THE
Handbook
OF
Health & Safety
AT
Work

To Angie, Tess
and Mary

THE *Handbook* OF *Health & Safety* AT *Work*

Mike Bateman
Brian King
Paul Lewis

First published in 1996

Kogan Page Limited
120 Pentonville Road
London N1 9JN

British Library Cataloguing in Publication Data

A CIP record for this book is available from the British Library.

ISBN 0 7494 1241 0

Typeset by Northern Phototypesetting, Co Ltd, Bolton
Printed and bound in Great Britain by Biddles Ltd, Guildford and King's Lynn

Contents

List of Figures 7
List of Tables 9
Preface 11
List of Abbreviations 13
List of Cases 16

Part I: The Law of Health and Safety at Work 19
1 The Legal System 21
2 The Role of the European Union 36
3 Enforcement of Health and Safety Law 50
4 Civil Liability 74
5 Changes and Developments in the Law 95

Part II: Managing Health and Safety 105
6 Principles of Health and Safety Management 107
7 Organising for Health and Safety 128
8 Risk Assessment, Prevention and Control 153
9 Accident Investigation and Reporting 184
10 Emergency Procedures 200
11 Audits and Inspections 217
12 Systems of Safety Control 250
13 Occupational Health and Hygiene 276
14 Employee Involvement 295
15 Selection and Management of Contractors 320

Part III: Practical Health and Safety Issues 347
16 Workplace Health, Safety and Welfare 349
17 Machinery Safety 367
18 Workplace Transport 390
19 Ergonomics, Manual Handling and VDUs 420
20 Electrical Safety 446
21 Fire Safety 476
22 Noise, Vibration and Radiation 515
23 Hazardous Substances 538
24 Asbestos and Lead 573

Contents

25 Personal Protective Equipment 589
26 Construction and Demolition Work 619
27 Social Factors and Health and Safety at Work 653
28 Environmental Aspects 667
Index 683

List of Figures

1.1 The Path to Statute Law 22
1.2 Comparison of Criminal and Civil Law Cases 25
1.3 The Courts System in Simplified Form 29

2.1 Simplified Description of the Role of the ECJ in Interpreting and Enforcing EU Law 38
2.2 Simplified Description of the EU Legislation Process 40
2.3 The Process of Implementing EU Directives in the UK 41

6.1 Outline of Three Accident Ratio Studies 109
6.2 The Domino Theory of Accident Causation 113
6.3 The Causes of an Accident 114–15
6.4 Topics Possibly Requiring Detailed Arrangements 121

7.1 Typical Management Health and Safety Responsibilities 130–1
7.2 Objectives and Performance Standards 136

8.1 Risk Assessment – Motor Vehicle Repair 178
8.2 Risk Assessment – Builders' Merchant 179
8.3 Risk Assessment – Home Helps 180
8.4 Risk Assessment – Solicitor's Office 181
8.5 Risk Assessment – Direct Works Depot 182

9.1 Incident Report 187

10.1 Bomb Threat Checklist 206
10.2 Dealing with Bombs and Suspect Packages 207

11.1 Rated Audit Subjects 225
11.2 Health and Safety Survey Reporting Format 228
11.3 Health and Safety Monitoring 229
11.4 Safety Inspection Topics 230
11.5 Health and Safety Inspections 231
11.6 General Safety Observations 234
11.7 People to be Considered for Inclusion in an Audit or Inspection 239
11.8 Examples of Plant and Support Units to be Considered in Audits and Inspections 240

11.9 Scope of Audits and Inspections 242
11.10 Managerial Implementation Checklist 248

12.1 Task Procedure 255
12.2 Permit to Work 265

13.1 Possible Health Problems 282

14.1 A Definition of Safety Culture 314

15.1 The Range of Contracted Services 321
15.2 Contractors: The Selection and Management Process 328
15.3 Contractor Selection Criteria 329
15.4 General Information for Contractors 337
15.5 Information to be Given to Contractors about Special or Unusual Risks 338
15.6 Contractor's Work Completion Report 345

16.1 Examples of Safety Signs 361

17.1 Guarding Characteristics 370
17.2 Moving Machinery Injury Types 375
17.3 Suggested Layout of Area Guarding 379
17.4 Reaching In and Through Elongated Openings with Parallel Sides 380–1

19.1 Key Stages of the MHO Regulations 1992 425
19.2 Specimen Form for Recording MHO Risk Assessments 428
19.3 Provision of Information Under the DSE Regulations (1) 437–8
19.4 DSE User Assessment Checklist 439

21.1 Fire Safety – An Overview 477
21.2 Risk Elements of Industrial Fires 482
21.3 Steps and Stages in the Development of a Fire 489
21.4 Fire Triangle 490

22.1 Noise Assessment Report 524

23.1 COSHH Assessment 554
23.2 COSHH Assessment 555
23.3 COSHH Assessment 556
23.4 COSHH Assessment 557

25.1 Attestation Procedures 595

25.2 PPE Standing Instruction 610
25.3 PPE Pocket Card 611
26.1 Application of the CDM Regulations 623
26.2 Particulars to be Notified to the HSE Under Regulation 7 of the CDM Regulations 626
26.3 Contents of the Health and Safety Plan 627–9
26.4 Ladders Checklist 634
26.5 Scaffolding Checklist 635

27.1 Employers' Policies on Drug Abuse at Work 661

28.1 Clean Air Legislation in Operation 669
28.2 Notice of Application of Authorisation in Respect of Prescribed Processes 670
28.3 Institute of Management: Example of an Organisational Environmental Policy 680

List of Tables

20.1 Physical Effects of Alternating Current 448
20.2 Electrical Inspections and Tests 467

22.1 Typical Noise Levels 517

Preface

The cost to the economy of accidents and work-related illnesses could be up to £16 billion a year, or 3 per cent of national income, according to the HSE's report *The Costs to the British Economy of Work Accidents and Work-related Ill-health* (1994). The direct cost to employers is estimated at between £4 and £9 billion, equivalent to 5–10 per cent of all British-based company profits, or £170–360 per person employed. These figures include costs of compensation, damage to equipment, goods and materials and the costs of replacing injured staff. The HSE believes that 70 per cent of the costs of accidents and work-related ill-health could be eliminated.

Against these startling figures needs to be set the improvements which have been achieved in health and safety in the UK. Thus, the number of deaths from accidents at work has continued to fall, although the decline in employment in high-risk industries such as coal, shipbuilding and steel will have been a factor. Moreover, there is now a much stronger emphasis on actively managing health and safety, rather than just responding to events. The risk assessment has become a key management tool as well as the cornerstone of legal requirements. The legal framework itself has become stronger, also more complex, but the HSC's review of the legislation promises a consolidation and thinning-out: health and safety law is set to become more user-friendly. Perhaps there is also a growing expectation about health and safety standards: public pressure arising out of some of the disasters of recent years has encouraged charges of manslaughter for health and safety breaches. There seems to be a national desire for those who cause death and injury to be made accountable.

The balance of health and safety is also changing. The historical emphasis on physical safeguards and physical injury is increasingly giving way to considerations of human behaviour and ill-health. Stress-related illnesses figure prominently here and one of every five adult sufferers of asthma is thought to have developed asthma because of work. Technology is also having a major impact, and in particular the microcomputer. The growth of work-related upper limb disorders, (repetitive strain injuries), associated with activities such as keyboard work, threatens to make this area the biggest health and safety issue in the UK. Both this and stress-related injuries may be the source of a considerable amount of future litigation.

The authors have tried to take account of all these developments in producing this *Handbook*. The approach has been to focus on the need to manage health and safety as effectively as possible in the interests of the business, its employees and society at large. The aim has been to integrate management principles and techniques with the legal requirements and the health and safety hazards themselves. The end product, it is hoped, is something which will provide practical assistance to managers in developing health and safety at work.

It is intended that our readers are principally managers who are not health and safety specialists, although we hope the latter may also find our work useful. We hope that union safety representatives might obtain assistance from what we have written and that perhaps students and teachers and trainees and trainers will benefit when involved in health and safety education and training.

It is hoped that the book will be useful for those in any size of organisation. Inevitably, some of the practices and procedures described have their origins in large companies, but adaptation is possible for use in small and medium-sized enterprises.

We have attempted to state the law as at 1 June 1995. Where possible we have incorporated some later changes, and in particular we have treated the Workplace Regulations as fully operational, which they are from 1 January 1996.

We have been greatly assisted by a number of people. We owe a debt to those who were kind enough to read various parts of the text. Among others, our thanks go to Jenny Coleman, Donald King, Joe McCarty and Ernie Walker. We also wish to thank our publisher, Pauline Goodwin, for her tolerant and flexible attitude, and for her encouragement and support.

A very big debt is owed to Norma Tuff who word-processed the text. Dealing with three busy authors separated by a distance of 60 miles cannot have been easy. She is to be congratulated not only on the high quality of what she produced but also for maintaining a strict timetable, a calmness and good humour. Finally, thanks are due to our wives, Angie, Tess and Mary, for their patience, support and encouragement during our work. We dedicate this book to them.

Mike Bateman
Brian King
Paul Lewis

June 1995

Abbreviations

ACOP	Approved Code of Practice
AIDS	acquired immune deficiency syndrome
APAU	Accident Prevention Advisory Unit
ASH	Action on Smoking and Health
BA	breathing apparatus
BASEEFA	British Approvals Service for Electrical Equipment in Flammable Atmospheres
BATNEEC	best available techniques not entailing excessive cost
BERBOH	British Examining and Registration Board in Occupational Hygiene
BS	British Standard
CDM	The Construction (Design and Management) Regulations 1994
CE	Communities European
CEN	European Committee for Standardisation
CHASM	Contractors' Health and Safety Manual
CHIP	Chemical (Hazard Information and Packaging for Supply) Regulations 1994
CIMAH	Control of Industrial Major Accident Hazard Regulations 1984
CNE	combined neutral and earth conductor
COP	code of practice
COSHH	Control of Substances Hazardous to Health
CPD	continuing professional development
dB	decibel
DNV	det norsk veritas
DSE	display screen equipment
DTI	Department of Trade and Industry
EAT	Employment Appeal Tribunal
EC	European Community
ECJ	European Court of Justice
EH	environmental hygiene
EHSA	European Health and Safety Agency
EIA	environmental impact assessment
EMAS	Employment Medical Advisory Service
EML	estimated maximum financial loss
EPA	Environmental Protection Act 1990

Abbreviations

EP(C)A	Employment Protection (Consolidation) Act 1978
EU	European Union
FA	Factories Act 1961
FEVI	forced expiratory volume in one second
FOC	Fire Officers' Committee
FPA	Fire Precautions Act 1971
FP(PW)	Fire Precautions (Places of Work) Regulations 1995
FVC	forced vital capacity
GATT	General Agreement on Tariffs and Trade
GMO	genetically modified organisms
HAVS	hand-arm vibration syndrome
HAS	health and safety
HASP	health and safety plan
HAZOP	hazard and operability studies
HMP	Her Majesty's Inspectorate of Pollution
HMSO	Her Majesty's Stationery Office
HSC	Health and Safety Commission
HSE	Health and Safety Executive
HSWA	Health and Safety at Work etc Act, 1974
Hz	Hertz
IEE	Institute of Electrical Engineers
IEHO	Institution of Environmental Health Officers
ILCI	International Loss Control Institute
IOSH	Institute of Occupational Safety and Health
IP	Index of Protection
IPC	integrated pollution control
IR	infra-red
ISRS	International Safety Rating Systems
LEC	Local Enterprise Council
LEV	local exhaust ventilation
LGV	loaded goods vehicle
LPC	Loss Prevention Council
MEP	Member of European Parliament
MDHS	methods of determination of hazardous substances
MEL	maximum exposure limit
MEWP	mobile elevating work platform
MHO	Manual Handling Operations Regulations 1992
MIOSH	Member of Institute of Occupational Safety and Health
NEBOSH	National Examination Board in Occupational Safety and Health
NIG	National Interest Group
NRA	National Rivers Authority
NVQ	national vocational qualification
OEL	occupational exposure limits

Abbreviations

OES	occupational exposure standard
OSRPA	Offices, Shops and Railway Premises Act 1963
PAT	portable appliance tester
PC	principal contractor
PDB	personal danger board
PLC	public limited company
PPE	personal protective clothing
PS	planning supervisor
PUWE	Provision and Use of Work Equipment Regulations 1992
RCD	residual current device
RF	radio frequency
RIDDOR	Reporting of Injuries, Diseases and Dangerous Occurrences Regulations 1985
ROSPA	Royal Society for the Prevention of Accidents
RPE	respiratory protective equipment
RSP	Registered Safety Practitioner
SEA	Single European Act 1986
SMS	safety method statements
SRSC	Safety Representatives and Safety Committees Regulations
SWP	safe working procedures
SWPL	safety working party leader
TEC	Training and Enterprise Council
TQM	total quality management
UK	United Kingdom
UV	ultra violet
VWF	vibration white finger
WBV	whole body vibration
WDA	Waste Disposal Authority
WRULD	work related upper limb disorder

List of Cases

Case abbreviations

AC	Appeal Cases
All ER	All England Reports
ICR	Industrial Cases Reports
IRLR	Industrial Relations Law Reports
KB	King's Bench Reports
KIR	Knight's Industrial Reports
LSG	Law Society Gazette
LT	Law Times
QB	Queen's Bench Reports
SJ	Solicitors' Journal
WLR	Weekly Law Reports

Armour v Skeen [1977] IRLR 310 68
Ashcroft v Harden (1959) CA 63 423
Austin Rover Group Ltd v HM Inspectorate of Factories [1989] IRLR 404 58
BAC Ltd v Austin [1978] IRLR 332 92
Bath v British Transport Commission [1954] 2 All ER 542 77, 365
Baxter v Harland and Wolff [1990] IRLR 516 79
Bell v Arnott and Harrison (1967) 2 KIR 825 79
Berry v Stone Manganese Marine [1972] 1 Lloyd's Rep 182 515
Birnie v Ford (1960) The Times 22 November 79
Bland v Stockport Metropolitan Council (1993) The Independent 28 January 665
Bourman and ors v Harland and Wolff plc 76
Boyle v Kodak [1969] 2 All ER 439 88
Bracebridge v Darby [1990] IRLR 3 82
British Transport Commission v Gourley [1956] AC 185 86
Broughton v Lucas (1958) CA 330 78
Bux v Slough Metals [1974] 1 All ER 262 78
Cambridge Water v Eastern Counties Leather [1994] 1 WLR 53 678
Campbell v Harland and Wolff [1959] 1 Lloyd's Rep 198 87
Century Insurance Co v Northern Ireland Road Transport Board [1942] AC 509 82
Chapman v Oakleigh Animal Products (1970) 8 KIR 1063 82
Chipchase v British Titan Products Ltd [1956] 1 QB 545 365
Clarkson v Jackson (1984) The Times 21 November 79
Clifford v Challen and Son Ltd [1951] 1 KB 495 365
Conway v Wimpey [1951] 2 KB 266 82

Cook v Square D Ltd and ors [1992] ICR 262 77
Cotterell v Stocks [1840] Liverpool Assizes 50
Coult v Szuba [1982] ICR 380 54
Crookall v Vickers Armstrong [1955] 2 All ER 12 78
Crouch v British Rail Engineering Ltd [1988] IRLR 404 77
Davie v New Merton Mills [1959] AC 604 79
Donoghue v Stevenson [1932] AC 562 75, 678
Donovan v Cammell Laird [1949] 2 All ER 82 87
Driver v Willett [1969] 1 All ER 665 88
Dryden v Greater Glasgow Health Board [1992] IRLR 469 659
Ebbs v James Whitson and Company Limited [1952] 2 QB 877 365
Edwards v National Coal Board [1949] 1 KB 704 22, 57
Ferner v Kemp (1960) CA 176 78
Foster v Flexile Metal (1967) 4 KIR 49 87
Franklin v Edmonton Corporation (1965) 109 SJ 876 85
Fricker v Perry (1974) 16 KIR 235 423
Gallagher v Dorman Long [1947] 2 All ER 38 87
Garrard v Southey [1952] 2 QB 174 83
Groves v Lord Wimborne [1898] 2 QB 402 51
Hall v McLaren (1958) CA 110 423
Hardaker v Huby (1962) 106 SJ 327 423
Hawes v Railway Executive (1952) 96 SJ 852 77, 364
Hawkins v Ross [1970] 1 All ER 180 79
Heasmans v Clarity Cleaning [1987] IRLR 286 82, 83
Herald of Free Enterprise 52, 71, 103
Hobbs v C G Robertson Ltd 33
Hopwood v Rolls Royce (1947) 176 LT 514 87
Hudson v Ridge [1957] 2 QB 348 80
Ioannou v Fisher's Foils (1957) CA 216 423
Iqbal v London Transport (1973) The Times 7 June 82
Irving v The Post Office [1987] IRLR 289 82
James v Hepworth and Grandage [1968] 1 QB 94 78
Johnstone v Bloomsbury Health Authority [1991] IRLR 118 655
Jones v Minton (1973) 15 KIR 309 83
Knowles v Liverpool City Council [1993] IRLR 6 33, 79
Latimer v AEC [1953] AC 643 77, 364
Lea v British Aerospace plc (1990) The Guardian 17 December 86
Lindsay v Dunlop Ltd [1980] IRLR 93 93
Lister v Romford Ice and Cold Storage [1957] AC 555 82
Lloyd v Grace Smith [1912] AC 716 82
Lyme Bay 52, 71, 103
Manwaring v Billington [1952] 2 All ER 747 88
Marshall v Gotham and Co Ltd [1954] AC 360 57
McCarthy v Daily Mirror [1949] 1 All ER 801 365
McDermid v Nash [1987] 3 WLR 212 88
McGuinness v Key Markets (1972) 13 KIR 249 87

McWilliams v Arrol [1962] 1 All ER 623 78
Middleton v Elliott Turbomachinery Ltd and anor (1990) The Guardian 22 November 86
Morris v Breaveglen Ltd [1993] IRLR 350 76
Mullard v Ben Line Steamers [1971] 2 All ER 424 87
Nahhas v Pier House (1984) 270 EG 328 82, 83
Nicholson v Atlas Steel Ltd [1957] 1 All ER 776 365
Overseas Tankships v Morts Dock (The Wagon Mound) [1961] AC 388 87
Pape v Cumbria County Council [1992] ICR 132 77
Paris v Stepney Borough Council [1951] AC 376 76
Parry v Cleaver [1970] AC 1 86
Pepper v Hart [1993] IRLR 33 33
Piggott Brothers & Co Ltd v Jackson and ors [1991] IRLR 309 93
Piper Alpha 69, 103
Qualcast v Haynes [1959] AC 743 78, 87
R v Boal [1992] 1 QB 591 68
R v Derby City Council, Bardon Contractors and City Plant Hire 322
R v Swan Hunter Shipbuilders and Telemeter Installations Ltd [1981] IRLR 403 57, 322
Ransom v McAlpine (1971) 11 KIR 141 79
Ready Mixed Concrete v Minister of Pensions [1968] 2 QB 497 83
Reid v Westfield Paper Co Ltd [1957] SC 218 365
Rose v Plenty [1976] 1 All ER 97 82
Ryan v Cambrian Dairies (1957) 101 SJ 493 80
Rylands v Fletcher (1868) LR 3 HL 330 678
Scholem v New South Wales Health Department 658
Sime v Sutcliffe Catering (Scotland) Ltd [1990] IRLR 228 85
Simmons v Walsall Conduits unreported, 1962 87
Smith v Davies and anor (1968) 5 KIR 320 77
Smith v Scott-Bowers [1986] IRLR 315 78
Smith v Stages and Darlington Insulation Co Ltd [1989] IRLR 177 81
Smoker v London Fire and Civil Defence Authority [1991] IRLR 271 86
Stokes v GKN [1968] 1 WLR 1776 82
Taylor v Coalite [1967] 3 KIR 315 83
Thompson v Ship Repairers (1984) 81 LSG 741 79
Uddin v Associated Portland Cement Manufacturers Ltd [1965] 2 QB 582 80, 81
Walker v Northumberland County Council [1995] IRLR 35 104, 653, 665
Walsh v Holst [1958] 3 All ER 33 83
Waugh v British Railways Board [1980] AC 521 85
Westwood v Post Office [1974] AC 1 81
White v Pressed Steel Fisher Ltd [1980] IRLR 176 60
Williams v Hemphill [1966] SLT 259 82
Woods v Durable Suites [1953] 2 All ER 391 78

Part I

The Law of Health and Safety at Work

1

The Legal System

INTRODUCTION

This chapter describes the main sources and types of law and explains some important concepts and doctrines which are part of the UK's legal fabric. It goes on to examine the system of courts and the personnel and procedures of the law.

The general legal context is important because health and safety law is not a completely separate branch of law. It adopts general legal principles, such as the duty of care, and cases are heard in mainstream legal institutions such as Magistrates' Courts or the High Court. However, health and safety law does have some of its own specialist institutions, notably the Health and Safety Commission (HSC) and its Executive (HSE), and these are discussed in chapter 3. The general legislative framework for health and safety is also set out in chapter 3, starting with a brief historical outline. The present chapter sets the scene by describing the legal system in general terms: chapter 2 adds the European Union (EU) dimension.

SOURCES AND TYPES OF LAW

Sources of law

The sources of law in the health and safety field are common law, statute and the law of the European Union.

Common law

The body of common law contains concepts and principles determined by judges through the process of deciding cases. The central legal relationship in the employment field – the contract of employment – is a common law concept. There is no Act of Parliament which says that the relationship between an employee and an employer is governed by a contract of employment, yet it is because this is what judges have determined. Case law, therefore, can be the source of major legal concepts.

It is also very important in the interpretation of the laws passed by Parliament. For example, Parliament has laid down that "it shall be the duty of every employer to ensure, so far as is reasonably practicable, the health, safety and welfare at work of all his employees"[1] but the legislation – the Health and Safety at Work etc Act 1974 (henceforth HSWA) does not define the term "reasonably practicable". Therefore, the courts have established the detailed rules through case law, for example in *Edwards v National Coal Board* and other cases. (See chapter 3 for an analysis of what is meant by the term.)

The judges are active, therefore, both in the creation and the interpretation of the law.

Statute law

Statute law is derived from Parliament. It comprises Acts of Parliament, regulations and orders made under such Acts and delegated legislation, for example, bye-laws made by local government under powers given to them by Parliament. The steps involved in the making of statute law are shown in Figure 1.1.

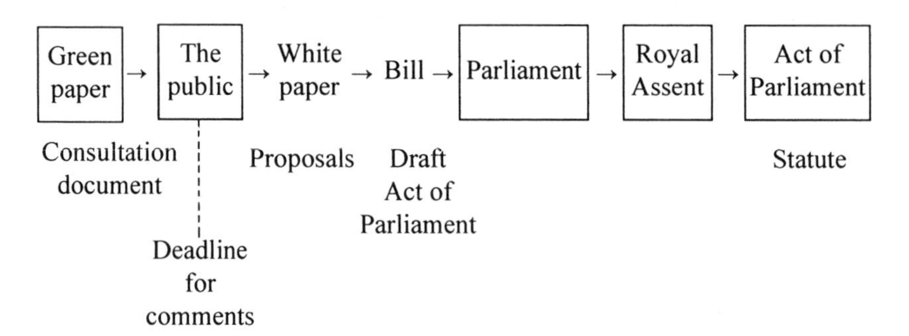

Figure 1.1 *The path to statute law*

Where law is expected to need extension, or change is anticipated or thought possible, an Act can give the relevant Minister the power to issue orders or regulations to give effect to such changes. For example, HSWA places a duty upon an employer to prepare a written statement of his safety policy, "except in such cases as may be prescribed".[2] Prescribed means "prescribed by regulations made by the Secretary of State"[3] and the regulations so far enacted provide exemption for employers with fewer than five employees.[4] For an example of an Order, see p26.

Regulations are often used where an Act lays down only general provisions, but where detailed law is required for particular circumstances. HSWA is a good example of this. It lays down very general duties, notably in section 2 (s.2). Regulations under the Act then cover specific circumstances such as the control of substances hazardous to health. Both orders and regulations have the full force of law and are of the same status as the Act under which they are made.

Parliament may also delegate legislative power to persons other than the appropriate Minister. In particular, such power may be delegated to local government authorities to be used through the medium of bye-laws. In the health and safety field this is relevant in respect of the employment of children: bye-laws may control (for health and safety and other reasons) where, when and for how long children are employed.[5] It is also necessary to mention what may be called quasi-law which is found in the form of codes of practice. These and guidance notes are discussed in chapter 3 in the context of the role of the HSC and HSE.

Law of the European Union

A third source of UK law is statute and case law from the European Union. The relationship between EU and UK law is described in chapter 2 both in general terms and specifically as regards health and safety at work.

Types of law

Law may be classified in a number of ways. The principal ones are: according to source (as above), that is, common law and statute; criminal and civil law (discussed below); and public and private law, ie public administration and citizenship as distinct from private transactions.

Criminal law

Parliament has decided that those who do not accept certain rules of conduct shall be punished as a deterrent to others. A crime, therefore, is a breach of society's rules, as determined by Parliament, and is punishable by the State. It follows that the criminal law is concerned with preventing breaches of society's rules and with punishing offenders. Those responsible for such breaches commit crimes – such as theft, murder, drunken driving and so on – and an agent of the State will prosecute them if there is sufficient evidence. In most cases that agent is the Crown Prosecution Service, although in very serious cases it may be the Director of Public Prosecutions. In the health and safety at work field the relevant health and safety inspector is responsible for prosecution in the magistrates' courts.[6] The standard of proof in criminal law is that the case is established beyond all reasonable doubt. This is a high standard of proof reflecting the fact that guilt has associated with it a stigma, possible financial penalty (a fine)[7] or loss of liberty (a custodial sentence), although probation, community service and so on are increasingly common forms of punishment. Criminal cases are normally dealt with in the magistrates' courts, although more serious or contested cases may or must (depending upon their nature) then be sent to the Crown Court for trial.

Civil law

Meaning of civil law

The civil law is concerned with settling disputes between private parties, for example between individuals, between individuals and organisations and between organisations. Thus, a dispute between a householder and his or her neighbour about nuisance created by the neighbour's noise would be a civil matter. So would a dispute between two individuals over a debt; between a worker and an employer over wages due; between a customer and a shop over a faulty product; or between a retail firm and one of its suppliers over late delivery. In the health and safety at work field the principal type of civil case is a dispute between an employee and his or her employer over the employer's alleged negligence leading to the employee suffering a personal injury (see chapter 4). A person or organization wishing to pursue a civil claim – the plaintiff – has to establish, on the balance of probabilities, that they have suffered some loss as a result of the defendant's unlawful act, and show the extent of that loss in a claim for damages. Sometimes an order (or injunction) may be sought to stop the unlawful act (for example, an employer wanting to stop a strike, a householder wanting to stop a neighbour's noise).

Categories of civil law

There are three categories of civil law relevant to the employment field – contract, trust and tort. A contract exists where two or more parties agree to make an exchange, but it need not in every case be in writing. There must be consideration – something provided by one party in exchange for what is received from the other. In most cases this is money, such as wages for work or money for consumer items. Contracts are entered into after an offer is made by one party and accepted by another. Where there is a contract, a breach by one party may give rise to a legitimate claim for damages for breach of contract by the other. The law of trusts operates in areas where a person is charged with looking after someone else's money. A failure to properly discharge such a role can lead to claims for damages for breach of trust. This would be possible, for instance, in relation to pension fund trustees. A tort is simply a civil wrong (as opposed to a criminal wrong, that is, a crime). There are a variety of different torts, largely the product of judicial creation. Examples are:

- Nuisance
- Defamation – libel and slander
- Negligence
- Trespass – person, land, property
- Inducing a breach of contract
- Breach of statutory duty.

Whereas a claim for damages for breach of contract can be pursued only where a contract exists, the law of torts applies generally, although different rules apply to different torts. Unions do much work, for example, in assisting members' claims for damages where an employer has been negligent. Here the rules are that there has to be, first, a duty of care; second, a breach of that duty and third, a loss suffered as a result. The duty of care can be established by reference to the contract of employment or more generally, and indeed statute law also lays down such a duty (in HSWA) although this cannot be used as a basis for civil actions. By failing in his or her duty of care the employer has committed a tort – the tort of negligence.

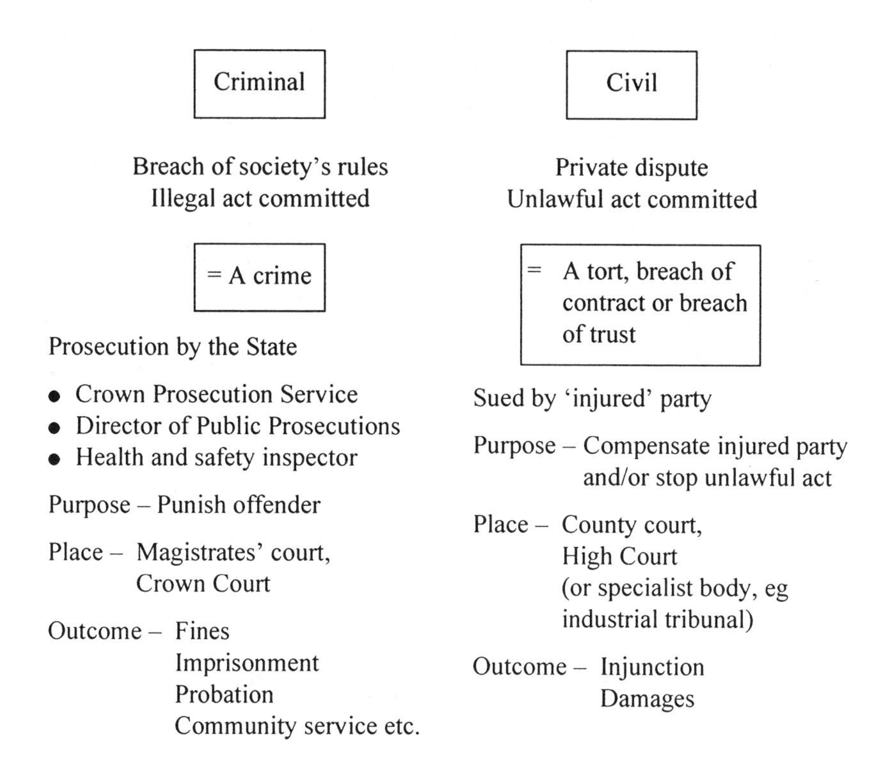

Figure 1.2 *Comparison of criminal and civil law cases*

Relationship between criminal and civil law

In general there is little relationship between the two systems. As already noted, they have different standards of proof. Moreover, cases take place in different courts and the atmosphere as well as the outcome is different. However an important factor is that one event can give rise to both criminal and civil proceedings.

EXAMPLE

A hole in the floor of the workplace is left uncovered, without being cordoned off and without any warning signs. A worker falls into it, injures his or her back and is absent from work as a result, so suffering financially and otherwise. The worker may pursue a civil case to claim for damages for losses occurring because of the employer's negligence. At the same time the factory inspector may feel that the employer's behaviour constitutes a crime – a failure to perform the duties laid down in HSWA – and may decide to prosecute. The civil case will be heard in the county court or High Court, the criminal case probably in the magistrates' court. The criminal system tends to be less slow than the civil system so these cases will be heard at different times as well as in different places. The civil court will decide whether there has been negligence, and if so what damages are appropriate. The criminal case will decide if the employer is in breach of HSWA. Where damages are a relatively minor matter, they may be dealt with in the criminal proceedings.

The extent of the law

The word extent refers to the geographical coverage of legislation. For example, Part I of HSWA, including the various general duties, applies only to Great Britain (ie England, Wales and Scotland, but not Northern Ireland).[8] However, a Ministerial Order (technically an Order in Council by Her Majesty) may extend the legislation to Northern Ireland. This has in fact been done.[9]

CONCEPTS AND DOCTRINES

Torts and crimes

The principal areas of civil law relating to employment were mentioned earlier and the concept of a tort, or civil wrong, was introduced. The law of torts applies generally, the State defining what conduct constitutes a tort and determining the remedies available. A tort may be seen as a breach of a duty – that duty having been created by the general law rather than by agreement (ie a contract) – for which the injured party may obtain damages. Negligence is one of the principal torts and is of great relevance in the field of health and safety at work.

The criminal law is the concern of the State with misconduct – crime – which affects the whole community. The general aim is to punish and deter. By contrast, the civil law, encompassing the law of torts, provides individual citizens with the opportunity to enforce their rights against other citizens who owe a

duty to them. Usually the aim is to provide damages which compensate the injured person so as to restore them to their former position.

Contracts

It has been seen that a significant difference between a crime and a tort is that the former is a public concern and the latter a private one. But what are the differences between tort and contract? A major difference is the fact that a contract is derived from agreement while a tort is based upon a breach of a duty imposed by the general law. Contractual duties are undertaken by choice, although the respective bargaining power of the parties is rarely considered, and can be enforced only where there is a contract, while duties under the law of torts are imposed and do not require the presence of a contract. In the law of torts the State determines the duties; in contract the parties determine the duties and the courts, historically, will intervene for enforcement purposes (typically to award damages) rather than to pronounce on the fairness or otherwise of what the contracting parties agreed.

Legal personality and liability

In law, the word person may be used to indicate an individual human being or a group of human beings with their own existence in law. The former are 'natural' legal persons, the latter 'artificial' ones. Legal personality involves acceptance of rights and obligations given effect by the law. Special rules apply to certain categories of natural legal person, such as children, people of insane mind, prisoners, bankrupts and aliens. Artificial legal persons are known as corporations. Those bodies which are not corporations are known as unincorporated associations. Corporations, which include limited liability companies and local authorities, may be established by Royal Charter or by statute. The latter method includes Acts establishing specific corporations as well as the general legislation (the Companies Acts). The achievement of legal personality means a corporation can sue and be sued in its own name, and may be the defendant in criminal proceedings.

In HSWA duties are placed upon employers (and others). Where there is an offence under the Act, the employer will be guilty. If the employer is a body corporate, that body will be guilty, although by s37 of the Act an individual "director, manager, secretary or other similar officer" may also be guilty. Where the employer is unincorporated the individual proprietor will be liable. In a partnership, the partners will be jointly and severally (ie collectively and individually) liable.

If the action is a civil one (see chapter 4) the employer will be personally liable, if, for example, there is a breach of the duty of care by that employer. If the breach is by an employee of the employer, rather than the employer himself, the doctrine of vicarious liability operates to make the employer liable for acts and omissions during the course of the employee's employment (again, see chapter 4). An employer's vicarious liability for the independent contractor is much weaker than the liability for employees; therefore the distinction between these two categories of worker is of considerable importance.

It should also be noted that HSWA places duties upon persons other than employers, notably employees, the self-employed, controllers of premises, designers, manufacturers, importers and suppliers. In addition, the civil law sets out the duties of occupiers towards those who come onto their premises (see chapter 4).

Employment status

As noted, employment status can be an important factor in determining liability. Rights and obligations differ both at common law and under statute. For example, employers' vicarious liability for the acts of contractors is less than it is for employees (see chapter 4); and only employees may use legislation such as the unfair dismissal provisions relating to health and safety (see chapter 4). Professional legal advice may be needed in order to determine employment status.[10] (The subject is dealt with briefly on p83).

Agency

As noted, a worker may be an employee or an independent contractor. Such a person may, at the same time, be an agent. In essence, a principal acts through an agent, the agent bringing the principal into legal relationships with third parties. In such relationships, the principal is deemed to have acted in person. The duties of the agent will depend upon any contract of agency and the implied terms of agency, which include the duty to exercise care and skill. The principal is under a duty to indemnify his agent against any losses, liabilities and expenses incurred in the performance of his undertaking. The general rule is that the agent is not liable under nor entitled to enforce a contract made on behalf of his principal, although there are exceptions to this rule, such as where the principal is undisclosed. The importance of agency is that it may affect the allocation of liability.

The duty of care

The duty of care may be found in contract, statute and in the general law of negligence. The last-mentioned imposes upon all of us, in certain situations, a duty to take reasonable care. The concept is explored at some length in chapter 4.

The correct application of the law

Law distinguishes between matters of law and matters of fact, although there can be a mixture of the two. The significance of the distinction is that in many circumstances appeals are permitted only on points of law. That is, the appeal must be on the basis that the courts have applied the law incorrectly – for example, by applying the wrong test, or by applying the right test incorrectly – or have made a finding of fact that is not compatible with the evidence.

An important doctrine, therefore, holds there to be a correct way of applying the law to a given set of facts. This is at its most obvious where the issue to be tried is the correct construction of a written document such as a contract. Here there should be a 'correct' answer, with extraneous material disregarded. Where such a contract is part written and part oral, there is less certainty because the courts need to establish facts based on oral evidence before they can decide the issue of interpretation.

THE LEGAL SYSTEM IN ACTION

The system of courts in England and Wales

The system is divided into criminal and civil sections, although in practice there is a degree of overlap. The jurisdiction of the different courts is set out briefly below: the roles of the HSC and HSE are described in chapter 3.

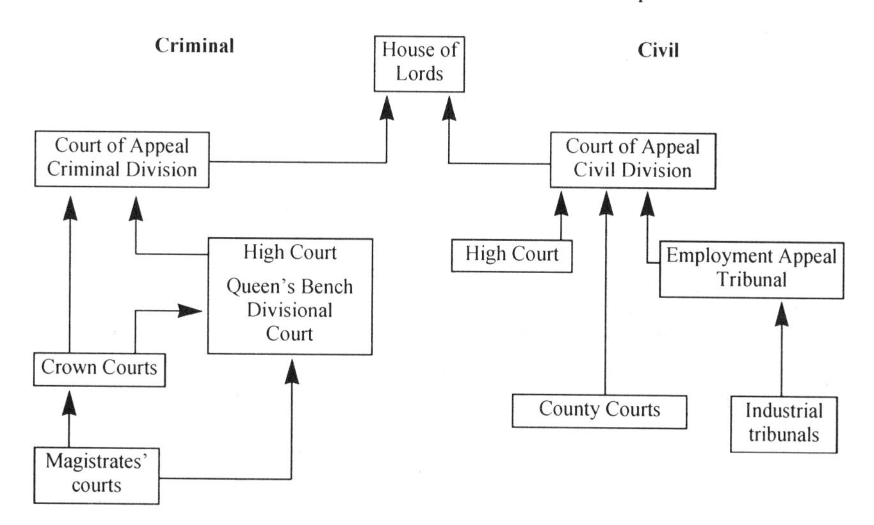

Figure 1.3 *The courts system in simplified form*

Magistrates' courts

These deal with the vast majority of criminal cases. The magistrates also have responsibility for some civil matters, for instance, maintenance orders and certain other aspects of family law, and licensing.

Crown Court

The Crown Court hears appeals on questions of fact from summary cases in the magistrates' courts and deals with indictable offences, that is, where the offence is more serious and the charges are set out in detail. Except on appeal, a jury is present if the defendant pleads not guilty.

County courts

These have a wide range of functions and are the main civil court. A large majority of civil cases start here, although fewer than five per cent result in actual trials, cases usually being settled by the parties. Cases in contract and most types of tort can be dealt with in the county courts, as well as adoption, bankruptcy, trust disputes, the winding up of companies and the land and estates of deceased people. Since 1973 the county court has operated a small claims procedure in which individuals can represent themselves. As a result of the Courts and Legal Services Act 1990, there has been a significant extension of the county courts' jurisdiction[11] Thus, claims for damages in respect of personal injuries must now be commenced in the county courts if the value of the action is less than £50,000 (previously £5,000). Actions with a value of less than £25,000 will normally be tried in the county courts and those valued at £50,000 or more in the High Court. Various criteria (such as complexity of the case) are set out for determining where actions involving sums between these figures should be heard. Many claims for damages for personal injuries are now tried in the county courts rather than in the High Court.

High Court

This comprises three divisions – Queen's Bench, Chancery and Family. The Queen's Bench Division hears all cases in contract and in tort. The Chancery Division's work includes land, financial matters (such as taxation) and company law. The Family Division's remit takes in cases involving children and those involving matrimonial matters. Each of the three divisions has a 'divisional' court. The Queen's Bench Division divisional court hears criminal appeals from the Crown and magistrates' courts. It also exercises the supervisory jurisdiction of the High Court over inferior courts, tribunals and public officials, including local authorities. This is known as judicial review.[12] In addition, the Divisional Court of the Queen's Bench Division hears appeals against the decisions of industrial tribunals in relation to contested improvement and prohibition notices (see chapter 3).

Appeals

A major feature of the system is the right of appeal. This is granted by the body which has just decided the case or by the appeal body itself. Appeals go to the appropriate division of the Court of Appeal, where they will be heard by three judges, and may go to the House of Lords which constitutes a court comprising five law lords. On matters involving EU law an appeal to the European Court of Justice may be possible.

Precedent

In deciding a case, attention will be paid to whether similar cases have already been decided by the courts. Precedent can be ignored if there is some difference between the cases being considered, that is, one case is distinguished from another. Otherwise precedent will usually apply unless the court has the authority to overturn an earlier decision on the grounds that it was wrong. Courts are normally bound by the precedents set by other courts which are of higher authority, for example, the Court of Appeal is bound by House of Lords' decisions and the High Court by Court of Appeal decisions.

The personnel of the law

Solicitors

Solicitors have to pass the examinations of their professional body – the Law Society – and obtain a certificate if they wish to practise. Usually they are self-employed, often in partnerships. They tend to be generalists rather than specialists, although as law firms grow in size specialisation is occurring. In general, solicitors are not allowed to act as advocates in the higher courts (for example, Crown Court, High Court and above), although recent reforms have started to alter this.[13]

Barristers

Until recently, barristers had a complete monopoly of representation in the higher courts. They tend to be specialists both in the practice of advocacy and in terms of particular areas of law, although the latter is not always true. They need to pass their professional body – Bar Council – examinations in order to practise. They do not deal directly with the client. The client deals with the solicitor, who then hires (and pays) the barrister. Barristers are self-employed, working in chambers, but are not allowed to form partnerships with each other.

Judges

Judges are drawn primarily from the ranks of practising barristers. Most of them are assigned to geographical circuits (hence circuit judges) taking in both county court and Crown Court work. Judges are appointed by the Queen on the advice of the Lord Chancellor. Senior judges, such as those who sit in the Court of Appeal, are appointed by the Queen on the advice of the Prime Minister, but the Lord Chancellor is consulted. Below the level of the circuit judge there are junior judicial posts in the Crown and county courts respectively. Immediately senior to the circuit judge (Judge Wisely) is the High Court judge (Mr. Justice Wisely, or Wisely J). In the even higher echelons are the judges of the Court of Appeal (Lord Justices – Wisely LJ) and the Law Lords (simply, Lord Wisely).

Magistrates

Magistrates are one of the most visible examples of 'ordinary' people being involved in the administration of the law. They are appointed by the Lord Chancellor on the advice of local advisory committees. They are often people prominent in local life, for example, in political parties or other voluntary bodies such as charities or trades unions. They are appointed for their soundness of judgment and responsible attitude rather than any legal knowledge. In large cities there is often a full-time, salaried (stipendiary) magistrate.

The burden of proof

As noted, the standard of proof in criminal law is that the case is proven beyond reasonable doubt, while in civil proceedings the less strict test is that the case is established on the balance of probabilities. The general legal rule is that it is for the plaintiff to prove his or her case. However, statute may vary this rule and there is a particularly important variation in the health and safety field. Where there is a duty or requirement to do something so far as is practicable or so far as is reasonably practicable, "it shall be for the accused to prove (as the case may be) that it was not practicable or not reasonably practicable to do more than was in fact done".[14] In addition, in criminal proceedings, a breach of an approved code of practice will lead to the prosecution case being "taken as proved" unless the defendant can show that he fulfilled his duty in some way other than by observance of the code.[15] Moreover, in civil cases there is what amounts almost to a presumption of negligence once the plaintiff establishes a prima facie case. (On this, see chapter 4.) Employers may be able to use their risk assessments as a means of showing that they have met their duties. (See chapter 8: Risk Assessment.)

Statutory interpretation

The courts are assisted in the matter of statutory interpretation by the Interpretation Act 1978. This states, among other things, that unless the contrary is shown masculine includes feminine (and vice versa) and singular includes plural (and vice versa). In addition, individual Acts of Parliament (and sometimes Parts of Acts) have interpretation sections. The Health and Safety at Work etc Act 1974 has four parts, three of which have interpretation sections. Interpretation of Part I is found in section 53 which defines 37 words or phrases (in alphabetical order), including "employee", "personal injury", "plant", "premises" and "substance". In addition, section 52 defines what is meant by "work" and "at work".

There are also rules of statutory interpretation derived from case law. Two examples are given below.[16]

EXAMPLE

Where specific words are followed by general words, the rule is that the general words are to be interpreted in the light of the specific words. Thus in *Hobbs v C G Robertson Ltd* the Court of Appeal had to construe the Construction (General Provisions) Regulations 1961 in respect of the provision of goggles. These had to be provided where there was "breaking, cutting, dressing or carving of stone, concrete, slag or *similar materials*" (emphasis added). Was brick, which splintered causing an eye injury, a similar material? No, held the Court of Appeal: thus provision of goggles was not compulsory.

EXAMPLE

Connected with this is the rule about lists of specific matters which are not followed by general words. Here the interpretation is that only the matters so listed are covered by the legislation in question, although a broader interpretation may occur where the list is preceded by the word "includes". Thus, s1 (3) of the Employers' Liability (Defective Equipment) Act 1969 states that "equipment includes any plant and machinery, vehicle, aircraft and clothing". But does it include the material with which the employee is working? In *Knowles v Liverpool City Council*, where the material was a paving stone, the Court of Appeal answered this question in the affirmative, saying that the word "equipment" should be interpreted broadly.

Historically the courts may not use extrinsic sources as an aid to interpretation of statutes. Therefore, sources such as the record of Parliamentary debates and White Papers have not been available. Since 1993, however, following *Pepper v Hart* (a tax case) such sources may be used in certain circumstances.

Intrinsic sources have been used as an aid historically. These include the long title of an Act: the long title of the Health and Safety at Work etc Act 1974 is shown in the box below.

HEALTH AND SAFETY AT WORK ETC ACT 1974

1974 Chapter 37

An Act to make further provision for securing the health, safety and welfare of persons at work, for protecting others against risks to health or safety in connection with the activities of persons at work, for controlling the keeping and use and preventing the unlawful acquisition, possession and use of dangerous substances, and for controlling certain emissions into the atmosphere; to make further provision with respect to the employment medical advisory service; to amend the law relating to building regulations, and the Building (Scotland) Act 1959; and for connected purposes.

Where the legislation being considered is relevant to any legislation of the European Union, the courts should, where possible, construe the domestic legislation in such a way as to give effect to the European provisions. (See chapter 2 for elaboration of this point.)

Notes

1. HSWA, s2(1)
2. HSWA, s2(3)
3. HSWA, s53(1). The Secretary of State's power to make such regulations is found in s15(1)
4. The Health and Safety Policy Statements (Exception) Regulations 1975
5. Children and Young Persons Acts 1933–69
6. HSWA, s39
7. Section 17 of the Criminal Justice Act 1991 provides that the maximum amounts specified in the standard scale of fines shall be as follows:

Level	£
1	200
2	500
3	1,000
4	2,500
5	5,000

   On increased fines under certain sections of HSWA, see chapter 3.
8. HSWA, s84
9. The Health and Safety at Work (Northern Ireland) Order 1978. Further extensions have been made through The Health and Safety at Work etc Act 1974 (Application Outside Great Britain) Order SI 1995/263.

10. Guidance is given in Lewis, P, *Practical Employment Law*, Blackwell, 1992, pp54–6. A more uptodate picture will be found in Lewis, P, *Law of Employment*, Kogan Page, 1996.
11. Courts and Legal Services Act 1990. See The High Court and County Courts Jurisdiction Order, SI 1991/724.
12. See Lewis, P, *Practical Employment Law*, Blackwell, 1992, p13 or Lewis, P, *Law of Employment*, Kogan Page, 1996.
13. Courts and Legal Services Act 1990
14. HSWA, s40
15. HSWA, s17(2). Approved codes are issued by the HSC under s16(1)
16. Another example is that where two or more words follow each other, they are taken to be related for the purposes of interpretation. Thus "health, safety and welfare at work" in s2(1) of HSWA clearly means health at work and safety at work, as well as welfare at work.

2

The Role of the European Union

THE EUROPEAN UNION LEGAL SYSTEM

Legal basis

The legal basis of the European Union lies in the founding treaties. Technically, there are three European Communities: the Coal and Steel Community (set up by the Treaty of Paris, 1951), the Economic Community (Rome, 1957) and the Atomic Energy Community (Rome, 1957). Operationally there is a unified structure combining all three communities, and the term European Community (EC) is commonly used to refer to this.

The name European Union (EU) is now widely used following the Maastricht Treaty on European Union. EU applies principally to the new areas of foreign policy, security and home affairs which are not within the jurisdiction of the EC but are within the jurisdiction of the EU. The original communities exist alongside the wider Union. The term European Union when used here should be taken to include the EC. For consistency the term is used even when describing events prior to the implementation of the Maastricht Treaty.

Each measure enacted by the EU must be based – that is, justified by reference to – a provision in an EU treaty, otherwise the European Court of Justice (ECJ) may rule it unconstitutional. Proposals from the European Commission thus specify the treaty article on which the proposal rests and this determines aspects of the procedure to be adopted eg unanimous or majority voting or whether the co-operation procedure involving the European Parliament has to be used (see below). Unless specifically provided for elsewhere in the treaties, Article 100 tends to be the base. This involves unanimous voting.

The Single European Act (SEA) of 1986 introduced some fundamental changes. In particular, the Treaty of Rome was amended to include new articles, of which 100A and 118A are especially relevant to health and safety at work. Both allow qualified majority voting. Article 100A has made possible a quickening of the pace of EU activity in the field of harmonisation of technical standards (see pp 42–3) which has implications for health and safety at work through matters such as safety standards for the design and construction of machinery.

Article 118A has been the base used for a number of health and safety at work directives and makes a specific commitment to improved health and safety at work on the part of the EU.

Sources of EU law

The sources of EU law may usefully be divided into three categories. First are the primary sources, comprising the founding treaties, Community acts (such as the SEA 1986) and further treaties (such as Maastricht or accession treaties). These provide the constitutional foundations of the EU as well as establishing fundamental policies (such as the operation of the Single Market) and principles (such as free movement of workers and equal pay). Next come secondary sources – regulations, directives and decisions – by which the EU implements policy in more detail. Finally, there are non-legally-binding sources – opinions and other non-treaty acts (memoranda, guidelines, resolutions, communications). In the health and safety field there is an important recommendation on policies to avoid sexual harassment.

By far the most important source of EU law in relation to health and safety at work is the directive. The way directives work is discussed below, and shown in diagrammatic form in Figure 2.3 (see p41).

Main institutions

European Council

This is a 'summit' meeting of the heads of government of Member States. It is held two or three times a year in order to discuss major issues and give general direction to the Union.

Council of Ministers

Now formally the Council of the European Union, this is the primary decision-making institution of the EU and takes final decisions on the laws to be applied throughout the Community. It comprises government ministers from each Member State, although the ministers actually present will vary according to the subject under discussion. Thus employment ministers will attend a Council where employment matters are being discussed, finance ministers for budgetary matters and so on.

European Parliament

The Parliament is elected directly by the voters in each Member State. The elections take place every five years. Parliament is consulted by the Council of Ministers but has increased powers as a result of the Maastricht Treaty.

European Commission

The Commission is the executive of the EU. It comprises full-time, salaried commissioners and their staff. The Commission proposes legislation which is then considered by the Council of Ministers, upholds the treaties and acts to ensure adherence to EU laws. Since Maastricht, the appointment of the Commission is subject to approval by the Parliament, which also has the power to dismiss the Commission by a vote of censure.

The Court of Justice[1]

The European Court of Justice (ECJ) is composed of judges from all Member States. It passes judgment in disputes involving the application and interpretation of EU law, and since Maastricht, it can fine Member States for being in breach of EU legislation. A Court of First Instance now deals with an increasing proportion of EU cases, with the ECJ judging any appeals. A simplified description of the ECJ's role in interpreting and enforcing EU legislation is given in Figure 2.1.

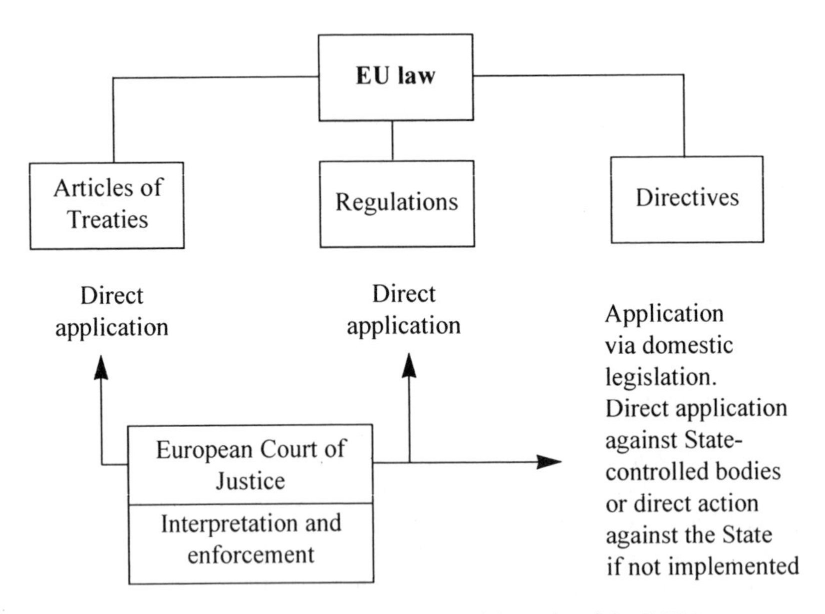

Figure 2.1 *Simplified description of the role of the ECJ in interpreting and enforcing EU law*

Other institutions

The Court of Auditors is responsible for checking the management of the EU's finances. There are also consultative bodies. The Economic and Social Committee has to be consulted on proposals relating to economic and social matters. It is an advisory body made up of representatives of employers, trades unions and consumers. The Maastricht Treaty set up a Committee of the Regions in order to shape and influence EU policy and expenditure in favour of the regions. In the main it comprises representatives of local and/or regional government.

An important committee in the present context is the Advisory Committee on Safety, Hygiene and Health Protection at Work. Formed in 1974, and bringing together representatives of governments, employers and trades unions, it discusses, proposes and reviews new health and safety developments. The Commission consults the Committee before it proposes health and safety measures.

Finally, there is an Ombudsman and a right of citizens of the Union to petition the European Parliament.

Legislative process

The European law-making process is understandably more distant and complex than that of domestic parliaments. Figure 2.2 sets it out in simplified form. The process starts with a proposal from the European Commission which is discussed in the Council of Ministers. The European Parliament is consulted. The Parliament is in fact consulted a second time (under what is called the co-operation procedure) after the Council has reached a common position ie an agreement in principle. The Parliament may reject, amend or accept the Council's position. In cases of rejection, the Council will need a unanimous vote in favour if it is to go against the Parliament's expressed wishes. Where the Council's position is amended, the proposal is reviewed by the Commission before being returned to the Council.

Most of the provisions of the Single European Act 1986 use this procedure and are subject to qualified majority voting in the Council. One of the exceptions, however, is the area relating to the rights and interests of employed persons, which includes the Community Charter of Fundamental Social Rights (The Social Charter). Provisions relating to technical standards are not an exception and are likely to result in a number of changes in the health and safety field.

Moreover, the general improvement of health and safety at work is emphasised as an EU objective and is subject to qualified majority voting.

The scope and method of EU decision-making has been altered as a result of the Treaty agreed at Maastricht in 1991 and the protocols attached to it. It seems likely that EU influence will increase in areas such as economic and monetary policy and that there will be greater inter-governmental co-operation among Member States over matters such as foreign policy, defence, immigration and crime prevention. Provision is made for the wider use of qualified majority voting. Parliament now has the right to override Council for the first time.

The EU Member States agreed to a Social Policy Protocol being annexed to the Maastricht Treaty. This Protocol allows the fourteen Member States apart

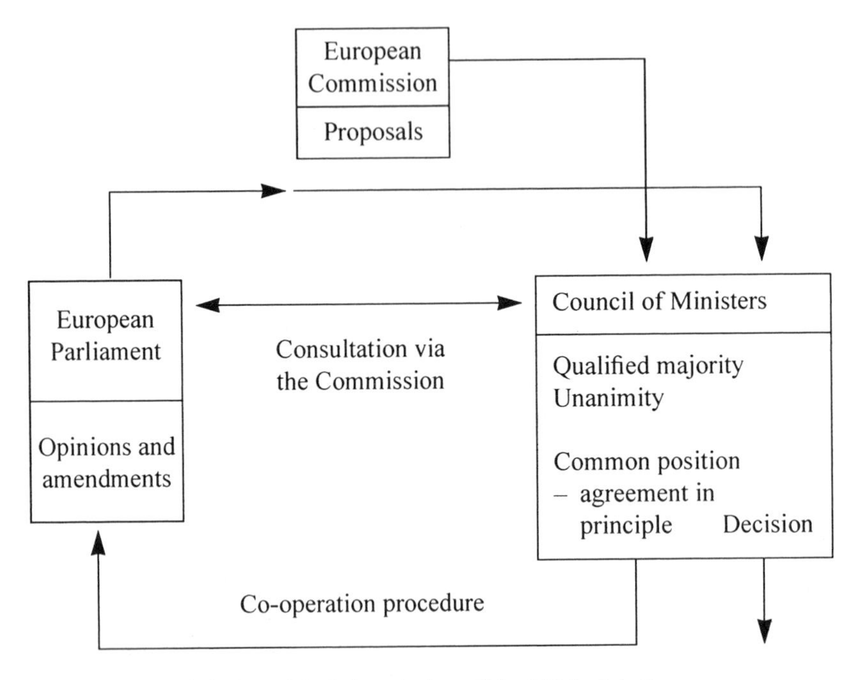

Figure 2.2 *Simplified description of the EU legislative process*

from the UK to use a procedure for implementing the Social Charter through qualified majority voting. The procedure applies to matters of equal treatment, working conditions and information and consultation of the workforce. The UK will not be bound by decisions made through the use of this procedure and unanimity will still be required in relation to terms and conditions of employment more generally, with the exception of health and safety at work.

Bodies representing management and labour are being encouraged to negotiate agreements which might meet the EU's requirements as well as taking into account the traditions of each Member State.

An agreement was reached at Maastricht in relation to occupational pensions. Equality will apply only to benefits accruing from and including 17 May 1990 unless proceedings had been commenced before that date. The European Court has subsequently confirmed this approach.

Relationships between EU and domestic law

The UK's membership of the EU is expressed through the European Communities Act, 1972. There are three principal methods by which EU law is effected under the 1972 Act:

- by direct effect
- by subordinate legislation under the 1972 Act

- through interpretation of existing legislation.

In addition, Parliament may enact new statutes or issue regulations under Acts other than the European Communities Act of 1972. For example, substantial amounts of EU law have already been given effect through regulations under HSWA.

Regulations and decisions of the EU have direct effect – that is, no UK legislation is necessary to give effect to them – and so too do articles of EU treaties where they are clear and precise, unconditional and unqualified and do not require further implementing measures by Member States. Where the EU issues directives, member countries must pass their own legislation in order to comply. The process is illustrated in Figure 2.3. The famous 'six pack', comprising six sets of health and safety at work regulations, all operative from 1 January 1993, is an example of the process in action.[2] Where a directive is not implemented by a Member State, or not fully implemented, complaints may be taken on the basis of the directive itself against organisations made responsible by the State and given special powers for providing a public service under the control of the State.

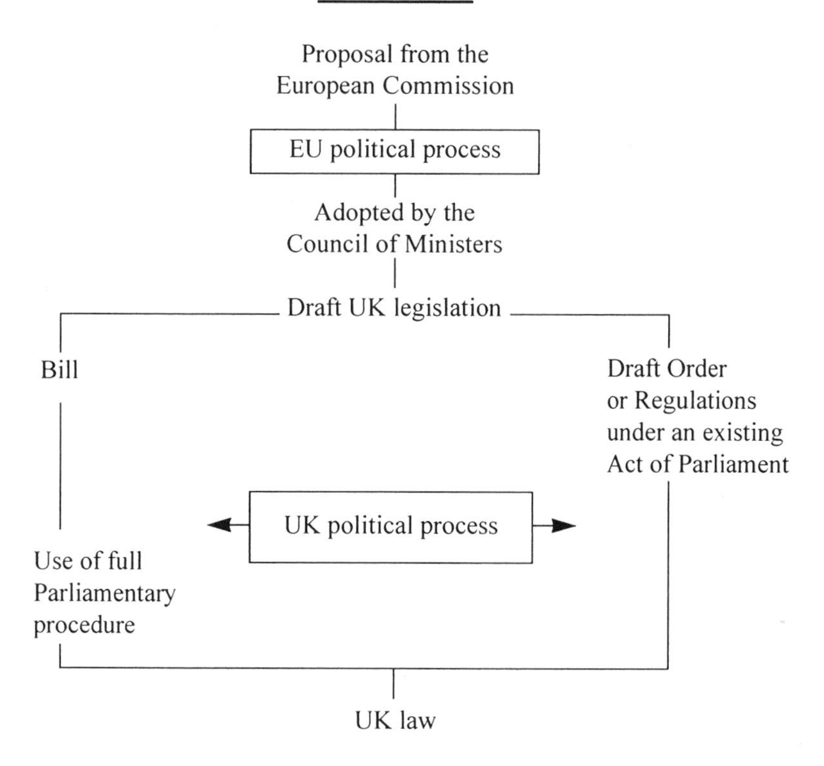

Figure 2.3 *The process of implementing EU directives in the UK*

Where the organisation is not a public body the proceedings can be taken directly against the State – effectively an action for damages suffered as a result of the State's failure to fully implement the directive. Where there is a conflict between UK and EU law, the latter must prevail and, in general, UK law is to be interpreted in such a way as to give effect to EU provisions. EU recommendations (eg in relation to the prevention of sexual harassment) may provide guidance where legislation is insufficiently clear.

THE EU AND HEALTH AND SAFETY AT WORK

The emphasis on health and safety in the EU objectives

Because of changes introduced by the Single European Act 1986, health and safety has become an important area of European legislation. Two aspects of the SEA are of particular significance.

Article 118A

First, the introduction of article 118A into the Treaty of Rome gives health and safety a particular prominence in the objectives of the European Union. Article 118A states that particular attention shall be paid to encouraging improvements in the working environment as regards the health and safety of workers in order to harmonise conditions within the Union. To help achieve this, the qualified majority voting procedure introduced by SEA can be used to adopt directives. This means that health and safety matters are less likely to get held up in the EU legislative machine since unanimity is no longer required. The vehicle of a 'framework' directive (p46) has been used to give effect to the objectives of article 118A.

Technical standards

The second aspect of the SEA that is particularly relevant to health and safety is article 100A. This allows the Council to adopt by a qualified majority, in cooperation with the European Parliament, measures to further the establishment and functioning of the internal market. While article 100A excludes matters of taxation, free movement of people and provisions relating to the rights and interests of employed people, it has been held to *include* technical standards and safety requirements for specific products.

The EU is agreeing a single set of rules on technical standards to replace the legislation in individual Member States. The main means of achieving this will be a set of directives ('new approach' directives). A new approach directive sets out the range of products covered by the directive and the essential requirements eg in relation to safety, which must be met before the product can be sold in the EU. There has to be attestation – a demonstration that the essential requirements have been met. This can be done by:

- a manufacturer's declaration (backed up by test results)
- a certificate from an independent body
- the provision of test results by an independent body.

The CE mark may then be displayed. Conforming with a European standard shows that the EU's requirements have been met. Such standards are prepared by two EU committees.

A number of new approach directives have been adopted and new UK legislation introduced:

Toy safety
Simple pressure vessels
Construction products
Electromagnetic compatibility
Gas appliances
Personal protective equipment
Machinery safety including mobile machinery and lifting equipment
Active implantable medical devices
Non-automatic weighing instruments
Telecommunications terminal equipment

The health and safety aspects of the 'new approach' directives may be far-reaching. For example, the machinery directive – relating to the design and construction of machinery – has extensive health and safety requirements (see chapter 17).

In addition to developments arising out of the introduction of articles 100A and 118A which use the qualified majority voting procedure, there are health and safety provisions in the Charter of Fundamental Social Rights (The 'Social Charter') dealt with below, and in the EU's environmental protection programme (see chapter 28).

The EU's social action programme

The Community Charter of Fundamental Social Rights (the 'Social Charter')

The Social Charter was adopted at the European Council in December 1989 by eleven Member States, excluding the UK. The Charter is essentially a declaration; it is not an internationally binding treaty and it has no legal force. An extract relating to health and safety at work is reproduced below.

EXTRACT FROM THE SOCIAL CHARTER

Health protection and safety at the workplace

19 Every worker must enjoy satisfactory health and safety conditions in his working environment. Appropriate measures must be taken in order to achieve further harmonisation of conditions in this area while maintaining the improvements made.

These measures shall take account, in particular, of the need for the training, information, consultation and balanced participation of workers as regards the risks incurred and the steps taken to eliminate or reduce them.

The provisions regarding implementation of the internal market shall help to ensure such protection.

As can be seen, part of the Social Charter is devoted specifically to health and safety at work and makes a commitment to an improvement while seeking to ensure that the process of harmonisation of conditions between the Member States does not undo that improvement. In addition, other parts of the Social Charter may have a bearing on health and safety at work, in particular:

- information, consultation and participation rights for workers, eg where there are technological changes affecting working conditions
- limitations on the employment of children and young persons.

The Social Action Programme

The Social Charter has been converted into a programme of measures – the 'Social Action Programme' – which was adopted by the Commission in 1990. This programme is currently going through the EU's legislative process.

Health and safety measures already adopted are discussed on pp46–8.

Because article 118A allows qualified majority voting, it has been found politically expedient to use it as a base for directives in the employment field even where the health and safety aspects are secondary. For example, a directive on the limitation of working time was adopted under article 118A (but the treaty base is subject to a challenge at the ECJ). Moreover, health and safety issues occur alongside other matters in a number of directives. Deciding what is a health and safety provision, therefore, is not always straightforward. Various aspects of the Social Action Programme have not yet been adopted principally because matters concerned with the terms and conditions of employment of workers have been subject to unanimous voting (this has not been the case with health and safety at work). However, the

Social Policy Protocol agreed at Maastricht gives Social Policy a legal status and will allow some employment matters to be subject to qualified majority voting, although the UK will not be bound by EU decisions arrived at through this procedure.

The main legislative device for implementing the programme is the directive, which requires Member States to enact their own legislation within a given time period. In addition to the Social Action Programme there is a continuing process of improving existing legislation. Thus in 1992 the Council of Ministers adopted a directive on collective redundancies to strengthen the provisions of an earlier directive.[3]

The EU's approach to health and safety at work

Apart from the Social Charter, the EU has formulated separate action programmes for health and safety at work. The Commission's third action programme, agreed in 1988, focused upon safety and ergonomics (including the high-risk sectors of construction, agriculture and shipping) and health and hygiene (including carcinogens, biological agents and noise). Particular stress has been laid upon information and training, especially in relation to work with chemicals. Some initiatives were targetted at small and medium-sized enterprises and at developing a dialogue between employers' and workers' organisations.

The approach of the EU to health and safety at work contrasts with that of HSWA. The latter adopted an approach based upon the idea of broad, generally-applicable duties (see chapter 3) while the EU legislation tends to set out the various obligations in considerable detail. The result is that the HSWA now provides an umbrella for numerous sets of EU-inspired regulations. An important feature of the EU approach has been the emphasis on health and safety management, and in particular upon the risk assessment as a basic management tool (see chapter 8). Risk assessments have now become a major requirement under the UK law.

The EU has published the framework of its fourth action programme. This covers the years 1994–2000. The emphasis is somewhat different from that of the previous period, where there was a strong legislative impetus. The emphasis for 1994–2000 reflects the EU's increasing concern to create jobs and be competitive on the wider world stage. Thus, the next period will see:

- consolidation of legislation and implementation by all Member States
- an evaluation of legislation already implemented
- attempts to achieve consistency of enforcement
- an awareness of the impact on small firms and
- a thrust towards investigation and guidance rather than legislation.[4]

THE EU'S IMPACT ON HEALTH AND SAFETY AT WORK IN THE UK

Earlier EU initiatives

A large number of EU health and safety initiatives were connected with the formation of the Single European Market in 1992, but by the late 1980s there had already been a considerable amount of significant EU legislation. The results of this in the UK were important sets of regulations governing, among other things:

- the control of substances hazardous to health
- electricity
- noise
- pressure systems.

The duties laid-down under the UK law in relation to these areas are dealt with in detail in Part III. A brief comment on their impact is made below (p48).

Developments associated with the formation of the Single Market

As noted earlier, the Single European Act provided an impetus for the development of health and safety legislation through the process of qualified majority voting. The result has been the adoption of a broad, framework directive and of numerous specific directives.

Directive on the introduction of measures to encourage improvement in the safety and health of workers at work (the 'framework directive')

The 'framework' directive adopted in 1989 can be seen as a EU equivalent of the HSWA. It lays down broad duties akin to those contained in s2 of HSWA and embodies the principle (found in British law, in, for example, the COSHH and Noise Regulations) that the risks inherent in work activities should be assessed, and appropriate control (including preventive) measures introduced. The directive has been given effect in Great Britain through the Management of Health and Safety at Work Regulations 1992 and an approved Code of Practice (see chapter 3) henceforth, the Management Regulations.

Workplace health and safety

This directive is given effect through the Workplace (Health, Safety and Welfare) Regulations 1992 and an approved Code of Practice (see chapter 16).

Use of work equipment

This directive is given effect through the Provision and Use of Work Equipment Regulations 1992, which are accompanied by guidance (see chapter 17).

Use of personal, protective equipment (PPE)

This directive is given effect through the Personal Protective Equipment at Work Regulations 1992 and accompanying guidance (see chapter 25).

Manual handling of loads

This directive is given effect through the Manual Handling Operations Regulations 1992 and accompanying guidance (see chapter 19).

Display screen equipment

This directive is given effect through the Health and Safety (Display Screen Equipment) Regulations 1992 and accompanying guidance (see chapter 19).

Temporary workers

This directive is given effect through the Management Regulations.

Protection of pregnant women at work

This directive deals with a number of employment issues, such as maternity leave, but also contains health and safety provisions. A duty is placed upon employers to assess the risks to the health and safety of pregnant women and to take health and safety measures in the light of the assessment. This requirement is given effect through amendments made to the Management Regulations (see chapter 27).

Other directives adopted or proposed

Among the other directives recently adopted are those covering:

- carcinogens
- asbestos
- biological agents
- health and safety on temporary or mobile construction sites (this directive has given rise to an important set of new regulations – the Construction (Design and Management) Regulations – see chapter 15)
- extractive industries
- safety signs at the workplace
- protection of young people at work
- health and safety on fishing vessels
- the organisation of working time.

There is also a recommendation that there should be an EU-wide schedule of prescribed industrial illnesses and diseases.

There remain a number of proposed directives which have yet to be adopted. Included are those dealing with:

- physical agents
- medical assistance on ships
- amendments to the Use of Work Equipment directive
- chemical agents
- industrial agents
- risks from physical agents
- transport safety.

Chapter 5 deals specifically with likely future developments, including the proposal to establish a European Health and Safety Agency.

Assessing the impact

Historically the UK approach to health and safety was a piecemeal one because legislation was an ad hoc response to problems arising out of specific hazards in particular types of workplace. The Robens Report argued for a simpler, all-embracing approach which was given effect through HSWA.[5] As noted earlier, the EU legislation has reintroduced a great amount of detail into UK legislation, in sharp contrast to the Robens approach.

It is difficult to assess the impact of the EU-derived law but the starting point is to recognise that EU law has to be implemented in the Member States. The UK has been one of the most active countries in implementing EU health and safety legislation[6] and the impact seems to have been as follows.

First, much of the EU law makes explicit those duties which already existed in a general form. Thus, the general duties under HSWA, s.2 could be said to imply that a risk assessment should be carried out.

Second, some areas are completely new: eg the new construction management regulations. The effects are difficult to separate from those caused by a general decline in the high-risk mining and manufacturing sectors, and from those caused by existing legislation, but progress appears to be being made, particularly on the safety front.[7] On the other hand, there is increasing awareness of the burden being placed on employers, (felt disproportionately by small and medium-sized employers) by a very substantial increase in detailed regulation. The HSC has recently completed a review of the legislation in which individual sets of regulations were subjected to a cost/benefit analysis (see chapter 5).

Another concern is the differential rate of implementation between Member States and the unevenness of enforcement regimes across the Community.[8] These concerns are visible in the framework of the EU's fourth health and safety action programme.

Notes

1. The ECJ should not be confused with the European Court of Human Rights (ECHR) set up under the European Charter of Human Rights 1950 by the wider, Council of Europe. Alleged breaches of human rights are filtered by a European Commission of Human Rights and cases which have merit are then heard by the Court which sits in Strasbourg.
2. The 'six pack' comprises:

   - The Management of Health and Safety at Work Regulations 1992
   - The Manual Handling Operations Regulations 1992
   - The Workplace (Health, Safety and Welfare) Regulations 1992
   - The Personal Protective Equipment at Work Regulations 1992
   - The Health and Safety (Display Screen Equipment) Regulations 1992
   - The Provision and Use of Work Equipment Regulations 1992

For details of where these are dealt with in the text see the index.

3. 75/129 of 17 February 1975 (OJ, 1975 L48) and 56/92 of 24 June 1992 (OJ, 1992 L 245).
4. *HSC Newsletter*, No 93, February 1994, p8
5. *Safety and Health at Work*, Report of a Committee, chairman Lord Robens, 1972, Cmnd 5034
6. *European Parliament News*, UK edition, Dec 13–17 1993, p3
7. *HSC Annual Report 1992–3* See HSC Newsletter, No 92, December 1993
8. *Harmonisation and Hazard: Workplace Health and Safety in the European Community*, Baldwin, R and Daintith, T (eds), London: Graham and Trotman, 1992

3

Enforcement of Health and Safety Law

The purpose of this chapter is to describe the general legal framework of health and safety at work in terms of the criminal law and its enforcement. The civil process is described in chapter 4, while legislation relating to particular hazards or processes is dealt with by subject in Part III. Practical guidance as to how to stay within the law is given in Parts II and III.

INTRODUCTION

Background[1]

The first legislation on health, safety and welfare was at the beginning of the nineteenth century, the focus being on health and welfare rather than safety.[2] In particular, there were restrictions on hours of work and provisions for cleanliness and ventilation. The legislation had limited effect: not only was it restricted in its scope, but there was no proper mechanism for enforcement. Voluntary, unpaid inspectors were appointed by local magistrates. Subsequent legislation prior to 1833 suffered from the same defects.

The Factory Act 1833 empowered the government to appoint paid factory inspectors but the numbers appointed were limited and the Act applied only to textile mills. Again, the legislation dealt with health and welfare, but not safety. However, the new inspectors quickly became aware of the problem of accidents and realised that the failure to fence and guard machinery was a major factor. The government called for a report on the issue and the inspectors provided this in 1841, arguing the case for safety legislation. Social reformers were in support of the case and attention was drawn to the issue indirectly in 1840 when Lord Shaftesbury took successful legal action for damages in negligence on behalf of an employee injured in a factory accident (*Cotterell v Stocks*).

In fact it was the Coal Mines Act of 1842 which contained the first safety provision. However, the principal development was contained in the Factories Act 1844. This included prohibition of the cleaning of machinery that was in motion, required the fencing of certain machinery and provided for the reporting of accidents. Opposition by employers led to the 1844 Act being weakened by an amending Act in 1856 in respect of its fencing requirements.

The Inspectorate developed in strength but from the outset adopted the method of working by persuasion and conciliation that remains one of its hallmarks. Legal action was used only when everything else had failed. Over many years there was a general extension of legislation to cover different types of workplace and different types of process. However, the effect was separate legislation for mines and quarries, agriculture and factories, and gaps in the legislation meant that not all employees had protection. There was a multiplicity of legislation even as regards factories, the first consolidation not occurring until the Factory and Workshop Act of 1901. By this time it was apparent that the inspectors were adopting a cautious approach to prosecution, and that in any event prosecution on a large scale would be administratively impracticable. Against this background, the development of the civil law was particularly important. Employees became able to use the statutory duties as a basis for civil actions for damages. The first successful action for damages for breach of statutory duty was in 1898 (*Groves v Lord Wimborne*). (See chapter 4 on civil liability.)

There were, of course, problems of interpretation such as: what was "dangerous" machinery within the meaning of the Factories Acts? Major reforms removing outdated distinctions – such as that between textile and non-textile factories – were effected through the Factories Act 1937. This and subsequent amending legislation was then consolidated in the Factories Act 1961, parts of which are still operative. The first legislation relating specifically to shops was 1886 and offices 1960.[3] The Offices, Shops and Railway Premises Act 1963 consolidated the legislation. The principal legislation governing mines and quarries is the 1954 Act.[4]

Finally, there have been two major developments in the last 25 years. First, the patchwork history of health, safety and welfare legislation that extended over 172 years was reformed in 1974 by HSWA. The new legislation was all-embracing in its coverage, removing the distinctions between different types of workplace and placing general duties upon employers, employees and others (see below). The intention was to replace the old, workplace-specific legislation with regulations issued under the 1974 Act, and to encourage employers, employees and others to take more responsibility for health and safety at work. Consequently, there has been increased emphasis on the *management* of health and safety at work. The 1974 reforms were based upon an analysis of the shortcomings of the old legislation carried out by the Robens Committee of Inquiry during the early 1970s.[5]

The second major development has been caused by the UK's membership of the European Union and the commitment of the Union to the improvement and harmonisation of health, safety and welfare in the Member States (see chapter 2). One effect of this has been to modify the more generalist approach of the HSWA by requiring detailed, prescriptive regulation. To date, this has been achieved mainly by issuing regulations under HSWA.

The current statutory provisions are examined later in this chapter. Prior to this, however, it is necessary to consider some of the key words and phrases used in the legislation. An understanding of these is required if the duties laid down in the legislation are to be translated into actual management practice. The first step is to describe the legal framework.

The legal framework

Breach of statute law in this area gives rise to criminal offences, with prosecutions taken by the Health and Safety Executive or the appropriate local authority, and to claims for damages by injured parties for breach of statutory duty. Civil claims for such breaches are not, however, permitted on the basis of the general sections of HSWA nor on the basis of the Management of Health and Safety at Work Regulations.[6] Nevertheless injured parties might be able to sue for damages for negligence (see chapter 4). Much of the legislation laying down specific duties is in the form of regulations made under one of the major Acts, predominantly HSWA, although some specific duties are laid down in the major Acts themselves. Regulations have the full force of law and are frequently backed by approved codes of practice. These have the object of providing practical guidance on the requirements contained in ss2–7 of HSWA, or in regulations, or in any of the existing statutory provisions. Approved codes of practice have a specific status (see p54).

In addition to the statutes, the common law imposes duties upon employers (and employees). A major source of these duties is the law of torts (civil wrongs) under which an employer has a duty of care arising out of his or her position as employer and as owner or controller of premises Thus, where the employer is also the owner or controller of premises his duty extends to people who are not employees. The duty of care towards employees is also found as part of the employment contract. It should be noted that criminal proceedings may be taken under the common law. Where negligence is gross, or behaviour reckless, a prosecution for manslaughter may occur (see p71). The prosecution would need to establish that the accused was aware of an obvious and serious risk as well as being guilty of the act or omission itself. Where senior employees who are in control of the organisation were aware of such a risk and allowed such acts or omissions to occur, a charge of corporate manslaughter may accompany the charges against individuals. This happened in the prosecution following the capsizing of the ferry *Herald of Free Enterprise* and also in the *Lyme Bay* canoeing case, although the prosecution was successful only in the latter.

A number of features of legislation in the health and safety field are worth stressing. First, the law is often invoked, particularly in respect of the more serious cases. Second, one incident often gives rise to two separate cases – a civil case, pursued in the High Court or county court, and a criminal case, usually prosecuted in the magistrates' (but sometimes Crown) court. Third, different parts of the legislation set different standards. Thus, some requirements are mandatory (that is, there is an absolute duty) and others need to be met as far as is practicable or as far as is reasonably practicable. Finally, health and safety is an area where EU law plays an increasingly important role. EU health and safety legislation is subject to the qualified majority voting process in the Council of Ministers (see chapter 2).

LEGAL TERMINOLOGY

Standards of care

Absolute

Where the statute states that something 'shall' or 'shall not' be done, there is an absolute requirement to comply. No test of practicability, nor indeed any other test is to be applied. For example, The Provision and Use of Work Equipment Regulations 1992 state that: "every employer shall ensure that work equipment is maintained in an efficient state, in efficient working order and in good repair."[7] Similarly under the Management Regulations, "every employer shall make a suitable and sufficient assessment ..."[8]

Practicable

If the requirement is that a duty is to be fulfilled so far as is practicable, it is the responsibility of the person on whom the duty is placed to meet their obligation as far as the current state of knowledge and invention allows. The burden of proving that the standard was met lies with the accused.[9] An example taken from the Management Regulations is as follows: "so far as is practicable", employees exposed to serious and imminent danger must be informed of the hazard and the protective measures.[10] A variant of this standard is the requirement to use the "best practicable means". The meaning of practicable is dealt with on p57.

Reasonably practicable

Practicable implies a standard that is less strict than the absolute, and reasonably practicable is less onerous than practicable. The quantum of risk is placed in one scale and the cost, inconvenience etc of preventive measures placed in the other. The onus of proof lies with the accused, but the test is the balance of probabilities. The general duties of employers in HSWA are an example of the use of this standard.[11] The meaning of reasonably practicable is dealt with on p57.

Work and employment

The issue and significance of employment status was mentioned in Chapter 1. A second issue which arises is whether the act or omission giving rise to the injury occurred at work, and if at the workplace, whether during the course of work. HSWA defines work to mean "work as an employee or as a self-employed person".[12] An employee is at work when "in the course of his employment".[13] The definitions leave a lot of room for interpretation and the issue has been much considered in cases concerning civil liability (see chapter 4). These are not necessarily authority for criminal law and in one case a person driving to his place of work but already on the employer's internal road system just under 20 minutes before the start of his shift was held to be not in the course of his employment (*Coult v Szuba*).

Other definitions

Personal injury

This is defined to include "any disease and any impairment of a person's physical or mental condition".[14] This is a very wide definition and one that is of considerable importance in view of the increasing incidence of stress-related illness arising out of overwork.

Competent person

Where there is a requirement to appoint such a person, the specific regulations and any associated materials (ACOPs; Guidance Notes) may provide assistance as to what is needed. There is no universal definition. (See p62 and chapter 8.)

Codes of practice and guidance notes

Approved codes of practice

A code of practice is defined to include "a standard, a specification and any other documentary form of practical guidance".[15] The HSC is empowered under HSWA to approve and issue codes of practice.[16] The aim of such codes is to provide practical guidance on how to comply with the statutory requirements. The HSC must obtain the consent of the Secretary of State, after consulting appropriate bodies including government departments. Codes do not have the full legal status of Acts, Regulations or Orders (see chapter 1) but neither are they merely advisory. Rather, they can be seen as quasi-legal. A person will not be liable simply because they failed to observe the provisions of a code and this will be the case under both criminal and civil law. However, in criminal proceedings the breach of an approved code of practice (ACOP), where the person is alleged to have committed an offence in respect of a matter to which the code relates, will cause the code to be admissible in evidence. Moreover, breach of the code will be taken as proof of a failure to comply with the statutory provisions unless the accused can show that compliance was achieved by some other equally satisfactory means.[17] Thus a breach of an ACOP creates a presumption of guilt which the accused must then rebut to establish their innocence. It seems unlikely that anyone following the provisions of an ACOP would be successfully prosecuted. Nothing is said about the status of ACOPs in civil proceedings but it seems likely that a breach would provide prima facie evidence of negligence. A successful defence would require the defendant to bring evidence to the contrary.

Guidance notes

In contrast to codes, guidance notes have no legal standing. However, they contain detailed practical guidance on how to comply with the statutory provisions, and are generally regarded as very useful supplementary material for those who have statutory duties. An important consideration is their use as evidence of the state of knowledge. Since the guidance is authoritative, being produced by the HSC, it is likely to be regarded as good evidence of what is known about particular risks and preventive measures at the time of issue. Thus it would be difficult for an employer to succeed with the argument that he was unaware of a risk if that risk and suitable preventive or control measures were dealt with in a set of guidance notes. For ease of access, the HSC publishes regulations, any ACOP and guidance notes in one booklet. The example below is taken from The Workplace (Health, Safety and Welfare) Regulations, ACOP and Guidance Notes, 1992 (see chapter 16 for discussion of the contents).

EXAMPLE OF INTEGRATED REGULATIONS, ACOP AND GUIDANCE NOTES

Regulation 7

"(1) During working hours, the temperature in all workplaces inside buildings shall be reasonable."

ACOP

"43 The temperature in workrooms should normally be at least 16 degrees Celsius ..."

Guidance

The Guidance Note contains details of relevant sources and indicates other, related statutory duties.

GENERAL STATUTORY DUTIES

Health and Safety at Work etc Act 1974

General duty of employers to their employees

The philosophy behind this legislation is that the duties of employers should be set out in broad terms covering all sorts of employment situations – shops, offices, factories etc. The Act covers all persons at work except domestic servants in private households. Sets of regulations issued under the Act then deal with different types of work situations or different health and safety issues.

It should be noted that all those receiving training or work experience from an employer in the workplace, and who are not employees, are deemed to be employees for the purposes of health and safety legislation.[18]

The essence of HSWA is a general duty imposed upon employers which requires them "to ensure, so far as is reasonably practicable, the health, safety and welfare at work" of all their employees.[19] This general duty includes health and safety in relation to the:

- provision and maintenance of plant and systems of work
- arrangements for use, handling, storage and transport of articles and substances
- maintenance of the place of work and access to and egress from it
- working environment

and includes the duty to provide:

- adequate welfare facilities
- necessary information, instruction, training and supervision

and to consult with union safety representatives if there are any. A notice containing the main provisions of the Act must be posted.[20]

Employers must prepare a written statement of their policy with respect to the health and safety at work of their employees, and show it to an inspector if requested to do so. The statement must indicate the organisation and arrangements in force to give effect to the policy and must be revised as often as is appropriate. The statement, and any revision of it, must be brought to the notice of all the employees.[21] There is a provision for exceptions, but the only exemption granted to date is for employers who employ fewer than five employees.[22] The number of employees relates to the undertaking rather than the site or establishment, and trainees count as employees for the purposes of these and other health and safety requirements. The HSC has produced a leaflet *Writing a Safety Policy Statement: Advice to Employers* (HSC6) (see chapter 7).

By virtue of s235 of, and schedule 7 to the Companies Act 1985, the Secretary of State has power to prescribe by regulations information to be contained in directors' reports about a company's arrangements for ensuring the health, safety and welfare of its employees, but this power has not yet been invoked.

Where, under HSWA,[23] a safe system of work for the employer's own employees involves giving information and instruction to persons other than his or her own employees (the employees of a contractor, for instance) it must be given (*R v Swan Hunter Shipbuilders Ltd and Telemeter Installations Ltd*). Thus, employers have a duty to provide relevant safety information to the employees of other employers working on their premises.

Meaning of practicable and reasonably practicable

'Practicable' means something less than physically possible. It means feasible – possible in the light of current knowledge and invention. Where the requirement is that something should be as safe as is practicable this means that the duty must be performed unless it is unreasonable to do so (*Marshall v Gotham and Co Ltd*). The word 'reasonably' requires that a comparison be made between the risk of injury (including the severity of any injury which might occur) and the time, trouble and expense of preventive action. If there is gross disproportion between them (that is, the risk is insignificant compared with the preventive measures needed), the defence that an employer has done what was reasonably practicable will succeed (*Edwards v National Coal Board*). Because reasonably practicable involves weighing the risks against the costs of prevention, employers will need to adduce evidence about the risks and costs involved, so influencing what is reasonably practicable. For example, in assessing the likelihood of risk a relevant consideration would be the length of time the employee is exposed to the risk. When considering preventive measures, doubts about the efficacy of those measures in reducing risk may be of relevance. The onus of proof lies upon the accused to show that it was not practicable or reasonably practicable to do more than was done.[24]

Duties owed to non-employees

The HSWA operates where people are at work rather than by specifying particular types of workplace. It therefore covers all premises including vehicles, movable structures and off-shore installations. Under the Act employers have a responsibility to show the same standard of care towards non-employees (such as visitors and contractors) as they are required to show to their employees and, as noted, this duty includes provision of information and instruction (*Swan Hunter*).[25] The duty is not restricted to the employer's premises. A self-employed person has similar duties. Both the self-employed and employers may be required to provide information to non-employees about the way in which their undertakings might affect such people's health and safety.

Persons who have "to any extent, control of premises"[26] (but not domestic premises) or control access to or egress from such premises, have duties to persons other than employees and must take such measures as are reasonable for a person in that position to take, as far as is reasonably practicable. This does not extend to taking measures to guard against unexpected events (*Austin Rover Group Ltd v HM Inspectorate of Factories*). Whether precautions in relation to someone else's use of the employer's premises are reasonable will depend upon:

- the employer's knowledge of the anticipated use of the premises
- the extent of control and knowledge of actual use.

In *Austin Rover*, contractors' employees were in breach of safety rules and the controller of the premises (Austin Rover) could not reasonably have been expected to foresee and to guard against this.

The Act also covers control of emission into the atmosphere of noxious or offensive substances from prescribed premises. These are premises and substances laid down in the Health and Safety (Emission into the Atmosphere) Regulations 1983 as amended.[27] The duty imposed is to use the 'best practicable means' for preventing the emission and to render harmless or inoffensive any substances emitted. (Part I of the Environment Protection Act 1990 and regulations thereunder introduce new controls over air pollution.) (See chapter 28)

Duties of designers, manufacturers and others

Duties are also placed upon designers, manufacturers, importers and suppliers to require the safe design and construction of articles, testing and examination and the provision of information indicating the designed use and precautions needed to avoid risks. Designers and manufacturers must carry out necessary research.

Those erecting or installing articles are also responsible for safety. Parallel requirements exist in relation to manufacturers, importers and suppliers of substances. The standard of care is what is reasonably practicable to achieve safety and the absence of risks to health. This section of the Act has been strengthened by the Consumer Protection Act 1987.[28]

Manufacturers and others must now consider reasonably foreseeable risks. Moreover, these risks are to be considered in relation to handling, maintenance and storage, as well as use. The requirement to provide health and safety information is also widened so that it covers revision (for example, in the light of new knowledge), applies not just to use, and covers situations such as foreseeable errors by users. Manufacturers and others must also take account of non-domestic premises other than workplaces to which they supply substances. As a result of the Control of Substances Hazardous to Health (COSHH) Regulations (see chapter 23) there is now a much clearer onus upon employers to obtain health and safety information, and data provided by suppliers are a major source.

Duties of employees

It is noted below that employees have a common law duty to go about their work with reasonable care. This is supplemented by the 1974 Act which lays upon employees a statutory duty of reasonable care towards themselves and those who may be affected by their acts or omissions. There is also a statutory duty to co-operate with the employer or any other person in meeting the statutory requirements. Reasonable care includes the use of equipment provided for employees' safety. The employer must make sure that employees know of the equipment and must take steps, as far as is reasonably practicable, to get them to use it. Standards for personal protective equipment are now set down by an EU Directive (see chapters 2 and 25) so that it is important for the equipment to meet these standards if a successful defence is to be raised by an employer. The duties of employees include not intentionally or recklessly interfering with or misusing anything provided for health, safety and welfare purposes in pursuance of the statutory provisions. (This is a duty placed upon all persons and not just employees.)

Individual employees can be, and occasionally are, prosecuted by the inspectors under these provisions. Moreover, as noted below (see p68) senior managers, as well as the organisation, may be liable under HSWA where they consented to or connived at the commission of an offence or neglected their duties.[29]

No charge must be made for anything done or provided in respect of any *specific* requirement of health and safety legislation.[30] Anything necessary as a result of an employer's PPE assessment is a specific requirement (see chapter 25).

Safety representatives[31]

Regulations under the Act allow a recognised trades union to appoint safety representatives, with specified functions:

- to investigate potential hazards, dangerous occurrences and the cause of accidents
- to investigate and process members' complaints about health and safety at work
- to deal with management over general questions of health and safety
- to carry out workplace inspections
- to represent members in discussions with inspectors
- to receive information from the inspectors
- to attend meetings of health and safety committees.

The Safety Representatives and Safety Committees Regulations have been extended by the Management Regulations of 1992. There is now a specific duty placed upon employers in respect of consultation with SRs and the provision of facilities and assistance. The duty to consult, as laid down in the

Management Regulations, covers: any change which may "substantially" affect employees' health and safety; the appointment of a competent person to assist with health and safety; the appointment of a person responsible for implementing emergency plans; information and training requirements; and the introduction of new technology.[32]

A code of practice and guidance notes give further detail. As far as is reasonably practicable a safety representative (SR) should have been employed by their employer for the previous two years, or have had at least two years' experience of similar employment.[33] Employers must provide SRs with facilities for inspections, including the opportunity for independent investigation and discussion. However, employers are entitled to be present during inspections.[34] SRs can have access to any documents kept by their employer for statutory health and safety purposes, except those relating to the health and safety of individuals, and in general are entitled to information which it is necessary for them to have to perform their duties. SRs are not entitled to information where:

- disclosure would be against the national interest
- disclosure would result in contravening a statutory prohibition;
- it relates to an individual (unless the individual gives their consent)
- apart from health and safety effects, the information would cause substantial injury to the employer's undertaking, or where the information was supplied to the employer by some other person, to the undertaking of that other person
- information was obtained for the purpose of any legal proceedings.

Disclosure is restricted to documents or parts of documents which relate to health, safety or welfare.

Safety representatives are the appropriate people to receive information from the inspectors about particular occurrences in the workplace, and any action that inspectors take or propose to take. The safety representatives are given a right to time off in order to perform their functions, and also for training.

A separate code of practice covers time off for training.[35] This states that basic training should be provided as soon as possible after appointment. There should also be further training for any special responsibilities or changes in legislation or work circumstances. Training should be approved by the TUC or the independent trades union to which the SR belongs. However, the code does not have full legal status: the law requires such training as may be reasonable in all the circumstances. While such circumstances include the code, the ultimate test is reasonableness. Thus in *White v Pressed Steel Fisher Ltd* the employer was reasonable in refusing time-off for union SR training when it preferred its own in-company course.

The code states that basic training should include the role of the SRs, and safety committees. These should be related to the legal requirements, the nature and extent of workplace hazards and precautions and the employer's health and safety policy. The training should also develop skills, for example, accident and incident investigation, conduct of inspections and use of legal and official sources. Management should be provided with a copy of the syllabus on request. Normally, a few weeks' notice should be given of the names of SRs nominated to attend a course. The numbers should be reasonable taking into account the availability of courses and the operational requirements of the employer. Union training for SRs should be complemented by employer training which should focus upon the technical hazards of the workplace, relevant precautions for safe methods of work and employer organisation and arrangements for health and safety.

A safety representative may complain to an industrial tribunal if refused reasonable time off or refused pay for such time off. Where at least two SRs request it in writing, an employer is required to establish a safety committee. The overall function of such a committee is to keep health and safety measures under review.[36] (See chapter 14)

Management of Health and Safety at Work Regulations 1992[37]

Introduction

With the exception of merchant shipping, these regulations apply to all workplaces and workers (including the self-employed). They flesh out some of the basic principles already established in the Health and Safety at Work Act and are accompanied by an Approved Code of Practice (ACOP) which gives some detail of how to comply with the regulations. The management implications of these regulations are dealt with fully in Part II.

Reg 3 Risk Assessment (see chapter 8)

Employers (and the self-employed) must make "suitable and sufficient" assessments of the risks to employees and others who may be affected by their work activities in order to identify the measures necessary to comply with the law. Significant findings must be recorded where there are five or more employees.

Unfortunately the ACOP does not give illustrations of risk assessments but it does stress that it should normally be a straightforward process with specialist skills and quantitative techniques only necessary in the more complex situations eg chemical plants, nuclear installations.

Reg 4 Health and Safety Arrangements

Arrangements must be made through planning, organisation, control, monitoring and review for putting the measures identified under reg 3 into effect. Again these must be recorded where there are five or more employees.

Reg 5 Health Surveillance

This may be required if problems are likely and there are valid techniques for detecting them.

Reg 6 Health and Safety Assistance (see chapter 7)

Employers must appoint one or more 'competent' persons to assist them in undertaking measures to comply with the law. The numbers, time available and resources must be adequate for the size and risks of the undertaking.

'Competence' is regarded as a knowledge and understanding of the work involved and health and safety principles and practices *together* with the capacity to apply this. (See chapter 8)

Formal competence-based qualifications provide a guide but in simple situations an understanding of relevant practices, a willingness and ability to learn and an awareness of one's own limitations may be sufficient.

Small employers may appoint themselves if they are competent.

Reg 7 Procedures for Serious and Imminent Danger and Danger Areas (see chapter 9)

Procedures must be established and put into effect for foreseeable emergencies including sufficient competent persons to implement evacuations if necessary. Where specific danger areas are identified (eg due to electrical or chemical risks) access to these must be restricted.

Reg 8 Information for Employees (see chapter 12)

Employees must be given 'comprehensible and relevant information' on risks they are exposed to, appropriate countermeasures, emergency arrangements and other special procedures.

Reg 9 Shared Premises (see chapter 15)

Employers sharing workplaces (temporarily or permanently) must co-operate, co-ordinate and inform each other on health and safety matters.

Reg 10 Persons Working in Host Employers' or Self- Employed Persons' Undertakings (see chapter 15)

The host must give visiting employers comprehensible information on risks arising out of, or in connection with, the host's undertaking and measures taken by the host to comply with the law (where this relates to visiting employees).

Hosts must also ensure visiting workers are given instructions and information on risks and emergency/ evacuation procedures.

Reg 11 Capabilities and Training (see chapter 12)

Employees must be given adequate health and safety training following recruitment, transfer, responsibility change or the introduction of new or changed equipment, technology or systems of work.

Account must be taken of health and safety capabilities in entrusting tasks to employees.

Reg 12 Employees' Duties

Employees must use machinery, equipment etc in accordance with their training and instructions. They must also report dangers or shortcomings which affect them or arise out of their work.

Reg 13 Temporary Workers

Information must be given to contract staff or agency employees on special safety related qualifications or skills needed and any health surveillance required before they commence duties.

Agencies must be given such information and pass it on to their employees.

Regs 14 – 17

The remaining regulations allow exemption certificates to be issued in certain defence-related situations; exclude civil liability; and modify the SRSC Regulations as indicated earlier. The regulations apply to Great Britain and to certain premises and activities outside Great Britain.

Recording and Reporting

Introduction

The occupier of a workplace – usually the employer – is required to keep particular registers and must display certain forms and notices under the various safety statutes. The HSE issues a comprehensive, free catalogue of its publications[38] and a complete list of forms can be obtained free of charge from any HSE enquiry point. The requirements include the following items.

- Display HSWA poster or distribute equivalent leaflets.
- Notification of accidents form. Employers will probably need a supply of these. Copies of the form can be used as a book.
- Accident book, for DSS purposes.[39]
- An up-to-date certificate of insurance. This must be displayed by virtue of the Employers' Liability (Compulsory Insurance) Act 1969. (see p85)

In addition, there are specific requirements under particular sets of regulations as well as under the main Acts themselves. The Factories Act, for instance, requires registers showing the testing, inspection and examination of chains, ropes and lifting tackle. A failure to display notices, submit forms or keep registers is a criminal offence.

Under the Factories Act 1961 and the Offices, Shops and Railway Premises Act 1963 an employer will need to register his occupation of premises with the relevant authority using the prescribed form.

Reporting of Injuries, Diseases and Dangerous Occurrences Regulations 1985 (RIDDOR)[40]

The requirements of these regulations are dealt with in chapter 9.

ENFORCEMENT OF THE LAW

The Health and Safety Commission (HSC)[41]

Constitution

The HSC is a body corporate established by the HSWA. It consists of a chairman and 6–9 other members, chairman and other members being appointed by the Secretary of State for Environment. Before appointing the members (other than the chairman) the Secretary of State must consult organisations of employers (in respect of three members), organisations of employees (in a further three cases) and local authorities and other bodies concerned with health, safety and welfare (with regard to the remaining three). The HSC began to operate in 1974.

Duties[42]

The Commission is under a duty "to do such things and make such arrangements as it considers appropriate" to give effect to the Act's general purposes. Principally, these are:

- to secure the health, safety and welfare of persons at work; and
- to protect those not at work from the activities of those at work.

More specifically, it has a duty:

- to assist and encourage persons concerned with matters relevant to HSWA's general purposes to further those purposes
- to make arrangements for and encourage research, publication of research findings, provision of information and training in relation to the Act's purposes
- to provide an advisory and information service for those concerned with the Act's purposes
- to make proposals for regulations.

The HSC plays a major role in proposing and shaping new legislation, particularly in the context of EU directives.

The Commission reports to and is under the direction of the Secretary of State for Environment. In 1993, the government directed the Commission to undertake a review of health and safety legislation to determine whether the legislation could be reduced and simplified. The results and implications of this review are discussed in chapter 5.

Powers[43]

The Commission has the power to do anything (except borrow money) which "is calculated to facilitate, or is conducive or incidental to" the performance of its functions. This includes appointing persons or committees to provide it with advice. A number of such committees have been formed covering particular classes of hazards and particular industries, and ad hoc committees may be established to deal with specific problems.

The HSC may issue codes of practice and guidance notes as described earlier (see pp54–5). It may also set up investigations and enquiries, for example where there has been an accident or occurrence, if this will assist it in its statutory duties. For there to be an enquiry, the consent of the Secretary of State is required. The report of an enquiry or investigation may be made public in part or in whole. An enquiry is a formal device with a statutory procedure.[44] The matter under consideration is likely to be of public interest.

In contrast, an investigation may be less formal and the investigator has freedom to determine the procedure. Nevertheless, rules of natural justice must be followed and evidence should be taken from anyone able to make a positive contribution.

The Health and Safety Executive (HSE)

Constitution and functions

The HSE is a body corporate consisting of three people one of whom is appointed by the Commission, with the approval of the Secretary of State, to be Director. The other two members are also appointed by the Commission with the approval of the Secretary of State, but after consultation with the Director.[45] The Executive exercises the functions of the Commission and is directed by the Commission. Thus, the HSE is the operational arm of the HSC. However, the Commission cannot direct the Executive on the enforcement of the law in any particular case.[46]

Structure of the HSE

Its basic job of inspections and work in the field is done by its Field Operations Division which incorporates the former Factory, Agricultural and Mines and Quarries Inspectorates and the Employment Medical Advisory Service (EMAS). The HSE is now responsible for the work formerly done by the

Railway Inspectorate and for offshore oil and gas safety. Other divisions of the HSE (eg health policy) provide back-up services in relation to both policy and operations. National interest groups for particular industries exist to provide specialist information.

Method of working

A major feature of the inspection process is the advice and encouragement given to employers to improve health and safety. Although the inspector carries a big stick, the day-to-day role is very much one of reasoning and persuasion. The aims are to establish what is going on, to encourage improvements, find out how the legislation is working and to assess new technological developments. A principal consideration is whether there is a commitment to health and safety – has management organised a system, and is it monitored and, if necessary, modified? The statutory written health and safety policy is the framework within which the HSE carries out its inspections. An assessment of management's ability to organise and maintain a safe workplace is formally recorded at the end of the inspection. Where there are problems, the improvement notice (see below) is a useful device. The average manager is likely to come into contact with an inspector at some stage: it could be through a routine inspection, because of an accident or complaint, or because the HSE is carrying out a special study or survey (eg of small firms).

A recent report by the National Audit Office suggests that the HSE is curently understaffed.[47] The report shows that the Executive had failed to assess 132 out of 331 reports on major hazard sites in the UK. Some 300 further reports are expected to be submitted in the next two years, threatening to increase the backlog.

Employment Medical Advisory Service

Next, there is the medical arm of HSE, the Employment Medical Advisory Service (EMAS).[48] This has a field staff of Employment Medical Advisors (doctors) and Employment Nursing Advisors (nurses). For statutory purposes the Employment Medical Advisors have the same powers of entry and investigation as inspectors. The role of EMAS includes research, provision of health advice and information and the medical examination of people working on hazardous operations. An EMA cannot force a person to be examined against his will.

The public information role of the HSE and HSC

Finally, there is the public information role of the HSE and HSC. They play a major role in the supply of information about health and safety at work. Their very large output ranges from detailed technical publications to free basic information leaflets. The HSE has public enquiry points and a fax information service.

The criminal process

Notices and prosecutions

Enforcement is principally by the HSE, although local authorities are responsible for enforcement in certain premises and for enforcing the implementation of general fire precautions. The Commission retains responsibility for guiding enforcement agencies which are outside the HSE. The allocation of enforcement responsibilities is determined by the Health and Safety (Enforcing Authority) Regulations 1989, issued under HSWA.[49]

These increased the local authorities' coverage by re-allocating premises including those used for leisure and consumer services, churches and other places of religious worship, and premises used for the care, treatment and accommodation of animals.[50] Certain lower risk construction work carried out in premises where local authorities already had enforcement responsibilities was also re-allocated. The 1989 regulations allocate to local authorities responsibilities for enforcing HSWA and relevant statutory provisions, subject to certain specific exceptions, in all premises where the main activity is listed under schedule 1 of the Act. Broadly speaking, local authorities have responsibility under the 1989 regulations for: premises where the main activity is the sale or storage of goods for retail or wholesale distribution; office activities; hotels and catering; places of entertainment. It should be noted, however, that the 1989 regulations allocate to the HSE sole responsibility for the enforcement of s6 of HSWA in relation to articles and substances for use at work including at those premises where the local authority would enforce the remaining provisions of the 1974 Act.

In terms of local authority enforcement, the principal role is performed by district councils and usually by the department which has an environmental health function. In addition, county councils (in England and Wales) and regional councils (in Scotland) have responsibilities such as those in relation to petroleum licensing, certain explosives, including fireworks, and the packaging and labelling of dangerous substances in consumer premises. Enforcement in these areas is carried out by trading standards officers or fire authorities as appropriate. Some of these 'county' functions, however, are performed at district level, for instance where the district is a metropolitan one or a London borough. If there is doubt about whether enforcement is in the hands of the HSE or a local authority (or which local authority), employers can check with the HSE. In each HSE area there is an enforcement liaison officer.

If there is a breach of duty an inspector may serve an improvement notice, or a prohibition notice, or may prosecute. An improvement notice will apply where the inspector alleges a breach of statutory duty. The inspector will describe in the schedule to the notice the practical steps which should be taken. The notice will specify a time period within which the improvement(s) must be made, although this is often negotiable. An employer may appeal to an industrial tribunal within 21 days against either the allegation of a breach or the shortness of the time period if, for example, this would cause undue difficulty for the production process. The appeal has the effect of suspending the notice. Very few employers appeal, and of those who do, very few succeed. This is not surprising: the HSE inspector is not only a health and safety expert but also has the full technical, medical and research backing of the HSE at his disposal. Where an inspector believes that there is a risk of serious personal injury he may serve a prohibition notice, which can have immediate effect. There is again a right of appeal but this does not suspend the notice unless an industrial tribunal so directs. The normal industrial tribunal costs rule does not apply.[51] In these cases costs may be awarded against the losing party at the tribunal's discretion. Because the Crown cannot prosecute the Crown, HSE issues Crown notices where improvement or prohibition notices would be used, (in the Civil Service for example) although these have no legal force.

Any breach of the general duties in the 1974 Act or of specific duties under other health and safety legislation is a criminal offence, as is obstructing an inspector or giving him false information. An employer will be liable for the acts of his or her employees during or arising out of the course of their employment. Where an offence is committed with the consent or connivance of, or due to the neglect of any director, manager, secretary or other similar officer the individual as well as the organisation is to be held guilty.[52] Thus, in *Armour v Skeen* the inspector prosecuted a senior employee because he had been given the responsibility for drawing up the health and safety policy, but had not done it. However, in *R v Boal* it was held by the Court of Appeal that the assistant general manager of a bookshop was not a 'manager' within the meaning of s23 of the Fire Precautions Act 1971.[53] Within the Act, the term manager, and the same can be said about HSWA, means managers who have corporate responsibility and are thus directing the policy of the organisation.[54] A manager in such a position may be individually liable. The other basis for the individual liability of managers lies in the duties placed upon employees, and here the duties of a manager may be more onerous because his acts and omissions may be capable of causing greater injury than an 'ordinary' worker (eg by instructing subordinates to use incorrect or faulty equipment, or unsafe methods of working). The legislation does not lay duties upon managers as a category: rather the duties are laid upon employers, employees and others.

Where there is a breach of an approved code of practice issued under HSWA this will be taken (in criminal proceedings) as conclusive proof of a contravention of a requirement or prohibition unless the court can be satisfied that there was compliance by some other method.[55] Prosecution for breaches of HSWA is limited to inspectors save for permission otherwise by the Director of Public Prosecutions; nor can the Act be used for the purposes of civil claims, that is for breach of statutory duty. Regulations under HSWA and legislation enacted prior to HSWA can be used for civil purposes unless they state otherwise.

A successful prosecution for a breach of HSWA requires proof beyond all reasonable doubt, as in criminal cases more generally. Thus, the relevant enforcement authority may decide not to prosecute if they conclude that the evidence is not strong enough. This was the case in relation to the *Piper Alpha* oil platform disaster in the North Sea in 1988, which resulted in the death of 167 men.

Powers of inspectors[56]

The powers of the inspectors are substantial. They have the right to:

- enter premises
- make examinations and inspections
- take samples
- take possession of articles and substances
- take measures and photographs
- make recordings
- keep things undisturbed
- have something tested, removed or dismantled
- obtain information and have answers to questions
- inspect and copy any entry in documents to be kept under statute.

Facilities and assistance must be provided for inspectors in connection with any of the above. Inspectors have any other powers necessary to fulfil their responsibilities. Where there is 'imminent danger of serious personal injury' an inspector may seize an article or substance and render it harmless.[57] An inspector does not need to give prior notice of a visit but an employer or occupier of premises is entitled to see proof of an inspector's identity prior to entry. An inspector may disclose relevant information to anyone who is a party in civil proceedings.[58]

Where an inspector exceeds his powers, there can be a civil action against him. Enforcing authorities have the power to indemnify their inspectors in such cases.[59]

Powers of the courts

Maximum penalties for different types of offence are laid down in HSWA. On summary conviction the maximum is £5,000 (ie level 5 – see chapter 1) except for those offences set out in the Offshore Safety Act 1992. Despite its specific title, this Act makes some generally applicable changes to penalties for certain offences under HSWA.

Magistrates' courts (Sheriff Court in Scotland) Fines of up to £20,000 may be imposed for breaches of:

- ss2–6 of HSWA
- an improvement notice
- a prohibition notice
- a court remedy order.

According to the HSE, 25–33 per cent of prosecutions are for such breaches.[60]

There can be imprisonment of up to six months for breach of a notice or a court remedy order. The time limit for commencement of prosecution for a summary offence is six months.

Crown Court There are no time limits for commencement of prosecution for indictable offences. There can be imprisonment for up to two years for certain offences including breach of a notice or court remedy order. A company director convicted on an indictable offence may be disqualified from being a director under the Company Directors Disqualification Act 1986.

In the Crown Courts, on indictment, the fines are unlimited. In addition, the courts can order specific forms of action in order to ensure safety.

All courts A fine relates to a particular offence. It is quite common for an incident to give rise to more than one offence. Fines totalling £750,000 have been recorded (BP, 1987) and, unlike in civil cases, employers cannot insure against criminal liability. An example of multiple penalties is given below (prior to the 1992 increases in fines).

Fines and imprisonment are not necessarily alternative penalties: they may be used together.

EXAMPLE OF MULTIPLE PENALTIES

Company	£
Machine not securely fenced	25,000
Failure to register a factory	500
Breach of a prohibition notice	1,000
Managing Director	
Machine not securely fenced	15,000
Failure to register a factory	500
Charge of manslaughter to lie on file	–
Another Director	
Machine not securely fenced	5,000
One year prison sentence suspended for two years	–
	47,000

Source: Works Management, April 1990, p 5

Employer defences

An employer's defence will be that there is no breach of statutory duty. For example, it might be maintained that there is little or no risk, therefore what has been done in the way of prevention is all that is reasonably practicable. An employer is unlikely to be successful in arguing that financial difficulties facing the firm mean that safety improvements cannot be afforded. Some of the statutory provisions provide for a defence of due diligence. For example, under the Control of Substances Hazardous to Health (COSHH) Regulations 1994 (see chapter 23) it is provided that "it shall be a defence for any person to prove that he took all reasonable precautions and exercised all due diligence" to avoid committing an offence.[61] The burden of proof will lie with the accused; the test of proof will be the balance of probabilities. The issue will be a matter of fact for the court to decide.

Manslaughter

The criminal process is not restricted to enforcement of the statutory provisions: criminal cases can be based on breaches of the common law. In particular, where there is gross negligence at common law there can be a prosecution for manslaughter and the accused may be an individual or a corporation. There have been a number of successful manslaughter prosecutions of individuals, but only one of a corporation: in the *Lyme Bay* canoe case. The successful prosecution of a corporation requires it to be shown that at least one of the company's 'controlling minds' was grossly negligent about an obvious risk of death or injury. It is not sufficient to show that there was *collective* negligence in the sense that a number of people were each responsible for part of the negligence. This helps to explain the unsuccessful prosecution of P & O European Ferries following the sinking of the *Herald of Free Enterprise*.

Notes

1. A useful source, and one that has been relied upon here, is: *The Development of Factory Legislation*, Royal Society for the Prevention of Accidents, undated (but no earlier than 1963), General Information Sheet No 8
2. The Health and Morals of Apprentices Act 1802
3. Shop Hours Regulation Act 1886 and Offices Act 1960
4. Mines and Quarries Act 1954
5. See: Report of a Committee, Chairman Lord Robens, 1972
6. HSWA, s47 and Management Regulations, Regulation 15
7. Provision and Use of Work Equipment Regulations, SI 1992/2932 Regulation 6(1)
8. Management Regulations, Regulation 3(1)
9. HSWA, s40

10. Management Regulations, Regulation 7(2)
11. See, for example, HSWA, s2
12. HSWA, s52(1)
13. HSWA, s52(1)
14. HSWA, s53(1)
15. HSWA, s53(1)
16. HSWA, s16
17. HSWA, s17
18. The Health and Safety (Training for Employment) Regulations, SI 1990/1380
19. HSWA, s2(1)
20. The poster is entitled "Health and Safety Law: What You Should Know". The requirement to display it (or to distribute leaflets with the same title) arises out of The Health and Safety (Information for Employees) Regulations, SI 1989/682.
21. HSWA, s2(3)
22. The Health and Safety Policy Statements (Exception) Regulations 1975
23. s2(1) – general duty, s2(2) (a) – safe system of work, and s2(2) (c) – information and instruction
24. HSWA, s40
25. s3(1)
26. s4(2)
27. SI 1983/943 as amended by SI 1989/319. The Alkali etc Works Regulation Act 1906 and other provisions require that certain works must be registered with the Secretary of State for the Environment. The registration system was modified recently by SI 1989/318.
28. HSWA, s6; Consumer Protection Act 1987, s36 and Schedule 3
29. HSWA, s37
30. HSWA, s2
31. Safety Representatives and Safety Committees (SRSC) Regulations, SI 1977/500. There are currently draft regulations and guidance on consultation with employees who are not members of a group covered by trades union safety representatives. Separate arrangements apply offshore: the Offshore Installations (Safety Representatives and Safety Committees) Regulations 1989.
32. Management Regulations, Regulation 17 and Schedule
33. SRSC Regulations, Regulation 3(4)
34. SRSC Regulations, Regulation 5
35. *Code of Practice on Time Off for the Training of Safety Representatives*. This was issued under Regulation 4(2) (b) of the SRSC Regulations. London: HSC, 1978.
36. HSWA, s2(7)
37. Management Regulations, SI 1992/2051
38. *Publications in Series*, London: HSE, 1990

39. Social Security (Claims and Payments) Regulations, SI 1979/628. The HMSO book is form BI510. Other documents are acceptable if they contain the same information.
40. RIDDOR, SI 1985/2023
41. HSWA, s10 and Schedule 2
42. HSWA, ss1, 11 and 12
43. HSWA, ss11, 13, 14 and 16
44. The Health and Safety Inquiries (Procedure) Regulations 1975
45. HSWA, s10 and Schedule 2
46. HSWA, s11(4)
47. *Enforcing Health and Safety Legislation in the Workplace*, Report from the National Audit Office, London: HMSO, 1994
48. HSWA, Part II
49. HSWA, s18
50. The Health and Safety (Enforcing Authority) Regulations 1989
51. The normal rule is that costs will be awarded only where a party acts frivolously, vexatiously, abusively, disruptively and otherwise unreasonably in bringing or conducting their case (rule 12, Schedule 1 of The Industrial Tribunals (Constitution and Rules of Procedure) Regulations 1993). Another procedural difference is that because the criminal law is involved an appeal against the decision of an industrial tribunal lies with the Divisional Court of the Queen's Bench Division of the High Court (see chapter 1) instead of with the EAT.
52. HSWA, s37
53. He was not a "manager of the body corporate" (Fire Precautions Act 1971, s23) because he had no responsibility for corporate policy and strategy. The Court of Appeal had a duty under s2(1) (a) of the Criminal Appeal Act 1968 to allow an appeal if a conviction was unsafe or unsatisfactory. (See: [1992] IRLR 420.)
54. HSWA,.s37
55. HSWA, s16
56. HSWA, s20
57. HSWA, s25
58. HSWA, s28(9)
59. HSWA, s26 Actions against inspectors, the HSE and HSC may be possible through the Parliamentary Ombudsman and through an application for judicial review.
60. HSE Press Notice E189: 92, October 1992
61. COSHH Regulations 1994, Regulation 16

4

Civil Liability

INTRODUCTION

In chapter 3 it was seen that the cornerstone of health and safety law is the framework provided by HSWA, and in particular the general duties it imposes. It was seen that HSWA establishes a criminal law system for dealing with health and safety matters but does not provide a basis for civil actions. Civil actions are likely to be founded in the common law of negligence or breach of statutory duties other than in HSWA. In addition, there is an overlap between health and safety and the wider law of employment in respect of employee protection on health and safety grounds and questions of health and safety discipline. It is all these areas of civil law which form the substance of the present chapter. The starting point is the civil process under common law, beginning with the different causes of action that are available.

DUTIES OF EMPLOYERS AND OTHERS AT COMMON LAW

Types of action

An injured person may have three types of possible legal case under the common law, all of which involve a claim for damages:

1. Breach of statutory duty (this constitutes a tort)
2. Breach of contract (through negligence)
3. Negligence in tort.

In the absence of provision to the contrary, civil action can be taken for breach of a duty contained in a statute by someone injured as a result of the breach. In fact, there is such provision to the contrary in HSWA so that breaches can be subject to legal action by the enforcement agencies but cannot be used as a basis for civil claims.[1] The same is not true of earlier legislation still in force, such as the Factories Act 1961. Thus employees may take a case on the basis of breach of statutory duty if a claim on the basis of negligence is ruled out, and vice versa. It should be noted that employer defences in the statute are in respect of criminal proceedings. They relate only indirectly to civil cases.

The basis of a worker's claim for damages arising out of an employer's negligence is as follows:

- The employer owes the worker a duty of care
- The employer has not fulfilled that duty
- As a result, the worker has suffered injury or damage to health.

The origin of the duty of care is found in the law of torts (civil wrongs – see chapter 1). The liability of the employer which arises in this respect is in tort. However, a duty of care also flows from the contract of employment, so an action for damages for breach of contract would also be possible in cases where the relationship is governed by such a contract. Nevertheless such actions are rare because the basis for the action is narrower. Actions for damages in tort are not limited to parties to a contract and encompass provision for greater damages. Even where there is a contract of employment the duty of care in the law of torts does not arise from the fact of contractual relations; it arises from the fact of the employee working for the employer. The employer's liability goes beyond situations where the employee is acting in the course of their employment. The test is whether the circumstances were within the control of the employer.

The duty of care

The concept of reasonable care

Employers have a duty of reasonable care for the safety of their employees, and the responsibility extends to the premises of third parties to which they send their employees. They also have a duty of care towards independent contractors, although the degree of care is less in such cases. The duty of reasonable care is part of a general duty which the common law imposes upon everyone in order to provide protection against injury. The word injury here tends to mean physical injury or harm to property – protection against economic loss (such as loss of livelihood) is much more restricted unless it flows from injury to the person or his or her property. The employer's duties are a particular application of the general duty to take reasonable care.

What is meant by reasonable care? In general it means avoiding acts and omissions which a person can reasonably foresee would be likely to injure their neighbour (*Donoghue v Stevenson*). A neighbour is anyone who is so affected by my acts or omissions that I ought reasonably to have these effects in mind. Liability arises, therefore, where there is not reasonable care to prevent reasonably foreseeable risk. Foresight has to be assessed on the basis of what is known (or should reasonably be known) at the time. Therefore, courts may set a date by which a reasonable employer would have been aware of a particular risk and taken precautions (1 January 1978 in the case of vibration white finger – *Bourman and ors v Harland and Wolff plc*). The employer cannot delegate his duty of care – it is a personal duty and he will be liable even if the job of fulfilling the duty has been given to an employee (see vicarious liability, pp81–3). An employer in breach of the duty of care at common law will be committing the tort of negligence, for which the usual remedy is an award of damages (see pp84–6). The key requirement is that the employer (or other duty-holder) must be at fault. There will be no liability if there is no negligence – for example, if the risks could not have been foreseen, or were not under the duty-holder's control. (The position is different where the action is founded on a breach of statutory duty: see, for example, *Morris v Breaveglen Ltd* – employer liable even though employee was under the control of a temporary employer.)

The test is an objective one – what the reasonable person would have done – rather than a subjective one (what the judge thinks is reasonable). The approach is to weigh the risks of injury against the costs and other drawbacks of the preventive action. It follows that the degree of care required is determined by balancing risk against the actions necessary to prevent or reduce it. Risk is judged in terms of how likely it is that an injury will occur and how serious it would be. The cost of preventive action will figure highly in the equation. The degree of care may vary according to the worker, for example how experienced they are, because this affects the likelihood of the injury occurring. Furthermore, in a case where the employee had only one eye, it was held that the duty of care was heavier because the seriousness of any injury to that person's eye was greater than for the normal employee (*Paris v Stepney Borough Council*). If the injury is caused by a manufacturing fault, the employer will not, ultimately, be liable if they could not reasonably tell that something was wrong. However, if the employer knew that there was something wrong and kept the machine in use they will be in breach of their duty of care. The Employers' Liability (Defective Equipment) Act 1969 provides that where an employee suffers injury in the course of their employment in consequence of a defect in equipment provided by their employer and the defect is wholly or partly attributable to the fault of a third party, the injury will be deemed to be attributable to the negligence of the employer. This leaves the employer to recover damages from the third party, for example the manufacturer of a machine.

The duty of care may be divided into a number of specific parts:

- safe premises
- a safe system of work
- safe plant, equipment and tools
- safe fellow workers.

Safe premises

An employer is under a duty to provide a safe place of work but not absolutely, since the approach already discussed involves weighing risks against costs and inconvenience in order to determine the standard of care that is reasonable. Thus, although there is an obvious danger of electrocution where employees work near live rails, it would not be reasonable to require the current to be switched off every time there were minor repairs to be done, since this would cause substantial disruption to the working of the railway (*Hawes v Railway Executive*). Where exceptional expenditure is needed in order to achieve a safe workplace, it will not be required if the risk is slight (eg the danger of slipping on an oily, wet floor – *Latimer v AEC*) but is likely to be required if there is imminent risk of death or serious injury (eg the danger of falling from a height – *Bath v British Transport Commission*).

If the employees' workplace is unsafe because of a third party and the employees' supervisor is aware of it but does nothing, the employer as well as the third party may be liable. This might occur where the employer is a sub-contractor on a building site and his employees are working in unsafe conditions because of an act or omission by the main contractor, as in *Smith v Davies and anor*. The duty of care is not removed because the employee is working on someone else's premises, nor because these premises are in another country (*Cook v Square D Ltd and ors*). Rather, what can reasonably be expected of the employer is less onerous. If the work is lengthy or poses an unusual hazard, it may be reasonable to require an inspection.

If an employer knows of a hazard, or should have known about it, and fails to take precautions in a reasonable time, he may be in breach of his duty of care. An adequate maintenance and housekeeping system, therefore, is an important consideration in ensuring that the duty of reasonable care is discharged. The duty to provide a safe place of work extends to providing safe access and egress.

A safe system of work

Part of a safe system of work is the provision of safety equipment. In *Crouch v British Rail Engineering Ltd*, the duty of care was not fulfilled by making safety goggles available for collection from a point about five minutes away. The lack of immediate availability encouraged risks to be taken. In *Pape v Cumbria County Council*, the employers provided gloves for cleaners handling chemical cleaning materials but the duty of care went beyond this; it extended to warning the cleaners about handling such materials with unprotected hands and instructing them as to the need to wear gloves at all times. The COSHH Regulations now require these steps to be taken (see chapter 23).

In general, a safe system of work means that working practices should be safe. This implies that there should be adequate training, safety equipment and supervision, and may require the employer to provide incentives or take disciplinary action. (On psychiatric harm, see pp104 and 653–4).

Supervisors cannot be expected to watch all of their employees all of the time. Nor are they expected to tell the employee what is obvious and commonsense in terms of precautions (*Ferner v Kemp*). Thus, an employer was not liable for not issuing new safety boots – it was the employee's duty to know that his old ones needed replacing (*Smith v Scott-Bowers*). However, where there is a danger which the employee cannot reasonably be aware of, the employer should reduce the risks and/or warn the employee.

Where employees ignore safety precautions (eg because they slow the work and reduce the bonus) employers may be liable if they do not enforce discipline or make adjustments to pay or time (*Broughton v Lucas* – the employer was 75 per cent liable where the employee ignored safety precautions). Similarly, where employees refuse to wear safety clothing (eg because it is uncomfortable), the employer must take all reasonable steps to get them to co-operate, especially if the risks are not obvious (*Crookall v Vickers Armstrong*; *Bux v Slough Metals*). This might include personal instruction, personal issue of the equipment, periodic checks as to use in practice, incentives, disciplinary action and safety awareness training. As a result of the PPE Regulations, the employer has a statutory duty not only to provide safety equipment but also to take all reasonable steps to ensure it is used (see chapter 25).

On the other hand, there is no duty to stand over an experienced worker, so once instructions are given and proper equipment provided there may be nothing more that an employer can be expected to do (*Woods v Durable Suites*). Nor, where the risk is obvious and the employee knows that safety clothing is available and will make the work safer, is there any liability on the part of the employer for not telling the employee what he already knows if he chooses not to wear the safety clothing and is injured (*Qualcast v Haynes*; followed in *James v Hepworth & Grandage*). The rule that an employer does not have to tell an experienced, skilled worker to wear safety equipment (here a safety belt when working at a height of almost 100 feet above the ground) was also upheld where the employer did not have the equipment available. This was because when belts had been provided in the past they had not been worn. On these facts, the provision of belts and attempts to secure their use would not have affected the outcome and the employer was not held to be liable (*McWilliams v Arrol*). This case illustrates that the injury has to be caused by the employer's failure. Here there was no causation because the belt would not have been worn even if provided.

In general, an employer should ensure that:

- the employee is warned about risks
- the employee is instructed as to the precautions to be taken
- safety clothing or equipment is available
- the employee is issued with the clothing or equipment or knows it is available; and that
- reasonable steps are taken to secure employees' co-operation in using safety equipment or clothing.

It is submitted that these factors apply as much to safe working procedures (SWPs) as they do to clothing and equipment. (On SWPs, see chapter 12). Establishing SWPs, instructing employees that they must be followed, instructing them in their use and checking periodically for actual use may demonstrate that an employer has taken reasonable steps to secure a safe system of work.

Safe plant, equipment and tools

The question is again one of balancing risk against expense and effort. The precautions generally taken in the particular industry will provide some evidence of what is appropriate, but employers will not be able to hide behind a general practice of inaction (*Thompson v Ship Repairers*; *Baxter v Harland and Wolff*).

Part of the duty of care will be met by adequate testing and maintenance, and ultimately renewal of plant. Thus, employers should periodically check the functioning of plant, equipment and tools (for example, the functioning of an electric drill – *Bell v Arnott and Harrison*).

Another aspect of the duty of care in respect of plant, equipment and tools is the need to keep up to date with relevant health and safety knowledge, especially that published by trade journals and official sources such as the HSC (*Ransom v McAlpine*). The significance of this stems from the fact that the availability of knowledge determines the date from which the employer should have known about a risk and taken appropriate precautions.

Where there are hidden defects, the employer would not be liable at common law but the manufacturer probably would be (*Davie v New Merton Mills*). However, the Employers' Liability (Defective Equipment) Act 1969 places strict liability on the employer (ie no negligence is required), leaving the employer to seek redress against the manufacturer or other negligent party (*Clarkson v Jackson*). Equipment includes plant, machinery, clothing, vehicles, ships and aircraft. It also extends to materials with which the employee is working – paving slabs in *Knowles v Liverpool City Council*.

Safe fellow workers

Fellow workers will generally be safe if they are competent to perform the work and are properly instructed and supervised. Fulfilling the duty of reasonable care means, therefore, that an employer should:

- select suitably qualified people to do the job (*Birnie v Ford*)
- provide adequate training, and
- provide instruction and supervision which is competent in terms of safety.

The requirements for training, instruction and supervision may be greater where the worker is young and inexperienced and employers need to take extra care in order to be sure that immigrant workers have understood what they must do (*Hawkins v Ross*).

Although a worker may be competent and properly instructed, he nevertheless may constitute a risk because of general behavioural characteristics. Employers may be liable if they know that someone presents such a risk but take no action (*Ryan v Cambrian Dairies* – bullying; *Hudson v Ridge* – practical jokes).

Similarly, in assessing the risks, an employer should not limit himself to contemplating the risks faced by the prudent, alert and skilled workman going about his task. The risks facing the careless worker may also also be reasonably foreseeable (see *Uddin v Associated Portland Cement Manufacturers Ltd* for an example of where this principle has been applied).

Negligence

Negligence occurs where there is a breach of the duty of care. In terms of the above definition of the duty of care this means that the actions taken by an employer to prevent or reduce risk were inadequate for the level of risk which was reasonably foreseeable. It should be noted that an employer cannot use a contract term or a notice to exclude or restrict liability for death or personal injury resulting from negligence.[2] The onus of proof of negligence lies with the plaintiff, who must also show causation, that is, that the negligence caused the injury, and that the employer had a reasonable and safer alternative to what they did. However, in the absence of an explanation of the cause of the injury there is what almost amounts to a presumption of negligence. This is because courts may infer from the immediate facts of the case that negligence occurred. It is particularly likely where *res ipsa loquitur* – the thing speaks for itself. Therefore, the plaintiff can establish a prima facie case on the immediate facts even if there is no direct proof that a negligent act or omission caused the injury. This discharges the plaintiff's burden of proof, leaving the defendant to rebut the inference of negligence by showing that the cause of the injury was not negligence.

If the action is against the defendant in the capacity of employer it will need to be established that the plaintiff was acting in the course of their employment when they received their injury. Where the injury was sustained in working hours on the employer's premises this is likely to prove difficult to challenge, especially if the employee was performing tasks in the employer's interest. Acting in the course of employment means carrying out acts which have been authorised by the employer, as well as some that have not been authorised, in order to fulfil contractual obligations (see p82). The latter, however, must be connected with authorised acts in that they can be regarded as ways of performing such acts. Anything normally and reasonably incidental to a day's work, such as going to the toilet, to the canteen or to collect tools or materials, will also be included.

In contrast to the above, there was breach of statutory duty under the Offices, Shops and Railway Premises Act 1963 even though an employee had no authority to be where he was. The legislation applies regardless of whether an employee is in the course of his employment. Since there was no warning of danger there was no contributory negligence on the part of the employee (*Westwood v Post Office*). Similarly in *Uddin* there was a breach of s14 of the Factories Act 1961 where the employee was injured on unfenced machinery while trying to catch a pigeon. More recently, in *Smith v Stages and Darlington Insulation Co Ltd*, the House of Lords set out guidelines for helping to decide when travel associated with work is in the course of employment. An employee paid wages (rather than a travelling allowance) to travel in their employer's time to a workplace other than their regular workplace will, prima facie, be acting in the course of their employment. The guidelines are as follows:

- Travelling from home to a regular workplace will be in the course of employment if the employee is compelled by the contract of employment to use the employer's transport, unless there is an express condition to the contrary.
- Travelling in the employer's time between workplaces will be in the course of employment.
- Receipt of wages (as distinct from a travelling allowance) is indicative of travel being in the course of employment.
- Travel in the employer's time from home to a workplace other than the regular workplace will be in the course of employment.
- Incidental deviation or interruption to a journey taken in the course of employment will not take the employee out of the course of employment, but anything more would do.
- Return journeys are to be treated on the same footing as outward journeys.

Where the action lies against the defendant in the capacity of occupier of premises the test is again whether there has been a breach of the duty of care. As a result of the Occupiers' Liability Act 1957, a statutory duty of care is owed to persons authorised to be on the premises.[3] The occupier's actual degree of control over the premises will be an important factor in determining liability. The Occupiers' Liability Act 1984 defines the duty owed to trespassers.

Vicarious liability

It is a rule of law that an employer is liable to persons injured by the wrongful acts of his employees if committed in the course of their employment. This is known as vicarious liability. The injured party may sue the employer and/or the employee, knowing that employers are required to have insurance cover as a result of the Employers' Liability (Compulsory Insurance) Act 1969.

Where the injured person sues the employer directly, negligence on the part of the employer must be demonstrated. Where the injured person sues the employer vicariously, negligence on the part of the employee must be shown. Once shown, the employer is strictly liable, ie no fault on the part of the employer needs to be demonstrated. The right of non-employees to sue an employer vicariously is long-established in common law; the employee's right to sue directly stems from the Law Reform (Personal Injuries) Act 1948. Where an employer is sued vicariously he may obtain an indemnity from the employee who committed the wrongful act (*Lister v Romford Ice and Cold Storage*).

A pre-requisite for vicarious liability is the committing of a legal wrong – a tort, breach of contract or crime. An error of judgement, an Act of God or some unexpected event will not fall into this category. The employee's wrongful act must be in the course of his employment. This means that the employee must be engaged in performing his contractual duties, albeit in an unauthorised way. Acts which are normally and reasonably incidental to an employee's work – such as going to the toilet or to a canteen – are also included. The following selection of cases illustrates where the dividing line is drawn in this matter.

Acts which were in the course of employment

- A tanker driver causing an explosion by smoking while unloading petrol, despite a clear prohibition – the employee was performing his duty to unload petrol (*Century Insurance Co v Northern Ireland Road Transport Board*).
- A works doctor failed to give proper advice to his employer and another employee died from cancer as a result (*Stokes v GKN*).
- A solicitors' clerk swindled his employer's clients under cover of his duties (*Lloyd v Grace Smith*)
- A driver took an extended route and caused an injury while doing so – he was employed as a driver (*Williams v Hemphill*). (But note that such an act may be regarded as the employee engaging in a 'frolic of his own'.)
- A milkman took on a boy to help him with his round: he was going about his contractual work (*Rose v Plenty*).
- A resident porter who stole tenants' property – it was his job to look after the property (*Nahhas v Pier House*).

Acts which were not in the course of employment

- A driver employed to carry his employer's employees carried the employees of other employers, despite a prohibition, and injured them – it was not his job to carry other employer's employees (*Conway v Wimpey*).
- A bus conductor drove a bus in the depot and caused injury – not his job to drive buses (*Iqbal v London Transport*).
- A postman wrote insulting messages on letters – not his job to do so (*Irving v The Post Office*).
- A cleaner made unauthorised telephone calls while on a client's premises – not part of the employee's job to make telephone calls (*Heasmans v Clarity Cleaning*).

Practical jokes are likely to be outside the course of employment, but may not be if carried out by an employee abusing his authority over a subordinate (*Chapman v Oakleigh Animal Products; Bracebridge v Darby*).

Vicarious liability is a principle extending beyond personal injury. There can be liability for breaches of contract by an employee unless the employee enters into the contract without actual or apparent authority – in which case the employee would be liable for breach of warranty of authority, or for fraud if there is an intention to deceive.

There is also vicarious liability for crime. Usually the employee is convicted while the employer is liable for damages. Examples include the delivery driver who exceeds the speed limit in the course of his deliveries and the resident porter who steals the tenants' property instead of protecting it (*Nahhas*: see above), but there was no liability for the cleaner who made telephone calls from the client's office (*Heasmans*: see above).

The vicarious liability of employers is stronger in respect of employees than in relation to independent contractors. This is explained historically by the fact that an employer can instruct his employee as to the manner of the work but cannot so instruct an independent contractor. Therefore there is less control and accordingly, less responsibility. This was the origin of the 'control' test to determine whether a worker was an employee or an independent contractor. In practice, under modern-day conditions, the distinction is less easy to draw. Professional employees determine their own methods of work, while small contractors working for a major business organisation may be instructed in detail. The legal test has developed therefore into a multiple test: first, is there a requirement for personal service; second, what is the degree of control; and third, are the other characteristics of the relationship consistent with a relationship based upon a contract of employment (*Ready Mixed Concrete v Minister of Pensions*).[4]

Vicarious liability for contractors arises when the contractor is employed to do work which creates a danger for the public and when the work is done negligently (*Walsh v Holst*). This might arise, for example, where a contractor is brought in to construct a new building and does so negligently. However, there will be no liability for contractors where the injury is incidental to rather than inherent in the work: for example, the employer's employee stands on a plank with a nail in it, left by the contractor (*Taylor v Coalite*). Negligent selection or supervision of contractors will give rise to direct rather than vicarious liability. (On safety in relation to contractors, see chapter 15, and in relation to construction, chapter 26.)

Duties of occupiers and suppliers

Just as the employer's duty of reasonable care is part of the wider responsibility laid down in law, so too are the duties of occupiers and suppliers. These are duties owed to non-employees, eg visitors and users. Where the occupier or supplier has substantial control, the duty of care may be as great as that of an employer (*Garrard v Southey*), but generally their duties are less strict because visitors and users are not their employees (*Jones v Minton*).

The civil liability of occupiers is laid down in the Occupiers' Liability Acts of 1957 and 1984 (in Scotland, an Act of 1960). The legislation covers land, buildings and structures (fixed or movable, such as vehicles and scaffolding) and in England and Wales distinguishes between the lawful visitor and the trespasser. An occupier is the person in immediate occupation or control, although control (and hence liability) may be shared. The occupier is not necessarily the owner.

The duty of the occupier is to take reasonable care to ensure that lawful visitors are reasonably safe in using the premises for the agreed purposes of their visit. Business occupiers cannot exclude liability for death or injury by negligence (Unfair Contract Terms Act 1977). Historically this was done through exclusion notices. Typically the occupier must warn the lawful visitor about known hazards which the visitor is unaware of or cannot avoid. The degree of care must be what is reasonable in the circumstances.

Where the hazard is apparent, the visitor will be under a duty to take more care, and where the visitor is a contractor, the occupier is entitled to assume that the contractor will take all appropriate precautions against the risks of his own trade. Providing the contractor has been selected with reasonable care (and properly supervised if the occupier is the main contractor), the occupier will not be liable for the contractor's negligence (except in Scotland).

Where the visitor is a trespasser, the law imposes a lesser duty upon the occupier (except in Scotland). If the occupier knows or ought to know of the danger, knows someone may be near it and knows that he ought reasonably to offer some protection, he must take reasonable steps to prevent the trespasser being injured. Warnings are one such step. Trespassers whose presence is condoned are likely to be regarded as lawful visitors.

The duties of suppliers are parallel to those of occupiers. There will be liability if all reasonable care is not taken eg for defective design, faulty workmanship or inadequate information, advice or warning for users. The liability may be reduced or removed if the product is used for an unintended purpose or where the user or his employer may be expected to test the goods himself. Anyone suffering personal injury as a result of a defect in a product may sue under the Consumer Protection Act 1987. Liability attaches to the producer: that is, the manufacturer, 'own brander' or, if outside the EU, the importer. There is strict liability unless the defendant can show that the state of scientific knowledge at the time did not enable him to discover the defect.

LITIGATION

Actions for damages

The Law Reform (Personal Injuries) Act 1948 gave employees the right to sue their employer by abolishing the principle that employees in common employment accepted the risks of their work. In the process, vicarious liability was allowed. The general rule now is that employers (principals) will be liable for the acts and omissions of their employees (agents); that is, they will be vicariously liable. The limiting factor is that the employee's act or omission must be in the course of employment or linked with it in the ways described earlier. Moreover, the courts have held that the person entitled to tell the employee how to do work would generally be liable for that employee's negligence, even if the employee is an employee of another company. (See *Sime v Sutcliffe Catering (Scotland) Ltd.*)

Vicarious liability does not mean that individual employees, whether managers or not, cannot be sued, since under the contract of employment (and in tort) the employee has a duty of reasonable care. In practice, however, the likelihood of a civil action is quite small since the sums involved in damages claims are larger than an individual could normally pay. The position would be entirely different if the individual had insurance cover. Most employers accept vicarious liability as a fact of life and the necessary insurance as an unavoidable cost of production.

The fact that a worker has not complained does not prove that the workplace is safe or that he thinks it is safe. Nor, on the other hand, does a complaint prove that it is unsafe. Complaints or their absence may help or hinder an action for damages, but are unlikely to be conclusive. If a specific complaint is made the employer must investigate, and if it is found to be justified must take action if he is to avoid liability. If there is a clear complaints procedure an employee may be partly responsible for his own injury if he does not use it properly and as a result the employer does not take the necessary precautions (*Franklin v Edmonton Corporation*).

The time limit for commencement of claims is three years from the time of knowledge of cause of action.[5] The time of knowledge will be the date on which the cause of action occurred or the date on which the plaintiff knew or should have known there was a significant injury and that it was caused by the employer's negligence. Where the person is fatally injured, the three years can be applied from the time of death or from the date of the personal representative's knowledge, whichever is later. Other exceptions are possible on grounds of equity.

Employers must be insured against such actions as a result of the Employers' Liability (Compulsory Insurance) Act 1969. The insurance must cover injury and disease arising out of and during the course of employment. Cases will be heard in the county courts or High Court (see chapter 1 for details of the allocation criteria). Convictions (for instance, under HSWA) are admissible as evidence: therefore, they can be cited as indicating civil liability. A breach of an approved code by an employer also substantially assists the plaintiff in discharging the burden of proof. A plaintiff may want to see the accident report in an attempt to establish their case. Such reports will be protected from disclosure by legal professional privilege only where use in possible litigation was the 'dominant purpose' of the report (*Waugh v British Railways Board*). The dominant purpose of accident reports is usually to establish why the accident happened, so there may not be protection against disclosure. The plaintiff may also obtain information from an inspector.[6] Where the employer seeks a medical report, the provisions of the Access to Medical Reports Act 1988 may apply and where the employee seeks medical information from a health professional, the Access to Health Records Act 1990 may be relevant.

Finally, the Civil Liability (Contribution) Act 1978 allows the plaintiff to take action against any of the tortfeasors (those who have allegedly committed torts), leaving the defendant to claim from others who contributed to the tortious act.

Damages are divided into special and general. The former covers provable loss to the date of the trial – loss of earnings, damage to clothing and so on. The loss of earnings is net loss. The general damages cover the remaining forms of loss such as pain and suffering before and after the trial and future loss of earnings. The broad aim is restitution – putting the injured person back where they were before the accident happened, in as much as money can do this (*British Transport Commision v Gourley*). Where appropriate, a money value will be put on disfigurement, loss of enjoyment of life, nursing expenses and the inability to pursue personal or social interests.

Under the Fatal Accidents Act 1976, certain dependent relatives can claim for the financial loss they suffer (but not in England and Wales) for the shock, grief and so on that they experience. Since 1985 a plaintiff may seek provisional damages and may return to apply for further damages in due course. However, the courts cannot declare that if a person dies before making that further claim their surviving dependents will have an entitlement to claim under the 1976 Act (*Middleton v Elliott Turbomachinery Ltd and anor*). Under s17 of the Judgments Act 1838 interest is payable on damages, but only from the date of that judgment in which the amount of damages is determined. Where there has been a split trial, interest is not backdated to the judgment on liability (*Lea v British Aerospace plc*). A number of statutory social security benefits are deductible from awards of damages. This principle does not extend, however, to benefits under an occupational pension scheme (*Smoker v London Fire and Civil Defence Authority*). The principle laid down by the House of Lords in *Parry v Cleaver* is that the fruits of money set aside in the past, through private insurance, cannot be appropriated by the tortfeasor. Nevertheless, employers and insurance companies are at liberty to draft pension schemes in such a way as to negate the effect of this principle.

Defences

The burden of proof of liability lies upon the employee but the employer may reduce or remove liability by succeeding with one or more defences. Despite the employer's duty of care, there is also a duty on the part of the employee, and where the employee is partly to blame there is said to be contributory negligence.

Contributory negligence

An employer may argue that the injured person was careless or reckless and was solely or partly to blame for their own injuries by, for example, ignoring clear safety rules. Contributory negligence might include contribution to the seriousness of the injury by the failure to wear available safety equipment. Where the injured person is partly to blame, the Law Reform (Contributory Negligence) Act 1945 requires that the damages be reduced rather than the claim defeated.

Courts will apportion liability according to the degree of fault. Accidental errors are distinguished from failure to take reasonable care (*Hopwood v Rolls Royce*) and recklessness or disobedience tend to be viewed seriously. Where the action is in respect of a breach of strict statutory duty rather than negligence, the impact of the employee's conduct (in terms of reduced damages) may be less (see below).

There was no contributory negligence where an employee misjudged whether he could safely lift a load (*Gallagher v Dorman Long)*, nor where a lapse of concentration resulted in the employee falling through the top of an unguarded tank (*Donovan v Cammell Laird*). Nevertheless, there may be contributory negligence where employees fall over objects on the floor or otherwise act carelessly, work in obviously dangerous conditions (*Campbell v Harland and Wolff*) or would not have used safety equipment even if it had been provided (*Simmons v Walsall Conduits*). The most serious contributory negligence is likely to be a breach of safety rules (see *Qualcast v Haynes*).

In actions for breach of statutory duty the test for contributory fault is stated as being recklessness rather than mere carelessness (*Foster v Flexile Metal*; *McGuinness v Key Markets*), but in practice it is difficult in a number of contributory negligence cases to see how the employee has been reckless (eg *Mullard v Ben Line Steamers*).

Injuries not reasonably foreseeable

An employer may argue that the type and/or extent of injuries sustained by the plaintiff were not reasonably foreseeable as a result of the employer's breach of duty of care. The plaintiff has the burden of proof in demonstrating that such injuries could be foreseen as arising out of that particular negligence. The employer's defence may be that the injuries were caused by an Act of God, that is, something beyond normal expectation and control. An employer may argue remoteness – that the injury was not in a category which could have been foreseen, the reason being that some unforeseen factor(s) intervened in the situation. There must be reasonably foreseeable cause and effect (*Overseas Tankships v Morts Dock [The Wagon Mound]*).

Delegation

An employer cannot exclude liability by delegating his duty to an employee

(*Driver v Willett*). However, delegation might be a complete defence where the employee delegated to perform the duty is solely to blame for his own injury (*Manwaring v Billington*). In other cases liability is likely to be shared (*Boyle v Kodak*). Where the duty is delegated to contractors, the question is whether reasonable care was taken in selecting and relying upon that contractor (*McDermid v Nash*).

Voluntary assumption of risk

In extreme cases an employer may succeed in escaping liability on the grounds that the employee had consented to taking risks as part of the job – risks other than those inherent in performing the job as safely as is reasonably practicable. This defence – *volenti non fit injuria* (one who consents cannot complain) – is argued on the basis of the plaintiff's knowledge and acceptance of the likelihood of the occurrence of a tortious act. It can apply to negligence but cannot apply to breach of statutory duty because the effect would be to allow the employee to contract out of Parliamentary protection.

Injury not sustained in the course of employment

It may be argued that the injury was not sustained in the course of employment. This could be because:

- the injury was not sustained at work at all; or
- it was sustained at work and during working hours, but the employee was performing unauthorised acts, which could not be regarded simply as unauthorised ways of performing authorised functions. A defence to liability here will be an express prohibition. Such a prohibition will not however, remove liability as regards unauthorised methods of performing authorised tasks.

Absence of vicarious liability

An employer may try to avoid liability by arguing that they are not vicariously liable. In cases of alleged vicarious liability it has to be established: first, that the employee is liable; and second, that the employer is vicariously liable for the employee's act(s) or omission(s). There may not be liability for the actions of an employee in the following situations.

- Where the employee is 'lent' to another employer and is working for, and under the control of that other employer. Vicarious liability may be transferred in such cases, the question of control being paramount.
- Where the employee knows (or should know) that what they have done is expressly outside the limits of their authority.
- Where the actions of the employee are excessive, for example a security guard who uses excessive violence against an intruder.

EMPLOYMENT PROTECTION IN HEALTH AND SAFETY CASES

Right of employees not to suffer detriment or dismissal

There are important new provisions in the Employment Protection (Consolidation) Act 1978 (EP(C)A) which aim to protect employees from action by employers taken against them on grounds of health and safety at work.[7] The employee has a new right not to suffer detriment and the existing right not to be unfairly dismissed is strengthened by making dismissal in health and safety cases automatically unfair and by removing the 2-year continuous employment qualification. In addition, selection for redundancy on health and safety grounds is also automatically unfair and again there is no qualifying period of employment required before an employee can exercise his or her right.

The health and safety activities which are protected are:

(i) Carrying out health and safety activities having been designated by the employer to carry out such activities.
(ii) Performing or proposing to perform functions as a union safety representative or safety committee member when holding such a position.
(iii) Bringing to the employer's attention work circumstances which the employee reasonably believed were harmful or potentially harmful in situations where there is no safety representative or committee or where it was not reasonably practicable to raise the matter by those means.
(iv) Leaving or proposing to leave or refusing to return to work where the employee reasonably believed there was serious and imminent danger which he could not reasonably be expected to avert.
(v) Taking or proposing to take appropriate steps to protect himself or others from what he reasonably believed was serious and imminent danger.

Both (iv) and (v) are capable of being regarded by managers as a breach of discipline (eg refusing to work or to obey an order). See pp92–3 for the common law position. First-aiders are likely to be included under (i) above. In (ii) the representative or member must be either a statutory or employer-recognised representative or committee member. Under (iii), the employee, when bringing matters to the employer's attention, must use "reasonable means". In (iv) the right to refuse to return applies only while the danger persists and in (v) the "appropriate steps" will depend on all the circumstances applying, including the employee's knowledge and the facilities and advice available at the time. An employer has the defence that the employee's steps (or proposed steps) were (or would have been) so negligent that the employer reasonably responded in the way he did. The Minister may by Order extend these rights to trainees who are not employees and the government intends this to be done. Where the person dismissed is a safety representative, safety committee member or some other designated health and safety person, compensation will be at the enhanced level applied in cases of dismissal for union or non-union reasons. Interim relief is also available.[8]

The onus will be upon the employer to show why the dismissal or other detrimental act was taken and the provisions restricting the unfair dismissal rights of those taking industrial action are removed where the dismissal is on health and safety grounds.[9]

Suspension from work on maternity grounds

An employee is suspended from work on maternity grounds when she cannot work because of risks made apparent by a maternity risk assessment or arising from night work. Where there is suitable alternative employment the employee has a right to be offered it before any suspension. Where suspension occurs, the employee has a right to be paid during that suspension.[10]

Medical suspension or dismissal

Under the EP(C)A, an employee, or someone in Crown employment, who is suspended on medical grounds, has a right to be paid for up to 26 weeks.[11] They must be suspended under one of a limited number of sets of regulations specified in schedule 1 to the Act or under related codes of practice issued or approved under s16 of HSWA. The qualification is one month's continuous employment.[12] An application may be made to an industrial tribunal within the usual three-month period. The employee must be available for work and not incapable of work because of illness or injury.

Where a suspension does not fall under the EP(C)A provisions, the employee may be able to mount a challenge on the grounds of breach of contract, possibly resigning and claiming constructive dismissal. The employee would have much less chance of success in cases where the suspension was with pay. If a dismissal is ultimately effected it will be subject to the usual requirements of unfair dismissal law provided that the employee has two years' qualifying employment. (The qualifying period will be only one month if the dismissal is for a reason which would otherwise give rise to a medical suspension under EP(C)A.) First, is there a fair reason? Breach of statute is one of the reasons laid down. Second, has the employer behaved reasonably? In this respect proper procedure is important as is investigation of the possibilities of a transfer to other work if the organisation is sufficiently large.

In cases where the fitness of the employee is in dispute the provisions of the Access to Medical Reports Act 1988 may be of relevance. A further problem may arise where colleagues refuse to work with a person because of medical factors. AIDS is a particular example. What if, despite medical evidence to the contrary, employees will not work alongside an AIDS virus carrier? In the event of a dismissal a tribunal would have to be convinced that this was a substantial reason for dismissal and that the dismissal was altogether reasonable.

Industrial tribunals

Employment protection matters in relation to health and safety at work are dealt with by industrial tribunals. These were originally set up under the Industrial Training Act 1964 to hear employers' appeals against levies made against them under that Act. In 1965 the responsibility for redundancy payments disputes was added. The 1971 Industrial Relations Act introduced unfair dismissal, which operated from the end of February 1972. Since that time, unfair dismissal claims have made up the vast majority of the cases heard by tribunals. Subsequently equal pay, sex and race discrimination and other cases were also put within the tribunals' jurisdiction.

The jurisdiction of industrial tribunals is found in United Kingdom (UK) statutes, but it now appears that complaints may be pursued directly on the basis of European law in the absence of a UK statutory right. A common law jurisdiction was added in 1994 but this excludes claims for damages in respect of personal injuries.[13] The current jurisdiction of industrial tribunals on the basis of UK statutes, and in respect of health and safety at work is:

- appeals against improvement and prohibition notices
- time off work with pay for union safety representatives
- right to receive pay while suspended on medical grounds
- right to receive pay while suspended on maternity grounds
- right not to be dismissed or suffer other detriment in health and safety cases.

Appeals against industrial tribunal decisions can be made to the Employment Appeal Tribunal (EAT) in most cases only on a point of law (that is, not generally on the grounds that the tribunal got its facts wrong). The appeal will need to show some incorrect legal interpretation, approach or procedure by the tribunal, or that no reasonable tribunal could have made the decision it did on the basis of the evidence before it.

Despite its modest sounding title, the EAT has the status of the High Court. It comprises a High Court judge, a senior union-nominated person, and a senior employer nominee. Sometimes there may be a tribunal of five rather than three people. This might occur where the EAT is deciding an issue which has hitherto been the subject of conflicting EAT judgments. Appeals have to be made within 42 days of the publication of the industrial tribunal decision.

Appeals against industrial tribunal decisions on matters relating to improvement and prohibition notices lie, as noted earlier, with the Divisional Court of the Queen's Bench Division.

SAFETY DISCIPLINE

Disregard of safety rules or procedures

Employees who disregard safety rules and procedures may be subject to discipline, including dismissal, in the same way as if there were a breach of some other type of rule. All the usual requirements laid upon employers need to be met – rules must be reasonable, applied consistently and fairly, and communicated to employees and the procedure for handling disciplinary breaches must be reasonable. A dismissal for a safety breach would need to be defended in the light of the normal requirements of unfair dismissal law.

An employer may seek to change existing safety rules or introduce new rules. If the change is a direct result of new legislation, the employer will be under a duty to meet the statutory requirements. Where the change stems from a policy change rather than a change in legislation an employer will need to consider the existing terms of employment and whether the change can represent a breach of contract. The procedure for implementation (advance warning, consultation and so on) may be important. For example, employers introducing no smoking policies might justify their actions on the basis of their general duties under HSWA, provide employees with evidence to support this, give advance warning, perhaps survey employee opinion or consult in other ways and provide some facilities for smokers after the rule becomes operative. (See chapter 27)

Refusal to work on grounds of lack of safety

Apart from where the new EP(C)A provisions apply, (see above), what is the legal position if an employee refuses to work because what they are being asked to do seems to them to be unsafe? On the one hand the employee has a contractual and statutory duty to take reasonable care, so that they may feel that they will be in breach of such a duty. On the other hand, they are required by their terms of their contract of employment to obey the lawful commands of their employer. The critical question here is whether the employer's command is lawful. This means it must not only be within the terms of the contract, but also must not involve the employee engaging in a criminal act, or indeed being unable to act with due care. The problem is that opinions may well differ and neither party is likely to be able to wait for a legal ruling. The advice of specialists (safety managers, union representatives and perhaps even factory inspectors) is about as far as one can go. If there is a refusal to work the matter may then become a disciplinary one.

As part of their contractual duty to provide their employees with a safe system of work and take reasonable care of their safety, employers must investigate all bona fide complaints about safety brought to their attention by employees (*BAC Ltd v Austin*). Failure to do so could give grounds for an employee to resign and claim unfair constructive dismissal. If the employee chooses to stay but refuses to do the work, the question of whether or not the employer is in breach of a contractual term or a statutory duty will not be conclusive in determining the fairness or otherwise of any consequent dismissal. The test will be reasonableness (*Lindsay v Dunlop Ltd*). Relevant factors may include the attitude of other employees in the same position and what steps the employer is taking to deal with the safety problem. Employers might fare better if they did not treat refusal to work on safety grounds as straightforward disobedience. Dismissing for refusal to work before there has been a proper investigation of the employee's complaint and before the results of that investigation have been communicated to the employee may well be unfair. Any employee with a special condition ought to be treated as sympathetically as possible and alternative duties considered, especially if the condition is likely to be temporary. Pregnant women anxious about the effects of working at VDUs might fall into this category.[14]

In *Piggott Brothers and Co Ltd v Jackson and ors*, an employer's failure to get a definitive explanation of the cause of the employees' symptoms (experienced as a result of exposure to fumes) was held to amount to unreasonableness. This was so even though ventilation had been improved and the problem had been investigated by HSE inspectors who thought it was a 'one-off' that had ceased to exist. Dismissals for refusal to work were held to be unfair. The Court of Appeal would not disturb these findings – the decision was a permissible option on the facts.

Notes

1. HSWA, s47. There is a similar restriction on civil liability in the Management of Health and Safety at Work Regulations (henceforth the Management Regulations), regulation 15.
2. Unfair Contract Terms Act 1977, s2
3. Occupiers' Liability Act 1957; and Occupiers' Liability (Scotland) Act 1960.
4. On this see Lewis, P., *Practical Employment Law*, Blackwell, 1992, pp54–7
5. Limitation Act 1980
6. Under s28(9) of HSWA
7. Trade Union Reform and Employment Rights Act 1993, (henceforth, TURERA), s28 and Schedule 5. This inserts a new s22A into the EP(C)A (right not to suffer detriment) and a new s57A (making dismissal for health and safety reasons unfair).
8. On the general provisions in relation to unfair dismissal, including interim relief, see Lewis, P, *Practical Employment Law*, Blackwell, 1992, chapter 7.
9. This is done by TURERA, sch 8, paras 76-77 amending the Trade Union and Labour Relations (Consolidation) Act 1992, ss 238 and 254.

10. These are new sections 45-47 of the EP(C)A (inserted by TURERA s25 and schedule 3) in conjunction with Management Regulations, regs 13A (3) and 13 (B).
11. EP(C)A, s19
12. On the rules for continuous employment see Lewis, *op cit* pp 85–88 and EP(C)A, sch 13, as amended by the Employment Protection (Part-time Employees) Regulations SI 1995/31. There is no longer any hours requirement for the accrual of continuous employment. Thus, part-time employees will now qualify for various statutory rights.
13. The Industrial Tribunals Extension of Jurisdiction (England and Wales) Order, SI 1994/1623 (SI 1624 for Scotland).
14. On this, see (1990) 433 IDS Brief 10. VDU health and safety is dealt with below on pp434–45.

5

Changes and Developments in the Law

INTRODUCTION

The purpose of this chapter is to alert managers to possible, likely and planned changes in health and safety legislation and to consider the ways in which the body of case law is developing. As regards legislation, a number of sources of change can be pinpointed, although these may overlap to some degree:

- the HSC's 1994 Review of Health and Safety legislation (in the context of the 'deregulation debate')
- developments in the health and safety at work programme of the European Union
- the continuing policy of replacing old legislation with new provisions under HSWA (the 'modernisation' of the law post-Robens), and
- the evolution of health and safety law in response to the emergence of new or increased hazards.

These are dealt with in turn before a final section of the chapter briefly examines the ways in which case law is developing and paints a broad-brush picture of tomorrow's legal framework.

THE HSC REVIEW OF HEALTH AND SAFETY LEGISLATION

The deregulation debate

The HSC's review of health and safety regulation needs to be seen against the backcloth of the wider debate about legal regulation in the UK. The essential argument on the one side is that legal regulation, including in the health and safety field, has gone too far and is making UK firms uncompetitive internationally by adding to their costs and restricting their ability to respond quickly and flexibly to market requirements. The opposing argument holds that this position is overstated and that the law is merely providing rights which should be regarded as basic in any advanced, democratic society.

The movement in favour of deregulation has been marshalled by the Department of Trade and Industry (DTI) which issued a set of proposals at the beginning of 1994.[1] These proposals gave rise to the Deregulation and Contracting Out Act 1994. The Act contains both specific deregulatory provisions and a general facility for the Minister to repeal legislation by statutory instrument.

Employment in general, and health and safety in particular, have been areas where the government thought there was substantial scope for deregulation. The introduction of the 'six pack' (see chapter 2) at the beginning of 1993, following closely on earlier requirements in other areas such as hazardous substances and noise, meant that the late 1980s and early 1990s had seen a tidal wave of health and safety legislation, and it seems certain that this played a part in fuelling the fires of deregulation.

Given the above background, it was not surprising that the government required a review of the whole of health and safety legislation. Accordingly, in December 1992, the HSC was 'invited' to conduct such a review.

The government's general approach to legislation is to weigh the compliance and enforcement costs against the risks the legislation is intended to reduce or eliminate. This has been the method adopted by the HSC in its review of health and safety legislation.

The HSC review

The report and recommendations of the HSC's review of regulations were published jointly in May 1994 by the HSC and the Employment Department.[2] Seven task forces, with members drawn from employers and trades unions, assisted in the review. Support facilities were provided by the HSE.

The HSC's terms of reference were to focus upon the costs and benefits of existing health and safety legislation, and to recommend reforms, while maintaining health and safety standards. The overall conclusion is that there is general support for the present system and its standards, based upon HSWA, but that health and safety law is criticised for its volume, complexity and fragmented nature. Accordingly, the HSC will reduce the volume and simplify and clarify that which remains. Since the HSC will not be proposing anything which reduces standards, the reforms will not be restricted by HSWA, s1(2): this requires regulations or orders under Part I of HSWA to maintain or improve health and safety standards. Thus, reforms can be achieved by regulations and ACOPs issued under HSWA. The reform programme will face two principal constraints, however: the requirements of EU directives and the need to alter legislation at a pace which enables businesses to keep up to date and meet the costs of change (such as training costs).

The HSC recommends that:

- Over 40 per cent of the legislation should be removed, including seven pieces of primary legislation and over 100 sets of regulations. The primary legislation to be repealed includes the remaining parts of the Factories Act 1961 and the Offices, Shops and Railway Premises Act 1963. Indeed, all pre-1974 legislation is likely to be removed.
- It should provide clarification of:
  — the respective roles of legislation, ACOPs and guidance;
  — many EU-derived provisions which it feels are misunderstood; and
  — the respective duties of employers, contractors, suppliers and designers.
- There should be a simplification of form-filling, record-keeping and other paperwork.
- There should be greater consistency in enforcement.
- Communication between inspectors and businesses, especially small businesses, should be improved. The HSC found little support for exempting the self-employed and small firms from the legislation but recommends that they be given more information and guidance on compliance.

Recommendations of the HSC based upon its review of health and safety regulations*

- *The overall 'architecture' of regulation:*

Recommendation 1: The Commission will continue to base its revision of the law on the approach set out by Robens, involving general duties in the main (1974) Act, goal-setting legislation which applies across the board, sector – or hazard – specific legislation and ACOPs where appropriate, and clear, targeted guidance.

Recommendation 2: The Commission will continue to seek to achieve the implementation of EC directives within that structure, while recognising that over time the EC will have a decisive influence on the form of much UK health and safety law.

Recommendation 3: Alongside this, the Commission will continue its domestic programme of rationalising and modernising health and safety legislation.

Recommendation 4: When proposing new legislation, the Commission will in future explain its place within the 'vertical' structure of health and safety legislation (in relation to the requirements of the general duties and of other across-the-board provisions) and the 'horizontal' links with the controls on specific categories of hazard or risk.

Recommendation 5: The Commission will re-examine the current portfolio of ACOPs, including their coverage, style, content and practical value to industry. The aim will be to return to a situation where if ACOPs are used they give *practical guidance* on specific hazards or to key sectors of industry on the implementation of legislation, especially legislation which applies across the board. A consultation document will be issued.

---

* Source: *HSC Review of Health and Safety Regulation*, HSE Books, 1994, pp9–17 © Crown Copyright reproduced with the permission of the Controller of HMSO.

Recommendation 6: The Commission will ask HSE to re-examine the wording of *all guidance* to ensure that its legal status is made clear and that in its coverage and content it complements the other components of the regulatory framework.

Recommendation 7: The Commission will ask HSE to publish a short guide to health and safety regulation which will set out the basic legislative framework; explain the respective roles of EC directives, domestic legislation – both goal-setting and prescriptive – ACOPs and guidance; clarify the relationship of these to the risk assesments and control measures required of employers; explore the relationship between government regulation and self-regulation by industry; and seek to set out a path for legislative change in the medium term. The Commission will itself use indicative criteria to bring greater consistency to its deployment of the various legislative and other instruments available.

- *The Commission's continuing work to simplify, clarify and modernise the law*:

Recommendation 8: The Commission proposes to examine further how best to ensure that there is an up to date and authoritative data-base of all current workplace health and safety legislation (and related legislation) in Great Britain; Approved Codes of Practice; and guidance. This might appropriately be done in partnership with one or more commercial providers.

Recommendation 9: The Commission proposes an intensive programme of consultation, which will necessarily stretch over three years or more, in order to secure the removal of outdated legislation.

Recommendation 10: The Commission proposes to issue a consultation document with proposals on the simplification or clarification of the risk assessment and control, and other common provisions, of modern health and safety legislation.

Recommendation 11: In response to the recommendation on the 'six pack' and COSHH Regulations made by its Task Groups and the government's deregulation Task Forces, the Commission proposes to revise guidance and produce new guidance; to consult on the scope for incorporating into COSHH the lead, asbestos, and other regulations on hazardous substances; and to keep the possibility of legislative change in view.

Recommendation 12: In addition, the Commission proposes to carry out its own evaluation, and to seek the earliest possible evaluation by the EC, of the benefits and costs of the Display Screen Equipment Directive (90/270/EEC) and the possible need for changes.

Recommendation 13: The Commission believes that its new guidance will be effective in reducing the impact of the misrepresentation of the Electricity at Work Regulations 1989 but proposes to keep this under review.

Recommendation 14: The Commission proposes to simplify and improve the requirements and procedures on the reporting of work-related accidents, injuries and disease in the light of responses to its recently published consultation document.

Recommendation 15: The Commission proposes to issue a series of consultation documents over the next 3–5 years with proposals on the reform of legislation in various high-risk areas, which include work in confined spaces, work with highly flammable substances, and compressed air working.

Recommendation 16: The Commission proposes to consult on the revocation of miscellaneous regulations, for example on metrication, which no longer have any practical effect.

Recommendation 17: The Commission recommends to the government that section 1(2) of the Health and Safety at Work etc Act 1974 remain unchanged.

- *Exemptions for small firms and for the self-employed*:

Recommendation 18: The Commission recommends to the government that there should be no general exemptions from health and safety law for small firms.

Recommendation 19: The Commission proposes to issue a discussion document in order to facilitate a public debate about the implications for health and safety regulation of recent trends in the structure of employment, including the growth of self-employment.

- *Regulatory paperwork*:

Recommendation 20: The Commission proposes to issue a consultation document with proposals for streamlining the requirements on the display of health and safety information and facilitating the use of non paper-based systems in support of health and safety management; and to remove other unduly prescriptive requirements for paperwork as the opportunity arises.

- *Enforcement practices*:

Recommendation 21: The Commission proposes to raise its own profile on enforcement policy by developing a statement building on the existing statements of approach. This will incorporate the principles of proportionality, consistency, transparency and targeting and will be embodied in new instructions from the Health and Safety Executive to HSE staff, and included in new guidance (which must be followed) for local authorities. The allocation of enforcement responsibilities between HSE and local authorities will also be clarified.

Recommendation 22: The Commission also welcomes HSE's plans to improve consistency of inspection, by means including a process of 'peer review', and proposes to encourage local authorities, through the Health and Safety

Executive/Local Authority Enforcement Liaison Committee (HELA), to adopt a similar strategy.

Recommendation 23: The Commission proposes to develop new guidance on the competencies and training which it considers appropriate for HSE and for local authority inspectors.

Recommendation 24: The Commission proposes to discuss with those responsible for policy on regulation related to workplace health and safety, such as food safety and environmental protection, the scope for more commonality of approach to enforcement policy.

Recommendation 25: The Commission proposes to ask local authorities for an annual analysis of complaints received on enforcement action, and of the action taken.

Recommendation 26: The Commission recommends to the government that new statutory powers are introduced to require each local authority to provide information on the effectiveness and efficiency with which it performs its enforcement functions.

- *Further help for businesses, especially small firms*:

Recommendation 27: The Commission proposes to ask HSE to ensure that there is simple and practical guidance about all key areas of risk – on what to do and what not to do – available to small firms.

Recommendation 28: The Commission will ask HSE to produce a catalogue and classification system designed to help small firms in any sector to identify easily what guidance they need.

Recommendation 29: The Commission proposes to review, at least annually, the advice, information, training and other support available to business, and especially to business start-ups, on health and safety.

Recommendation 30: The Commission proposes to evaluate all the current central and local initiatives to assess the most cost-effective means of closing the information gap – with small firms in low-risk sectors in particular.

Recommendation 31: The Commission proposes to evaluate and implement the best practicable means of providing rapid and accessible technical advice to business, especially to small firms, for example through a telephone 'help line'.

Recommendation 32: The Commission proposes to issue a discussion document on the strategy it should adopt towards improving compliance by small firms and self-employed people, drawing on proposals suggested by its Task Groups.

- *Reporting progress*:

Recommendation 33: The Commission proposes to report regularly, in its published Annual Report, on progress with this programme designed to increase the effectiveness of health and safety regulation.

The government's response

The government has accepted all the proposals and recommendations in the report of the HSC's review, noting that safety standards themselves are not burdensome but that the way in which these standards are put into law and enforced may be. The government's 1994 White Paper on competitiveness proposes six sets of regulations to replace the primary and secondary legislation identified as redundant by the HSC.[3]

THE EU's HEALTH AND SAFETY PROGRAMME

The EU's fourth action programme

The role of the European Union in the regulation of UK health and safety at work was dealt with in chapter 2. It was noted that the EU's fourth action programme on health and safety covered 1994–2000, and the emphasis was to be less on new legislation and more on consolidation, evaluation, consistent enforcement and assessment of impact. The thrust is towards investigation of problems, and guidance.

The European Health and Safety Agency

In 1994, the EU agreed to set up the European Health and Safety Agency (EHSA).[4] It will be located in Bilbao, Spain, and should become operational shortly. The expected annual cost is 5 million ecus (currently £4.2 million) but no final decision has been taken in this matter. The EHSA will be run by an administrative board of 27, with rotating membership for the social partners.

The objective of the EHSA will be to promote improvements in European health and safety by:

- supplying information to Member States, EU institutions and those involved in the health and safety field
- sponsoring an information network between Member States, through which information will flow in both directions between the Agency and Member States
- encouraging an exchange of experts between Member States
- encouraging conferences and seminars, and by
- contributing to the development of EU health and safety programmes.

The Agency will also be responsible for assessing the impact of legislation upon business, especially small and medium-sized enterprises.

Health and safety in the context of the EU's social policy

As noted in chapter 2, health and safety initiatives from the EU may come from different sources, including the EU's social policy. The social policy programme to the end of the century, agreed by the Commission in July 1994, is therefore of some interest. The emphasis is upon enforcement and consolidation of existing directives rather than on new legislation. The policy paper proposes completion of the implementation of the 1989 social action programme, including legislation to protect part-time and temporary workers and to enhance maternity and paternity rights (see chapter 2).

THE EVOLUTION AND MODERNISATION OF UK HEALTH AND SAFETY LEGISLATION

As noted, HSWA and parallel legislation in Northern Ireland were designed to modernise UK health and safety regulation. The intention was to have a generally applicable Act with regulations and ACOPs issued under it in order to replace the older legislation. In the early 1970s, the drive for modernisation had its roots in the need for a more effective health and safety system: today, much of the impetus is derived from the need for economy and a reduction in the burdens on business.

Two recent examples illustrate the trend. The Chemical (Hazard Information and Packaging) Regulations, which took effect at the beginning of September 1993, deal with the packaging and labelling of dangerous substances. These regulations replace five previous sets of regulations, at a saving of £8 million per annum after initial costs of £6 million.[5] New legislation on explosives, shafts and windings in mines (operative from April 1993) enabled the repeal of six sets of regulations and parts of others. It is claimed that the new regulations reduce risks and save money.[6]

The emphasis of the HSC's work in 1994/5 has been upon beginning to fulfil the recommendations of its review of regulation. The aim is to achieve a clearer, better-focused and more economical system of health and safety legislation.[7] Such an approach suggests that the modernisation process will continue, that the principal concern will be to implement the review recommendations and that there will be few new requirements originating from within the HSC.

In its review of regulation, the HSC notes the need to coordinate regulation where different authorities are involved and to consider the implications of the changing structure of employment (more self-employment, temporary work etc). It confirms its continuing commitment to encouraging the improvement of health and safety management. All of this can be seen as part of the evolution and modernisation of health and safety, and the importance of this is reflected in the fact that a third of the HSC review's recommendations relate to it.

The changing occupational structure has been an important factor in the evolution of health and safety law. Primary (extractive) and manufacturing occupations have declined in number while clerical, administrative, technical, managerial and professional jobs have increased. Consequently the focus of health and safety law is increasingly upon office-based and professional work hazards such as stress. Another key factor is technological change. The use of computers is a major influence, and it seems to be giving rise to substantial health and safety problems in the form of repetitive strain injury (see chapter 19).

TOMORROW'S LEGAL FRAMEWORK

The general slowing of EU intervention and the clear line emerging from the HSC's review suggests that a picture can be painted showing tomorrow's legal framework. First, it will remain founded in the general duties of the 1974 Act. That is, it will be based upon goal-setting legislation which applies across the board. Second, there will continue to be hazard-specific legislation, with ACOPs where necessary, and with guidance. The balance between general and specific legislation (much of the latter coming from the EU in recent years) will be of considerable importance. The specific, EU approach is seen by some health and safety professionals as a return to a pre-HSWA situation, where effective enforcement proved very dificult. Third, there will be less health and safety legislation, but what there is should be simpler and clearer. Old, and familiar, legislation such as the Factories Act 1961, is likely to disappear, to be replaced by a small number of sets of regulations issued under HSWA. Finally, the statutory provisions are likely to be of increasing relevance in civil cases. The specific duties of employers and their general duty to assess risks will help injured employees in providing evidence of the standards to be applied and of a failure to reach those standards.

Inevitably, it is more difficult to predict developments in case law, but perhaps there could be speculation about three areas. First, the sequence of disasters, including *The Herald of Free Enterprise* and *Piper Alpha*, and the general movement in favour of a more socially responsible corporate sector, may lead to the wider use of the charge of corporate manslaughter, as well as the wider use of manslaughter charges against senior managers and directors. Here, the *Lyme Bay* canoeing case, the first successful prosecution for corporate manslaughter, may signify an important development. Second, the delayering of management structures in the 1980s and 1990s may help to establish the law in the area of stress-related personal injury. It seems possible that a series of cases in which professionals have suffered nervous breakdowns, heart attacks and other injuries as a result of employer-induced or enforced overwork may emphasise the importance of health issues in general and stress in particular, within the UK health and safety system.

The case of *Walker v Northumberland County Council* is particularly significant. Here the High Court ruled that the employer owed his employee a duty of care not to cause him psychiatric damage through the volume or character of the work that was required. This was the first time that the duty to provide a safe system of work, by taking reasonable steps to protect the employee from reasonably foreseeable risks, had been applied to hold an employer liable for mental as opposed to physical injury. Where it is clear that the employee is at risk – in the *Walker* case, after he had already suffered a nervous breakdown – employers will find it difficult to argue that the risk was not reasonably foreseeable.

The third area about which there can be speculation is repetitive strain injury (RSI). It does appear that the hidden costs of ill-considered work design have yet to be paid. (See chapter 19 on ergonomic approaches.) Computer operators with strains induced by repetitive keyboard work, awkward movement and bad posture seem likely to generate considerable litigation, much of which has so far been successful. The same is likely to be true of those performing other types of short, repetitive tasks. There may be arguments over precisely what medical conditions are involved, and over what to call it, but RSI as it is popularly known seems likely to generate a substantial body of case law in the years to come.

Notes

1. *Deregulation*, London: Department of Trade and Industry, 1994
2. *HSC Review of Health and Safety Regulation*, HSE Books, 1994
3. See: *Employment News*, No. 231, July 1994
4. Regulation adopted by the Council of Ministers in June 1994
5. *Deregulation*, p35
6. *Deregulation*, p35
7. *HSC Newsletter*, No. 96, August 1994, p1

Part II

Managing Health and Safety

6
Principles of Health and Safety Management

This chapter deals with the general principles of health and safety management, developing in particular the 5-step approach recommended by the HSE. The causes of accidents are also reviewed since an understanding of these is essential to managing their prevention. However, the chapter starts by examining the reasons why health and safety should be managed effectively.

WHY HEALTH AND SAFETY SHOULD BE WELL-MANAGED

The reasons why health and safety should be well-managed fall under three main headings:

- humanitarian
- financial
- legal.

Humanitarian considerations

In a typical year:

- 400 people are killed in accidents at work
- 30,000 people suffer major injuries (broken limbs, amputations etc.)
- 135,000 people are absent from work for more than three days due to accidents
- 2,000 people die as a result of work diseases
- 10,000 deaths are partially due to occupational illness
- 500,000 people suffer long term health problems due to work.

The economic costs are considered later but it is clear that this scale of human suffering should not be acceptable in an advanced, civilised society. Fortunately the picture is improving, partly due to a gradual raising of health and safety standards. However the contraction of the country's manufacturing and industrial base has also played a part.

Financial costs

The financial costs of accidents provide a second reason for effective management of health and safety. A broad definition of the term 'accident' is 'an undesired event that results in harm to people, damage to property or loss to process'. Clearly this definition will incorporate cases of occupational ill-health as well as injuries at work. Various studies of accidents at work have established a numerical relationship between serious injuries, minor injuries and various other 'undesired events'. Several of these were drawn together in the HSE book *Successful Health and Safety Management*[1] – see Figure 6.1. Each study clearly demonstrates that the number of serious injuries is very small compared with the number of non-injury accidents. The importance of investigating non-injury accidents as an accident prevention measure is dealt with fully in chapter 9. Few, if any, organisations have any clear idea of the costs they incur as a result of all of these accidents. Some of these costs are potentially measurable whilst others are hidden.

Measurable costs

Some employers count the total days lost from work due to accident injuries and then calculate a cost on that basis. However, calculating days lost due to occupational illness is much more difficult, particularly separating this from normal sickness absence. Others have access to figures for compensation paid to employees in the form of damages and sometimes the related legal and administrative costs. However, insurers are often reluctant to release this information and the picture is further complicated by the lengthy delay between the accident and settlement of the claim. Although these payments are made by the insurance company, in the long run the insurance premium paid will inevitably reflect the claims history of the employer. Since insurance arrangements are usually made by the finance department, senior operational managers and even safety specialists may be unaware of the size of the premium being paid.

Repair costs can also be measured although few employers try. More commonly the costs of repairing accident damage are hidden in general repair and maintenance costs. A more proactive approach to the prevention of accidental damage can often pay big dividends in the reduction of the cost of what previously was categorised as routine 'wear and tear'.

Despite their imperfections at least these costs can be identified with a little effort. The 'hidden costs' of accidents are either impossible to quantify or their quantification would be totally impracticable. However the 'hidden costs' are usually far greater than the 'measurable costs', particularly when the large numbers of minor injuries and non-injury accidents are taken into account.

Heinrich (1950)
From data available to him on the frequency of potential injury accidents Heinrich estimated that in a unit group of 330 accidents of the same kind and involving the same person there would be:

1 major or lost time injury
29 minor injuries
300 no injury accidents
(*Industrial accident prevention* 3rd edition – Heinrich 1950.)

1
29
300

Bird (1969)
From an analysis of 1 753 498 accidents reported by 297 co-operating organisations in the USA, representing 21 different types of occupational establishment and employing 1 750 000 people who worked more than 3 billion man hours during the exposure period analysed, F E Bird Jnr drew up the following ratio:

1 Serious or disabling injury
10 minor injuries (any reported injury less than serious)
30 property damage accidents (all types)
600 incidents with no visible injury or damage.
(*Practical loss control leadership* – F E Bird Jnr and G L Germain 1985.)

Tye/Pearson (1974/75)
Based on a study of almost 1 000 000 accidents in British industry Tye and Parson drew up the following ratio:

1 fatal or serious injury
3 minor injuries – when the victim would be absent for up to 3 days
50 injuries requiring first-aid treatment
80 property damage accidents
400 non-injury/damage incidents or 'near misses'
(*Management safety manual* – British Safety Council 5-Star Health and Safety Management System.)

1
3
50
80
400

Source: *Successful Health and Safety Management*, HSE, 1991 © Crown Copyright reproduced with the permission of the Controller of HMSO.

Figure 6.1 *Outline of three accident ratio studies*

Hidden costs

The following are all possible 'hidden costs' resulting from an accident:

- *Injuries to employees* (and others)
  First aid
  — time spent receiving it
  — time spent by others administering it
  — cost of materials
  After-effects
  — may not be capable of full duties
  — may be slowed down
  Unable to continue
  — time arranging a replacement
  — high cost of replacement
  — time training replacement
  — poorer performance of replacement

- *Damaged equipment or plant*
  Cost of repair (if not separately quantified)
  Cost of removal for repair
  Unavailability during repair – production delays/losses
  Temporary replacement costs
  Disruption during repair
- *Damaged product or materials*
  Cost of replacement, repair or recycling
  Product delay – customer dissatisfaction
  Clean-up time
- *Other costs*
  Time spent by other employees either attending the scene or discussing the accident
  Investigation time
  Civil claim administration costs
  Possible union action
  Demoralising effect on workforce

The HSE publication *The Costs of Accidents at Work*[2] includes five case studies in which the hidden costs of accidents were examined in detail. These show the ratio of insured to uninsured costs to vary between 1:8 and 1:36. Similarly high ratios have been found in previous research.

Legal sanctions

The possibility of legal sanctions provides a further reason for managing health and safety properly. It is clear from earlier chapters that there is considerable legislation on the subject, some general in application and some specific. Much of it can give rise to claims for damages, but all could result in action by the relevant enforcing authority.

The effect on the organisation's image

The fact that an inspector has chosen to use one of the legal powers open to him means that he, at least, considers it a serious matter. Often the most damaging effect of legal sanctions can be on the organisation's own image of itself – someone in authority doesn't consider it to be as law-abiding, as caring or as well-managed as it previously believed.

The organisation's external image can also be significantly damaged by a successful prosecution or even, on occasions, an unsuccessful one. Local and regional newspapers feast upon health and safety prosecutions, particularly if there has been a serious injury involved and there may also be interest from the national press, radio or TV.

Such reports will be seen by customers, local politicians, council officials, neighbouring companies and by actual and potential employees. Indeed the newspaper report of the facts may be the only version that existing employees receive. An effect on employee morale will be inevitable with recruitment and sales also likely to be influenced.

The costs of legal sanctions

As noted in chapter 3, the fines that can be levied as a result of successful prosecutions are unlimited with £750,000 the highest so far. Even the local magistrates' court can impose fines of up to £20,000 for each separate offence. However, legal fees can often match or even exceed the normal size of fines meted out, particularly in the event of a 'not guilty' plea. Added to this there will be company time spent preparing for the case and also in actually attending court. In many cases the tripling or quadrupling of the fine would give a truer estimate of the cost of the case to the organisation.

In some situations the imposition of a Prohibition Notice could prove much more expensive than a prosecution. If a piece of equipment, a process or a building is suddenly taken out of action then the costs that result will depend upon how critical it is to the operation. In one instance a Prohibition Notice resulted in the loss of 3000 man-hours in a single day. It is of course quite open to an inspector to issue a Prohibition Notice and follow it up with a prosecution – double jeopardy from a cost point of view, with the additional possibility of a damages claim.

Generally, Improvement Notices will be less expensive than Prohibition Notices but the fact that action has to be taken within an imposed time-scale, rather than managed normally, could result in additional costs, eg overtime, premium payments to contractors or charges for rushed special orders. With expenditure concentrated into a narrow period there could also be budgetary difficulties or even real cash-flow problems for smaller companies.

THE CAUSES OF ACCIDENTS

In studying how health and safety should be managed it is important to have an appreciation of how accidents (using the broad definition given on page 108) are caused.

The loss causation model

HW Heinrich, a pioneer in safety thinking, originated the loss causation model. This has since been developed by a variety of other authorities and is depicted in Figure 6.2. In the accident sequence the five dominoes topple over in turn, the final one representing the loss. This could be an injury to an employee (or someone else), ill-health or damage to equipment or materials. The loss resulted from the incident, the causes of which need to be investigated. (Chapter 9 covers investigation techniques in some detail.) The cause or, more likely, causes emerging from the investigation are usually immediate causes – for example poor technique, lack of attention, faulty equipment. However, it is important to realise that underlying these are much more basic causes. Poor technique could result from lack of training, inattention from a poor attitude or inadequate supervision, faulty equipment from insufficient maintenance.

All of these basic causes are indicative of a lack of management control. In the well-managed organisation training arrangements, attitudes of employees, supervisory standards and maintenance programmes should be effectively controlled and losses due to accidents thus greatly reduced.

Immediate and basic causes

The differences between immediate and basic causes are illustrated in Figure 6.3 which describes the possible causes of a hypothetical accident in which a fork lift truck has crashed into a wall. Clearly not all the causes will necessarily apply and some possibilities may well have been omitted.

A review of the basic causes will quickly identify those areas where management control needs to be exerted – many of these are of course dealt with more specifically elsewhere in this book. In the example these include:

- training
- supervision
- employees' attitudes
- medical screening
- control of visitors and contractors
- maintenance
- pre-use check
- defect reporting and rectification
- planning
- housekeeping.

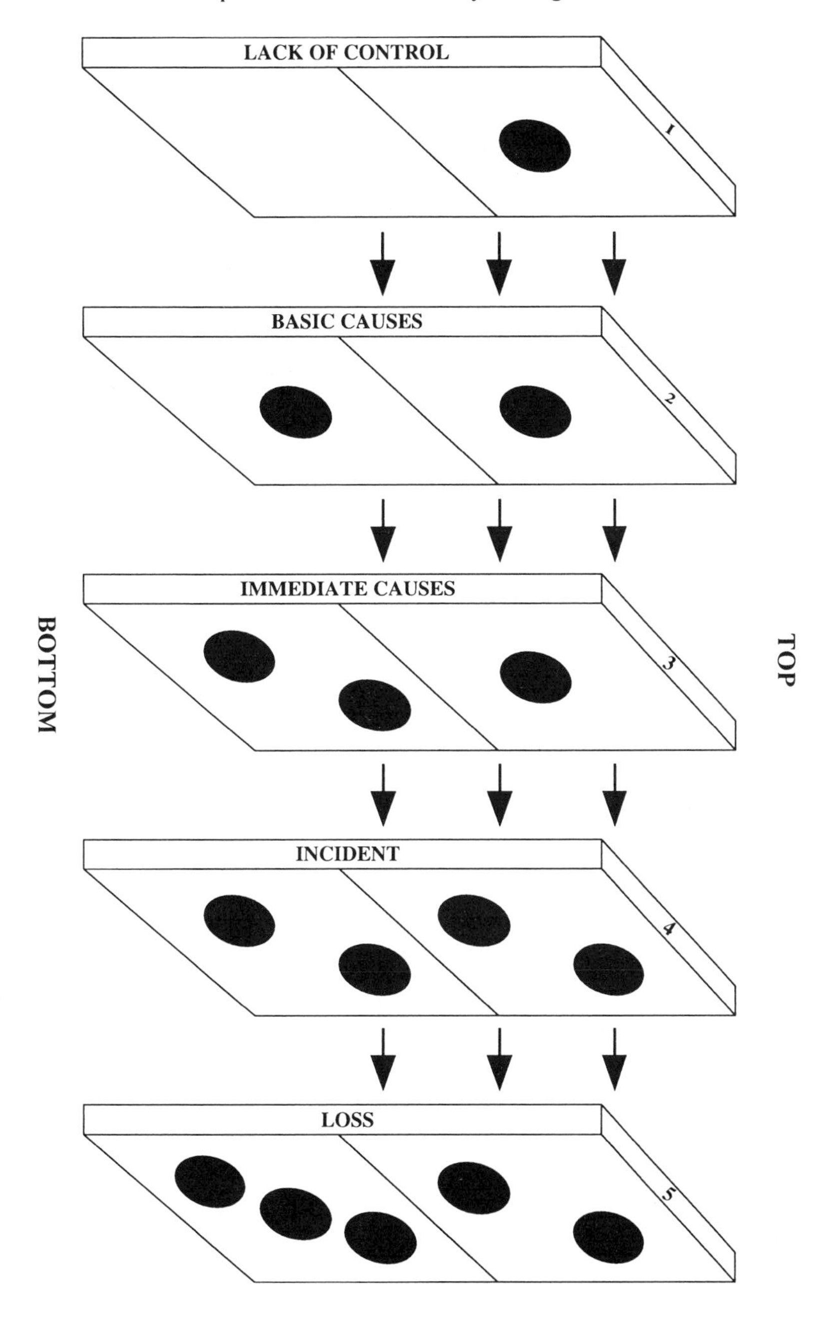

Figure 6.2 *The loss causation model*

Source: Reprinted from *Practical Loss Control Leadership*, Frank E. Bird, Jr. and George L. Germain. Copyright © 1996 Det Norske Veritas Industry, Inc., Loganville, Georgia, USA.

This figure details the possible immediate causes and the related basic causes of an accident in which

A FORK LIFT TRUCK CRASHES INTO A WALL

POSSIBLE IMMEDIATE CAUSES	RELATED BASIC CAUSES
Involving the DRIVER Speeding Careless driving Affected by drink or drugs Tired Not authorised to drive	 Driver's attitude Lack of supervision
Poor technique Unfamiliar with the working environment	Lack of training
Poor eyesight Blackout Heart attack	Absence of medical screening Lack of supervision Attitude of the driver (and possibly others)
Involving OTHER PEOPLE Wandering into the path of the fork-lift truck	 Lack of training Poor control of visitors/ contractors
Distracting the driver Leaving obstructions on the route	Employee's attitude Lack of supervision

(Continued opposite)

POSSIBLE IMMEDIATE CAUSES	RELATED BASIC CAUSES
Relating to the FORK LIFT TRUCK	
Defective brakes	Poor maintenance
Defective steering	Absence of pre-use check
Other mechanical failure	Failure to report defects (driver's attitude)
	Failure to rectify defects when reported (lack of procedure, attitude of others)
Unsuitable for the task for which it was being used	Lack of planning Lack of training Lack of supervision
Relating to the LOAD	
Unsuitable for the fork lift	Lack of planning
Not sufficiently secure	Lack of training
	Lack of supervision
Relating to the WORKING ENVIRONMENT	
Poor floor surface	Lack of maintenance
Inadequate lighting	Inadequate housekeeping arrangements
Oil or water on the floor	Employees' attitudes
Restricted visibility	Lack of supervision
Storage encroachment	Lack of planning
Poorly designed route	
Absence of barriers	
Effects of weather	

Figure 6.3 *Analysis of possible causes of an accident*

The four categories of causes*

Both immediate and basic causes of accidents can be analysed using four categories.

- PEOPLE
  - EQUIPMENT
    - MATERIALS
      - ENVIRONMENT

As explained more fully in chapter 8 a Risk Assessment must take proper account of each of the four categories, with the corresponding management control measures detailed in the same way.

People

Here it is necessary to consider not just the people who have the accidents or are the immediate causes of them, but all of the people who may influence the business.

- *Senior managers* – who set the policies and procedures, decide what types of activities the organisation will be involved in, determine the budgets etc.
- *Middle managers* – who also set procedures and are responsible for implementing them, select staff and place them in specific jobs.
- *Supervisors* – who organise, lead, train and control employees.
- *Engineers* – who plan production systems and processes and specify equipment.
- *Purchasing staff* – who purchase materials, equipment, personal protective equipment (PPE) and services.
- *Maintenance staff* – who should be keeping the equipment and the environment in good working order.

Equipment

Again it is necessary to look at a wide range of equipment.

- *Fixed machines* – power presses, baking ovens, printing machines.
- *Portable machines* – welding sets, spray units, word processors.
- *Materials-handling equipment* – cranes, conveyors, hoists.
- *Vehicles* – fork lift trucks, vans, tractors.
- *Hand tools* – spades, hammers, scalpels.
- *PPE* – helmets, gloves, respirators.

Materials

The full range of materials which might be encountered must be considered.

* These categories were previously used in *Practical Loss Control Leadership*, published by DNV, Author: Frank Bird.

- *Raw materials* – used in the process.
- *End products* – there may be risks from these.
- *Intermediate materials, by-products, waste materials.*
- *Maintenance and cleaning materials.*

All these types of materials may cause problems in a variety of different ways, for example they may be

Toxic	Flammable
Corrosive	Explosive
Heavy	Hot
Sharp	Cold

Environment

All aspects of the working environment must be considered:

- *Buildings* – their walls, roofs, floors and fixtures within them.
- *Surfaces* – roadways, walkways, storage surfaces.
- *Surroundings* – public roads, waterways, drains, neighbouring workplaces, domestic property.
- *Physical factors* – noise, radiation, heat, cold.
- *Climate* – rain, snow, ice, sun, wind.

There will obviously be overlap and interaction between the four categories but this approach is a useful aid in ensuring that no elements are overlooked in developing management control measures.

THE FIVE STEPS OF SAFETY MANAGEMENT

Five Steps to Successful Health and Safety Management is the title of a free leaflet published by the HSE.[3] The five steps described in the leaflet are:

(1) Set your policy
(2) Organise your staff
(3) Plan and set standards
(4) Measure your performance
(5) Learn from experience: audit and review

Each of these steps will be explained in greater detail in the remaining sections of this chapter.

Successful Health and Safety Management

The Five Steps leaflet has its origins in the book *Successful Health and Safety*

Management[4] published in 1991 and based on many years' practical work in the field by the HSE's Accident Prevention Advisory Unit (APAU).

The book's introduction goes on to point out that "commercially successful companies often also excel at health and safety management precisely because they bring efficient business expertise to bear on health and safety as on all other aspects of their operations".

Whilst somewhat academic in its presentation, the book nevertheless gives a comprehensive description of all of the key elements of successful health and safety management. However, a vital comment is made by the Chief Inspector of Factories who states:

> The path described is neither easy nor short. There are no short cuts to successful health and safety management. It cannot be sidelined. It must not be delegated out of sight. The clearest lesson from practical experience is that the starting point is the genuine and thoughtful commitment of top management.[5]

HSE's guidance on the route to follow is firmly based upon established principles of Total Quality Management (TQM). This theme was developed by Andrew Deacon in his award-winning essay for the Safety and Health Practitioner[6] where he demonstrated the links between successful health and safety management and the quality approach taken by WE Deming and others.

The Management Regulations

Under the Management of Health and Safety at Work Regulations 1992[7] it has become a legal obligation to manage health and safety effectively.

Regulation 3 of these regulations requires both employers and the self-employed to carry out an assessment of the risks to their employees and others in order to identify the measures necessary to comply with health and safety legislation. (The process of Risk Assessment is dealt with in detail in chapter 8.) Regulation 4(1) then states that "Every employer shall make and give effect to such arrangements as are appropriate, having regard to the nature of his activities and the size of his undertaking, for the effective **planning, organisation, control, monitoring and review** of the preventive and protective measures." (emphasis added)

This requirement closely matches the approach of the *Five Steps* leaflet. Regulation 4(2) also requires the "arrangements" to be recorded by undertakings with five or more employees.

International Safety Rating System™

The International Safety Rating System™ (ISRS) was developed by the International Loss Control Institute (now part of DNV) as a tool for auditing management safety performance. Although ISRS originated within the USA and Canada it is now extensively used in other parts of the world including the

UK and other European countries.[8] Auditing techniques are dealt with in more detail in chapter 11.

The Management Control System on which ISRS is based uses the acronym ISMEC, taking a similar Five Step approach:

I – IDENTIFICATION of the loss control activities necessary to achieve the desired results
S – STANDARDS – establishing performance standards for each of these activities
M – MEASUREMENT of what performance is being achieved in practice
E – EVALUATION of how well the standards are being met (and the reasons why if they are not)
C – COMMEND AND CORRECT – commending good performance and taking constructive action to correct sub-standard performance

POLICY AND ORGANISATION

Set your policy (Step 1)

The requirement for organisations with five or more employees to have a health and safety policy dates back to s.2(3) of HSWA. Regulation 4 of the Management Regulations reinforces this by requiring those with five or more employees to record their health and safety arrangements. However, although many employers have produced policy documents, these have often been filed away and forgotten, and bear little relationship to what is actually happening in the workplace.

A well prepared and up to date policy can provide a clear demonstration of the organisation's commitment to health and safety which can act as a reference point for employees at all levels. What is included in the policy should very much depend on local circumstances – the organisation's activities, the risks involved and how it is structured.

Although various standard 'off the peg' policies are available, these are not recommended. The HSE produce their own version *Guide to preparing a safety policy statement for a small business*.[9] This contains an arrangements checklist which, although helpful, consists predominantly of blank pages on which users must fill in their own text.

Another HSE document, *Writing a safety policy statement – Advice to employers*,[10] contains much useful material for those producing a policy from scratch to meet their own organisation's needs. Such a policy is likely to consist of three main sections:

- a general statement of policy
- who is responsible for putting it into effect
- detailed arrangements for its implementation.

The likely content of these sections is described below together with:

- arrangements for the policy's communication and keeping it up to date.

The statement

This usually consists of a general statement of good intent on health and safety matters, often linked to a commitment to comply with relevant legislation, for example:

> The Company is committed to achieving high standards of health and safety in all of its operations, not only in respect of its own employees but also in respect of all others involved in or affected by its activities. In particular we will comply with all relevant legislation which we will regard as setting a minimum standard.
>
> Detailed responsibilities and arrangements for implementing this policy are contained elsewhere in this document. The policy will be reviewed on an annual basis.

The statement should be signed by a partner or senior director and dated. Such a statement can then form the frontpiece to the complete document and can be prominently displayed around the workplace.

Responsibilities

The good intentions of the statement are of little value unless it is clearly identified who is responsible for carrying them out. How responsibilities are allocated will depend upon the size and the structure of the organisation. Some detailed advice on this is contained in chapter 7 (see figure 7.1).

Detailed arrangements

The policy itself need not contain all the company's health and safety arrangements in detail. Cross references to other documents are perfectly acceptable. The topics included will depend very much on the organisation concerned, hence the earlier recommendation against 'off the peg' policies. Typical topics which may be included are listed in figure 6.4. (Many of these are explored in more detail in other chapters of this book.)

Communication and updating

All employees must be familiar with and understand the policy, particularly their own responsibilities under it. They may be provided with their own copy (possibly as part of an employees' handbook or similar document) or alternatively copies could be prominently displayed around the workplace. In either case this should be supplemented by a personal briefing on the contents – this would usually be included in an induction programme.

The policy should be regularly reviewed to ensure it remains up to date in the light of changing circumstances within the organisation (eg new processes, revised management structures) or outside changes (eg new legislation). It could be subjected to the documentation and review standards of BS5750.

Health and safety training	– general	– managers
	– task related	– supervisors
Operational procedures		
Personal protective equipment	– general provision	
	– specific standards	
Maintenance and repair	– general	– defect reporting
	– electrical	– isolation/permit to work
Inspection arrangements	– statutory inspections	
	– pre use checks	
	– safety inspections	
Safety representatives		
Health and safety committee		
Accident and incident investigation		
Fire prevention		
Fire and other emergency procedures		
First aid		
Occupational health		
Hazardous substances		
Purchasing controls		
Design of new equipment and processes		
Control of contractors and visitors		
Work at other locations		
Employee involvement and motivation		

Figure 6.4 *Topics possibly requiring detailed arrangements*

Organise your staff (Step 2)

The detail of how to organise for health and safety is dealt with in chapter 7 but it is important to set down here some of the important principles in this key step in the management cycle.

1. The term 'health and safety culture' is increasingly being used to describe not only the organisational structure but also the involvement and commitment of everyone in the organisation to high standards of health and safety. These will never be achieved just by the health and safety professionals but must actively involve people at all levels.
2. Importance of senior management actions
   The wholehearted commitment of senior management is essential in creating this culture by means of actions such as:
   - *Allocating responsibilities*, through job descriptions and in other ways, to flesh out the general responsibilities identified in the policy.
   - *Providing resources* through adequate specialist support and in budgetary provision for health and safety related activities.
   - *Setting objectives* by developing the intentions of the policy to set detailed standards for their achievement (see step 3).
   - *Emphasising the importance* of health and safety through its integration into performance appraisal systems and making it a real factor in personal development and career progression.
   - *Setting the example* through their own conduct. Employees always remember the senior manager who didn't conform to PPE requirements or failed to respond quickly to a fire drill.
   - *Demonstrating an interest* in what is happening on the subject both formally through a presence at health and safety and informally in day to day contacts. Everyone becomes interested in something that interests their boss.
   - *Action on failures* – whether by reviewing policies or procedures or taking action with individuals responsible (managers as well as shopfloor employees).
3. The 'Four Cs'
   HSE emphasise in their Five Steps leaflet (3) the 'Four Cs' essential to achieving a "positive health and safety culture".
   - *Control* Through the types of senior management actions listed above.
   - *Competence* Ensuring the availability of the necessary competence throughout the organisation through the assessment, recruitment and training etc of staff.
   - *Communication* Ensuring an adequate flow of information into the organisation (legal, technical, practical etc) and within it (policies, objectives, standards, factual and performance).
   - *Co-operation* Participation encourages ownership. The knowledge and experience within the organisation needs to be pooled for the common

good. This can be done formally through safety committees and problem-solving teams or informally by face to face contact.

The Four Cs will be considered more fully in chapter 7 whilst chapter 14 deals more specifically with employee involvement.

STANDARD SETTING AND PERFORMANCE MEASUREMENT

Plan and set standards (Step 3)

Define objectives and standards

Having set general objectives through the policy (Step 1) and created the culture for carrying them out (Step 2) there is now a need to define objectives in much greater detail and set standards for their achievement.

The objectives and standards must relate to positive actions to be taken to improve the health and safety performance. The setting of goals based purely upon numbers of accidents or accident frequency rates is just wishful thinking unless the actions which it is hoped will produce these figures are clearly identified.

"You get things right by doing right things"

Anon

The standards used must be:

- *Measurable* – a defined activity within a designated time or to a specified frequency by identified individuals or groups.
- *Realistic* – there must be a realistic expectation of their achievement by those individuals or groups within the time frame, bearing in mind their own personal situation (eg knowledge, experience, time) and other constraints within the organisation (eg budgets, availability of others).

Examples of standards

Typically, general policy objectives might include:

- the provision of necessary *training* for employees
- the carrying out of regular *safety inspections* of the company's premises

Specific objectives and standards might be developed as follows:

- *Training* Having become aware of the requirements of the Manual Handling Operations Regulations 1992 (see chapter 19) a specific objective could have been set to: "train a team of ten assessors, drawn from all departments, in the content and implications of the regulations and practical assessment techniques during the first quarter of 1993." This could have been allocated to a training officer.

From this could have developed further specific objectives
— carrying out assessments within a defined period (agreed by the assessment team)
— for task-related manual handling technique training (identified as a result of the assessment process)

- *Safety inspections* Standards for the safety inspection programme might be specified in a slightly different way:
  — The premises split into clearly identified and manageable inspection units eg the main stores building, the maintenance workshop etc.
  — An inspection frequency agreed (relating to the risks in the unit) eg every two months.
  — Those carrying out inspections identified eg the supervisor and safety representative (or their deputies).
  — A method for reporting and taking action specified eg standard report form completed and sent to the manager (to initiate future action) with a copy to the safety officer.

(Safety inspections are covered in greater detail in chapter 11.)

Measure your performance (Step 4)

The importance of measurement

"What gets measured gets done" goes the old adage and it applies to health and safety in the same way as other aspects of the organisation's performance – production, sales, quality. There is little point in setting those measurable and realistic objectives and standards in Step 3 and then not checking whether they have been achieved.

How this measurement and comparison is carried out will depend upon the structure of the organisation and the aspect of performance being measured. It is likely to involve one or more of the following:

Senior managers
Departmental managers
Supervisory staff
Administrative support staff
Safety specialists
Other specialists
Safety committee

As well as the measurement of proactive activities, health and safety can also be measured by reactive means – by identifying unwanted or sub-standard performances through the investigation of accidents or during simple informal observations in the day-to-day course of events.

In addition to actually measuring the performance it is important to have an appreciation of the reasons for it so that the necessary corrective action can be taken or commendation given.

Examples of measurement

Applying this approach to the examples used in Step 3:

- *Training* Whether the training of manual handling assessors had been carried out and had met the stated needs would be a simple matter of fact – possibly established by the training manager or the safety committee some time soon after completion of the relevant time-frame.

  However, the possible reasons for it not being carried out could be quite varied eg
  - — indolence on the part of the training officer
  - — unavailability of a suitable tutor
  - — unwillingness of some managers to nominate participants
  - — failure of several nominees to attend.

  Using reactive measurement at a later stage in this example, an increasing number of manual handling accidents could indicate that this training or some other part of the assessment programme and resultant action had not been very successful.
- *Safety inspections* Compliance with the laid-down safety inspection programme could be measured by the relevant department manager, the safety specialist or the safety committee. As well as checking that the inspections had actually been carried out, the quality of the reports could also be monitored as could the effectiveness of remedial action taken – whether by a formal feedback system, sample checking or simply the absence (or presence) of repeated faults.

  Again, the reasons for poor performance could be varied:
  - — lack of commitment by the supervisor or safety representative
  - — absence due to sickness or other duties
  - — poor quality inspections due to lack of hazard spotting training
  - — no remedial action due to inaction by the manager

  Similarly an informal visit to the department could reveal faults that should have been obvious during an inspection (eg unstable racking, defective ladders) indicating that the inspections are not being carried out as thoroughly as they should be or corrective action is not being taken.

LEARNING FROM EXPERIENCE (Step 5)

Much of the earlier effort will be wasted if the experience accumulated – whether by establishing through proactive monitoring that standards have or have not been met or as demonstrated through events (accidents or observations) – is not put to good use. For short-term matters action of the commend or correct type will be necessary, whilst in the longer term periodic auditing and review of safety performance becomes important.

Commendation

Commendation is something that managers in the UK are not very good at, especially when compared with organisations across the Atlantic. Flag-waving and 'Employee of the Month' awards may be somewhat alien to British culture but a well chosen word of praise can do wonders for employee morale and therefore future performance.

Corrective action

Much of the corrective action necessary to achieve satisfactory performance will be relatively straightforward. The action (or inaction) of individuals can often be corrected by the right word in the right ear. Whether it is the individual's own ear or that of his boss will depend on circumstances (and tactics). In a true health and safety culture there should seldom be the need to resort to formal discipline.

In the earlier examples:

- alternative sources of assessor training could be suggested
- substitutes could be arranged for absentees from the safety inspection programme.

Auditing

An audit is part of the learning process. It should look at what is being done during each of the five steps:
– POLICY – ORGANISING – STANDARDS – MEASURING – LEARNING.
It should probe for weaknesses in the existing procedures and systems and whether these are actually producing the right results. A safety audit is far more comprehensive than measurement of a simple parameter or a routine safety inspection (to which the term is often incorrectly applied). It can be carried out by someone within the organisation or an outsider – a combination of the two is often particularly effective. (Audits are dealt with in much greater detail in chapter 11.)

Review

Review is the detailed examination of one or more parts of an organisation's health and safety arrangements.

- Is there a policy laid down on the topic or is the existing policy adequate?
- Is the organisational structure capable of implementing it?
- Do standards exist or are the present standards right?
- Can performance be measured more effectively?
- Why do the same old problems keep recurring?

For example, a review of a regular failure to meet safety inspection standards could reveal:

- An organisational structure that did not allow enough time to carry them out.
- A need to increase (or decrease) their frequency.
- A general need for hazard spotting training.

Periodic reviews of arrangements are a good discipline in any case but a much more extensive review is likely to be necessary after an audit. The need for improvement on many fronts may have been identified during the audit, the review must decide how these recommendations should best be addressed.

Any review will throw up action points, all of which will need to be prioritised alongside other health and safety and business activities, with measurable and realistic standards being set for their completion and those responsible for taking action clearly identified.

Notes

1. *Successful Health and Safety Management* – HSE 1991 (HS(G)65)
2. *The Costs of Accidents at Work* – HSE 1993 (HS(G)96)
3. *Five Steps to Successful Health and Safety Management* – HSE (IND(G)132L)
4. *Successful Health and Safety Management* (see note 1)
5. *Ibid.*, pV
6. *The Safety and Health Practitioner* – Institution of Occupational Safety and Health, January 1994 p18
7. Management of Health and Safety at Work Regulations SI 1992/2051. The Regulation, ACOP and guidance are available in one document (L21) from HSE Books.
8. *International Safety Rating System* – Det Norske Veritas (DNV)
9. *Guide to preparing a safety policy statement for a small business* – HSE 1990 (C100)
10. *Writing a Safety Policy Statement – Advice to Employers* – HSE 1994 (HSC6)

7

Organising for Health and Safety

This chapter is concerned with how the principles described in chapter 6 can be put into practice. It is structured around the HSE's four Cs of positive health and safety culture (see p122).

- CONTROL
- COMPETENCE
- COMMUNICATION
- CO-OPERATION

Co-operation is only dealt with briefly here as it receives more detailed coverage in chapter 14.

CONTROL

All methods of successfully controlling health and safety centre around three areas:

- THE KEY PEOPLE
- KEY ATTRIBUTES OF THOSE PEOPLE
- KEY HEALTH AND SAFETY SYSTEMS

The key people

The list of key people below may need to be expanded in the case of large organisations (eg taking account of national, regional and local levels) or compressed in the case of smaller ones. For very small organisations the roles may well be combined into one or two posts.

Senior management responsibility

A member of the board or senior management team should be clearly identified as responsible for health and safety related matters in the same way as individuals carry responsibility for production, sales, finance etc.

The allocation of adequate resources (in terms of manpower and finance) to carry the good intentions of the health and safety policy into practice is a major issue which can only be addressed properly at senior management level. Typical senior management responsibilities (which could be incorporated into a health and safety policy) are illustrated in figure 7.1.

Line management

Despite the vast amount that has been written on health and safety theory there are still far too many managers and supervisors who believe that these topics are the sole province of the safety officer or the health and safety committee, or even the safety representative, and that their responsibility is simply to produce widgets, repair machines or count beans.

No organisation will be successful in the long term in controlling health and safety unless responsibilities for these aspects are properly integrated into management responsibilities at all levels.

The types of responsibilities which should be carried by managers and supervisors are illustrated in figure 7.1 and the levels of competence required for managers and supervisors are examined later in the chapter.

Health and safety specialists

Every employer now has a legal duty under Regulation 6 of the Management Regulations to appoint one or more competent persons to assist him in undertaking the measures he needs to take to comply with the requirements and prohibitions imposed upon him by or under the relevant statutory provisions – ie to assist in complying with the law.

The ACOP accompanying the Management Regulations then goes on to point out that small employers (sole traders or members of partnerships) may appoint themselves (or other partners) for this purpose, so long as they are competent.

What constitutes competence in this context is developed further in the ACOP and is examined in more detail later in this chapter, although clearly the level of competence required will reflect the level of risk in the work being carried out and, to a lesser extent, the size of the organisation.

The Management ACOP allows for the use of external support such as consultants in the provision of competent assistance, although it points out that such services will usually be advisory and that their activities will need to be co-ordinated by the employer (or his appointee). The HSE booklet *Selecting a Health and Safety Consultancy* gives much useful advice in this respect.[1]

The employer still retains all his responsibilities under health and safety legislation, even where external services are used or one or more employees are appointed. He cannot pass these on to his appointee, although the competent person still has his general duties as an employee and will be expected to carry out his role in a diligent and responsible manner.

Set out below are the health and safety responsibilities which might be held by individuals at different levels in an organisation. These would normally be detailed in a health and safety policy with reference also made (possibly in an abbreviated form) in job descriptions.

Senior manager (responsible for health and safety)
(board or senior management team level)

- Demonstrating a commitment to, and interest in, health and safety
- Bringing health and safety matters to the attention of the board
- Ensuring adequate resources are available to carry the policy into effect
- Monitoring the effectiveness of the policy and related arrangements
- Clearly allocating health and safety responsibilities to others within the organisation
- Setting health and safety objectives for the organisation and individuals within it
- Monitoring how well responsibilities are carried out and objectives achieved
- Participating in (probably chairing) the health and safety committee

All managers (within their own area of responsibility)

- Demonstrating a commitment, to and an interest in, health and safety
- Ensuring that health and safety aspects are fully integrated into procedures and arrangements
- Monitoring the effectiveness of health and safety policy and related arrangements
- Ensuring that subordinates carry out their health and safety responsibilities
- Ensuring that employees are adequately trained, particularly in health and safety matters
- Reviewing accident and incident reports and ensuring remedial action is taken
- Overseeing the health and safety inspection programme

Supervisory staff (these could include supervisors, team leaders, charge-hands or junior engineers)

- Ensuring that employees follow laid-down procedures, particularly in respect of health and safety aspects
- Reporting shortcomings in, or the absence of, health and safety-related procedures
- Allocating tasks only to those employees who have been adequately trained for them

(Continued)

- Taking appropriate remedial action where employees commit unsafe acts
- Ensuring high standards of housekeeping
- Investigating and reporting on accidents and incidents
- Participating in health and safety inspections

Figure 7.1 *Typical management health and safety responsibilities*

The level and type of specialist support necessary will depend upon the level of risk present in the work, the size of the workforce and the structure of the organisation. For example, a large national employer with many small branches will need to decide what support should be provided at national, regional and local level. This might involve a full-time specialist with a national role, part-time specialists combining health and safety with other duties at regional level, and the appointment of a senior manager at each local branch, providing of course that these managers had received an appropriate level of training (see section 2 of this chapter).

In smaller organisations a part-time health and safety specialist may combine these duties with other work eg personnel, administration, facilities management.

The Construction (General Provisions) Regulations 1961 contain a requirement similar to that in the Management Regulations in respect of contractors and employers normally employing over 20 persons (even if they work on different sites). They must appoint an experienced and suitably qualified person or persons to advise them on safety requirements, to exercise "general supervision" of the observance of such requirements and to promote "safe conduct of the work generally".[2]

Even if other members of the management team have health and safety fully integrated into their responsibilities, the specialist still must play a pivotal and very proactive role – promoting, monitoring, prompting and doing. The exact nature of the role will depend upon the size, structure and needs of the organisation but responsibilities are likely to include the following:

Promoting

- Clearly demonstrating his own commitment.
- Utilising the health and safety committee to promote and communicate.
- Publicising health and safety messages eg poster campaigns, in-house magazines, newsletters etc.
- Providing information and resource material for others.

Monitoring

- Checking the effectiveness of the key systems such as accident and incident reporting, health and safety inspections.
- Monitoring other procedures and arrangements.

- Measuring the achievement of organisational and individual objectives.
- Auditing.
- Observing.

Prompting

- Identifying shortcomings and liaising with the senior manager (and others) to rectify them.
- Identifying training needs.
- Nudging along reluctant or dilatory members of the management team.
- Identifying future health and safety issues.

Doing

- Playing a functional role in the health and safety committee (acting as secretary is believed by many to be the most appropriate role).
- Liaising with HSE or local authority inspectors, insurers etc.
- Conducting some health and safety training.
- Providing advice and assistance at all levels.
- Developing and writing health and safety related procedures (or assisting in the process).

In larger organisations the specialist may head a multi-disciplinary team and his role may extend to include:

- Co-ordinating and supervising the other team members.
- Setting objectives for the team and individuals within it.
- Identifying training needs of team members.
- Budgeting for the team's expenditure.

Related specialists

Large organisations may employ other specialists, particularly in the health and hygiene fields but in the case of small companies such posts are unlikely to exist at all, expertise being hired in from consultancies whenever specific needs arise. For medium-sized employers the roles may be combined with other functions eg with that of the general health and safety specialist or be carried out on a part-time basis, possibly shared with other employers in the area.

Occupational hygienists are involved in the recognition, evaluation and control of risks to health, from hazardous substances, noise etc. Their role is examined in greater detail in chapter 13.

While the hygienist is concerned principally with the environment in which employees work, occupational physicians (medical officers) and occupational health nurses are much more involved with the employees working in that environment. Chapter 13 also deals with their role and those of other health specialists.

Training staff have an important part to play in ensuring health and safety is properly managed within their organisations. Health and safety should generally be an important element in induction programmes and safety aspects need to be properly integrated into other training activities that are taking place. Trainers need to be aware and take account of health and safety legislation containing specific training requirements.

The key attributes

To develop a positive health and safety culture, those involved must possess some important attributes.

A shared philosophy

Although the health and safety policy statement is a starting point for a shared philosophy, it really must penetrate into the organisation's psyche. There must be a common and genuine belief that health and safety is much more than keeping out of trouble and complying with the law. Management at all levels within the organisation must believe that health and safety needs to be managed systematically in the same way as all of the organisation's other activities.

Commitment and interest

The need to demonstrate commitment and interest was specifically highlighted within the responsibilities of several of the key players listed earlier. A genuine commitment and interest among senior management is essential in developing the same level of commitment amongst those further down the structure.

This must go far beyond formal statements and periodic declarations of commitment to them. To be recognised as genuine by others the commitment must shine through in day-to-day behaviour and example. The higher the individual's position, the more important these factors become – everyone eventually gets interested in those things that interest their boss.

Another major factor in overcoming the scepticism barrier within an organisation is the willingness to provide the necessary resources to meet health and safety objectives. Provision of financial resources and the appointment of an appropriate number of specialist staff are just part of the picture here.

Availability of both management and employee time for health and safety-related activities is a very important feature. The cancellation of a health and safety training course or a routine inspection because staff cannot be released will be a significant demotivating factor.

A sense of realism

Providing the resources necessary to meet the declared objectives has to be part of maintaining a sense of realism. Realism is also necessary in setting the objectives in the first place – the 'quick fix' is seldom successful in health and safety. Cultural changes involving increased commitment to health and safety at all levels are likely to take several years to achieve. Indeed the process may never be complete with aspirations steadily climbing higher and higher. However, periodic bouts of retrospection can sometimes be quite morale-boosting in enabling the organisation to realise how far it has progressed over the years.

Progress to the eventual objectives will be encouraged by the setting of short-term objectives and standards which are both measurable and realistic (as described on p123). These should be based upon positive actions to be taken to improve health and safety performance and not upon accidents or accident frequency rates.

An integrated approach

One result of a shared philosophy on health and safety should be that its management is properly integrated into other aspects of running the organisation rather than it being treated as something special and apart or, worse still, as an afterthought.

Procedures, whether to meet the requirements of BS 5750 or otherwise, should take health and safety into account along with operational instructions, product quality requirements, environmental considerations etc (see chapter 12).

A similarly integrated approach should be taken to the writing of job descriptions, the setting of personal objectives and the appraisal of performance in meeting those objectives. Likewise health and safety should be a regular feature within management meetings and employee briefings.

Key health and safety systems

There are a number of areas where formalised health and safety systems (or arrangements) are likely to be necessary, whatever the size of the organisation or the nature of its activites. Where the system or arrangements are set-out in writing this could be included within the health and safety policy or within some other document.

The list below is not intended to be exhaustive and certainly some of the topics will not apply to all employers. Most of the topics are covered elsewhere in this book but a few will be discussed later in this chapter.

Topic	**Chapter(s)**
Health and safety policy	6 and 7
Setting of objectives	6 and 7
Recruitment	7
Training	7 and 12
Communications	7 and 14
Assessments	8
Accident and incident investigation	9
Emergency procedures	10
Health and safety inspections	11

Health and safety audits	11
New equipment and processes	12
Purchasing controls	12, 23 and 25
Health and safety-related procedures	12
Hazard reporting	12
Occupational health and hygiene	13
Health and safety committees	14
Contractors	15
First-aid	16
Personal protective equipment	25

Health and safety policy

Chapter 6 provided guidance on the contents of a health and safety policy. The organisation for putting the policy into effect will be based upon the key people referred to earlier in this chapter. Normally their responsibilities will be spelled out in the policy document and (possibly in an abbreviated form) in individual job descriptions.

How many of the detailed systems or arrangements are actually contained in the policy and how many are located in other documents is a matter of choice and convenience, although the policy should contain references to those other documents.

Setting of objectives

As already noted, objectives should be measurable and realistic and based on positive actions to be taken to improve health and safety performance rather than on accidents or accident frequency rates.

While the organisation may identify its long-term objectives, it will generally need to break these down into component actions if it is to achieve them. These components will then result in shorter-term objectives, both for the organisation as a whole (and possibly its constituent departments) and for individuals within it.

Commitment to achieving objectives is vital, and is more likely to be present if those charged with achieving them have been involved in their formulation and accept them as realistic and important. Regular monitoring of performance together with feedback and review will stress this importance.

Organisational objectives could be set within the framework of management meetings or the health and safety committee. There may be advantages in larger organisations of convening a special health and safety planning meeting in the same way as meetings are often held to specifically prepare the annual financial plan.

The senior manager with special health and safety responsibilities and the health and safety specialist are likely to be very actively involved in this process, particularly in carrying out the groundwork to identify which objectives are relevant and how they might be achieved. Individual objectives should then cascade through the structure with the objectives being agreed between the individual and his immediate superior, utilising the general performance appraisal system where one exists.

Some objectives may be ongoing in which case they may better be described as performance standards. Examples of objectives and performance standards are illustrated in figure 7.2 (see also pp123 and 124).

Long-term objective
Achieve a score of 75% on the emergency procedures section of the Group Health and Safety Auditing System by the end of 19—.

Short-term objectives

Organisation
- each department to publish a procedure for dealing with one potential major emergency and train employees in its content by the end of the year

Department
- finalise and publish an emergency procedure for dealing with leaks of chemical X by the end of the second quarter
- give 80% of employees a one-hour briefing in the procedure by the end of the third quarter
- give all employees a one-hour briefing by the end of the fourth quarter

Individual
- draft an emergency procedure for dealing with leaks of chemical X by the end of the first quarter

Performance standards
Where an employee requires first-aid treatment as a result of an accident at work, his or her supervisor must complete a company incident report form within 24 hours and send copies to the department manager and the health and safety officer.

All new employees must receive the standard health and safety briefing within their first day at work and must complete the company induction programme within their first two weeks.

Figure 7.2 *Objectives and performance standards*

COMPETENCE

Whilst the need for competence in health and safety matters is more obvious in respect of health and safety specialists, managers and supervisors, a level of competence in these areas is necessary for all employees.

What is competence?

The requirement for competence, and particularly for a "competent person" appears in many places in health and safety legislation. However, statutory definitions of the term competent person are relatively rare and even when they appear they are not always helpful. For example Regulation 2 (1) of the Pressure Systems and Transportable Gas Containers Regulations 1989 states that "competent person means a competent individual person (other than an employee) or a competent body of persons corporate or unincorporate; and accordingly any reference in these regulations to a competent person performing a function includes a reference to his performing it through his employees".[3] (On these regulations more generally, see chapter 16.)

However, the ACOP accompanying these regulations gives much more practical guidance, setting out not only the attributes required by the competent person but also clearly stating that different levels of competence will be appropriate in different circumstances – in this case devising schemes of examination for pressure systems of differing complexity and capacity.

This more helpful trend has continued in the ACOP to the Management Regulations where regulation 6 requires every employer to appoint one or more competent persons to assist him in complying with the law. Whilst the guidance in the ACOP is related to the specific attributes required for this purpose (ie providing specialist advice and services) the principles followed can be adapted to other areas where competence is required. Broadly speaking these can be split into two.

Knowledge and understanding

The ACOP refers to:

- The work involved.
- Principles of risk assessment and prevention.

To these can probably be added:

- Relevant legislation.
- Technical guidance (from HSE and other sources).

The capacity to apply this in practice

The ACOP specifies several areas of practical application:

- Identifying health and safety problems.
- Assessing the need for action.
- Designing and developing strategy and plans.
- Implementing strategies and plans.
- Evaluating their effectiveness.
- Promotion and communication.

For true competence both the theoretical knowledge and the capacity to put it into practice in the work situation are essential. Whilst professional and other qualifications, both inside and outside the developing National Vocational Qualification (NVQ) structure, are increasingly based upon practical demonstrations of competence, there are nevertheless still some very well-qualified individuals who are incapable of putting theory into practice.

In fact the ACOP states that in simple situations formal qualifications are not an automatic requirement. What is relevant is:

- An understanding of relevant current best practice
- Awareness of one's own limitations
- The willingness and ability to learn

An awareness of the limitations of one's own knowledge and experience is very important in determining competence and this is also of relevance in carrying out risk assessments (see chapter 8).

Competence requirements for all employees

Identifying the requirements of the job

All employees must be competent to carry out the tasks they are expected to perform. Regulation 11(1) of the Management Regulations imposes a general obligation: "Every employer shall, in entrusting tasks to his employees, take into account their capabilities as regards health and safety". Many other sets of regulations also contain requirements for competence or training. Such regulations include those relating to electricity, hazardous substances, noise, manual handling, work equipment, personal protective equipment and display screens. Moreover, the requirement to provide training may form part of the employer's duty of care at common law (see chapter 4).

Any occupation will require some general competencies – for example an electrician needs to be knowledgeable about electricity and its dangers, isolation procedures, live testing precautions and wiring standards, as well as being skilled in using test equipment and other tools. However, his competence must also relate to the task in hand, including a knowledge of the equipment to be worked on and any special risks it might present. (On electrical safety, see chapter 20.)

For some activities specific training requirements have been laid down in statutes or in ACOPs. These include abrasive wheel mounting,[4] power press setting,[5] operating woodworking machines[6] and driving rider operated lift trucks.[7] In other cases detailed guidance on training standards has been published by the HSE. Examples here include the operation of chain saws[8] and crane driving and slinging.[9]

Orders made under the Factories Act[10] and Offices, Shops and Railway Premises Act[11] list a variety of dangerous machines on which persons must not work unless they have been fully instructed as to the dangers arising and precautions to be taken and either have received sufficient training in work at the machine or are under adequate supervision by a person who has a thorough knowledge and experience of the machine. The Factories Act Order applies only to young persons (under 18) but the same approach should really be taken for employees of all ages.

Assessing employees' capabilities

Capabilities (or competence) may need to be assessed informally in the workplace if an employee is to be assigned to a task for the first time, or in the more formal situation of an interview for a vacancy. In both cases not only the present capabilities of the individual must be considered, but also how likely these are to be further developed through future training and experience. Factors to be taken into account in assessing capabilities are:

- Formal qualifications eg NVQs, apprenticeships.
- Attendance at relevant training courses.
- Experience of similar work elsewhere.
- Awareness of relevant risks and precautions.
- Attitude (including an awareness of one's own limitations).
- Physical attributes eg strength, dexterity, health.

Detailed questioning on practical aspects of the work is important in identifying exaggerated or falsified claims relating to qualifications or experience.

Action after selection

Most new recruits and existing employees allocated to new tasks will need to be brought up to the standard required through a combination of formal and informal training and the acquisition of relevant experience. Records should be kept of training undertaken and assessments made of employee competence together with any documentation issued.

General health and safety induction

All new employees, whether permanent or temporary, full-time or part-time, will need some type of induction. Regulation 11 (2) (a) of the Management Regulations requires all new recruits to be provided with adequate health and safety training. In large organisations an induction may be necessary for existing employees if they are transferred to new departments or locations.

The duration of the induction and whether it is delivered on a formal or informal basis will depend on local circumstances. Common topics to be included in most inductions are:

- Health and safety policy
- Personal responsibilities
- Fire procedures

- Other emergency procedures
- First-aid arrangements
- Accident and incident reporting procedures
- Washing, changing and eating facilities
- Basic manual handling techniques
- General PPE requirements
- Prohibited areas, tasks or equipment.

Local induction
Newcomers will also need to be briefed on the risks and precautions associated with the tasks they are expected to perform. Such briefings will usually be carried out by their supervisor or an experienced co-worker. A checklist of items to be included is often invaluable in ensuring that nothing important is missed out. Some of the specific statutory training requirements referred to earlier in this chapter could be covered at this stage.

Further training
Training will usually be required both in the short-term and the long-term to gradually bring the employee up to the full requirements of the job. Until that point is reached it will be necessary to restrict the range of duties that he is allowed to perform to avoid either him or others being put at risk. Such training may be through attendance on formal courses or through a structured programme of coaching by competent employees. The provision of competent supervision during this training is particularly important in the case of under 18s. The employee may well be on probation until a specified level of competence has been achieved.

Changing circumstances
Regulation 11 (2) (b) of the Management Regulations requires employers to provide adequate health and safety training for employees being exposed to new or increased risks because of:

- Transfer
- Changed responsibilities
- New or changed work equipment
- The introduction of new technology
- New or changed systems of work.

Many sets of regulations place similar requirements on employers to provide training where circumstances change.

Additionally Regulation 11 requires training to be repeated periodically where appropriate. Such refresher training may be necessary where employees have become out of practice at certain tasks or where bad habits have crept into their work.

Needs of the health and safety specialist

What level is appropriate?

In considering this question the employer must be mindful of his duty under regulation 6 of the Management Regulations to "appoint one or more competent persons to assist him" in complying with his statutory duties. The ACOP points out that it is open to small employers to appoint themselves, whereas large employers may have a specialist health and safety department. Consultants or other external specialists may also be appointed to provide advice.

Sub-paragraph 3 of the regulation states that the number of persons appointed, the time available for them to fulfil their functions and the means at their disposal must be adequate having regards to:

- the size of the undertaking
- the risks to which employees are exposed
- the distribution of those risks.

Guidance as to competence can be given by:

- membership of a professional body
- possession of an appropriate qualification
- achievement of a relevant NVQ.

However, competence may also be achieved through a programme of personal study or even lengthy experience in a particular field. Knowledge and understanding of safety principles must be accompanied by the capacity to put them into practice.

In a small business with relatively low risks the attendance of a manager or the employer himself at the type of course described on p143 may be all that is required. Full-time specialists or those with a substantial part-time role are more likely to need courses leading to professional qualifications.

Specialist qualifications

The National Examination Board in Occupational Safety and Health (NEBOSH) developed from the main professional body for safety and health practitioners – the Institution of Occupational Safety and Health (IOSH). Achievement of NEBOSH qualifications earns eligibility for different grades of IOSH membership.

The NEBOSH Certificate

The NEBOSH National General Certificate is a basic qualification aimed at newly-appointed safety advisors without previous experience, managers or supervisors with part-time responsibilities and safety representatives. The course involves 60 to 100 hours of study culminating in a written examination in two parts (each of two hours) together with a practical assessment requiring the identification of hazards in a workplace and the preparation of a short report for management attention and action.

Certificate courses are offered by many colleges and private training companies. They may involve day release or block release attendance and can also be run in-house for clients. Some open learning packages are also available. There are specialist versions of the Certificate catering for the needs of industries such as construction and offshore oil and gas exploration.

Successful candidates are eligible for associate membership of IOSH who also accept other qualifications such as the Diploma in Safety Management (issued by the British Safety Council) and the Diploma in Safety and Health (ROSPA) as having equivalent status.

The NEBOSH Diploma

The National Diploma course is directed at those intending to become professional safety and health practitioners. It normally requires 400–500 hours of study – between one and two academic years of attendance for one day per week, although block release courses are available as are open learning programmes.

The course consists of four modules – Risk Management, Health and Safety Law, Occupational Health and Hygiene and Safety Technology – there is a three-hour examination paper on each module. These can be taken either together or separately (in any order or combination). Finally a case-study paper of two hours must be taken in which candidates are expected to utilise their knowledge from the four modules in a hypothetical but practical context such as an accident scenario.

Successful candidates with three years suitable practical experience in the field (at least one of which is post-qualification) become eligible for corporate membership of IOSH and can then use the letters MIOSH after their names. Various universities offer degree courses or their own Health and Safety Diplomas and many of these are accepted by IOSH as equivalent to the NEBOSH Diploma.

IOSH maintains a Register of Safety Practitioners for corporate members who are currently active within the profession. These members may use the letters RSP after their names. From 1 January 1995 in order to maintain their status as RSPs, members must participate in IOSH's Continuing Professional Development scheme (CPD). Through the scheme they must demonstrate that they are maintaining and enhancing their knowledge and continuing to accumulate experience and skill.

The future status of bodies such as IOSH, NEBOSH, the British Safety Council and ROSPA in accrediting specialist qualifications is unclear because of developments taking place within the NVQ system. Specialist qualifications in the fields of occupational health and hygiene are referred to in chapter 13.

Needs of managers and supervisors

All managers and supervisors should have an understanding of the application of good management principles to health and safety and a broad awareness of the requirements of health and safety legislation. They must also understand their own role in the management process. Most nationally recognised courses cater for these general needs but individual managers and supervisors may also need training to address the more specific needs of their own workplace.

IOSH – Working Safely

This course is aimed at those who have had little or no previous systematic health and safety training. Whilst it is intended to be for all employees, it may be of use to a newly appointed first-line supervisor. It lasts approximately 6 hours and is assessed using a 45-minute written test together with a workplace-based safety assessment.

IEHO – Basic Health and Safety Certificate

The Institution of Environmental Health Officers aims this course at employees in general. It involves a minimum of six hours teaching time and a 40-minute multiple choice examination.

IOSH – Managing Safely

This course addresses the needs of managers and supervisors much more specifically. A total of approximately 30 hours of directed study is required. Candidates must take a 1¼-hour examination and carry out a work-based assignment during the course.

IEHO Advanced Health and Safety Certificate

As with IOSH's Managing Safely course, this caters much more for the needs of those in managerial and supervisory positions. At least 40 hours of teaching time is required and candidates are assessed upon a project, at least 2 course assignments and a 3-hour written paper.

Other courses

There are many other training courses covering the general needs of managers and supervisors and also catering for specific needs. Colleges and TECs (LECs in Scotland) are often active in this field as are many independent training organisations.

In-house training

Training for managers and supervisors can also be carried out in-house, either using the organisation's own health and safety specialist or through external consultants. This gives the opportunity to tailor the content of the training to more closely match local needs.

As well as covering topics of general importance such as legislative requirements, accident causation and safety management techniques, the course can concentrate on those risks and precautions existing within the workplace. Local procedures such as those for incident investigations, emergencies or inspections can be covered in detail.

Identifying new needs

The need for training because of changing circumstances was identified earlier. There the focus was on changes initiated internally such as transfers, new equipment, new technology etc. However, there are also changes totally outside the employer's control which will necessitate training, possibly at a variety of different levels within the workforce. Foremost amongst these are changes in legislation. To take the Manual Handling Operations Regulations 1992 as an example, training in a large organisation could be required as follows:

- *Health and safety specialist*
  — detailed knowledge of the regulations and their implications.
- *Medical officer, managers and supervisors*
  — appreciation of the main themes and principal implications.
- *Selected individuals*
  — training as assessors.
- *Occupational health nurse*
  — training to give advice and possibly training in techniques
- *Relevant employees*
  — general principles of manual handling and specific handling techniques.

Similarly changes in ACOPs, HSE Guidance, British Standard requirements or industry-based codes or guidance could necessitate training.

With so much changing in the world of health and safety the need for health and safety-related training should be reviewed on an annual basis, preferably as part of an overall analysis of training needs.

COMMUNICATION

Communication has been defined as "What we do to give and get understanding".[12] The process of communication is much more than the giving out of information – the message must be both received and understood.

The what, who and how of communication

What to communicate

There is much that management needs to communicate in the field of health and safety.

Philosophy and commitment
All concerned with the organisation need to understand that health and safety is regarded as an important issue and that management is committed to high standards. Management must gain the commitment and co-operation of employees in order to achieve their overall goals and objectives.

Standards
The detailed standards set for achieving these objectives must also be communicated. The workforce needs to be aware of relevant standards whether they are for quarterly fire drills, monthly safety inspections or all incident investigation reports to be completed within 24 hours.

Responsibilities
Individuals will be allocated a variety of personal responsibilities. Some of these may be of a general nature arising from the health and safety policy. Others may relate to their involvement in organisational procedures (eg incident investigation) or the allocation to them of a specific task (such as carrying out a manual handling assessment in a particular area).

Such individuals need to know that they carry these responsibilities, to understand how to carry them out and to feel capable of doing so.

Hazards and precautions
Employees must be fully aware of the hazards present in their work and of the precautions necessary to minimise the risk – the know why and the know how of health and safety. In some cases this may be very simple but in others the knowledge of detailed procedures or other arrangements may be necessary.

Performance
The workforce need to know how well the organisation, their department or their section is performing. This could involve negative indicators (such as accidents) or positive ones such as compliance with safety inspection schedules or the speed of correction of safety defects.

Thoughts and concerns
Both managers and their employees may have concerns about poor performance in certain areas of safety activity, a 'gut feeling' that things are not as they should be, or ideas on directions in which to move in the future. It is important that there is upward as well as downward communication of these thoughts and concerns.

With whom to communicate

Incoming communication

The organisation needs to receive a constant flow of information in order to keep abreast of developments in the legal and technical fields or in good management practice. There also needs to be an awareness of mistakes that have been made and problems that have arisen elsewhere.

Among the many means of keeping up to date are:

- health and safety periodicals
- HSE publications
- health and safety information systems (computer-based, microfiche or specialist manuals)
- trade associations
- local employers groups
- informal networks
- sister companies.

Within the organisation

In most organisations the emphasis is upon downward communication but upward communication must also be encouraged on health and safety matters. All employees must feel capable of effectively communicating with management, in particular when:

- the risks of an activity are greater than is generally realised
- precautions are impractical or inadequate
- performance indicators are measuring the wrong thing (or are being 'doctored').

There must also be effective communication across the organisation between:

- production departments
- production and maintenance
- specialists (including the health and safety specialists) and operating departments.

Those outside

HSWA places duties on each employer in respect of "persons not in his employment".[13] Those whose health and safety may be adversely affected by the employer's activities include:

- contractors
- visitors
- customers
- immediate neighbours
- emergency services
- the public in general.

Communication in respect of contractors is covered in chapter 15 of this book and the need for communication in respect of emergency situations is dealt with in chapter 10.

However, communication may also be necessary to:

- warn visitors of risks they may encounter on the premises and precautions they need to take
- tell neighbours about unusually heavy traffic movements expected
- ensure local youngsters are aware of the dangers of trespassing on waste tips

The need to feed information into the various associations, groups and networks referred to earlier must not be overlooked.

How to communicate

Communication must be understood in order to be effective. What is going to be most effective will depend upon the circumstances – the message to be communicated, the target audience and the prevailing environment. Communication can be categorised as written or oral and can take place in a formal or informal manner.

More than one means of communication may be necessary to ensure that the message goes home and the message may need to be reinforced through further communication and management actions.

Written communications

By their very nature written communications are generally more formal than oral communications. The information to be communicated can be presented in a very precise and detailed form if wished – and it can be proved that it was presented in this way. However, what is not so easy to prove is that it will be read and understood or even that it is capable of being understood. Too much written material is prepared more as a means of covering the issuer's back rather than as an effective form of communication.

Written communications have their place but often need to be supported by formal or informal oral methods, including the opportunity for questioning. Common types of written health and safety communications are:

Employee Health and Safety Handbooks

These might contain the organisation's health and safety policy statement together with details of those responsible for carrying it into effect. Some of the more general health and safety related procedures (eg fire procedures, accident/incident investigation) may also be included together with general advice on some of the more common hazards present and the precautions to be taken against them. There may be a need to include cross references to other health and safety documents.

Such handbooks should be freely available to all employees and preferably issued on an individual basis. There is likely to be a need for periodic updates and revisions. Opinion is divided on whether loose-leaf or bound formats are preferable for this type of handbook. Loose leaf is cheaper to update but it is difficult to ensure everyone inserts the revision sheets. With a bound format all employees can be personally issued with the latest copy.

Procedures and procedural manuals

Employees need to be made aware of specific health and safety procedures such as those for reporting accidents and dealing with emergencies, many of which are dealt with in other chapters of this book. Health and safety aspects of work activities also need to be effectively communicated and this may involve the use of written operating procedures.

The form in which procedures are kept will be determined by their content and many local factors. Manuals containing relevant procedures can be kept in managers', supervisors' or team leaders' offices whilst in other cases procedures may be issued on a personal basis. It may be appropriate for procedures or abstracts of key features of them to be posted on or near the individual item of plant or equipment they relate to.

Notice boards

Notice boards may be used in two ways. One is to display permanent items of major importance, for example:

- key actions in case of fire
- how to get first aid treatment.

They are not the place to display detailed procedures, even those of some potential urgency such as fire procedures – they will just not be read.

The main use of notice boards should be to pass on information of relatively short-term importance or interest such as:

- causes and effects of recent accidents or incidents
- general health and safety performance
- special achievements – by the organisation, departments or individuals
- evidence of good accident prevention eg eye injuries prevented by wearing eye protection
- periodic reminders of hazards
- emphasising the need to follow specific procedures
- changes to procedures or other arrangements
- minutes of safety committee meetings
- safety inspection reports.

Notice boards must be kept up to date, with obsolete items removed, otherwise employees will not bother to search them for new material. Electronic message boards may be effective in passing on short-term information, providing they are in situations where employees have time to actually read them, such as canteens, rest areas or entrance foyers.

Posters

As with notice boards it is important that posters are changed regularly, otherwise they become ineffective. In a large site it may be sufficient to move them to a different location. Poster sites must be prominent, well-lit and well-frequented. They may be adjacent to notice boards or contained in an area on a notice board.

Not everyone's attention will be caught and stimulated by the same poster. Some guidelines to follow in using posters (as well as permanent safety slogans or safety stickers) are:

- Posters are not a substitute for good health and safety management.
- Posters influencing general attitudes to safety are likely to be effective in most locations.
- Try to link posters with other current safety activities.
- Make sure posters with specific advice are relevant to where they are displayed (guarding posters are not of great interest in an office).

Newsletters

Health and safety newsletters or sections of general newsletters provide an opportunity for getting over a more detailed message to employees than can be achieved through notice boards or posters. However, this opportunity will be wasted if the articles are not read – they need to capture the interest of the potential reader. Ways of doing this include:

- Making the format visually attractive with photographs, illustrations etc.
- Using the human interest angle (achievements, milestones, personal views).
- Highlighting injuries prevented by correct precautions.
- Using humour in a way that emphasises the message.
- Including some controversial topics.
- Keeping employees up to date on planned activities or events.

Personal memos

These are often used to draw the attention of managers or supervisors to:

- Changes in procedures or arrangements.

- The need to follow procedures or enforce standards.
- Accidents or incidents which have occurred.

Personal memos may also be used to communicate with the workforce generally – a process made easier where electronic mail systems exist. Important messages could even be posted to employees' home addresses for greater impact.

Suggestion schemes

A suggestion scheme offers employees the opportunity to communicate upwards within the organisation. Ideally health and safety should be included within general suggestion schemes rather than treated as something apart. Suggestions need to be properly evaluated, possibly by a committee or an ad hoc group of individuals and a response should always be given to the employee submitting the suggestion, even if it cannot be implemented.

Employees should not regard the suggestion scheme as a substitute for a hazard reporting system. The emphasis should be on providing practical solutions to difficult health and safety problems rather than identifying simple actions needed to deal with obvious hazards.

Many effective suggestion schemes reward employees submitting good suggestions, sometimes on the basis of a percentage of the costs saved or profits increased. Because the benefits of health and safety suggestions are often difficult to quantify there is a danger that their value is underestimated. Awards should be of the same order of magnitude as those given for other types of suggestions.

Oral communications

Oral communications are potentially much more effective than written communications since they facilitate two way communication. The giver of the information can be questioned about it, leading to greater understanding, and he in turn has the opportunity to test how much of the message has been understood.

In many cases oral communications are accompanied by written communications with participants being given a discussion document, a handout or a new procedure as a record of the subject matter that has been discussed.

Oral communication with employees

Some managers fight shy of oral communications, particularly at shopfloor level, since they provide opportunity for difficult questions, arguments and for pet hobby-horses to be ridden. However, an unwillingness to engage in oral communication with the workforce will only reinforce us and them attitudes.

Several methods of communicating verbally with employees are reviewed in chapter 14 which deals with employee involvement. This includes consideration of the important role of safety committees and safety representatives.

Communication within the training context was dealt with earlier in this chapter and is also covered in chapter 12.

Management meetings

Health and safety items should be regularly on the agenda of management meetings so that all members of the management team are kept up to date on the organisation's objectives, standards and performance, together with any issues which may have arisen.

On occasions it may be necessary for health and safety specialists to attend senior management meetings to ensure that they are briefed on the implications of new legislation or of planned health and safety activities within the organisation. Specialist meetings may also be necessary to discuss major health and safety issues.

Communication through actions

"Actions speak louder than words" goes the old adage and certainly the manager who says "Do as I say, not as I do" (other than in jest) forfeits his credibility.

The important health and safety role of senior management was emphasised earlier. Senior managers can also demonstrate their commitment to health and safety through:

Conducting health and safety inspections

This may be as part of a formal inspection programme or just through an occasional informal inspection tour.

Health and safety committees

Committees should normally be chaired by a senior member of management. Even if the most senior manager does not personally sit on the committee his support for it and interest in its activities should be clearly apparent.

Incident investigations

Whilst senior managers may not routinely be involved in investigations they should be actively concerned in the investigation of major accidents and incidents and should demonstrate their interest in the activity generally through relevant questions and comments.

Questions

Regular questioning by senior managers of their subordinates on health and safety related matters will not only demonstrate their own interest but will stimulate an interest among the subordinates so that they are better prepared to respond to the enquiries.

Recognising good performance

Senior managers can recognise good health and safety performance in a variety of ways.

- Presenting safety awards.
- Participating in similar recognition events.
- Through the performance appraisal system.
- By informal praise – both public and private.

CO-OPERATION

Commitment at all levels is essential in achieving a positive health and safety culture. Communicating with the workforce is an important part of gaining their co-operation but other steps need to be taken if they are going to actively participate in meeting health and safety objectives. These aspects of employee involvement are addressed in chapter 14.

Notes

1. *Selecting a Health and Safety Consultancy* – HSE 1992 (IND(G)133(L))
2. The Construction (General Provisions) Regulations 1961, regulation 5
3. The Pressure Systems and Transportable Gas Containers Regulations 1989, regulation 2(1)
4. The Abrasive Wheels Regulations 1970, regulation 9 and the schedule to the regulations
5. The Power Presses Regulations 1965, regulation 4 and the schedule to the regulations
6. The Woodworking Machines Regulations 1974, regulation 13
7. *Rider operated lift trucks – operator training: approved code of practice and supplementary guidance* – HSE 1988 (COP26)
8. *Training and standards of competence for users of chainsaws in agriculture, arboriculture and forestry* – HSE 1990 (Guidance Note GS48)
9. *Training of crane drivers and slingers* – HSE 1986 (Guidance Note GS39)
10. The Factories Act 1961, s21 and The Dangerous Machines (Training of Young Persons) Order 1954
11. The Offices, Shops and Railway Premises Act 1963, s19 and The Prescribed Dangerous Machines Order 1964
12. *Practical Loss Control Leadership* – FE Bird Jr and GL Germain
13. The Health and Safety at Work etc Act 1974, s3

8

Risk Assessment, Prevention and Control

The requirement for risk assessment is a feature of several recent sets of regulations. Whilst this chapter deals with the subject from the standpoint of the general type of risk assessment required under the Management Regulations, many of the principles can be applied to the more specific types of assessments necessary under other regulations. The chapter is structured as follows:

- the purpose and meaning of risk assessments
- carrying out a risk assessment
- recording the assessment
- acting on the assessment
- practical assessment examples.

THE PURPOSE AND MEANING OF RISK ASSESSMENTS

The need to make an assessment of the level of risk and of the adequacy of the associated prevention and control measures goes back much further than the recent legislation. It has always been necessary in order to determine what is reasonably practicable under HSWA and other legislation. Ascertaining whether the level of risk justifies the money, time or trouble necessary to avert it, and therefore whether the precautions are reasonably practicable, requires a knowledge of the level of risk and the effectiveness of the precautions.

Risk assessment means an end to bland claims of complying with relevant legislation. It is necessary for an organisation to demonstrate that it has reviewed how both general and specific legislation apply to its work activities and that it is satisfied suitable precautions are being taken. In most cases records must be kept to prove that these processes have taken place.

The general requirement for risk assessment is contained in regulation 3 of the Management Regulations and this chapter concentrates mainly on that requirement. However, there are other sets of regulations requiring more specific assessments and these are covered in some detail in other chapters of the book.

Control of Substances Hazardous to Health Regulations 1994
Noise at Work Regulations 1989
Manual Handling Operations Regulations 1992
Health and Safety (Display Screen Equipment) Regulations 1992
Personal Protective Equipment at Work Regulations 1992

Much of the advice within this chapter is based upon the contents of the ACOP which accompanies the Management Regulations[1] and upon the recommendations in the HSE leaflet *5 Steps to Risk Assessment.*[2]

The purpose of assessment

Paragraph (1) of regulation 3 of the Management Regulations states that:

> "Every employer shall make a suitable and sufficient assessment of –
>
> (a) the risks to the health and safety of his employees to which they are exposed whilst they are at work; and
> (b) the risks to the health and safety of persons not in his employment arising out of or in connection with the conduct by him of his undertaking,
>
> for the purpose of identifying the measures he needs to take to comply with the requirements or prohibitions imposed upon him by or under the relevant statutory provisions."[3]

Paragraph (2) of regulation 3 puts an identical requirement upon self-employed persons in respect of their own health and safety and persons not in their employment.

Under paragraph (3) assessments must be reviewed if it is suspected they are no longer valid or there are significant changes (see pp170 and 171). Where the employer employs five or more employees paragraph (4) requires him to record

(a) the significant findings of the assessment; and
(b) any group of his employees identified by it as being especially at risk.

(Assessment records are covered later in the chapter.)

The purpose of the risk assessment therefore is to determine what measures need to be taken to comply with "the relevant statutory provisions" ie to comply with the law. Since the general sections of HSWA contain such wide-ranging obligations (see chapter 3) the rule of thumb should always be – if the risk is significant, it needs to be assessed.

Hazard and risk

In common usage these two terms are usually regarded as interchangeable, but health and safety terminology has ascribed different meanings to them.

A hazard is something with the potential to cause harm. The term risk describes the chance of someone being harmed by the hazard. The extent of the risk relates to the number of people who may be harmed and the extent of the harm they may suffer ie a bruise, a fracture or a fatality.

EXAMPLE

A spanner left on a walkway above ground is a hazard: it can fall onto a person below or trip someone walking on the walkway.

The risk of it falling on to someone below will be determined by:

- the possibility of its disturbance by someone or something (eg vibration)
- the presence or absence of openings in the walkway
- the presence or absence of edge protection on the walkway
- the numbers of people below.

Realistically the spanner could only fall onto one person, the extent of the risk depending upon:

- the height of the fall
- the weight and shape of the spanner
- the likelihood of the person wearing head protection.

Most people have personal experience of the hazard of sharp paper edges as cuts from them occur quite often. However these injuries are generally very minor in nature and the overall extent of the risk is low.

In looking at what precautions might be reasonably practicable in these examples, the removal of the spanner would be fully justified whereas the provision of protective gloves for handling paper would be reasonably practicable only for those who must handle large quantities.

What is 'suitable and sufficient'?

In *5 Steps to Risk Assessment* the HSE stresses that risk assessments need to be suitable and sufficient, not perfect. Some organisations have tried to go into far more detail than is necessary. The Management ACOP makes some important points about the way the task should be approached.

Identify the significant risks

The exercise should identify the *significant* risks *arising out of the work*. Trivial risks and risks arising from routine activities associated with life in general can be ignored, unless the work activity compounds these risks or they are significantly relevant to the work. For example, there is always a risk of hitting one's hand with a hammer. A risk assessment would be necessary only if the use of a hammer formed a major part of work activities or the hammer had to be used in an unusual or particularly hazardous way.

Employers and the self-employed are expected to take reasonable steps to familiarise themselves with the hazards and risks in their work. The ACOP specifically mentions:

- HSE Guidance (of which much is published)
- the trade press
- company or supplier manuals.

To these should be added:

- accident and ill-health records
- liaising with others in the same work activity
- talking to employees
- observing the work

(The methodology of risk assessment will be developed later in this chapter.)

Identify the precautions necessary

The adequacy of the precautions already being taken should be examined and any additional measures necessary to comply with the law should be identified and prioritised.

Appropriate and remaining valid

The ACOP states that the assessment "should be appropriate to the nature of the work and such that it remains valid for a reasonable period of time".[4] This means that the assessment should not need to be revised every time that there is a minor change in the work and need not cater for every variation and nuance of the activities. For example, separate assessments need not be carried out for the production of type one and type two widgets. Any risks associated with differences between the models should be taken into account in the main assessment. Similarly a garage normally servicing Ford cars should not need to carry out a new assessment because a Vauxhall comes in. However, major change such as a switch from servicing diesel engines to petrol-driven ones or from cars to HGVs would justify a review of some parts of the assessment.

For more dynamic activities such as those involving:

- changing work activities
- changing or developing workplaces
- peripatetic workers

the assessment will need to concentrate more on the broad range of risks that might arise rather than attempting to assess each and every task. A good example of this type of work would be a local authority direct works department carrying out a wide range of building, installation and maintenance tasks in a variety of environments – building sites, council premises, residential property etc. Suggested ways of approaching these types of assessments are put forward later in this chapter.

CARRYING OUT A RISK ASSESSMENT

Who should make the assessment?

There are no hard and fast rules on who should carry out the assessment. It could involve:

- managers
- supervisors/team leaders
- health and safety specialists
- engineers
- safety representatives
- ordinary employees

or any combination of the above.

One combination particularly recommended is the insider and outsider working together. The insider should know the work activities, many of the risks involved and precautions available and be aware of most of the possible variations that may arise. The outsider can question the status quo and why things should be so, bringing his own knowledge and experience to bear on a new environment.

Even where a health and safety specialist is brought in from outside the department, or the services of external consultants are used, there will still need to be significant involvement of local management in the risk assessment process. Outsiders, even highly competent ones, cannot be expected to have enough detailed knowledge of local circumstances to carry out a "suitable and sufficient" assessment in isolation.

Ultimately management must be satisfied that those carrying out the assessment are competent for the purpose and in particular that they:

1. Understand what risk assessment requires – through attendance on a formal training course or briefing session or by personal study
2. Know what legislation applies – they do not need to be legal experts but the assessment has to decide what precautions are necessary to comply with the law
3. Have some knowledge and experience of this type of activity and its hazards
4. Have a clear idea of how the task is to be approached
5. Know where to turn to for help

The assessment process

Risk assessment involves deciding whether risks are being adequately controlled in work environments. Consequently a significant proportion of the process will involve visiting work situations – a "suitable and sufficient" assessment cannot be done sitting in an office. The approach recommended here closely matches that in the HSE's *Five Steps to Risk Assessment*, although slightly different headings have been used. (HSE's last two steps – Record and Review – are covered in later parts of the chapter.)

Identify the hazards

This will be done by a combination of careful thought, relevant research and visits to the workplace. Rather than simply attempting to list hazards in an unstructured way it is better to group them under logical headings such as transport, machinery, storage etc. Different parts of the workplace may need different lists with production departments separated from each other and an additional list for maintenance. Those carrying out work away from their home base may need separate lists too – one for the base depot or workshop and one for site work. Some examples of this approach are given later in the chapter.

The purpose of the exercise is not to list every trivial hazard and the assessor cannot be expected to anticipate hazards beyond the limit of current knowledge. He should be concerned with what is reasonably foreseeable (see chapter 4) and, in doing that, needs to be reasonably well informed. There are various ways of identifying reasonably foreseeable hazards.

Personal knowledge and experience

The assessor may well be able to draw upon a detailed personal knowledge and experience of the work activity, he may have worked in a similar environment or may be able to apply his experience of work in general. The knowledge and experience of others can also be utilised, whether by a brainstorming session or asking them to assist in the exercise. This could take place in the workplace and might involve employees at all levels in the organisation.

Accidents, incidents and ill-health records

If it has happened before it might happen again. The examination of records of accidents, incidents and ill health may reveal hazards in the workplace that had not previously been realised. Even for familiar hazards, the exercise may highlight related precautions which need to be checked out in detail at a later stage of the assessment.

Unfortunately many organisations maintain adequate records of only the more serious accidents (notifiable under RIDDOR) (see Chapter 9) with many minor accidents not even recorded in the accident book and little or no formal investigation of minor accidents or non-injury incidents.

The existence of a good accident and incident reporting and investigation procedure is considered to be an important element in managing health and safety effectively, the subject being covered in much more detail in chapter 9.

Specific legal requirements
While specific legal requirements are concerned mainly with the precautions required to overcome known risks, reference to them still may be useful in identifying hazards which might otherwise have been overlooked.

HSE guidance
The HSE produces many publications which may be of relevance in carrying out risk assessments. Separate listings of priced and free publications are available from the HSE Information Centre, Broad Lane, Sheffield, S3 7HQ (tel 0114 2892345). However, these listings need to be studied carefully as it is easy to miss relevant publications with unfamiliar titles.

Priced publications can be bought from local HSE stockists or directly from HSE Books, PO Box 1999, Sudbury, Suffolk, CO10 6FS (tel 01787 881165). Free publications are available from the HSE Information Centre or from HSE Area Offices. Various National Interest Groups (NIGs) based at HSE Area Offices publish separate advisory literature relating to specific occupational activities.

The HSE book, *Essentials of Health and Safety at Work*, may be particularly helpful in the risk assessment process, especially for smaller employers. It is reasonably priced and easy-to-read, identifying various hazards and risks together with recommended control measures. The chapter headings use logical groupings such as machinery safety, pressurised plant and systems and flammable and explosive substances. The book has additional value in that its references section relates to each individual chapter, highlighting other HSE publications which may be of use if the reader needs to find out more on a particular subject.[5] Some HSE booklets deal with particular occupational sectors such as motor vehicle repair[6] or kitchens[7] making them invaluable aids in risk assessment.

Trade information
Trade associations or other bodies connected with a particular work activity often produce publications of relevance to health and safety. These may be in the form of general health and safety manuals, codes of practice, journals, studies of particular problems, advisory booklets or leaflets, possibly just dealing with a specific subject. Such bodies may also provide information or advisory services and may even publish model risk assessments (see pp163 and 164).

Commercial and political rivalries are usually set aside in the interests of health and safety and there may be considerable advantages in similar businesses in an area or neighbouring local authorities working together as part of the risk assessment process.

Information from manufacturers and suppliers
Manufacturers and suppliers etc of articles and substances have a duty under s6

of HSWA to supply adequate information so that the article or substance can be used with safety. This should cover reasonably foreseeable risks and will also be of use at a later stage of the assessment process in deciding whether the risks are adequately controlled in practice.

Other sources of information

These may be other sources of information which are of use in risk assessment such as company health and safety manuals or reference books on the subject.

Observation

Observation of people at work or of their work environment or work equipment may reveal hazards which had not previously been appreciated or had been underestimated. At this stage it is also appropriate to consider who might be affected by these hazards:

- production workers
- maintenance employees
- office staff
- cleaners
- security staff
- delivery staff
- temporary employees
- co-tenants
- contractors
- customers
- visitors
- neighbours
- other members of the public.

Some types of people may be particularly vulnerable:

- young people
- inexperienced workers
- lone workers
- people with disabilities.

Evaluate the present risks

Having identified the hazards or potential risks, the level of risk actually present in practice must now be evaluated. This involves taking account of the level of precautions already being taken. Many of the sources of information used in identifying hazards will also be of use in evaluating the risks, as will details of operating procedures, training programmes and other documents. Much of this stage of the assessment will have to be carried out in the workplace, observing the work activities and talking to those involved in them. At the end of the exercise the assessor needs to be in a position to answer a series of questions.

What is the level of risk?

- Does a risk exist in practice?
- If not, why not?
- How likely is it that things will go wrong?
- If they do, what will be the consequences?

It may well be that the scale or manner of use of a hazardous material or a piece of equipment is such as to represent no significant risk. If this is the case then this should be stated. If not, further questions need to be asked.

Can the risk be eliminated?
Is it necessary to continue using a particular material, process or piece of equipment? Can open bladed knives be replaced by scissors? Could items be rinsed with cold water rather than steam cleaned? However, many risks cannot be eliminated entirely and therefore the standards of control need to be examined. Are enough precautions being taken?

What is the present level of control?
Controls could include precautions such as:

- procedures
- training
- physical guards
- PPE.

and several questions need to be asked about them:

- What is generally required/recommended in this situation?
- What controls are currently specified?
- Are they actually used in practice?
- Do they work effectively?
- What happens in non-routine situations eg breakdowns, maintenance, rush orders?

Is this enough?
The existing level of control needs to be reviewed.

- Does it achieve compliance with the law?
- Does it conform with HSE and other informed guidance?
- Is the risk controlled to an acceptably low level?

If the risk is not adequately controlled then the improvements required will need to be identified.

In the vast majority of cases the level of risk can be evaluated by considering the levels of risks which exist and how well they are controlled. Quantification of risk is usually required only in large and potentially high risk situations such as the nuclear industry or major chemical complexes. Some organisations have attempted to quantify risks to far too great a degree. Apart from the unnecessary time involved, this 'number crunching' approach often detracts from a proper evaluation of the control measures – what should be there, and is it actually in place? The systematic and well-informed application of common sense will normally be all that is required, although the assessor must always be aware of his own limitations and should be prepared to ask for assistance or a second opinion where necessary.

Identify any improvements needed

It is likely that the process of risk assessment will identify some necessary improvements in the health and safety arrangements. Changes could be required in several areas:

Enforcement of existing control measures

Control measures may not be being enforced as they should be. Supervisors or managers may need to be reminded of their responsibility to ensure that specified procedures are followed in all instances or to insist on compliance with PPE standards.

Modification of existing procedures

Changed circumstances may have made procedures obsolete or they may be incapable of being followed in practice. They may not cater for the full range of work activities that they are intended for or they may not produce an adequate level of control of the risks.

Creation of new procedures

It is not necessary to have formal procedures for all work activities. However, the risk assessment may identify situations where new procedures are required to ensure that an unsafe approach is not taken in the carrying out of a particular task.

Training of employees

Procedures and other control measures are of no value unless relevant employees are made aware of them. The assessment is likely to identify gaps in employees' knowledge and these need to be filled. Whether formal training sessions or informal coaching or training in the workplace is required will depend upon the circumstances.

Physical safeguards

Repairs, modifications or the provision of new physical safeguards could be required as a result of the risk assessment. These could involve the work environment generally, the fabric of buildings, the installations within them, work equipment or PPE.

Some improvements will have cost implications associated with them. Those carrying out risk assessments should beware of going 'over the top' in their recommendations. The objective is to achieve compliance with the law, which in many cases will mean doing what is reasonably practicable. However, on the other hand they should not be put off recommending control measures that they consider to be necessary, even if they believe that the necessary resources will not be available to carry them out.

The risk assessor must identify what control measures are necessary to comply with the law. The responsibility for actually achieving compliance rests with the employer (or the self-employed person). Nevertheless those carrying out the assessments may have an important part to play in helping the employer prioritise the recommendations.

In some cases the assessor may be uncertain as to exactly what measures should be taken. He may just indicate the need for something to be done or may put forward tentative suggestions. For example one of the sample assessments later in the chapter (p179) contains the recommendation "consider speed ramps" in order to control the problem of speeding on internal roadways. However, the introduction of speed ramps could make it impossible or dangerous for fork lift trucks to operate in the area and other traffic calming measures such as chicanes may be more suitable. This explains why the assessor adopted a tentative approach.

Assessments under other regulations

Where specific risk assessments have already been made under other regulations (COSHH, noise, manual handling etc) then there is no need to repeat these. However, it would be prudent to check that they were still valid or that they had been carried out to a satisfactory standard, particularly if they were carried out some time ago.

In a similar way, detailed procedures, training programmes and other documented arrangements do not have to be reproduced in full within the risk assessment records – a simple cross-reference to the appropriate document will suffice. However, an important part of risk assessment is to check that these arrangements are still appropriate and that they are being followed in practice.

Model assessments

Where employers control a number of similar workplaces containing similar work activities the ACOP suggests the use of a model risk assessment (sometimes described as a generic assessment) reflecting the core hazards and risks associated with these activities. This may be an extremely useful concept in a variety of situations for example:

- local authorities
  - — schools
  - — residential homes
  - — direct works depots
  - — fire stations
- retail chains
- chains of estate agents
- networks of tyre fitting and exhaust workshops

However the model assessment will need to be adapted to cover the circumstances in the individual location and, particularly, to ensure that the control measures referred to in the model are actually being followed in practice. In some cases it may be necessary to extend the basic model to take account of activities and risks that are peculiar to a location. Factors to be considered would include:

Premises

Schools or residential homes may be single storey or multi-storey. Retail outlets or estate agents may be in single occupancy or multi-occupancy premises. These factors would be very important in determining appropriate fire precautions.

Occupancy

Schools may be for the very young or exclusively for sixth formers. Residential homes may be for children, adults or the elderly, the disabled or the disturbed.

Location

Retail outlets may be located in areas where there is a prevalence of robberies and assaults. Local authority depots could be situated within residential areas, introducing risks to residents from depot traffic.

Equipment and activities

A fire station may contain the brigade's turntable ladder or a specialist accident response vehicle. One depot could carry out all the local authority's joinery work using a wide range of machines whilst another holds the vast majority of its building material stores. Many tyre and exhaust depots may cater exclusively for cars and light vans but a small minority may also service HGVs.

RECORDING THE ASSESSMENT

Practical examples of risk assessment records in five different work environments are given in figures 8.1 to 8.5 on pages 178–182.

The legal requirements

Paragraph (4) of regulation 3 of the Management Regulations states:

> Where the employer employs five or more employees, he shall record –
>
> (a) the significant findings of the assessment; and
> (b) any group of his employees identified by it as being especially at risk.

The accompanying ACOP states that the "significant findings" should include:

1. The significant hazards identified in the assessment ie those hazards which might pose a serious risk to workers and others if they were not properly controlled.
2. The existing control measures in place and the extent to which they control the risks. Here cross-reference could be made to other relevant documents eg works manuals.

3. The population which may be affected by these significant risks or hazards, including any groups of employees who are especially at risk.

How much detail is required

Although neither the ACOP nor the later *5 Steps to Risk Assessment* specifically says so, there is little point in stating the obvious. Taking the example in fig 8.5 on p182 as an illustration, it is already well established that the principal hazards associated with the use of ladders are falls due to the ladder failing or slipping. It is only where the hazards are less obvious that they need to be spelt out in detail.

However, rather more detail will normally be required on the control measures – these are the measures that the employer carrying out the risk assessment is actually taking to eliminate or reduce the risk. Whilst they may relate to legal requirements, codes of practice, guidance etc, the precise way in which these are complied with will vary constantly. For example, in the case of the ladder in figure 8.5 (p182) it is the responsibility of the *depot foreman* to carry out inspections on a *monthly* basis.

Again it can normally be assumed that all employees carrying out the work or using the equipment which is the subject of the risk assessment are likely to be affected, unless a particular group of employees or other people are singled out. In the examples in figures 8.2 and 8.4 (pp179 and 181) all those on the premises are identified as being at risk whereas in figure 8.3 (p180) the risks to the clients as well as the home helps are specifically identified. In figure 8.1 (p178) only the risk from the open pit is considered to be be significant to customers.

The ACOP states that the record should be sufficient to demonstrate that a "suitable and sufficient" assessment has taken place and to facilitate review if circumstances change. "Only in the most straightforward and obvious cases in which the risk assessment can be easily repeated and explained is a record totally unnecessary."[8]

Suggested formats for records

Types of record

The ACOP states that records will normally be in writing but could be recorded by electronic means, eg on a computer. However, it must be retrievable, whether for reference by management or examination by an inspector or safety representative.

Form design

The format of the record is left to those carrying out the assessment providing it contains the essential ingredients required by the ACOP. The use of simple forms is recommended, more complicated forms generally being associated with the quantification of the risk, which is unnecessary in most cases.

The forms used in the examples in Figures 8.1–8.5 on pp178–182 have just three main columns:

- hazards (and those particularly at risk)
- control measures
- recommendations.

The HSE's *5 Steps to Risk Assessment* uses four columns:

- hazard
- who might be harmed?
- is the risk adequately controlled?
- What further action is necessary to control the risk?

Although not specifically referred to in the ACOP the final column is an important part of the risk assessment in both cases. It recognises that the actuality will fall short of perfection and that something more will need to be done to exert adequate control.

Both the form in Figures 8.1–8.5 and the 5 Steps form include a space for signature by those carrying out the risk assessment and for the date. Although the ACOP makes no mention of these, they are an important part of proving that a suitable and sufficient assessment has been carried out.

Manuals and procedures

Within a small organisation it may be possible to record the risk assessment in the form of a manual which details the risks involved in the work activities, the arrangements for controlling them and the responsibilities for these arrangements. This would be easier to use as a training resource than a separate 'stand alone' risk assessment document.

In larger organisations it may be difficult to produce a single comprehensive manual although this may be possible on a departmental basis. Similarly written operational procedures may largely represent a risk assessment record providing they identify the risks and how they should be controlled. However, there must also be a clear demonstration that the required controls are actually being applied in practice. This could be provided by the existence of an effective task observation programme or health and safety inspection system. Even where there is a good system of operational procedures it should not be assumed that no further risk assessment activity is necessary. Within large operations there will be many significant areas of risk which are not covered by the procedures (which tend to concentrate on production processes), for example:

- traffic and transport on site roads
- machinery and tools used by maintenance staff
- computer staff working alone after hours.

Safety cases

Some highly hazardous plants (eg major chemical or nuclear installations) are required by law to present a safety case which is likely to include all the elements required of a risk assessment for the main processes. However, once again further risk assessment will be necessary in respect of the work activities not covered specifically by the safety case.

Other related documents

The risk assessment record need not reproduce details of documents which are available elsewhere. It is acceptable simply to record, where relevant, sufficient details of their existence to identify them so that they can be referred to as necessary, for example:

- contained in induction training programme
- precautions detailed in Operating Procedure No. ...

Documents which may be relevant in this respect are:

1. health and safety policy
2. health and safety manuals
3. health and safety related procedures, eg inspection procedures
4. health and safety rules
5. training programmes
6. operational procedures
7. maintenance or inspection schedules
8. record systems
9. notices or signs.

ACTING ON THE ASSESSMENT

Once the risk assessments have been carried out and recorded, the most important and productive phase of the process has still to take place – the recommendations will need to be acted upon. It must also be recognised that the assessment is not a once and for all exercise. There will be a need to keep it up to date to reflect changes within the organisation and the world around it.

Action on recommendations

Organisations, especially large ones, seem to have big black holes into which recommendations disappear. If the recommendations from the risk assessment process go that way then most of the time spent in risk assessment will have been wasted. A systematic approach is necessary to ensure this is not the case.

Communication and evaluation

The risk assessment records are not the personal property of those carrying out the assessment. They need to be communicated to others affected and the recommendations evaluated. In small organisations this might simply involve giving the boss a copy to read and then having an informal discussion about the contents. However, in other cases a more formal approach will be necessary. A special review panel could be created for the purpose or the task may be given to the health and safety committee.

The assessors should be involved in the review process, particularly if alternative methods of controlling the risks are being evaluated. Safety representatives should also be consulted. Those safety representatives appointed by recognised trade unions have a legal right to see the risk assessment records and to make representations about them.

Costing and prioritisation

It was stated earlier that risk assessors should not be deterred from recommending necessary control measures because they believe the resources required to carry them out will not be available. Nevertheless most recommendations will have resource implications and these need to be identified. There will need to be some sort of prioritisation of expenditure in relation to the risks involved. Time constraints may also influence the programme. For example, the lead time for obtaining equipment or materials may be lengthy; some actions may only be possible during plant shutdown; or the logistics of a retraining programme may necessitate it being spread over a long period.

This part of the process is much more likely to be carried out within the framework of a management group than within a joint health and safety committee, although those carrying out the risk assessments should be able to contribute, even if they are not involved in making the final decisions. Ultimately the decision-making and the responsibility for complying with the law rests with the employer.

Implementation

Once the decision has been made to carry out certain recommendations the responsibility for their implementation must be clearly allocated to individuals with the capability and the resources to carry them out. In most cases this will be a relatively straight-forward process but, in some, more detailed activities such as the preparation of an action programme may be required. It may be appropriate to include these actions as formal objectives for the organisation, for departments within it or for individuals.

Following-up recommendations

It is an important part of the assessor's role to carry out a follow-up in order to check what action has been taken on his recommendations and the implications of that action.

Planning

Both the assessment record form illustrated in Figures 8.1–8.5 and the HSE's 5 Steps version have a space for indicating when the next assessment review should take place. Clearly sufficient time should be allowed for action on the recommendations to have been completed. Whilst a three month period has been allowed in the examples this will not always be the case. Some items of major urgency may need to be followed up the next day or next week whilst others may need much longer periods.

There is clearly little purpose in carrying out a formal follow-up review if the assessor is already aware from other sources that no action has been taken. He may however wish to bring pressure to bear in other ways in order to voice concern about failure to act upon his recommendations.

Checking

Most people involved in the world of work are not surprised if they find that something on which action is stated to have been taken is still in exactly the same condition. The assessor will actually need to check on the ground by observations or discussions that the necessary action has been taken.

In some cases the action taken may necessitate a further assessment of the risk. For example it may have been decided to reduce the risks from using ladders for certain tasks by introducing a hydraulic platform. This would require an assessment of:

- its suitability for the purpose
- its condition
- maintenance arrangements
- operational procedures for its use
- training of relevant employees in its use.

Records

Ideally once the follow-up has been carried out, the 'Recommendations' column in the Risk Assessment form will be empty. Short-term problems will have been rectified and long-term controls will have passed into the 'Control Measures' column. Whether such improvements are recorded by simply annotating the original risk assessment with appropriate comments (preferably initialled and dated) or whether a new assessment form is prepared is a matter for local policy or even personal choice. What does need to be recorded is the level of control that has been achieved by a particular date.

Further follow-up

It is possible that all of the recommendations may not have been carried out by the time of the first follow-up and a further follow-up may need to be scheduled to check on subsequent action.

The need for review

Nothing stays the same and the risk assessments will need to be kept up to date. Paragraph (3) of regulation 3 of the Management Regulations requires review if:

- "there is reason to suspect that it is *no longer valid*; or
- there has been *significant change* in the matters to which it relates". (Emphasis added)

No longer valid

There may be a variety of reasons to suspect that an assessment is no longer valid.

- *Accidents or near misses occurring*
  — within the organisation's own activities or similar ones elsewhere.
- *Awareness of new risks*
  — faulty components discovered
  — research revealing new health hazards.
- *Problems observed*
  — unexpected failure of control measures
  — unexpected behaviour making it 'reasonably foreseeable' that the same could occur again.

Significant changes

These changes need to be significant in that they affect either the levels of risk or the previous levels of their control. Changes which may justify a review of the risk assessment are those involving:

- equipment
- materials
- substances
- processes
- procedures or systems of work
- allocation of duties/responsibilities
- manning levels
- location/environment.

Whether the change justifies a review of the whole assessment or just one element of it will depend upon the circumstances.

Routine review

Change is often gradual and unnoticed; as the ACOP states "it is prudent to plan to review risk assessments at regular intervals".[9] The frequency of such reviews would depend upon the nature and extent of the risks and a perception of how much gradual change in the work activity is likely. A period of between one and five years would seem appropriate to cater for the full range of possibilities.

The nature of the review is also likely to vary. In some cases an office-based appraisal may be sufficient to confirm that nothing of any consequence has changed whereas in others detailed observations and discussions in the workplace may be necessary to confirm that the level of risks and the standard of their control remain unchanged.

PRACTICAL ASSESSMENT EXAMPLES

In this section of the chapter some practical examples are provided to show how risk assessment techniques can be applied in several different working environments. The workplaces illustrated are:

- motor vehicle repair workshops
- builders' merchants
- domestic premises (in respect of home helps)
- solicitors' offices
- local authority direct works departments.

Hazard grouping

A good way of starting the risk assessment process is to list the hazards of the work activities (or part of the activities) under logical headings. In preparing this list, use should be made of all the sources of information mentioned earlier in the chapter. Which of these proves to be the most helpful will vary according to the situation. Those carrying out risk assessment on their own are strongly recommended to check their list of headings with someone else in case there is an important hazard area which has been missed. This approach is demonstrated below for the five sample working environments.

Motor vehicle repair

In this case the HSE booklet *Health and Safety in Motor Vehicle Repair*[6] is of great assistance. The section on Servicing and Mechanical Repair gives the following hazard groupings:

- lifting equipment
- electrical safety

- compressed air equipment
- vehicle inspection pits
- petrol fires
- brake and clutch linings
- wheels and tyres
- batteries and chargers
- used engine oils
- engine running
- rolling road and brake testing equipment
- moving and road testing vehicles
- vehicle valeting
- steam and water pressure cleaners.

Not all of these will necessarily apply to every garage of course. To this list should then be added any significant areas of hazard associated with the offices or stores and any of general application to the premises, for example:

- display screen equipment*
- racking and shelving
- general working environment
- manual handling*.

*These hazards and some in the later examples are subject to risk assessment under other, more specific regulations.

If the garage is involved in body repair or painting it will be necessary to include hazard groupings from those sections of the booklet. If a breakdown repair and/or recovery service is offered then again additional headings will need to be included: work on public roads and towing.

Builders' merchants

For this example we will consider a builders' merchant involved in selling (and possibly hiring) materials and equipment both to building trade customers and to the general public. Although there is no easy HSE list to follow in this case, application of thought based upon knowledge and experience of such premises can be used to produce a similar list.

- traffic and transport
- loading and unloading
- fork lift trucks
- manual handling
- racking, shelving and stacking
- woodworking machines
- hazardous substances
- flammable gases
- glass

- fire
- access routes
- general working environment
- display screen equipment.

Some hazards may involve outsiders, particularly customers, either as potential victims or as a potential cause, for example:

- hazards to customers (on premises)
- creche areas
- product hazards
- equipment hire
- violence to staff (including robberies)

Domestic premises (home helps)

Home helps provide a good example of peripatetic workers who must still be included in the risk assessment process. The vast majority of hazards to home helps are due to the environment they work in and the clients with whom they work, over both of which their employer can introduce very little direct control. Main hazard areas to be considered in respect of the home helps themselves are:

- access within clients' homes
- electrical installations and equipment
- hazardous substances (eg cleaning)
- burns and scalds
- kitchen equipment (including knives)
- fire
- manual handling (people and objects)
- violence (people and animals)
- hygiene.

Hazards to the clients from the activities of the home helps must also be considered although in some cases there will be an overlap with the previous list, eg fire, manual handling.

Solicitors' offices

The hazards within offices are generally much less than those within the other workplaces considered so far in this section. Nevertheless, risk assessment must still be carried out within an office environment, even if at the end of the exercise the risks are shown to be satisfactorily under control. Main risk groupings within a solicitor's office would typically include:

- fire
- electrical equipment

- manual handling
- record storage
- filing cabinets
- display screen equipment
- hazardous substances
- general working environment
- violence.

In a solicitor's office the volume of client records kept and hence the possible manual handling and storage problems may be considerable, while the risk of violence from clients or others must not be overlooked, particularly where staff may work alone.

Whilst the use of hazardous substances may only involve cleaning and office stationery materials it must be remembered that the public (possibly including children) will have access to parts of the premises, necessitating tighter controls than in other types of offices.

Local authority direct works departments

It would be impossible to carry out separate risk assessments for each and every work activity carried out by a council direct works department as the permutations are endless – the activity itself, the people carrying it out, the environment where it is taking place. The same problem will arise with a maintenance department carrying out work within a factory.

By applying the hazard grouping approach under a number of general headings the potentially *significant* hazards (ie the risks) associated with the activities can be separated out. The lists contained within this example are not complete but are intended to illustrate how the technique can be used in such situations.

Hazards within the depot

The exact range of hazards within the depot will vary considerably. Some depots may just provide a base from which outside work is done whilst others may house manufacturing activities, repair work, major storage facilities or significant numbers of administrative staff. The following list of main hazards to be considered is not untypical.

- traffic and transport
- pedestrian access
- storage
- access equipment
- manual handling
- glass
- flammable liquids
- compressed gases
- woodworking machines
- welding and burning

- hazardous substances
- fire
- noise
- electrical equipment.

General site hazards

A direct works department will carry out the majority of its work on site. Whilst some hazards will be peculiar to individual sites there are some general hazards which may be encountered on any site.

- ladders, trestles and boards
- scaffolding
- powered platforms
- access to roofs
- transport
- powerlines
- power supplies
- electrical equipment
- compressed air equipment
- underground services
- confined space entry
- lifting equipment
- asbestos
- noise
- equipment and materials storage/security
- work near water.

Hazards of specific sites

The range of sites where work is carried out will depend upon the types of services provided by the direct works department. Whilst some departments may carry out major building works (with a similar range of hazards to any building site), others may do little more than maintain council-owned buildings. Some illustrations of hazards related to particular sites are below.

- Domestic or other residential property
  - — hazards to residents (including children)
  - — violence from residents
  - — animals
  - — unhygienic premises

- Educational premises
  - — hazards to pupils or students
  - — hazards from educational activities

- Other council premises
  - — hazards to occupants
  - — hazards from activities within premises

- Highways work
  - hazards from traffic
  - hazards to traffic
  - hazards to pedestrians

Special hazards of trades or activities
Under this heading we seek to identify these special hazards associated with a particular trade or activity which have not been adequately identified using the three previous headings. The contents of this list will again depend upon the range of work done by the department. Some examples are:

- joiners – use of cartridge operated tools
- gas fitters
  - leak detection
  - isolation of supplies
- bitumen handling
- removal of fly tipped material
- spraying herbicides in public areas
- grass cutting.

Hazards can be grouped in a similar way for any work activity. There will inevitably be some overlap between the headings used. The important thing to ensure is that no important hazards fail to be identified at this stage.

Detailed assessment

A detailed risk assessment can then be carried out for each of the hazard groupings identified using the approach suggested in the previous section. This assessment should follow the methods set out on pp160–4. As already emphasised, it is essential to find out what is actually happening in the workplace rather than simply to review the procedures and other precautions contained in manuals and rulebooks.

Examples of detailed assessments are contained in figures 8.1 to 8.5 using one hazard grouping from each of the five illustrative work environments. The record format used is similar to that suggested in the HSE's *5 Steps to Risk Assessment* but alternative types of assessment records are equally acceptable. As can be seen from the examples, in some cases a significant number of improvements needed to be made whereas in others the existing arrangements were satisfactory.

Within risk assessments it is not necessary to state the obvious. For example in figure 8.2 dealing with traffic and transport within the builders' merchants, the vehicles, pedestrians and their areas of operation are identified. To then say that vehicles may collide with each other or run over pedestrians in these places would be superfluous.

The five examples used are:

Figure 8.1 Motor vehicle repair – vehicle inspection pits
Figure 8.2 Builders' merchants – traffic and transport
Figure 8.3 Domestic premises (home helps – electrical installations and equipment
Figure 8.4 Solicitors' offices – fire
Figure 8.5 Direct works department – access equipment

NUMBER 4	**HAZARD GROUPING: VEHICLE INSPECTION PITS**	**SHEET 1 OF 1**
OTHER RELEVANT RISK ASSESSMENTS	5 PETROL FIRES 10 ENGINE RUNNING	OTHER REFERENCES: COSHH ASSESSMENT HS(G)67
HAZARDS (AND THOSE PARTICULARLY AT RISK)	CONTROL MEASURES	RECOMMENDATIONS
VAPOURS IN THE PIT (FROM PETROL, PAINT, SOLVENTS)	PIT LIGHTS SEALED, WITH POLYCARBONATE COVERS	
– IGNITION BY ELECTRICAL SOURCE	AIR POWERED TOOLS NORMALLY USED IN PIT HANDLAMPS NOT USED IN PIT)	} STRESS THIS TO ANY NEW EMPLOYEES
– TOXIC EFFECTS	SEE COSHH ASSESSMENT	
ACCESS INTO PIT	GOOD CONCRETE STEPS PROVIDED LIGHTING ADEQUATE	
FALLS INTO PIT (EMPLOYEES AND CUSTOMERS)	LIGHTING ADEQUATE PIT EDGES MARKED WHITE BOARDS AVAILABLE TO COVER PIT WHEN NOT IN USE	REPLACE CRACKED BOARDS WORKSHOP FOREMAN TO ENFORCE STRICTLY
	CUSTOMERS SHOULD ONLY BE IN WORKSHOP IF PERSONALLY ESCORTED	REMIND ALL EMPLOYEES OF PROCEDURE
SIGNATURE(S)	NAME(S): J. SMITH	DATE: 21st DECEMBER, 1992
DATE FOR NEXT ASSESSMENT: 21st MARCH, 1993		FOLLOW UP OF RECOMMENDATIONS

Figure 8.1 *Risk assessment MOTOR VEHICLE REPAIR*

NUMBER 1	HAZARD GROUPING: TRAFFIC AND TRANSPORT	SHEET 1 OF 1
OTHER RELEVANT RISK ASSESSMENTS: 2 LOADING AND UNLOADING 3 FORK LIFT TRUCKS 14 HAZARDS TO CUSTOMERS		OTHER REFERENCES: HSE GUIDANCE NOTE GS 9 (ROAD TRANSPORT IN FACTORIES AND SIMILAR WORKPLACES)
HAZARDS (AND THOSE PARTICULARLY AT RISK)	**CONTROL MEASURES**	**RECOMMENDATIONS**
VEHICLES – DELIVERY WAGONS – TRADE CUSTOMERS' VEHICLES – PUBLIC VEHICLES – OWN VEHICLES – OWN FORK LIFTS	10 mph SIGNS AT ENTRANCE AND ROADWAYS (OFTEN IGNORED) PUBLIC CAR PARK AWAY FROM BAY AND YARD ONE WAY SYSTEM THROUGH CAR PARK (WITH PROMINENT SIGNS)	CONSIDER SPEED RAMPS REPAIR POT HOLES IN CAR PARK
PEDESTRIANS – STAFF – DELIVERY DRIVERS – TRADE CUSTOMERS – PUBLIC	SIGNS PROHIBIT UNACCOMPANIED PUBLIC ACCESS TO ROADWAYS, BAY AND YARD ROADWAYS OF ADEQUATE WIDTH AND IN GOOD CONDITION STORAGE MUST NOT ENCROACH BEYOND WHITE LINES	BIGGER SIGNS REQUIRED MORE EFFECTIVE 'POLICING' NEEDED REMIND STAFF
TRAFFIC AREAS – INTERNAL ROADWAYS – PUBLIC CAR PARK – LOADING BAY – YARD STORAGE AREAS	REVERSING VEHICLES ASSISTED BY YARD STAFF ALL EXTERNAL AREAS HAVE GOOD ARTIFICIAL LIGHTING	
SIGNATURE(S)	NAME(S): R. JONES	DATE: 16 DECEMBER, 1992
DATE FOR NEXT ASSESSMENT: 16th MARCH, 1993		FOLLOW UP OF RECOMMENDATIONS

Figure 8.2 *Risk assessment BUILDERS' MERCHANTS*

NUMBER 2	HAZARD GROUPING: ELECTRICAL INSTALLATIONS AND EQUIPMENT	SHEET 1 OF 1
OTHER RELEVANT RISK ASSESSMENTS:	5 KITCHEN EQUIPMENT 6 FIRE	OTHER REFERENCES: TRAINING MODULE 4
HAZARDS (AND THOSE PARTICULARLY AT RISK)	**CONTROL MEASURES**	**RECOMMENDATIONS**
ELECTRIC SHOCK ELECTRICAL FIRES FROM – CLIENT'S INSTALLATION – CLIENT'S EQUIPMENT – OWN EQUIPMENT	HOME HELPS INSTRUCTED ON – PRINCIPLES OF ELECTRICAL SAFETY – SAFE WORKING PROCEDURES – EQUIPMENT TO BEWARE OF – NOT TO USE SUSPECT EQUIPMENT AND TO REPORT IT TO THEIR ORGANISER (SEE TRAINING MODULE 4)	REFRESHER TRAINING NEEDED (NOT ALL STAFF SEEM FULLY AWARE OF RISKS) ORGANISERS TO MONITOR WHETHER THIS HAPPENS AS OFTEN AS IT SHOULD
RISKS TO HOME HELPS AND CLIENTS	ALL HOME HELPS PROVIDED WITH A PORTABLE EARTH LEAKAGE CIRCUIT BREAKER OWN EQUIPMENT INSPECTED AND TESTED BY COUNCIL ELECTRICIAN EVERY 6 MONTHS (RECORDS CHECKED)	ORGANISERS TO MONITOR WHETHER THESE ARE ALWAYS USED
SIGNATURE(S)	NAME(S): G. THOMPSON	DATE: 7th DECEMBER, 1992
DATE FOR NEXT ASSESSMENT: 7th MARCH, 1993		FOLLOW UP OF RECOMMENDATIONS

Figure 8.3 *Risk assessment DOMESTIC PREMISES (HOME HELPS)*

NUMBER 1	HAZARD GROUPING: FIRE	SHEET 1 OF 1
OTHER RELEVANT RISK ASSESSMENTS:		OTHER REFERENCES: FIRE CERTIFICATE (HELD BY LANDLORD)
HAZARDS (AND THOSE PARTICULARLY AT RISK)	CONTROL MEASURES	RECOMMENDATIONS
FIRE DANGERS TO STAFF, CLIENTS AND VISITORS FROM – FIRE IN OWN OFFICES – FIRE ELSEWHERE IN BUILDING	NO SMOKING IN OWN OFFICES ALL EXIT ROUTES CLEAR AND WELL SIGNED 'PANIC BAR' EXIT DOORS OPEN EASILY WEEKLY FIRE ALARM TESTS QUARTERLY FIRE DRILLS (BOTH INITIATED BY LANDLORD WHO KEEPS RECORDS – CHECKED AND UP TO DATE) NEW STAFF INSTRUCTED IMMEDIATELY IN FIRE PROCEDURE FIRE PROCEDURE SIGNS IN POSITION ADEQUATE SUPPLY OF WATER AND CARBON DIOXIDE FIRE EXTINGUISHERS (MOST STAFF INSTRUCTED IN THEIR USE)	
SIGNATURE(S)	NAME(S): P. WILSON	DATE: 9th DECEMBER, 1992
DATE FOR NEXT ASSESSMENT: 9th DECEMBER, 1995		ROUTINE REVIEW

Figure 8.4 *Risk assessment SOLICITOR'S OFFICE*

NUMBER 4	HAZARD GROUPING: ACCESS EQUIPMENT	SHEET 1 OF 1
OTHER RELEVANT RISK ASSESSMENTS: GS1 LADDERS, TRESTLES AND BOARDS		OTHER REFRENCES: LADDER INSPECTION RECORDS HSE GUIDANCE NOTE GS31 (SAFE USE OF LADDERS, STEPLADDERS AND TRESTLES)
HAZARDS (AND THOSE PARTICULARLY AT RISK)	CONTROL MEASURES	RECOMMENDATIONS
EXTENSION LADDERS STEP LADDERS ROOF LADDERS TRESTLES AND BOARDS THESE ARE ALL USED BOTH WITHIN THE DEPOT AND FOR SITE WORK (SEE ASSESSMENT GS1)	LADDERS SHOULD BE IDENTIFIED WITH REFERENCE NUMBERS AND INSPECTED MONTHLY BY DEPOT FOREMAN	ENSURE INSPECTIONS TAKE PLACE EACH MONTH (RECORDS SHOW SOME GAPS) INCLUDE HIRED LADDERS AND ALL TRESTLES AND BOARDS IN MONTHLY INSPECTION SYSTEM
	ALL EQUIPMENT SEEN (WITH ONE EXCEPTION) WAS IN GOOD CONDITION	SCRAP SMALL UN-NUMBERED METAL STEPLADDER (BADLY CORRODED) FOUND IN PAINTERS' WORKSHOP
	EXTENSION LADDERS TIED OR FOOTED WHEN IN USE	
	ROOF LADDERS ONLY USED BY TRAINED ROOFING SQUAD	
	STAFF INSTRUCTED TO VISUALLY CHECK ALL EQUIPMENT BEFORE USE	INCLUDE THIS IN EMPLOYEE INDUCTION PROGRAMME
SIGNATURE(S)	NAME(S): K. ROBINSON	DATE: 15th DECEMBER, 1992
DATE FOR NEXT ASSESSMENT: 15th MARCH, 1993		FOLLOW UP OF RECOMMENDATIONS

Figure 8.5 *Risk assessment DIRECT WORKS DEPOT*

Notes

1. Management of Health and Safety at Work Regulations 1992 and Approved Code of Practice – HSE 1992 (L21)
2. *5 Steps to Risk Assessment* – HSE 1994 (IND(G)163L)
3. Management Regulations, regulation 3
4. *Management ACOP*, paragraph 9(c)
5. *Essentials of health and safety at work* – HSE 1994
6. *Health and safety in motor vehicle repair* – HSE 1991 (HS(G)67)
7. *Health and safety in kitchens and food preparation areas* – HSE 1990 (HS(G)55)
8. *Management ACOP*, paragraph 26
9. *Ibid.*, paragraph 11

9

Accident Investigation and Reporting

INTRODUCTION

Accidents occur, even in well-managed operations, and an important part of health and safety management involves minimising both the personal and financial costs of such incidents and learning from them to prevent future occurrences.

Accidents and incidents

In chapter 6 the term accident was defined as "an undesired event that results in harm to people, damage to property or loss to process". However, in common usage accident is often restricted to situations involving injury to people. Within this chapter therefore, the phrase accident and incident is used to embrace property damage, process loss and near misses (or near hits as some people would prefer to describe them) as well as injury to people.

Reasons for investigation

Future prevention

A major reason for having an effective investigation procedure is the prevention of future accidents and incidents, possibly with more serious consequences. Whether a particular accident or incident results in physical injury, damage to plant or materials, a "dangerous occurrence" or just a near miss is often a matter of chance. For example, an item falling from a considerable height could cause:

- damage to the item itself
- damage to equipment or materials below
- slight injury from a glancing blow
- serious injury or a fatality.

Which of these actually results will very largely be a matter of chance, the potential is the same regardless of the actual consequences and the causes remain the same.

Unfortunately many organisations have investigation arrangements driven either by the statutory requirements to report certain accidents or incidents (The Reporting of Injuries, Diseases and Dangerous Occurrences Regulations [RIDDOR] – see pp194 to 198) or by the demands of insurers in relation to claims for damages. Such organisations are not using accident and incident investigation as a positive accident prevention tool, often represented by the slogan 'never waste an accident'.

Accident prevention can result in considerable cost savings to the employer (as noted in chapter 6) and the accident and incident investigation system can also provide important feedback into the process of risk assessment, prevention and control (see chapter 8). Accidents and incidents are evidence of the failure of control measures and what is reasonably foreseeable in respect of future events.

Recording the facts

There must be an investigation into the circumstances of an accident or incident if the facts are to be accurately reported as legally required under RIDDOR. Demands for information from those involved in litigation are quite considerable and are usually made some considerable time after the event (see chapter 4). If the facts are recorded accurately at or around the time of the occurrence then it will be much easier to meet these demands.

The investigator should always resist the temptation to slant the report towards one party or another with a view to possible future civil claims. The report could be analysed in great detail (possibly within a court) and, if it proves to be inaccurate or unfair, the investigator's credibility will be undermined. The lawyers and others involved in the claims process are trained to seek out and exploit the weaknesses in the opposition's case and the witness box can be a very lonely place. Honesty (and accuracy) will always prove the best policy.

Additional demands for the facts relating to accidents may come from accident (and incident) reporting requirements within large organisations or from related organisations such as trade associations.

INVESTIGATION PROCEDURES

The elements below are all considered to be important components of an investigation procedure for larger organisations. Whilst all of the detail and formality may not be appropriate for the smaller employer, the same principles should be applied.

Written procedure

Putting the procedure in writing provides an important point of reference for all involved in the accident and incident investigation process. This will also create various relevant performance standards as described in chapter 7 and illustrated in figure 7.2 on p136.

Prompt reporting

The procedure should require employees to promptly report *all* incidents/accidents, work-related illness, property damage, process loss and near misses to a designated person for investigation. This would usually be their immediate superior – a supervisor, foreman or team leader. In order for the procedure to be effective all employees must be aware of it and in particular understand the importance of reporting non-injury incidents as well as those causing physical injury. This creates a training need, probably best met during the induction programme. Means of encouraging prompt reporting are reviewed on pp192 to 194.

Prompt investigation

A realistic period should be set for the designated person to carry out an investigation and prepare a written report. Where it is not possible to complete the investigation within the timescale set (eg the injured person or another important witness may not be available) this should not be a deterrent to partial investigation and the preparation of an interim report.

Those required to conduct investigations should be given training in investigation and report writing techniques. Much of pp189 to 192 is likely to be relevant to such training.

The procedure may also provide for the involvement of safety representatives in the investigation process. Some organisations have two tier investigation procedures whereby there is a more detailed investigation into more serious accidents or incidents, often directly involving senior managers.

Written report

The investigation is intended to determine what happened, why it happened and how it can be prevented from happening again. The existence of a written report provides a permanent record of these aspects and can be used to check that the necessary actions have been taken. Use of a standard report form helps ensure that certain basic facts are recorded and that the causes are determined – ideally separating out the immediate causes from the more basic, underlying causes (see pp112 to 115). An example of a report form is provided in figure 9.1.

<table>
<tr><td colspan="2">Department</td><td colspan="3">Employee(s) involved</td><td colspan="2">Type of incident</td></tr>
<tr><td colspan="2"></td><td colspan="3">Surname First Name(s)</td><td>INJURY
DAMAGE
NEAR MISS</td><td></td></tr>
<tr><td>Incident Date</td><td></td><td>Time</td><td colspan="2">Place</td><td colspan="2"></td></tr>
<tr><td colspan="7">Details of any injuries or damage</td></tr>
<tr><td colspan="5">Description of incident

A sketch or photo may be helpful
Continue on a separate sheet if necessary.</td><td colspan="2">KEY POINTS
TO CHECK
Equipment
PPE
Access
Lighting
Training
Instruction</td></tr>
<tr><td colspan="7">Contributory causes</td></tr>
<tr><td colspan="7">Action already taken and recommendations for further action</td></tr>
<tr><td colspan="2">Investigated by</td><td colspan="2">Date</td><td colspan="3">Safety Rep (if involved)</td></tr>
<tr><td colspan="7"><u>MANAGER'S COMMENTS</u>

Date action completed____________ Signature ______________</td></tr>
</table>

Figure 9.1 *Incident report*

Distribution of report

The procedure should specify who must receive a copy of the investigation report – department manager, senior manager, health and safety specialist etc. Some discretion should be allowed for the wider distribution of the report eg copies to other departments operating similar equipment.

As well as reviewing the content of the individual investigation, those receiving investigation reports should use them as a means of monitoring the standards of investigation by individuals or within different sections of the organisation.

Actions and recommendations

A key element of the procedure is that action is taken to rectify the causes of accidents and incidents identified during the investigation process. Those carrying out the investigation should be encouraged to take corrective action where this is within their own control eg repairing defective equipment, clearing untidy areas or dealing with employees not following procedures. In other cases they may make recommendations which will need to be reviewed and acted upon by others eg changes to procedures or additional training. The investigation procedure should identify who is responsible for ensuring that this takes place, within the designated timescale.

Senior management involvement

Senior managers' commitment to the investigation procedure should be readily apparent. This can be demonstrated in several ways:

- comments on good (and bad) aspects of investigation reports
- participation in the review of recommendations
- interest in the progress of corrective actions
- personal participation in the investigation of serious accidents or potentially serious incidents.

Monitoring

The effectiveness of the implementation of the procedure can be monitored in a variety of ways and it may be appropriate to include some of these aspects in the procedure itself.

Relevant events investigated

Checks can be made on first aid treatment records to ensure that all accidents have been investigated. (Some organisations use first aid treatment points to initiate a form on which the investigation report must subsequently be completed.) Comparisons can be made between the numbers of investigation reports relating to accidents with those relating to non-injury incidents. These ratios may vary considerably between different sections of the same organisation.

Promptness of investigation

Managers should check that reports are being submitted within the period laid down in the procedure.

Quality of investigation and report

Both of these can be monitored by the recipients of reports.

Progress of remedial actions

Progress could be monitored by an individual identified in the procedure, eg the department manager or the health and safety specialist, or alternatively this could be done by a health and safety committee. In either case, checks should be made to ensure that the actions have actually been carried out, rather than simply accepting claims of their completion.

Analysis

Whilst not usually part of a written procedure, some analysis of investigation reports may be appropriate. However, much time spent in analysis is wasted producing statistics which are of little relevance, and on which no action is ever taken. The effort would be better applied in monitoring the implementation of the procedure, using some of the techniques described earlier.

INVESTIGATION TECHNIQUES

Good technique will develop with training and practice but there are a number of important points which must always be borne in mind whilst carrying out investigations.

Observe

Where possible a visit should normally be made to the scene of the accident or incident. The sooner this is done the less likelihood there is that important evidence has been disturbed or removed. In some cases the scene must be left undisturbed so that the inspector has the opportunity of viewing it (see p196).

The investigator should have some idea of the questions he will need to ask in order to determine what happened and what the causes were. It may be helpful to make a note of a few questions in advance. The questions will often be based on the investigator's own knowledge and experience but other sources may throw up useful pointers such as:

- operational procedures
- rulebooks or manuals
- risk assessment records
- legal requirements or HSE guidance.

Observation of the scene is more likely to reveal unsafe conditions but it may also provide an indication of unsafe acts by employees. Physical evidence may provide confirmation (or otherwise) of employees' versions of events.

There is often a temptation to jump to premature conclusions but this should be resisted. Most accidents have several causes and, even though one has been identified, there may still be others.

Interview

It will usually be necessary to interview those directly involved in the accident or incident, including the victim if there is one. Others may also be able to provide valuable circumstantial evidence, even if they did not actually see the incident. Interviewing witnesses at the scene will often be beneficial, particularly if they are not especially articulate. It makes it far easier for them to pinpoint the position of people, equipment or materials.

Questioning should be in a fair and friendly manner. This is far more likely to elicit information than adopting an accusing or aggressive tone. Open questions (usually prefaced by what, how or why) should be used in preference to questions demanding yes or no or other one word answers. The more the witness can be encouraged to talk, the more information he will provide.

Witnesses should preferably be interviewed separately. This can prevent some witnesses unduly influencing what others might say. Although the investigator is not an interrogator, he needs to avoid being given an 'agreed version' of the events. Whilst interviewing separately may not always reveal the truth it is more likely to reveal inconsistencies in accounts or between verbal and physical evidence.

Consult

Procedures often cater for the involvement of a safety representative in the investigation process. This will often be of great assistance, providing an extra pair of eyes and the insight of someone who possibly has more practical experience of the activity. However on occasions the safety representative may have different views of the incident (or even different motives) and the investigator will need to maintain a separate stance.

Others may be able to contribute to the investigation through their knowledge, experience or skills. Specialist engineers, chemists, analytical services or health and safety specialists may need to be called upon to assist in the process.

Report

The format of the report will depend upon whether a standard form is being used or not. Even if this is the case, it may be necessary to extend the report onto a continuation sheet – the investigator should not tailor the length of his report to fit box sizes. All reports should contain the following elements:

Basic details

Who was involved, where and when the incident took place, details of injuries or damage sustained.

What happened

A description of the incident. This should be accurate, particularly in identifying items of plant. If dimensions are given it should be clear if this is an estimate (eg he fell approximately 1.5 metres).

Jargon should be avoided as the report may need to be read and understood by people unfamiliar with local terminology. Where appropriate, sketches or even photographs may give a clearer indication of what happened.

In some cases witnesses may provide information about which the investigator has some element of doubt. If there is any doubt, the statement should not be reported as fact. Instead, a qualifying phrase should be used such as "he stated that" or "he reported that" eg "he stated that his attention had been distracted by a pig flying past the chimney".

Why it happened

The temptation to jump to early conclusions should be resisted. The investigator must try to identify *all* the causes – the immediate ones and the basic causes behind them.

Some investigators show a tendency to blame the individuals involved in the incident (including the victim, if there is one) rather than looking for underlying causes such as the lack of adequate training or supervision. Others show the opposite tendency – being overprotective of individuals and looking for equipment faults or procedural imperfections rather than identifying and dealing with unsafe acts.

Future prevention

The investigator should state what has already been done or what, in his view, should be done to prevent future occurrences. A sense of realism should be kept where recommendations are being made for others to implement. For example realignment of traffic routes to avoid drivers getting the sun in their eyes is not likely to be reasonably practicable but the use of sun visors, filters or blinds may well be.

EMPLOYEE REPORTING

The value of a well thought out investigation procedure backed up by well-trained investigators will be very much diminished if employees at shopfloor level do not report incidents in the first place. Some accidents and damage incidents will be so serious that reporting of them is unavoidable but the proportion of other incidents actually seen by investigators will be very small, so others need to be encouraged to join in the process.

Reasons for incidents not being reported

Before looking at how best to encourage reporting it is necessary to look at the factors which act as deterrents.

Fear of discipline

Often those involved in the incident will have either made a mistake or have failed to follow correct procedures and they are understandably concerned that reporting it will lead to disciplinary action against them (or others).

Personal image

Some employees do not want to admit to their own fallibility or perhaps have a 'tough guy' image that might be spoilt if they report a minor accident.

Spoiling records

Some organisations place undue emphasis upon the negative aspects of incident reporting, using them to make comparisons of performance. As a result individuals may be labelled accident prone or departments as unsafe, simply because they are more conscientious than others in reporting incidents.

Concerns over bureaucracy

Most incident investigation procedures involve some element of form-filling. This is always likely to act as a deterrent to reporting, especially if the individuals involved are hard pressed with other duties. Long, complicated forms will only make this worse.

Lack of understanding

Relatively few employees have a clear understanding of the value of incident reporting as an accident prevention measure. Reactions such as "no-one was hurt" or "it was only a scratch" are quite commonly advanced as reasons for not reporting. The potential of the incident is often not thought about at all.

Means to encourage reporting

Attitudes towards the reporting of incidents will not change overnight but several measures can be taken which should influence such attitudes in the long term.

Avoid 'knee-jerk' discipline

A much more tolerant attitude should be taken to those who have the honesty to report on their own failings than to those who are subsequently found out. Indeed there is a case for taking a tougher attitude with those found committing unsafe acts where no incident results than with those reporting on themselves. In some cases some form of disciplinary action may be inevitable but it should be made clear that this would have been more severe if the individual hadn't made the report.

Stress the value to others

Emphasise the teamwork element – reporting a minor incident may help prevent serious injury to a workmate. Well designed posters could assist in this respect.

Use statistics positively

Clearly an increase in serious accidents and damage costs is going to be of concern but an apparent increase in the numbers of minor injuries or non-injury incidents will often reflect improved reporting. This should be encouraged, especially if the number of serious accidents is going down. Presenting statistics in a different way, eg comparing ratios of non-injury incidents to serious accidents between departments, could be one way of doing this.

Increase the system's credibility

Designing forms that are relatively easy to complete will help reduce complaints about bureaucracy. Some forms are overly-complicated for (often spurious) reasons of statistical analysis or because of the demands of insurers.

However, there will always be those who are reluctant to fill in forms of any type. Their resistance can be diminished by demonstrating that the system is working. If it is clear that improvements are being made as a result of following the incident reporting and investigation procedure, the effort involved in reporting, investigating and filling in the form can be seen to be worthwhile.

Raise awareness

Employees should be made aware of the importance of reporting all incidents as a means of preventing future accidents. However, they are only likely to respond regularly if reporting is integrated into the culture of the organisation. Positive responses from supervisors to incidents reported to them and the use of incident reports in a public way (team briefings, notice boards, newsletters) to highlight safety problems or concerns are ways in which their value can be demonstrated.

Some companies have successfully operated Incident Report of the Month competitions in which reports are assessed as to the importance of the incident, the quality of the investigation, the clarity of the report and the degree of remedial action already taken. Both the shop floor employee reporting the incident and the supervisor investigating it and completing the report form receive a small prize and the attendant publicity.

RIDDOR REQUIREMENTS

General responsibilities

The Reporting of Injuries, Diseases and Dangerous Occurrences Regulations 1985 (RIDDOR) create a legal obligation to report certain incidents to the "relevant enforcing authority". Full details of the regulations and associated guidance are contained in an HSE booklet[1] whilst a synopsis is given in a free leaflet.[2]

Enforcing authorities

The "relevant enforcing authority" will generally be the environmental health department of the local authority in the case of accidents in the following sectors:

- commercial
- retail
- wholesale
- hotel and catering
- consumer/personal services
- leisure and entertainment
- residential accommodation (excluding nursing homes).

In other cases it will be the area office of the HSE (their address can be found in the telephone directory).

Duties of employers and others

Employers, the self-employed and persons in control of work premises all have duties under the regulations. The definition of employee includes trainees and school pupils receiving 'relevant training' as it does in other areas of health and safety law.[3] In some cases immediate notification is required followed by a written report within seven days whilst in others just the written report within seven days is needed.

Immediate notification

Notification must be made forthwith "by the quickest practicable means" (usually a telephone call) of:

- Fatalities or major injuries (see below) to employees *or other people* "arising out of or in connection with work".
- Dangerous occurrences listed in the regulations (see pp196 and 197).

This must be followed up within seven days by a report on a standard form F2508.

Other reports

Written reports must be sent within seven days in respect of:

- Other injuries to employees resulting in their absence or being unable to do their normal work for more than three days (see pp197 to 198) – again using F2508.
- Diseases listed in the regulations (see p198) – using F2508A.

Fatalities and major injuries

Meaning of major injuries

The definition of 'major injuries' includes:

- Fracture of the skull, spine or pelvis.
- Fracture of the arm, wrist, leg or ankle, but not a bone in the hand or foot.
- Amputation of a hand or foot, or a finger, thumb or toe where the bone or a joint is completely severed.
- Loss of consciousness resulting from lack of oxygen.
- Decompression sickness.
- Acute illness requiring medical treatment or loss of consciousness resulting from inhalation, ingestion or skin absorption of any substance.
- Acute illness requiring medical treatment believed to result from exposure to a pathogen or infected material.
- Any other injury resulting in the person being admitted to hospital for more than 24 hours.

Reporting of fatalities and major injuries

The requirement to report fatalities and major injuries includes any person injured "as a result of an accident arising out of or in connection with work"[4] and would include for example:

- a customer falling in a shop
- a patient in a hospital or home tripping over a cable
- a passer-by hit by a falling paint pot
- a member of the public hit by a building site dumper truck
- neighbours or visitors affected by a chemical leak

Employers are normally responsible for reporting fatalities and major injuries to their employees, although special reporting duties are placed on the owners of pipelines or quarries, together with mine managers and immediate providers of training. In other cases the person in control of the premises where the incident happened is responsible.

When fatalities and major injuries are reported during the working day, the enforcing authority is likely to give information as to when and how they intend to investigate. Inspectors may utilise their powers under the Health and Safety at Work etc Act to direct that the scene is left undisturbed until they arrive.[5]

Outside normal working hours telephone callers to the HSE are more likely to hear a recorded message inviting them to leave details of the incident or possibly requesting them to telephone another number. In the case of fatalities this is likely to include the Police (who must be informed of fatal accidents in any case) and the Police may in turn contact the relevant HSE or local authority inspector to inform them directly.

The telephone notification must in all cases be followed up within seven days by a written report on form F2508. Where an employee has suffered an injury as a result of an accident at work, and this is a cause of his death within one year of the accident, his employer must inform the enforcing authority in writing as soon as the death comes to his knowledge.

Dangerous occurrences

A full listing of dangerous occurrences is contained in Schedule 1 to the regulations but these include:

- failure of lifting machinery, hoists, lifts, powered access platforms etc
- failures of pressure vessels
- serious electrical short circuits or overloads
- escapes of hazardous or flammable substances
- scaffolding collapses
- collapses of buildings or structures (including unintended ones during construction, demolition etc)

- road incidents involving dangerous substances
- malfunctions of breathing apparatus
- accidental overhead electric line discharges
- collisions between trains and other vehicles.

Reference should always be made to the Schedule 1 listing in cases of doubt.[6]

The responsibilities and arrangements for notifying dangerous occurrences are very similar to those for fatalities and major injuries. In the case of road incidents involving dangerous substances, the company operating the vehicle is responsible for notification.

More than three days incapacity

RIDDOR regulation 3(3) states that:

> Where a person at work is incapacitated for work of a kind which he might reasonably be expected to do, either under his contract of employment, or, if there is no such contract, in the normal course of his work, for more than three consecutive days because of an injury resulting from an accident at work, a report must be sent to the enforcing authority on form F2508.

If an employee is absent from work a report must be made but where an injured employee remains at work a report could still be necessary. Where the employee is still capable of carrying out a limited range of his normal duties no report need be made but if the employee is switched to completely different work then the accident is reportable.

The period of more than three consecutive days excludes the day of the accident but includes any days which would not have been working days (rest days). If rest days are bracketed by periods of absence the position is straightforward but where an employee returns to work after rest days it may be difficult to determine whether he was incapacitated for those days. Where this presents difficulties to employers (especially those with unusual shift patterns) it may be advisable to discuss interpretation with the enforcing authority inspector. However, the general rule should be 'if in doubt, report'.

The same rule should be applied where there is doubt over whether the absence is really due to an accident at work or is attributable to a sporting injury, an accident elsewhere or just general malingering. Use can be made of the phrase 'he states that' or 'he reports that' in order to indicate that there may be some doubt on the matter and, if necessary, the body of the report in section H of the form F2508 can be headed 'without prejudice'. Whilst inspectors will investigate all fatalities and most major injuries, they only investigate a proportion of the 'more than three days' injuries and the report of a doubtful minor accident is likely to become just another statistic.

Responsibility for reporting these accidents is as for fatalities and major injuries, although the requirement only relates to employees (and trainees). An individual will need to be identified as responsible for submitting reports and to be provided with necessary information about employee absences. This is often one of the duties of the health and safety specialist or of his administrative support staff.

Diseases

Schedule 2 to RIDDOR lists a wide range of diseases which are reportable under the regulations. However, in most cases these diseases must only be reported if they occur in relation to a particular industry or work activity.

For employees (or trainees) a written statement must have been received by the employer (or training provider) from a registered medical practitioner diagnosing the disease as one of those specified in Schedule 2. Until this occurs there is no need to check out whether it relates to a specified work activity and therefore is reportable. In the case of self-employed persons they must be informed by a registered medical practitioner that they are suffering from a specified disease.

Once it is established that the disease is reportable then a report must be sent on form 2508A 'forthwith' – presumably first-class post or fax.

The general diseases covered are:

- certain poisonings (a full list of the chemicals causing these conditions is contained in Schedule 2)
- some skin diseases such as skin cancer, chrome ulcer, oil folliculitis/acne
- lung diseases including:
  occupational asthma, farmer's lung, pneumoconiosis, asbestosis, mesothelioma
- the following infections:
  leptospirosis, hepatitis, tuberculosis, anthrax, any illness caused by a pathogen
- other conditions such as:
  occupational cancer, cataracts, decompression sickness and vibration white finger.

Reference to the HSE RIDDOR booklet will give the full details of the scheduled diseases if required.

Records

Regulation 7 states that records must be kept of injuries, diseases and dangerous occurrences required to be reported under RIDDOR. These must normally be kept at the place where the work to which they relate is carried on and must be retained for at least three years. (Some employers may prefer to keep them longer.) The usual method is simply to keep a file of photocopies of completed F2508s (or F2508As).

Notes

1. *A guide to the Reporting of Injuries, Diseases and Dangerous Occurrences Regulations 1985 (RIDDOR)* – HSE 1986 (HS(R)23)
2. *Reporting under RIDDOR* – HSE 1994 (HSE 24)
3. The Health and Safety (Training for Employment) Regulations, SI 1990/1380
4. RIDDOR, regulation 3(1)
5. HSWA, s20(2)(e)
6. RIDDOR, Schedule 1. Changes to RIDDOR mean that from April 1996 employers will also have to report acts of violence against persons at work (see pp361–2).

10

Emergency Procedures

This chapter examines the procedures necessary for the effective handling of emergencies. Its starting point is the legal requirement for employers to have such procedures.

LEGAL REQUIREMENTS

The general need for emergency procedures "to be followed in the event of serious and imminent danger" is established in regulation 7 of the Management Regulations. Some large industrial plants will be subject to the Control of Industrial Major Accident Hazards (CIMAH) Regulations 1984 which also contain requirements for emergency procedures.[1]

Requirements of the Management Regulations

Paragraph (1) of regulation 7 requires every employer to:

> (a) establish and where necessary give effect to appropriate procedures to be followed in the event of serious and imminent danger to persons at work in his undertaking; and
> (b) nominate a sufficient number of competent persons to implement those procedures insofar as they relate to the evacuation from premises of persons at work in his undertaking.

The paragraph also requires the employer to:

> (c) ensure that none of his employees has access to any area occupied by him to which it is necessary to restrict access on grounds of health and safety unless the employee concerned has received adequate health and safety instruction.

This latter requirement to prevent access to danger areas might apply (the ACOP suggests) to areas where toxic gas or bare live electrical conductors are present. A risk assessment might identify danger areas such as those containing potentially dangerous equipment (eg robots) or animals (eg on farms or in zoos).

Returning to emergency procedures, paragraph (2) of regulation 7 requires the procedures to:

(a) so far as is practicable, require any persons at work who are exposed to serious and imminent danger to be informed of the nature of the hazard and of the steps taken or to be taken to protect them from it;
(b) enable the persons concerned (if necessary by taking appropriate steps in the absence of guidance or instruction and in the light of their knowledge and the technical means at their disposal) to stop work and immediately proceed to a place of safety in the event of their being exposed to serious, imminent and unavoidable danger; and
(c) save in exceptional cases for reasons duly substantiated (which cases and reasons shall be specified in those procedures), require the persons concerned to be prevented from resuming work in any situation where there is still a serious and imminent danger.

Separate legislation protects employees from dismissal or other detriment where they stop work or refuse to resume working in dangerous conditions (see chapter 4).

Meaning of "serious and imminent danger"

The identification of foreseeable situations of serious and imminent danger necessitating emergency procedures should be part of the overall risk assessment process. The employer's own experience together with that of others carrying out similar activities must be taken into account, as must HSE and other authoritative guidance.

Such situations might include the following:

Fire

Most employers have formalised fire procedures, often as a result of the requirements of their fire certificate. The certificate is likely to be primarily concerned with safe evacuation of buildings and there may be the need to extend it into other areas eg process shutdown or fire fighting by trained personnel. (Chaper 21 deals with precautions against fires and explosions.)

Bomb threats/suspect packages

This is a risk which has increased considerably in recent years, particularly for those employers in the public sector. If such a risk is foreseeable then a specific procedure will need to be established for dealing with the situation. This will be different from the fire procedure, and is discussed more fully later.

Hazardous substances

Leaks or discharges of substances in their liquid or gaseous form are most likely to present serious and imminent danger, although escapes of solids eg dusts, might also cause danger. Here there is likely to be an overlap with the requirements of the COSHH regulations and some substances will also be subject to the CIMAH regulations.

Process problems

Problems within industrial processes could also create situations of serious and imminent danger. Reactions could run away or there could be a failure of process controls leading to high pressure, high temperature etc.

Power failure

This could be the cause of process problems or could create dangers through the loss of ventilation equipment or lighting. Employees could be left in potentially dangerous positions from which safe access in the dark is not possible.

Violence

This is another risk that is increasing, especially within the public sector. The need for an emergency procedure should be considered wherever staff deal with the general public and especially in situations of potential conflict, eg complaints desks.

Employees handling cash are also likely to be subject to actual or threatened violence and those specifically dealing with potentially violent clients (eg the probation service or some sectors of the health care professions) will also require emergency procedures to be established. (On violence to staff more generally, see chapter 16.)

Crowds

Even peaceful crowds can create situations of very serious and imminent danger. Although regulation 7 requires the procedures to take account of "persons at work", it is impossible to separate their safety from safety of the public (to whom the employer has general duties anyway under s3 of HSWA). Employers in the fields of entertainment and sport will need to consider these risks – they may also be covered by other legislative requirements.

Animals

Paragraph 2(c) of regulation 7 may require access to be restricted to dangerous animals in the captive state, but their escape into other areas may cause serious and imminent danger to persons at work and others. All animals including farm and domestic animals could cause danger to those working with them, particularly if they are in a distressed state.

Weather

Severe weather in the form of wind, rain, snow, sub-zero temperatures or lightning could result in serious and imminent danger such as unstable buildings or scaffolding, flooding and hypothermia.

CIMAH requirements outlined

The Control of Industrial Major Accident Hazards (CIMAH) Regulations 1984 are intended both to prevent major chemical industrial accidents and to limit the consequences to people and the environment if accidents do occur. They apply to processes and storage involving a variety of chemicals listed in schedules to the regulations. These are mainly included because of their flammable, oxidising, explosive or toxic properties. Detailed guidance on the regulations is available in an HSE booklet.[1]

The regulations operate at two levels. Regulations 4 and 5 apply to a fairly wide range of processing or storage activities (as defined through regulation 4 and schedules 1 and 2) whereas regulations 7 to 12 contain much more detailed requirements for potentially higher risk sites (as defined through regulation 6 and schedules 2 and 3).

General requirements

- *Regulation 4 – Demonstration of safe operation*
  The person in control of the activity must be able to produce documentary evidence to show that he has

  (a) identified the major accident hazards; and
  (b) taken adequate steps to
    (i) prevent such accidents and limit their consequences to persons and the environment; and
    (ii) provide persons working on the site with the information, training and equipment necessary to ensure their safety.

- *Regulation 5 – Notification of major accidents*
  This requires major accidents to be notified forthwith to the HSE. Major accidents (fully defined in regulation 2) are occurrences, such as major emissions, fires or explosions, leading to serious danger to persons (inside or outside the installation) or to the environment.

Requirements for high risk activities

The more hazardous industrial activities to which these requirements apply are defined by regulation 6 and Schedules 2 and 3.

- *Regulation 7 – Reports on industrial activities*
  A written safety report must be submitted to the HSE, normally at least three months before the activity commences. Schedule 6 sets out the information which must be included in the report, including details of emergency procedures.

- *Regulation 8 – Updating of reports*
  Where material modifications are to be made to the activity concerned, a further report must be sent to the HSE at least three months in advance. Existing safety reports must be reviewed every three years (in the light of changes in technical knowledge or other relevant developments) and a report of the review sent to the HSE.
- *Regulation 9 – Requirement for further information*
  The HSE may require further information on any aspect of a safety report.
- *Regulation 10 – Preparation of on-site emergency plan*
  The person in control of the activity must prepare and keep up to date an adequate on-site emergency plan (in consultation with appropriate bodies such as workers' representatives, local authorities, the emergency services and the HSE). Everyone on the site who is affected must be given relevant information on the plan.
- *Regulation 11 – Preparation of off-site emergency plan*
  Local authorities must prepare and keep up to date off-site emergency plans for CIMAH sites. This must be done in consultation with the person in control of the activity, the HSE and other appropriate persons. (Under regulation 15 the local authority may charge a fee to the person in control of the activity for preparing the plan.)
- *Regulation 12 – Information to the public*
  The person in control of the activity must ensure that persons outside who are liable to be affected by a major accident are supplied with relevant information. Schedule 8 sets out the information to be communicated. This information should be prepared in consultation with the local authority whom it is expected will enter into an agreement to disseminate it.

COMPONENTS OF EMERGENCY PROCEDURES

The content of an emergency procedure will depend very much on the nature of the emergency. The ACOP to the Management Regulations clearly states that procedures should normally be written down and should set out the limits of actions to be taken by employees.[2] Some employees will be allocated specific responsibilities eg supervision of evacuation, emergency shut-down of equipment, firefighting or rescue. Where different employers share a workplace their emergency procedures need to be co-ordinated.

Most procedures will include some or all of the following components, although they may vary in order depending upon local circumstances or the type of emergency.

- *Discovery, alert and investigation*
  — what action the person discovering the emergency is expected to take
  — how others are to be alerted to the emergency

  - what action they should then take
  - what investigations should be made into the nature and extent of the emergency
- *External liaison and evacuation*
  - how and when the emergency services and others should be contacted
  - arrangements for safe evacuation
- *Control and rescue*
  - how the extent of the emergency may be controlled (or its eventual impact on the business or the community)
  - provision for rescuing casualties
  - when the emergency is deemed to be at an end
- *Information and training*
  - how those involved are to be informed about the procedure
  - what training is required, especially for those with key roles.

Each of these aspects is covered in greater detail in the remainder of the chapter.

DISCOVERY, ALERT AND INVESTIGATION

Discovery and alert

The person discovering what he believes to be a situation of serious and imminent danger must be given clear guidance on what action to take. The nature of this action and stage of the procedure which follows will vary according to the situation as the examples below indicate.

EXAMPLES

Situation	*Action*	*Leading to*
Suspect package	Report to supervisor	Further investigation
Telephoned bomb threat	Record specified details+ and report to designated warden	Further investigation Contact with Police
Process abnormality	Report to supervisor	Further investigation
Significant fire	Break fire alarm glass	*Evacuation
Personal attack	Operate personal alarm device	Despatch of assistance
Escape of animal	Telephone Police	Road closures

Notes:

* Some fire alarms can operate on two levels – one sound alerts occupants to the fact that a fire has been discovered (so they can prepare for evacuation), a different sound indicates evacuation must take place.

\+ See figure 10.1 for suggested questions.

KEEP CALM – DON'T HANG UP, TRY TO GET AS MUCH DETAIL AS POSSIBLE
INDICATE TO A COLLEAGUE THAT YOU'RE RECEIVING A BOMB THREAT
(it may be possible to trace it)

1. WHERE IS IT?
2. WHEN WILL IT EXPLODE?
3. WHAT DOES IT LOOK LIKE?
4. WHAT WILL MAKE IT EXPLODE?
5. WHEN WAS IT PUT THERE?
6. DID YOU PUT IT THERE?
7. WHY?
8. WHAT IS YOUR NAME?
9. WHERE ARE YOU?
10. WHAT IS YOUR ADDRESS?

EXACT WORDING USED:

CALLER'S SEX:

AGE:

ACCENT:

ANY BACKGROUND NOISES?:

REPORT THE CALL IMMEDIATELY TO: (insert local instruction)

Figure 10.1 *Bomb threat checklist*

Investigation

In many cases the emergency procedure will require a further investigation by a designated person (eg a supervisor or a warden) to determine whether a real emergency exists and, if so, its nature and extent. The bomb threat may be a hoax designed to disrupt or the process abnormality may in fact be an instrument fault.

A suspect package may need to be inspected by someone better equipped to decide whether it is an explosive or incendiary device – figure 10.2 gives some guidance in this respect. Even if a bomb threat is believed to be genuine, a check should be made of the proposed evacuation route and assembly point to ensure that they are safe. It is not unknown for terrorists to attempt to force people towards their devices so that they can have the maximum impact.

Investigation of a process abnormality may result in replacement of a faulty instrument, alterations to process controls or even the shutdown of the process.

GENERAL CHARACTERISTICS

Not expected to be there

Wires or batteries apparent

Watch or clock parts apparent

Greasy marks

Unusual smell – almonds or marzipan

LETTERBOMBS/PARCEL BOMBS

Incorrectly or badly addressed

Stamped rather than franked

Lots of small value stamps

Poorly packaged

Scruffy handwriting

Lopsided

Heavy for its size

DO NOT TAMPER WITH SUSPECT PACKAGES

GET OTHERS OUT OF THE ROOM

AVOID TOUCHING

PUT DOWN GENTLY ONTO FLAT, FIRM SURFACE (if you are already holding it)

OPEN WINDOWS (if possible)

LEAVE THE ROOM

SHUT THE DOOR AND PREVENT OTHERS ENTERING (if possible)

RAISE THE ALARM

Figure 10.2 *Dealing with bombs and suspect packages*

EXTERNAL LIAISON AND EVACUATION

External liaison

The type of external liaison required will depend on the nature of the emergency. Those with whom contact may be necessary will include:

Emergency services

Contact with the Fire and Rescue Service, the Police or the Ambulance Service is very likely to be required. This will usually be via a 999 call. In some cases agreed messages or detailed information may need to be given to indicate the nature and size of the emergency so that the Emergency Services can initiate their own emergency procedures, obtain necessary equipment or approach by a particular route.

Authorities

Local authorities or other bodies (HSE or HM Inspectorate of Pollution) may need to be notified about the emergency either because of a statutory obligation or because they are expected to render assistance, eg in mitigating the effects of the emergency, in advising others about it or in clearing up afterwards.

Neighbours

Neighbouring property could include workplaces, residential areas, schools or recreational buildings. The emergency procedure must take account of whether they need to be notified about the emergency, how they are to be alerted and what action they might need to take. In most cases neighbours can be notified by telephone or word of mouth but larger high risk sites may need to utilise alarm sounders. The emergency services may carry out this role.

Sources of assistance

The procedure may identify sources of assistance which can be activated in the event of an emergency. This may be off-duty employees, specialist contractors or neighbouring plants who may be able to supply essential equipment or specialist staff.

Senior management

Whether or not they are going to provide direct assistance (eg as emergency controllers or for liaison with the media) senior management away from the site (and possibly outside office hours) are likely to wish to be informed about any emergency situations which have arisen.

Evacuation

In some cases evacuation may be an automatic result of an earlier action, eg the breaking of a fire alarm glass to sound an alarm. However, in other cases it may result from a definite decision by a responsible person, possibly only after detailed investigation of the circumstances of the emergency. Aspects to be taken into account in evacuation arrangements are:

Place of safety

A place of safety must be identified for those being evacuated. In most fire procedures this assembly point will already have been identified but this may not be a place of safety in the case of other emergencies. In the case of a bomb alert, assembly in a car park may be ill-advised because of the possible presence of car bombs and assembly points in a street may be vulnerable to flying glass.

If the emergency results from a chemical escape then the assembly point will probably need to be upwind or, in the case of gases or vapours heavier than air, in a high building or on high ground. Some plants may even have refuges specifically provided.

This choice of a safe assembly point is an important one and it is essential that the person (or persons) responsible for making the choice is clearly identified in the procedure. These persons must receive appropriate training and guidance to enable them to make their choice wisely.

Safe evacuation route

Similar considerations apply in selecting the evacuation route. Whilst possible fire evacuation routes will probably have been designed into the building, in other cases routes will have to be chosen that keep occupants away from the source of danger. This may involve those in charge during a bomb alert detailing individuals to check that proposed exit routes are clear of suspect items. In the case of chemical leaks there may need to be some study of plant drawings or even maps in conjunction with weather information (particularly wind direction) in order to select a safe exit route.

Special considerations

Some occupants of the premises will warrant special consideration if an evacuation has to be carried out. Contractors or visitors are likely to be unaware or less aware of the place of safety and how to reach it safely. Whilst this difficulty may be overcome by briefing them in advance about emergency procedures it may still be necessary to allocate resources to liaise with and assist them.

Where occupants have disabilities, resources are also likely to be needed to assist them in evacuating safely. In the case of the physically disabled, help may be needed to guide the blind, push wheelchairs or even carry people to safety.

Various devices are available to assist in this process. Deaf people may also need special consideration such as visual as well as audible evacuation alarms.

However, the evacuation of people with learning difficulties or suffering from mental illness is likely to pose much more of a problem. Apart from the difficulty of ensuring that such people understand what action they are expected to take, there is the danger that faced by an emergency they will react in irrational ways, possibly panicking at the sounds of alarms. The resources to deal with this type of person in the event of evacuation may be greater than in catering for the physically disabled.

The evacuation signal

In the case of premises where the range of possible emergencies is limited, the evacuation may be signalled by a simple device – a fire alarm bell or an emergency siren. However, in more complex situations other means may have to be used.

Tannoy systems, telephone messages or even direct word of mouth are all worthy of consideration. Whichever is chosen, it is important that the message given is clear and concise:

- what is the nature of the emergency
- where is the place of safety to which occupants should go
- which route should they take
- any other essential information.

The consistency and reliability of such messages may be increased by preparing a simple written script for the deliverer to use. In some cases this may be able to be prepared in advance.

Sweeping out

If only employees are on the premises then there will probably not be a need to appoint employees to check through the building to ensure everyone has been evacuated. However, if others, particularly members of the public, are present this is likely to be an essential part of the evacuation process.

Places of entertainment or retail premises are particularly likely to include this in their procedures. People are often reluctant to leave goods or services that they have paid for; in one recent fatal fire in a large department store, customers in the cafeteria took a considerable time to leave the area after the sounding of the alarm.

There may also be considerable security implications if members of the public are allowed to remain in buildings which have otherwise been evacuated.

The roll call

The Fire Brigade or in-house emergency team will immediately need to know if anyone is still inside the building. If no-one needs to be rescued then they can deal with the emergency from a position of relative safety.

Sufficient members of staff need to be given the responsibility of carrying out a roll call, either for the premises as a whole or for their own particular section. The people to choose for this task will not necessarily be the most senior. They should be those who do not normally stray far from their work location so that they are likely to be there when the emergency arises and will have a good knowledge of who else is on the premises.

In relatively low-risk situations an oral roll call may be all that is required, but in higher risk locations written lists supported by formalised signing in and signing out procedures are likely to be necessary. Such procedures will also need to take account of contractors and visitors.

Where there is frequent public access to the premises it will not be possible to account for members of the public individually and reliance will have to be placed on 'sweeping out' arrangements. The individuals carrying out the roll call will normally be different from those sweeping out so that the arrival of the 'sweepers out' to confirm that the premises are clear is the final stage of a successful roll call.

Deputies will need to be appointed for those conducting the roll call (and those sweeping out) to cater for situations where they are unavailable due to holiday or sickness or even where they themselves have been affected by the emergency.

CONTROL AND RESCUE

The need for formalised procedures for controlling emergencies and possibly rescuing casualties will be influenced by the foreseeable nature of the emergency, its extent and perhaps its likely duration. The following are all likely to be components of the control part of the procedure.

Emergency controller

An individual must be clearly given the responsibility of acting as the emergency controller. This could be a named person or it could be defined by status eg the shift manager on duty. Arrangements should also be made for a deputy to take control in the case of absence. Provision may be necessary for the controller changing as the situation develops, possibly as more senior or more capable people arrive on site.

The role of the controller will need to be defined but it is likely to centre around the utilisation of the emergency resources (staff and equipment) available and liaison with others dealing with the emergency, particularly the public emergency services.

Control point

Allocation of resources and liaison with others is likely to be more effective if the emergency is controlled from a clearly defined point. This could be pre-selected in advance or it could be determined by the emergency controller.

The control point must be selected so as to be safe (upwind of any leaks or outside the range of possible fire or explosion). It should also be of an adequate size and allow for easy communication both with those off-site (eg by telephone) and for those dealing with the emergency on-site.

Pre-selection of the control point could allow it to be provided in advance with information (eg drawings or plans), communication equipment or other emergency equipment.

Emergency teams

It may be necessary to select (and to train) teams of employees to assist in dealing with the emergency. Their role may vary according to the nature of the emergency. It could involve control (of leaks, spillages or violent situations), rescue (of injured or threatened employees) or the emergency shut down of plant or processes.

Such teams will need to be given clear guidance on the level of risk they themselves are expected to tolerate in carrying out their duties, with the emphasis very much on preserving their own safety. Liaison with the public emergency services will be particularly important, possibly with clear demarcations being drawn on exactly who is expected to do what.

Emergency equipment

The type of emergency equipment which needs to be kept available will depend upon the type of emergency which is foreseeable. Typical equipment might include:

- *PPE* – protective suits, gloves, self-contained breathing apparatus or other respiratory protection
- *spillage-control equipment* – absorbent material, neutralising agents
- *tools* – hammers, cutters, burning equipment, jacks, torches, ladders, ropes etc
- *fire fighting equipment* – additional extinguishers or hosereels
- *communication equipment* – mobile phones, radios
- *animal or personnel restraint equipment.*

Restricting access

Some emergencies necessitate the restriction of access to affected areas. If restrictions are necessary within the work site itself then these can be implemented by staff (eg security personnel), possibly using barriers, tape or appropriate warning signs.

However, if public areas are affected (eg large scale leaks of chemicals, unsafe buildings or scaffolding or escaped animals) then there is likely to be a need for close liaison with the Police to ensure that unwary members of the public do not stray into dangerous areas.

Staff call-in

The need to inform senior management about the emergency was mentioned earlier but, as emergencies develop, there may be a need to call in members of staff with specialist skills. These may be required to help control the emergency or to assist in the restoration of normal operations once the emergency is over. Access to home telephone numbers of relevant staff will be an essential part of this process.

Additional resources

Emergencies may necessitate the obtaining of a variety of services from outside sources. Emergency generators might be required as might mobile compressors or temporary lighting rigs. Lengthy emergencies may result in a need for the supply of food and drink to those involved in dealing with them. Once again the telephone numbers of companies able to provide such specialist equipment or services should be kept readily available for use by the emergency controller.

Medical assistance

The general topic of first aid is covered in chapter 16 but the need for medical assistance must also be considered when planning for emergencies. An emergency could quite easily swamp an organisation's own capacity to provide first aid treatment, indeed, the first-aider could be one of the casualties.

The procedure must include provision for liaison with the Ambulance Service and local hospitals. Preparations can be made in advance so that they are aware of the likely numbers and nature of casualties and also of any dangers which they may encounter on site.

Contact with relatives

Arrangements will also be necessary to deal with relatives of employees or others affected. Telephone enquiries will have to be responded to sensitively and areas may need to be set aside for any concerned relatives who come to site. Where there are casualties, arrangements will need to be made to inform their relatives. For serious cases a personal visit will almost certainly be essential.

Personnel specialists are likely to be involved in this work, together with the public emergency services. Friends and colleagues of casualties are also likely to be able to make an important contribution.

Media liaison

Provision for media liaison should be made in order to prevent publicity which may damage the business. Uncontrolled media access to the site should be prevented. This may be necessary in any case to protect the media representatives from danger. An individual should be appointed to deal with media enquiries rather than leaving the media to feed upon the ill-informed or ill-considered statements of employees or others.

Clearly the public relations aspect of news management in such circumstances is a specialism in its own right but honesty will generally be the best policy. However, an effort must always be made to keep things in perspective – the emphasis should always be on what is being done or has been done to keep the emergency under control rather than on what could have happened if it had got out of control.

Termination of the emergency

At some stage the emergency will end and the procedure must state who makes the decision to end it. If an emergency controller is designated then he is likely to be the relevant person but in simpler procedures fire wardens or others could be given this authority.

They may need to make their decision in conjunction with others – emergency services, specialists etc. Then steps can be taken to restore the situation to normal, possibly in a phased manner. A decision to end the emergency is likely to be necessary before the following can happen:

- evacuated persons are allowed to return
- emergency teams are stood down
- external bodies or neighbours are informed
- barriers or signs are removed
- normal activities are resumed
- clean-up operations can begin.

INFORMATION AND TRAINING

Statutory requirements

Employers have a general duty under s2(2)(c) of the HSWA to provide necessary "information, instruction, training and supervision". Additionally under the Management Regulations they have a duty to provide their employees and others working in their undertaking with information about their emergency procedures (regulations 8 and 10).

Under regulation 11 of the Management Regulations, employers must take into account the capabilities of employees entrusted with tasks under their emergency procedures and provide them with adequate health and safety training when they are given these responsibilities. As noted earlier, the CIMAH Regulations may also require information to be provided to the public.

Information

The manner in which employees and others are informed about emergency procedures and how they are expected to act will vary considerably depending upon the types and potential severity of the foreseeable emergencies. Issuing procedures alone is unlikely to prove effective since employees will not have the opportunity to ask questions about parts they do not fully understand. Generally there should be some form of oral briefing allowing for two-way communication.

It is unlikely employees will be able to find the relevant parts of detailed procedures if an alarm is suddenly sounded and therefore the key elements of procedures should be prominently displayed on notice boards or in other suitable positions.

Training

Where employees have specific emergency roles – emergency controllers, fire wardens, telephonists, security staff, members of emergency teams etc – they must be provided with adequate training. They must understand their role and how best to carry it out in the full range of foreseeable emergencies.

Practice

In most cases the information and training will need to be augmented by practice. Periodic fire drills or emergency exercises may be required as a condition of the "fire certificate" or form an integral part of an emergency plan under the CIMAH Regulations. The holding of a full-scale drill may not always be practicable because it is so disruptive to normal operations or because it introduces dangers of its own.

This should not be an excuse for not practising at all however, as realistic 'table-top' exercises can be designed to both test the procedure as a whole and the capabilities and knowledge of those responsible for its implementation. Individual employees can be questioned on what steps they would take in designated emergency scenarios. In some cases detailed practice of specific skills (eg use of breathing apparatus, fire fighting, emergency shutdown sequences) can be arranged.

Experience has shown that if emergency procedures are not practised regularly then there is a far greater chance that they will prove partially or totally ineffective if they ever have to be put into operation to deal with a real emergency.

Notes

1. The Control of Industrial Major Accident Hazards Regulations, SI 1984/1902; and *A Guide to the Control of Industrial Major Accident Hazards Regulations 1984* – HSE 1990 (HS(R)21(Rev)
2. Management of Health and Safety at Work Regulations 1992 Approved Code of Practice – HSE 1992 (L21), paragraph 48

11

Audits and Inspections

INTRODUCTION

Historical perspective

By the beginning of the 20th century most of the industrial countries of Europe had established a form of statutory factory or mines inspectorate. In the UK, the statutory inspection process has been constantly extended and with this sort of intervention it was not long before the more responsible organisations began to consider and formulate inspection systems to match those of the state. For these companies, health and safety inspections have become a well-established and essential managerial activity.

The Factories Act 1961 assisted many organisations in facilitating and developing their internal safety inspection systems, since sections of the Act laid down detailed safety standards for plant and equipment. These standards were the inspectorates' views of all the main safety issues relevant at that time. The 1961 Act covered for the most part what can be termed the hardware aspects of health and safety. It set standards on aspects which included:

- *health* drainage, lighting, ventilation; cleanliness and overcrowding
- *safety* guarding, hoists, lifts; chains, ropes, lifting tackle; cranes, lifting machines; access, floors, passages, stairs; steam boilers, fire prevention means of escape
- *welfare* washing facilities, accommodation and first-aid.

Today many of these items form the basis of health and safety **inspection** programmes operating throughout the UK. The list previously detailed is far from complete and a brief examination of Parts I to IV in the Arrangement of Sections of the 1961 Factories Act can provide a valuable source of information to anyone wishing to find a starting point for the design of inspection systems.

It was not until 1972, with the publication of the Robens Committee Report[1] that the value of the health and safety audit became really apparent. In examining the role of management the report discussed methods for the control of safety emphasising:

> safety audits in which each aspect of workplace organisation and operation is subjected to a carefully planned and comprehensive safety examination.

The difference between inspections and audits then began to emerge and these techniques came to be employed more and more by safety professionals and company managers.

The advent of HSWA in 1974 assisted organisations in examining the non-hardware aspects of health and safety. This was because HSWA centred on items such as:

- provision of safe systems of work
- provision of information and instruction
- provision of training and supervision
- provision of a health and safety policy.

It can be argued that in 1974 there was a change in emphasis from hardware aspects of health and safety, mainly covered by inspections, to the managerial control aspects mainly covered by audits. Safety control systems are usually recorded in document form which can be termed the software of health and safety. HSWA recognised that it was no longer possible to continue with the approach adopted by the early reformers who saw legislative safeguards for particular hazards as the only way forward. It started to impose responsibilities on those who created risks at work and linked health, safety and welfare as an integral part of the process of management. If health and safety is an integral element of the management process, audits and inspections become important activities for management if it is to effectively discharge its duties. Moreover, the use of audits and inspections is one way of demonstrating that duties imposed by HSWA have been effectively discharged.

Legal considerations

The legal provisions most relevant to audits and inspections are those found in s2 of HSWA and in particular subsection (2) which requires the employer to ensure:

> (a) the provision and maintenance of plant and systems of work that are, so far as is reasonably practicable, safe and without risks to health;
> (d) so far as is reasonably practicable as regards any place of work under the employer's control, the maintenance of it in a condition that is safe and without risks to health and the provision and maintenance of means of access to and egress from it that are safe and without such risks;
> (e) the provision and maintenance of a working environment for his employees that is so far as reasonably practicable, safe, without risks to health, and adequate as regards facilities and arrangements for their welfare at work.

These requirements involve management in taking positive steps to ensure duties are met and to demonstrate compliance. One of the best studies to demonstrate the need for professional audits and inspections based on the application of principles outlined in this chapter is the HSE Report *Blackspot*

Construction.[2] The report concluded that out of 739 deaths examined during the report period (1981–5), positive action by management could have saved lives in 70 per cent of the cases. Many of these deaths could have been foreseen and proactive measures taken, in particular management of sites through the use of audits and inspections. Audits and inspections must be given the same magnitude of professional management attention as (for example) production costs and order completion dates.

Slipping on badly designed or oil covered steps is still a common accident today. Accidents caused by collapsed gratings and defective high level walkways still cause serious leg and back injuries. These categories are but two simple examples of areas in which accidents can be reduced by the application of audits and inspections. The provision of "safe plant with safe systems of work, a safe place of work with safe access and egress and a safe working environment with adequate welfare facilities" can be seen as a fundamental health and safety requirement which may be successfully achieved by the deployment of audits and inspections. Clearly, many more health, safety, welfare and environmental issues can be managed by the deployment of audits and inspections but the three statutory provisions set out above are paramount.

Regular audits and inspections of the workplace are excellent managerial tools to be included in an accident prevention policy. Post-accident action and rectification is still too prevalent an approach to health and safety today. Audits and inspections allow a proactive approach. Organisations which successfully manage health and safety display a number of common characteristics: one of the most significant is the deployment of regular health and safety audits and inspections.

OVERVIEW OF AUDITS AND INSPECTIONS

Subject definition

Throughout this chapter health and safety audits and inspections are mainly presented together as a generic labelling for all of the many types and classifications of this managerial technique. The term health and safety monitoring could equally be employed. However, while many of the functions of audits and inspections are similar and indeed complementary, for these techniques to be used effectively it is necessary to demonstrate their main differences.

Audits are seen by most health and safety professionals as a system for the total examination of an organisation's health and safety. An audit is a measured account of progress towards a company's health and safety targets. It is an in-depth appraisal of the health and safety management culture. An audit is a review of the control systems and procedures in place measured against the effectiveness of the control. Measurement and quantifiable data are the main hallmarks of an audit.

Inspections of health and safety are a careful look at a defined field of examination. They are a transient appraisal of actions and conditions at a given point in time and viewed at a given location or locations within an organisation. It is a technique for the monitoring of performance, a fact-establishing approach concerned with issues that can usually be corrected by supervisory and middle management. An inspection will produce an action list rather than an assessment sheet.

Health and safety auditing may be defined as:
management evaluation of company-wide health and safety control from which to develop **long-term** plans and financial provision.

A health and safety inspection involves:
identifying the conditions in a given workplace to enable action to be taken so that standards are maintained through ongoing programmes and budgets in both the **short and medium-term**.

In addition to the time span differences between audits and inspections another difference is that audits apply to strategic health and safety management while inspections relate to tactical questions.

Objectives and aims

The aims of audits and inspections are different in their detail and are examined later in this chapter. However, their objects are the same:

- FINANCIAL
- LEGAL
- HUMAN

Financial benefits

Financial arguments for the deployment of health and safety audits and inspections focus upon the loss avoidance area. Here, audits and inspections will, if employed correctly, highlight unplanned waste of materials in consumable stores, unnecessary contractor labour and poor supervision.

Damage is another major area where audits and inspections will highlight unnecessary costs. In particular damage to plant and machinery, to the product and to transport facilities (both external and internal) will be identified. At times even the examination of personal protective equipment by a health and safety audit or inspection will highlight waste. An inspection of out-of-the-way places and rubbish skips on site will reveal discarded PPE. With a minimum of training, audit and inspection teams will quickly identify cost savings within any type of enterprise. Another angle from which to approach this financial objective is to look at PPE purchasing and stores withdrawal records. Take, for example, ear protectors and obtain say two or three years' records showing how many have been issued. Then, look at the number of employees who are required to use such hearing protection at work. Do the facts fit or is there a leakage? This type of paper audit can be utilised as a supplement to normal audits and inspections and adds to their financial importance.

The principle can be expanded into more technical fields, eg examination of waste gas analysis from furnaces or boilers. Not only will such audits and inspections have beneficial effects on the environment, but they will also highlight possible savings in fuel consumption, reduce refractory damage and possibly extend the operational life of the plant. The ever increasing real cost of employers' liability insurance and the mainly unknown uninsured losses from accidents (generally thought to be 10 per cent of all losses) can and do have a devastating effect on an organisation's profit and loss. Expedient commercial presentation demonstrating the use of audits and inspections can play an important monetary role in reductions within this framework.

Meeting employers' legal duties

Audits and inspections are a powerful tool in allowing an organisation to fulfil, and to be seen to be fulfilling, both general and specific legal obligations in terms of health and safety. HSWA places some general statutory requirements on undertakings for which audits and inspections can and do provide practical instruments of managerial control. In particular s2(2)(a) requires the maintenance of safe plant and safe systems of work. In addition s2(2)(d) requires safe places of work and safe access and egress. These requirements are the very essence of audits and inspections and a major argument for their deployment as an instrument of health and safety management. The Management Regulations support this. Regulation 3 states:

> (1) Every employer shall make a suitable and sufficient assessment of –
>
> (a) the risks to the health and safety of his employees to which they are exposed whilst they are at work ...

There are also specific legal duties which require periodic inspections and records of inspection to be maintained. For example, the Pressure Systems and Transportable Gas Containers Regulations 1989 require records on scheme examination and the Control of Substances Hazardous to Health Regulations 1994 call for examination of local exhaust ventilation plant used in certain processes.[3]

The legal requirements should inform the management decision as to whether or not a system of audits and inspections is required and should prompt further review on points of application such as frequency and scope.

Human aspects

The human objective of audits and inspections can be defined as the prevention of injuries or ill-health on moral or human grounds. Most importantly it is the obligation to prevent fatal injury. Real and substantive improvements can be made in reducing the human suffering as a result of accidents by applying some of the most simple of audit and inspection routines. The view held by many health and safety professionals is that the vast majority of accident causation factors are related to poor management control. With a few exceptions the frequency of accidents can be reduced through the use of regular and efficient audits and inspections. Some organisations go beyond ensuring that periodic legal inspections are carried out and employ a system of internal examination, testing and inspection.

Aims of audits and inspections

The aims of audits and inspections are practical principle: these must be clearly distinguished from the objectives, which should be seen as forming a company policy. In particular, aims or principles can be said to have three main elements:

- monitoring
- reviewing
- assessing

In terms of technique, this three-stage process can employ both on-site examinations and a survey of records and other documents.

Monitoring

Monitoring is observing, checking and tapping into shop floor communication channels to evaluate, against laid-down standards, the safety status of a works, a plant, an identifiable area of equipment or an operation. It involves both examination and study. Monitoring is about the achievement of a safe system of work: this is a key aim of any health and safety audit or inspection. In essence, monitoring is about the collection of information on the subject or subjects by the observation of the work situation, its people and procedures of operation. This will either at the time or later require some study as to how these three aspects relate to each other.

Reviewing

Next, a review is required of the facts established at the monitoring stage. Reviewing against standard is an important management function. It involves measurement and consideration and is about revision and critical examination. It is a critique designed to establish a base of information on the health and safety subject or a number of subjects derived from the collection of information during the monitoring stage. Thus, a review cannot be effective until this monitoring stage has been undertaken and key elements are the preparation and processing of information.

Assessing

The final step in the threefold process is the assessment stage leading to action.

Assessment in this context is about evaluating and judging safety issues against set standards. Assessment of a health and safety subject or subjects is a final managerial evaluation involving a decision on the action to be taken. Managerial action is then carried out (where necessary) to rectify any unsafe acts or unsafe conditions. The assessment stage should result in action based on the health and safety information obtained from the previous two stages, monitoring and reviewing.

CLASSIFICATION OF AUDIT AND INSPECTION TYPES

Main grouping

Audits and inspections can be distinguished as indicated earlier and subdivided as shown below:

- *Audits* – health and safety audits
  - – health and safety surveys
- *Inspections* – scheduled safety inspections
  - – safety sampling
  - – safety tours
  - – safety observations.

These six types all have the same purpose, which is the prevention of damage, loss, injury and ill-health, but vary in their application. This section of the chapter examines the fundamental characteristics of each. In doing so a standard format will be used, comprising:

- a model remit
- location and routes
- personnel and leadership
- duration
- frequency and recording methods.

Health and safety audits

In basic scope a health and safety audit will be an in-depth appraisal. It should consider strategic health and safety subjects. Above all, health and safety audits must, like all other types of managerial audits, be measurable. The significance of measurement in audits can be illustrated by the Fennell Report on the King's Cross disaster which stated:

> It is essential that a system should be devised whereby safety of operation can be the subject of audit in the same way as efficiency and economy.

> If the internal audit has become the yardstick by which financial performance is measured then the safety audit should become the yardstick by which safety is measured.[4]

They should be designed so that data collected when processed can be compared with a standard. A health and safety audit must identify and target the strengths and weaknesses within an organisation and those areas vulnerable to major risk.

An open system of health and safety auditing is a powerful indicator to employees that management at all levels is professional. This style of auditing can provide vital synergistic advantages throughout a company. In addition general communications will improve. A positive competitive atmosphere can also be created through audit league tables.

Some of the strategic subjects for audit are set out in figure 11.1.

There are two main methods of measuring health and safety audits in industry and commerce, the rated audit (based on items in figure 11.1) and the evaluated audit. Rated audits are comprised of a systematic and structured set of standard questions each with a weighting which, when totalled, will present a comparable statement of health and safety performance as applied to a plant, department, works or site. The rated system is ideally suited to target-setting objectives as results can be assessed without too much in-depth knowledge of health and safety. Rated audits are a very effective managerial control tool.

Evaluated audits can be measured against a number of criteria. Some of those more commonly used in measurement are:

- unsafe issues breaching statute laws or regulations
- the number of unsafe issues observed as against the total number examined
- the estimated cost of rectifying unsafe conditions
- the level of risk associated with hazards observed (eg insignificant – low – significant or high)
- measuring against health and safety plans or programmes
- the loss effect (in financial terms) on the organisation should the hazards cause damage, production loss, injury or ill health etc.

The remit of an audit should specify whether or not it will include health or more importantly ill-health problems, such as effects of hazardous materials, for example oils and greases producing skin problems.

Should the audit include welfare and environmental considerations? The location of the area to be audited must be defined and the general route to and around the audit area outlined. Health and safety audits are best carried out by a small team of trained auditors. The team should include a member of production management or supervision, an engineer, a safety practitioner and any other specialist considered necessary. Most organisations consider an audit to be a formal managerial function but nothing should prevent employees being part of an audit. A team leader is vital for a health and safety audit and the production representative in most cases will be the best person for this role. The

(1) Accident and incident reporting and investigating procedures

(2) Auditing and inspection of health and safety – planned programmes

(3) Control of substances hazardous to health programmes

(4) Contractor control systems

(5) Emergency preparation plans

(6) Engineering maintenance controls and statutory inspection programmes

(7) Environmental and waste management control systems

(8) Fire prevention programmes

(9) Health monitoring of employees programme

(10) Employee safety evaluation systems

(11) Leadership in safety programmes

(12) Policy statement on health and safety together with implementation plan

(13) Personal protective equipment controls and standards

(14) Purchasing department controls on health and safety aspects

(15) Safety department organisation and operation

(16) Safety meetings and associated infrastructure groups

(17) Safe systems of work control procedures

(18) Standing (health and safety) instructions control process

(19) Task behaviour monitoring programmes

(20) Training policy and programmes for health and safety – management and employees

Figure 11.1 *Rated audit subjects*

team leader will, amongst other things, be responsible for ensuring the audit report is produced and in particular that an action list programme is part of that report. Team leaders should also ensure that the report is put into the hands of those who are responsible for undertaking the necessary control and rectification measures.

A professional health and safety audit would usually have a minimum duration of approximately four hours, but could last a number of days. Audits of the standard necessary to achieve levels outlined in this chapter should take place on an annual or even two-yearly basis. Depending on the audit area every six months may be feasible but anything more frequent is really a routine inspection.

Safety survey

A safety survey is a detailed, measured examination of a narrow or specific field of operation. Usually a single, given subject is the target of a safety survey. In many cases it will arise from a requirement to examine a particular issue throughout an organisation, possibly following an accident or an incident, or perhaps as a result of new legislation. An example would be a safety survey of visual display screen work stations. To be effective safety surveys must have quantifiable aspects. This is best illustrated by looking at an example.

EXAMPLE

Accessways to roofs and work areas at heights (such as steps, stairs and ladders) is a typical target subject for a safety survey. This would first require all accessways to be known in advance and to be detailed on a works map, drawing or register. A visual survey of all accessways would provide a quantifiable measurement (in percentage terms) of those meeting (for example) a ten-point standard. So, if out of 15 accessways surveyed ten meet the standard, that results in a 66 per cent success rate. Each of the 'failed' accessways could then be measured (again in percentage terms) against a ten point pre-set standard.

The remit of a safety survey, therefore, would be the evaluation of a single subject against a safety standard set by the organisation. Standards could be derived from legal requirements, national or international standards, supply or proprietory standards or from best industrial practice.

The location and especially the routes to be followed on safety surveys must be considerd very carefully as in many cases the survey subject would be of a hazardous nature. This would require careful planning. The staff to be employed on safety surveys should have the necessary knowledge and experience to evaluate the survey subjects against the various standards set. Because some typical examples of safety survey subjects are laser equipment, overhead cranes, confined spaces, pressure systems and immobilisation controls, the safety survey team members or individual examiners must in some cases be specialists in the field that the safety survey subject is to cover. Where a team of examiners is to be employed, the team leader should be a specialist. The duration and frequency of safety surveys will be dictated by the subject matter as the main aim is to achieve an in-depth study. Normal duration would be between two and three hours and frequency should be every three to four months. Recording of safety surveys will usually be in a tailored form again dependent on the safety survey subject. An example of a safety survey reporting format is shown in figure 11.2.

Scheduled safety inspections

A health and safety workplace inspection is probably the most commonly employed type of audit or inspection.

Safety inspections are a routine and scheduled examination of plant, factory, department or offices, or in fact any workplace unit. They check on all of the general day to day health and safety aspects common to many organisations throughout the UK. Inspections should take in the full spectrum of health and safety examining in detail the plant, people and procedures and their interface. Subjects to be examined would best be left to those taking part in the inspection. Hazards would be identified by the team and their colleagues. The remit would be to examine as many unsafe issues as possible and then put into action remedial measures that can be effected quickly and easily by the inspection team. Ownership of health and safety problem rectification is a central theme to effective inspections. To assist in keeping inspections professional, the inspection area must be clearly defined. Normally, to carry out detailed inspections people need to wander about and to examine certain plant and equipment. Dividing the inspection between ground level, above ground and below ground would assist. Safety inspections are best undertaken by a small team comprising approximately 50 per cent management/engineers and 50 per cent employees, with trade union members if applicable. As well as helping to secure employee acceptance of safety inspections, this mix will ensure that some of the unofficial aspects of an organisation's health and safety operations are seen and examined.

An important element of safety inspections is the use of the 'behind rule' – look behind locked doors, instrument and electrical panels, behind product and spares stocks, behind cupboard doors and behind the reluctance of some people to permit inspections of their 'patch'. Most safety inspections are a teamwork exercise and a respected leader is vital. This may be the safety specialist, an experienced and trained shop floor employee, a trade union safety representa-

<table>
<tr><td>HEALTH AND SAFETY SURVEY</td><td rowspan="2">SURVEY TEAM</td></tr>
<tr><td rowspan="2">1. REASON FOR SURVEY</td></tr>
<tr><td>SURVEY DATE</td></tr>
<tr><td colspan="2">2. SURVEY BASE REFERENCE (Detail Register, Schedule or Checklist)</td></tr>
<tr><td colspan="2">3. FINDINGS (detail)

(a) Did findings meet company standards?

(b) Did findings fulfil legal requirements?</td></tr>
<tr><td colspan="2">4. ACTIONS RESULTING FROM SURVEY</td></tr>
<tr><td>5. ACTION DATE:-</td><td>6. ACTION BY:-</td></tr>
</table>

Figure 11.2 *Health and safety survey reporting format*

tive or a member of line management. Whoever leads the team during a safety inspection must be made aware of their role and the importance of their duty.

The duration of a safety inspection should be between one and three hours. The frequency is ideally monthly, depending on risks. Inspections should dovetail with safety surveys. (See figure 11.3) Recording of safety inspections can be in a simple format showing date, plant, team and detailing findings under the following six main headings:

- area
- problem
- rectification action
- hazard/risk level
- action by
- target date

These proposed headings can be supplemented by other items of detail depending on a given organisation's requirements. Some of the most appropriate topics, often covered by safety inspections, are listed on p230 to form an initial checklist of work.

1. AUDITS – every 12 months
2. SURVEYS – every 4 months
3. INSPECTIONS – every month
4. SAMPLING – no fixed frequency
5. TOURS – every week
6. OBSERVATIONS – every day

Figure 11.3 *Health and safety monitoring: suggested frequencies**

* These are only suggested frequencies. The risk associated with each workplace must be the deciding factor.

The report format for a health and safety inspection should be based on a table type layout. An example is shown in figure 11.5. This report could be backed up with additional text reporting on the area covered, dates, names and responsibilities of team members. Most importantly, it could draw attention to points of major concern to the team or serious risk situations requiring urgent managerial attention.

The subjects on this list are not exhaustive but could be used to draw up an area-specific customised safety inspection format.

Access and egress routes	Immobilisation systems
Confined spaces operations	Lifting and slinging operations and equipment
Contractor operations	Lighting standards
Crane operations	Mobile plant
Dust and fume areas	Piped services and pressurised systems
Electrical plant and equipment	Power tools
Emergency preparedness	Permit to work systems
Environmental issues	Personal protective equipment
Fall conditions	Safety signs
Fire prevention and fighting equipment	Storage areas
First aid facilities	Safe working procedures
Gas areas	Ventilation systems
Guarding and moving machinery	Welfare facilities
Hazardous materials and substances	Working at heights operations

Figure 11.4 *Safety inspection topics*

Safety sampling

Safety sampling is a form of mini-safety inspection. It usually takes place in a small part of a work area. The area would form the unit for an inspection. This form of safety examination concentrates on those health and safety issues which injury and near-miss reporting would indicate may be a problem in relation to parts of plant or operations, or in respect of certain groups of people. It is suggested that safety sampling should be carried out on a weekly basis so it deals with health and safety issues that arise between scheduled safety inspections. Because it concentrates on small sections of an area, a number of safety sampling examinations should be taking place weekly to cover the gaps left or areas uncovered by monthly safety inspections. Safety sampling could be subject to a reference drawing to ensure all parts and elements of a particular area are included and are not overlapped or overlooked. Safety sampling locations can be detailed on the drawing with a reference list mapping the locations or tasks known to be problematic in safety terms.

The main remit of safety sampling is to identify safety problems in the early stages and make or request immediate rectification or action. It is a method of stopping minor safety problems becoming major or frequent safety problems. Frequent problems could include housekeeping and fire risks, guards left off,

<table>
<tr><td colspan="4">HEALTH AND SAFETY INSPECTIONS</td><td colspan="2">Date of inspection</td></tr>
<tr><td colspan="4">Department</td><td colspan="2" rowspan="2">Inspection team</td></tr>
<tr><td colspan="4">Management authorisation:</td></tr>
<tr><td>Item Ref</td><td>Location Detail</td><td>Non-Standard Condition or Feature</td><td>Action Required</td><td>Completion Required By</td><td>Item Accountability (By)</td></tr>
<tr><td></td><td></td><td></td><td></td><td></td><td></td></tr>
<tr><td colspan="6">Remarks</td></tr>
</table>

Figure 11.5 *Health and safety inspections*

unclear walkways and poor lighting. Experience has shown that safety sampling is most effective when carried out by groups of no more than two or three people, one perhaps a manager or engineer and one a shop floor employee. From time to time a safety practitioner may need to be involved in order to set and maintain standards of observation, examination and action. Leadership of a small group of this size is not usually of importance. However, the person responsible for reporting the safety sampling findings and recommendations for rectification should be identified from the outset. The report route for safety sampling findings should be firstly to the first line supervisor or manager of that particular part of the workplace. This should be by oral report immediately after the sampling exercise. This oral report should be then followed up by written notes and rectification recommendations. A safety sampling exercise would ideally take one to two hours.

Safety tours

In its most simple form the safety tour is a one-person safety evaluation. In the main, it serves the purpose of enabling the person conducting the safety tour to obtain a quick, general impression of the health and safety standards of a given plant or department. In most cases the tour keeps to a predetermined and standard route, usually via the main accessways through that plant or department or particular workplace. In some cases, it is undertaken by small numbers of people, typically senior managers. On occasions, it might include a member of an inspecting authority, or an insurance inspector. The purpose of a safety tour is to obtain an accurate picture at the time the tour is taken. It is a simple check on the facts as opposed to perceptions obtained from casual comments. Safety tour remits are fact-finding missions sometimes referred to as safety walk-throughs. A typical safety tour is limited to one or two hours or sometimes shorter, depending on the amount of information which needs to be obtained. As with all spot-checks no frequency should be applied as it is an unscheduled operation conducted to gain maximum effect.

Although unscheduled, it should be carried out as often as is required to obtain the appropriate results, as specified in the remit. Frequent safety tours by the same person will quickly allow them to make broad judgements as to whether or not a given plant or department is generally making progress on health and safety, staying the same or getting worse. Additionally, that person will observe which parts of the workplace demonstrate improvement and which do not. Such observations can be compared with written reports.

Safety observations

The safety observation has two variations and is one of the most simple and effective methods for checking health and safety.

General safety observation

The first variation is the general safety observation. Its purpose is to ensure operations do not go unchecked in the workplace. The remit intention is to highlight unsafe acts and conditions, operational failings, errors of judgement and incorrect attitudes. The aim is to make certain these safety issues do not remain unidentified and, in particular, unrecorded. The remit can be achieved by simple observation and examination of work methods and of the maintenance and operating conditions of plant and equipment, as well as by the checking and validation of procedures and the adherence to them. Safety observation is mainly about surveillance of safety systems of work: are they in place and in operation?

The duration of one safety observation should be five to ten minutes and with practice a two-minute observation can be undertaken. How can it be done? First consider the type of safety issues to be observed – people, plant or procedures. Then select a system, a job, an area of work, a site, a location, a piece of equipment, a procedure, a work team or an individual. Observe, evaluate and record. Some of the examples below can be used as an 'aide memoire':

- check workmanship of trade and craft practices
- check methods of safety communication
- ensure acceptable safety standards of engineering or production practice
- ensure errors and misunderstandings over safety are recorded
- establish the level and quality of safety supervision
- examine safety documents employed
- inspect housekeeping and site working conditions especially access, egress and lighting
- look into unsafe acts and purposeful omissions
- monitor potentially hazardous or dangerous operations
- observe safety resources and human resources
- review safe working procedures, safety method statements and permits to work (see chapters 7 and 12)
- survey safety teamworking, commitment and training.

Who should carry out safety observations? They can be carried out by anyone who is committed to health and safety and has been given a short safety briefing on the subject.

Records of safety observations can be kept on the data card or sheet shown in figure 11.6. A simple judgement, evaluation or marking method demonstrates the state of play in terms of whether or not there has been an improvement.

The recording of general safety observations could be kept in a wallet-size pocket book designed with replacement pads.

Safety attitude check

The second variation of safety observations is the safety attitude or safety behaviour check. A number of proprietory systems are available off-the-shelf. The underlying principle is that individual or work group safety attitudes are difficult to deal with, and therefore to change or improve, because they are impossible to see. However, safety behaviour can be seen and observed and therefore can be improved. In fact, many safety practitioners do this in an informal manner in day-to-day work. For example, an employee grinding with a portable power grinding machine and not wearing eye protection may demonstrate an attitude problem. If the employee has been clearly given safety information, understands it, has been supplied with eye protection and is effectively supervised, then a safety attitude issue could be at play.

The approach of this type of observation is formally to take time out to observe individuals or small groups at the workplace. The aim is to examine

Date	Time
Location	
1. Type of check People ☐ Plant ☐ Procedures ☐ Works check ☐ Contractors ☐ Others ☐ Detail:	
2. Observations	
3. Action Signature........................	
4. Evaluation –1 ☐ 0 ☐ ☐ 1 ☐ 2	

Figure 11.6 *General safety observations*

some key points linked to health and safety which are particular to that job or task. Key points could include the wearing of PPE, use of tools and equipment and the following of safety procedures. Observations might also be used where employees are in possible injury positions (eg in danger of falling). It is important to observe any change in practice which occurs while employees are under observation. The results of the observation should be recorded by observers on a score-card and discussed with the individual or group, pointing out both the positive and negative aspects of their performance against set safety standards.

Safety attitude checks should not be entered into lightly because of the work study, 'spying' connotation they can be seen to have. However through careful training of observers and briefing of the workforce, this type of safety observation is a valuable health and safety tool, especially for organisations whose current standards are high but wish to move off a plateau by developing a future health and safety policy.

AUDITS AND INSPECTIONS IN PRACTICE

Application

Every organisation will have its own needs to fulfil when undertaking various audits or inspections and these will be crystallised into objectives and aims which can be translated into actual management practice. A summary of management practice in this area is set out below. It is not proposed that all these points are checked during each and every audit or inspection. Rather, it is suggested that from time to time they are considered and applied, modified or updated, as necessary.

Formality

Is the audit or inspection to be carried out in a formal or informal manner? It is important to set the tone of audits and inspections before they commence. This is not to suggest that informal audits or inspections are somehow less valuable as both approaches have a contribution to make. Many organisations will see the informal route to be the most applicable, as it permits easy communication between those involved. Teamwork between those carrying out audits or inspections will be easier, allowing for full and frank observations and assessments, which will benefit the health and safety of the area under examination.

However, it may be that at times formal or very formal audits or inspections may need to be undertaken. An advantage of this approach is that the audits can be subject to quantifiable measurements in rated and comparable formats. This is valuable for measuring an organisation's or a department's progress in health and safety from year to year. Formal audits or inspections can be an excellent managerial tool for removing subjectivity from the evaluation. However, they do require time and also to some extent more highly-trained people to perform the task. Formal audits and inspections can and do facilitate the effective channelling of performance results to senior management. Reports to senior management can be an advantage as such documents demand action (where necesssary) by the senior management team, and are much better than opinion, or messages through the grape-vine. Where considerable expenditure is required, reports at this level are effective in all but the most unresponsive of organisations. The employment of formal audits and inspections can be used to impose a company or site-wide standard audit or inspection. This is useful when an organisation is required to be aware of the current state of play across all the company operations or in the case of a major site across the various plants and departments within that site. This approach may arise from a professional managerial decision to undertake periodic and major health and safety audits or it may be as a result of a positive reactive decision. For example, a positive approach may be as the result of a recent fatal or serious accident where certain identifiable unsafe conditions were seen as the main cause, possibly with wider implications across the enterprise. It may be necessary to apply this formal approach as the result of an adverse letter from a health and safety or environmental enforcing authority, or simply to ensure all corners of an enterprise are adhering to company policy as regards new legislation.

Task areas

Specifying areas of plant to examine may seem simple to those who have carried out audits and inspections. However, the number of audits and inspections that do not fulfil expectations by task members is high because of a failure to clearly define the task area limits prior to the commencement of the audit or inspection. Even when single person audits or inspections are to be carried out, area limits should be detailed before starting out. Areas to be covered are subject to many variations such as the physical size of a department, plant, equipment or building or perhaps a works size or layout. More important, however, are the parameters to be applied within a specified audit or inspection area given that the geographical border lines are agreed. From the experience of those who carry out health and safety audits and inspections, an issue which is highlighted over and over again is how they can be made effective. It is suggested that this is very much linked to the issue of the audit or inspection parameters within the area.

These parameters can be best expressed as

overhead – underground – behind and outside

Is the audit or inspection to include examination of plant and equipment overhead ie above ground level such as roofs, overhead cranes, lighting gantries or other high areas? Will the audit or inspection include underground (accessible and inaccessible) parts of plant ie cellars, electrical basements, tunnels, subways, cable and service trenches?

Does the audit or inspection team intend to look behind plant and equipment, especially panels, locked doors and access ways? This of course should be undertaken only when accompanied by authorised persons. It could include store rooms, motor rooms, hazardous plant, pump rooms and welfare blocks. It will quickly be realised by audit and inspection team members that this 'behind' category is a rich source of health and safety problems as, by and large, these out-of-the-way places or restricted areas tend to be seen on a regular basis only by the few who have normal business there. Unfortunately, such people can become too close to the problems and may not see them, therefore frequent audits and inspections can do much to ensure that conditions in these locations do not become unsafe by default. The 'outside' category is similar to the 'behind' category in that it is a reminder to think of examining areas just outside building, departments or plant boundaries, but within the accountability curtilage of a given departmental manager or engineer.

Organisations with medium and large factories, plants, sites or workshops may find that considerable confusion can arise from the issue of accountability curtilages. The principle is that of total single ownership of any given area bounded by an obvious feature or curtilage such as the kerb of a road, a boundary fence or similar. In terms of health and safety accountability single ownership should prevail. So for example if a locked electrical sub-station is in use within a saw mill then the senior production manager through his electrical personnel should be accountable for it in terms of health and safety, even though the unit is the operational responsibility of some electrical director at head office. Identifiable health and safety ownership of industrial and commercial kit in many organisations helps avoid misunderstanding as to who should or must carry out audits and inspections in these grey areas. Such misunderstanding can, and does, cause injury, along with plant and equipment damage, as a result of audits and inspections not being carried out. Audits and inspections clearly targetted on areas normally perceived to be outside a manager's responsibility will assist in reducing accidents in those places.

Scope and depth

Effectiveness requires an audit or inspection that:

- is produced on time
- is uncluttered by non-health and safety issues
- reports clearly on items specified in the remit
- clarifies in its findings the practical action necessary to rectify any hazards
- specifies the unsafe conditions or unsafe operations which have been observed (not just a safety snagging list)
- reflects the views of the team or individual who carried out the audit or inspection.

Scope

In attempting to achieve an effective audit or inspection what should be the scope of the examination? First, it must be decided whether the audit or inspection will cover welfare aspects, health issues, safety problems and environmental items. Will it be all four, just one or a combination? In each case the necessary scope should be determined. This could be a general examination, a serious attempt at reducing recurring accidents, a specialist examination or a risk assessment exercise. The breadth of examination will emerge from the remit team.

Depth

Having first focused on the breadth of audits or inspections, depth should then be considered. In effect this is what can or should be observed and/or examined. The simplest, and therefore possibly the best way to consider this issue is to look at it under the following three headings:

- PEOPLE
- PLANT
- PROCEDURES

People

In the main, the people involved are the organisation's employees, regardless of status. Contractors might also be included, both resident and transitory, again covering the full range of occupations. Transport contractors' employees in particular could be included, for example, where there are truck, lorry or van delivery or dispatch operations. Visitors should be included. There are two main types, commercial visitors and guests. Commercial visitors include sales representatives, contractors' agents, insurance inspectors and consultants. Guest visitors could be school parties, the general public, works visitors or visiting foreign or UK management. Finally, under the 'people' heading, consideration should be given to trespassers (ie unlawful visitors) which should lead to audit and inspection of security control, gatehouse procedures, identity cards and boundary fence conditions. Figure 11.7 presents a list, not exhaustive, of people to be considered for inclusion in or exclusion from audits or inspections.

Plant

Plant is the term applied throughout most of UK industry and commerce to mean the kit and support units that produce the company's product. In the case of service industries it could include offices, canteens, construction sites, warehouses, docks, airports and many other things. Plant can be seen as tangible hardware for the purpose of profitable service provision or product manufacture. This hardware can be fixed, moving or mobile. Usually alongside the term plant in the field of health and safety will go equipment. Here the audit and inspection task team can consider matters such as: moving machinery; pres-

1. **Production**
   Labourers
   Operatives
   Machine drivers
   Crane drivers
   Semi-skilled
   Line servicers
   Control room operatives

2. **Maintenance**
   Electricians
   Fitters
   Civil tradesmen
   Boilermakers
   Machine tool craftsmen
   Technicians

3. **Management**
   Inspectors
   Supervisors/foremen
   Engineers
   Stocktakers
   Managers
   Team leaders

4. **Miscellaneous**
   Transport staff
   Service department staff
   Contractors and sub contractors
   Visitors
   Security officers
   Trainees/apprentices
   Office staff
   Functional staff
   Cleaners
   Agency staff
   Technical support staff

Figure 11.7 *People to be considered for inclusion in an audit or inspection (industrial example)*

surised systems; transport; mobile tools and plant; hand and power tools; lifting tackle; electrical installations; process plant; stores and storage areas; fire-fighting equipment; hazardous areas (eg where there is gas or chemicals); and workshops. Raw materials of all kinds, and means of service provision or product manufacture such as fuel and power supply, along with materials of production, should be included in any audit or inspection under the heading of plant. Further examples of plant and plant support units are listed in figure 11.8.

Assembly areas	Offices
Building sites	Pipe lines
Basements	Pump houses
Canteens	Roof areas
Cellars	Stock grounds
Cleaning systems	Stores
Confined spaces plant	Switch rooms
Control rooms	Transformer rooms
Construction areas	Transport depots
Demolition site	Towers
Dismantling areas	Tunnels
Furnaces and stacks	Warehouses
Flues and flumes	Waste tip areas
Garages	Workshops
High level areas	Yards
Motor rooms	

Figure 11.8 *Examples of plant and support units to be considered in audits and inspections*

Procedures

Bringing people and plant together is process, best examined in health and safety terms under the heading of procedures. A safety procedure in relation to audits and inspections is usually a document which links plant and people at the workplace. A procedure in industrial or commercial terms is a mode or method of business. A safety procedure, therefore, can be seen as a safe course of action associated closely to the business mode or method of production. It should be noted that a safety procedure could be an unwritten safe practice. There are four main aspects to consider when examining a safety procedure:

- *requirement* – is a procedure needed?
- *adequate* – does a standard procedure exist?

- *instruction* – has the procedure been communicated?
- *operation* – is the procedure being followed?

Requirement Is a safety procedure required to be in place for a given operation, job or task? This assumes a written safety procedure: an unwritten safety procedure may be in place through custom and practice. Establishing the need for a written safety procedure can be developed from or by undertaking a risk assessment.

Adequacy The next consideration when examining a safety procedure is to make an evaluation as to whether or not it is adequate for its purpose. More importantly is it the best method in terms of cost effectiveness and safety?

EXAMPLE

An engineering slinging accident in a heavy industrial maintenance situation caused a serious leg crushing injury. When investigated it was found that two other shifts had not experienced any problems with this operation. They always used three men as per the safety procedure. However, the shift which had the accident was only employing two men on the operation to allow the third man to carry out preparation work on the next task. He was not working in line with the safety procedure so therefore the best (ie safest) method was established but was not in practice on that one shift.

Instruction Do the appropriate members of the workforce, or contractors, know about the safe working procedures and practices in force? Have they been instructed, trained or briefed? This can easily be established by asking a small sample of people:

- whether they know about a particular safety procedure; and
- asking them to briefly outline it.

Operation Is the safety procedure being operated, especially when management or supervision are away from the task location? This point deals with the very important subject of individual attitudes as demonstrated by safety behaviour. Safety procedures come in a number of types and formats; written, printed, oral and computer screen displayed. They can be in the form of information, instruction, guidance, training or control directives. Examples of the more common ones are a safety working procedure, a safety briefing sheet, a safety standing instruction, a health and safety guidance note, a training manual, a safety method statement and a company health and safety directive.

The full scope of audit and inspection can be best understood by use and development of the simple matrix presented in figure 11.9.

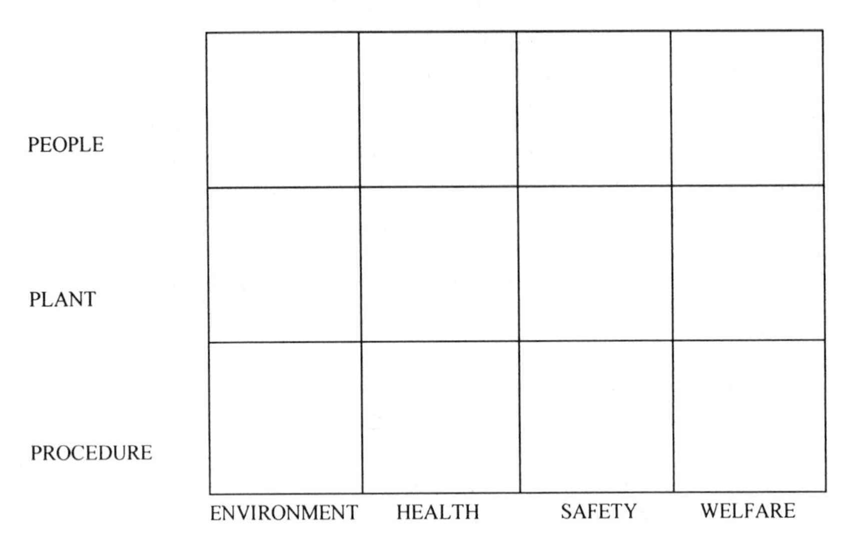

Figure 11.9 *Scope of audits and inspections*

Time factor

As all audits and inspections must be cost effective, it is important that they do not consume a disproportionate amount of time. Therefore, target duration times should be commensurate with the remit objectives and aims. Task teams or individuals should be made aware of this managerial parameter and oriented towards completing the audit or inspection on time. As health and safety must be a normal function of management, there is no reason why normal cost effective controls should not apply. But time allocation, especially when task teams are employed, needs to be made.

Another vital aspect of the time factor is the span of effective concentration or attention span. In the main health and safety audits and inspections need high levels of concentration. In some cases life and limb issues can be at stake along with potentially high costs or even disaster. Managerial planning of the time and duration of audits and inspection is critical, especially those with on-site requirements. The experience of many health and safety specialists is that for large operations four hours' duration is about the limit of the span of effective concentration without a break. Professional auditors and inspectors who are trained and experienced may be the exception to this rule. It is possible to continue for longer periods and organisations will do so, but depending on the industry there can be serious drawbacks. The *time v quality* equation is an important and usually little considered aspect of audits and inspections.

Team selection

When considering selection of people for audits and inspections, four factors should be considered:

- team size
- composition
- training
- briefing.

Team size

This will vary across industry. The balance is to have enough of the necessary skills to fulfil the audit or inspection remit while ensuring that the team can be effectively managed. It is of equal importance to ensure that some of the team members come from outside the area or conditions being examined. This helps to establish an objective point of view.

Composition

The make up of the team in terms of experience needs to be considered when more than a one-person audit or inspection is to be carried out. For example, should the team be all production staff or all engineering? Should it be line management only or safety specialists? The best results are usually obtained when teams are multi-disciplinary, coming from production, engineering, service and functional departments, have both managerial/supervisory staff and shop floor employees, and include trade union safety representatives. This type of mix should ensure all parts of the company's operations are examined in detail, thus avoiding the tendency to keep to visitor or senior management routes and to eschew examination of one's own and others' 'back yard'. The mix may also help to remove the embarrassment or reluctance to report on costly rectification or long-term health and safety problems 'in your own back yard'. If properly selected, team members can bring surprising inter-personal skills to play during an audit or inspection.

Individual qualities of team members should be checked out. The first check should be to determine whether they want to take part or have they been pressed into service. Do they have qualities to allow a balanced view on health and safety problems to be reported or do they have a score to settle? What sort of safety attitude will they bring to the task? Will they be looking for accident prevention issues that could affect anyone, or just those that affect their colleagues? Will their approach be proactive, looking for practical solutions to safety problems or will it be just a blacklisting of the problems? Will attitudes be business-oriented (ie cost effective, programme planned and realistic, with problem-solving contributions) or reflective of an approach which demonstrates minimal short-term input. Where only some of the necessary basic qualities are present in team members they can be built upon through training and briefing.

Training

Training for audits and inspections can, and should, be simple, short and cost-effective. A one-day, in-house course would in most cases be sufficient for inspections and three to five days for audits, with a follow-up refresher session every two or three years. Extra training could also be provided when major plant or conditions change, or when there are major changes in health and safety law which directly affect the organisation. The contents of this type of training could include:

- basic principles of accident causation and prevention
- major health and safety topics (hand tools – work at heights – electricity etc)
- the interface between people, plant, procedures
- overhead, underground, behind and outside issues
- local site health, safety and welfare policy
- fire prevention and
- environmental controls.

Personnel trained in audits and inspections can make all the difference in terms of effective feedback of information, data and solutions. A team of trained safety auditors or inspectors can transform an organisation's health and safety culture.

Briefing

The best use will be made of training if before each and every audit or inspection a short but concise briefing is carried out in the form of a team remit. This will set the overall objective of the audit or inspection and its key aims.

Reporting methods

Record of findings

There are three stages in the reporting phase of audits and inspections. In the initial stages of any type of health and safety audit or inspection someone will be required to keep a record of the findings. The usual manner is to keep written notes, either by individuals calling out items for the records to a central record keeper or by each person making records and handing them over to a records keeper for collation. Tape recordings could be considered instead or to assist with note taking, and for the more complex safety problem, consideration could be given to the use of a camera. It is always advisable to agree on who is to be responsible for the provision of the audit or inspection information at the start of each examination.

Final report

The next stage is to produce the audit or inspection information and records in a final report. This can be in various forms depending on company requirements and the type of audit or inspection undertaken. For example it could take the form of a typed report, ie:

- introduction
- remit details
- observations and examinations
- conclusions
- recommendations
- summary of key actions.

Presentation for audits and inspections can be on pre-formed standard layout sheet. These documents, both in text and standard sheet types, can form the permanent record of a particular audit or inspection.

Action-needed and action-completed mechanism

No audit or inspection system could be considered as being complete until the final and all-important stage is created, which is the action-needed and action-completed mechanism. All previous efforts will have been wasted if necessary actions are not identified and carried out. This function of the audit or inspection is a dynamic list of action points distilled from both the records and report stages. It presents in an easily understood and accessible manner the essence of that which needs to be done to rectify the unsafe acts, conditions and operations that were observed during the various stages of the audit or inspection. To close the loop and keep track of the change from unsafe acts, conditions and operations into safe situations, a tick-off system is necessary to confirm that the action needed has been carried out and completed.

Some organisations use computer systems to capture the changing information at relevant progress stages of this mechanism. The information could be recorded manually by the auditor on site and passed on for entry into the system, or scanned using an optical scanner. Alternatively, the auditor could type the information into a hand-held electronic data collector. The data could then be transferred electronically to the central computer on a periodic basis (eg daily/weekly). Once a record has been created, a printout of each item to be rectified could be produced, and passed to the person responsible for the work (eg engineer/manager).

Status reports could then be produced detailing items to be rectified, items which have been rectified in a specified period and items which have not been rectified by the target date. These reports could be produced and listed against a particular manager/supervisor/engineer or against a specific section of the site.

Reporting route

After the establishment of the reporting methods, the reporting routes and the scope of reporting should be determined. It should not be automatically assumed that all of the relevant managers will get copies of audit and inspection information or that they will check computer screens or be given oral information. To make full use of the information that can be collected and processed from audits and inspections, the reporting routes and the scope of reporting should be established as in any other management information system. Above all it must be a proactive system and be routed to the levels of management that have the resources to rectify the unsafe issues, or are empowered to obtain resources to carry out rectification (for example to draw stores, order spares, requisition contractor services or purchase equipment). The various functional sections within an organisation should not be overlooked in the scope of reporting. Unsafe acts by staff may require action by the training department so copies or relevant part copies of audits and inspections should be sent to the training department.

The practical rule of thumb on scope and reporting routes is to make sure audit and inspection reports are passed to the 'next-desk-up' in the management hierarchy, commensurate with proactive authority. In other words make sure that those who call the tune pay the piper. Management at all levels must know that health and safety audits and inspections raise matters that have resource implications and sometimes substantial ones. However, in practice many of the points arising from audits and inspections are related to rectification of equipment and in the main incur only modest expenditure.

Action authority

Following and closely linked to the reporting route is the need to ensure that action points are directed to the owner who has authority for rectification. In other words, it is necessary to direct matters to the manager, engineer or supervisor held by the organisation as being responsible for action on any health or safety topic highlighted by the audit or inspection. This will include repair, overhaul, modification, adjustment, procurement or implementation. The systematic inflow of audit and inspection summaries to managers in authority cannot be over-emphasised: it is the very source of proactive health and safety in any organisation. The resultant duties placed upon managers will need to be reinforced by senior management.

Remedial action to be set in train by managers must, where possible, be immediate and on the basis of total and permanent solutions. Therefore a simple but effective route is to make sure that accountable managers' names are allocated to each identifiable rectification item in an audit or inspection report with an agreed target date for completion. Communication and feedback are essential at all the post-audit or inspection stages. Implementation of health and safety audit/inspection control should normally progress in a systematic manner as with any managerial improvement programme. However, urgent action on any hazard likely to cause immediate and serious damage, loss or injury should not have to wait for a report. The general rule should be that all unsafe acts or unsafe conditions must be actioned there and then by members of an audit or inspection team if within their scope of competence and authority.

Progress review

One feature of health and safety audits or inspections often overlooked is continuous progress. Dates should be set for the audit or inspection team to review and check up on the action(s) necessary to confirm completion and 'close down' of a particular audit or inspection. This can only be done when all rectification items are accepted as completed. In the main, reviews can be carried out around a desk but at times could require revisiting the audit or inspection site or location. Action or rectification items may have to wait months for monies to be approved if considerable expenditure is needed. In these cases it may be that the audit or inspection can be closed by transferring the item or items to a long-term outstanding file along with other data from previous or future audits or inspections. A safety term often used to describe this file is a 'blacklist'.

Safety data referencing

A final operation that needs to be considered by everyone involved with audits and inspections is to carry out data reference comparisons. This will show trends, high risk areas, prevailing hazards and recurring health and safety problems across an organisation. Data-referencing will greatly assist senior management in planning long-term development programmes on health and safety. Policy decisions and strategic budgeting can also be derived from these exercises.

Cross audits/inspections

Most audits or inspections can be subject to a further and interesting dimension which can make significant in-roads into injury prevention, damage and loss prevention. Cross auditing (or inspecting) is a health and safety discipline whereby a safety team, from a department, works or section undertakes an audit or inspection of a different department, works or section from the one in which they normally work. In essence this would bring in new sets of eyes to measure standards and spot issues overlooked by the home team. This cross auditing/inspection principle provides an arena where centres of excellence can be observed and ideas and standards can be referred across established boundaries.

Summary of key points

The ten practical key points are:

1. Consider the formality of audits and inspections, set the tone from the outset.

2. Select and define the audit or inspection task area.
3. Establish the scope of the audit or inspection in terms of depth and breadth.
4. Allocate a time factor.
5. Select the team: consider experience, training and commitment.
6. Consider reporting methods and format.
7. Evaluate reporting routes.
8. Identify action authority.
9. Establish a review procedure.
10. Data reference the audit and inspection results.

CHECKLIST

Finally a set of key principles in checklist form is set out in figure 11.10 in order to summarise the common factors applicable to all health and safety audit or inspection information.

1. Identify the audits and inspections which are undertaken within the organisation. (This should include both internal and external sources.)
2. Ensure all audits and inspections have stated objectives, aims and practical remits.
3. Check the scope of audits and inspections are understood, along with standard of measurement.
4. Make sure all staff involved in audit and inspection are adequately trained in basic investigative techniques and are formally briefed before each audit or inspection.
5. Establish a senior managerial control system whereby audit and inspection results and action plans are developed, followed up, coordinated, monitored and communicated to all concerned, and accountability determined.
6. Manage practical and permanent solutions to audit and inspection results: avoiding recurrence must be the important management policy.

Figure 11.10 *Managerial implementation checklist*

Notes

1. *Report* of the Robens Committee, HMSO, 1972
2. *Blackspot Construction*, HSE, 1985
3. On The Pressure Systems etc Regulations, see chapter 16; on COSHH, see chapter 23
4. Fennell Report on the King's Cross disaster.

12

Systems of Safety Control

INTRODUCTION

The importance of systems of control in establishing a positive health and safety culture within an organisation was stressed in chapter 7. Some types of management control system, such as the safety inspection and audit or the arrangements necessary to manage contractors effectively, are reviewed in other chapters dealing specifically with those subjects. In this chapter several other types of safety control systems are described but firstly the concept of a safe system of work is examined.

Providing safe systems of work

There is a general duty placed on the employer under s2(2)(a) of HSWA to ensure 'the provision and maintenance of plant and systems of work that are, so far as reasonably practicable, safe and without risks to health'. The duty for employers to provide and maintain a safe system of work is one that has been a long-established principle of common law (see chapter 4). Although the law requires a safe system of work, it does not define what the term means.

A safe system of work can be defined as the work method resulting from an assessment of the risks associated with a task and the identification of the precautions necessary to carry out the task in a safe and healthy way.

Although this chapter mainly deals with the need for safe systems of work in an industrial context, the requirement applies to all types of work activity. For example, in managerial or professional work the risks created by volume of work, excessive hours or insufficient support facilities must also be addressed (see the case of *Walker v Northumberland County Council* in Chapter 4).

The degree of formality necessary in establishing safe systems of work will depend upon a variety of factors including:

- the level of risk involved

- how frequently the task is carried out
- the complexity of the task
- the variability of the task
- the capabilities of the employees concerned
- the complexity of the precautions required.

Some tasks will necessitate the preparation of written task procedures or the use of permit to work systems of the types described later in this chapter. However, in other cases safe systems of work can be established informally through custom and practice, providing this is based on sound health and safety principles. In some cases a short term system of work may need to be identified through an instant assessment of the task in hand.

The quality of both formal and informal systems of work should be assessed as part of the overall risk assessment programme (see chapter 8). An assessment may identify the need for the greater formalisation of procedures for carrying out certain tasks. Risk assessment should also take account of the capabilities of those employees who, to some extent, may be expected to devise safe systems of work themselves. This might be necessary for example in a 'one-off' repair task, where a written task procedure would be inappropriate.

Training is closely linked to the establishment of safe systems of work. Where formal procedures exist, employees must be given adequate training so that they can follow the procedures correctly. However, training is also essential where employees are expected to play a part in devising safe systems so that they are properly capable of assessing the risks of the task and identifying the necessary precautions.

Maintaining safe systems

The employer's duty is to provide and maintain safe systems of work. There is little value in having a safe system that exists in theory (or even on paper) if it is not followed in practice. Part of the risk assessment process is to check whether control measures are actually being implemented. However, management also has a duty to monitor or measure what is happening in practice on an ongoing basis. Supervision plays a key part in this process but there is also a need for other systems (such as the inspections and audits described in chapter 11) to ensure that supervision is effective and that safe systems of work are maintained.

WRITTEN TASK PROCEDURES

The need for written procedures

Written procedures are more likely to be required where the risks connected with the task are high. In addition to health and safety risks, there may be other risks, such as equipment or product damage or environmental pollution. The effort involved in producing a written procedure is much more likely to be justified for those tasks which are performed regularly, but some non-routine tasks may still involve such a high degree of risk as to also require written procedures. Some non-routine tasks, especially those in the maintenance field may involve a significant degree of complexity or variability and may best be dealt with using the type of isolation or permit to work procedure described later in the chapter rather than the step-by-step approach described here.

In circumstances where an overall task procedure does not exist, there may still be justification for documenting the necessary safety precautions. This is often described as a safe working procedure. However, the development of separate written procedures dealing with safety, operational aspects, quality etc should be discouraged in favour of the more integrated approach of the written task procedure.

What is a written task procedure?

A written task procedure is a step-by-step description of how to carry out a task correctly. It should be a 'one-stop' document taking into account all of the various aspects connected with the task.

The purpose

Rather than being simply a series of operational instructions, the procedure should contain a description both of the overall purpose of the task and, where relevant, of the individual steps within it.

Operational instructions

The necessary operational instructions need to be included in the procedure in a logical sequence. These should contain sufficient detail to prevent mistakes being made by those who are relatively unfamiliar with the task eg newcomers or those out of practice. Examples of the type of detail which might be needed are:

- locations of isolators
- clear designation of controls (by symbol or label)
- sources of necessary materials or equipment
- precision with parameters eg "when above 40°C" rather than "when warm enough".

Health and safety aspects

These aspects of the task should have been identified during either a general risk assessment or a specific assessment under relevant regulations. Where possible the health and safety aspect should be covered at the appropriate point in the operational sequence eg the employee should be instructed to take care when mounting the platform in case it is slippery. Some precautions may be necessary throughout the task eg an instruction that high visibility clothing must be worn at all stages because of the risks from passing traffic. There should always be some explanation of the reason for the precautions – the 'know why' as well as the 'know how'.

Quality considerations

Quality aspects of tasks, like health and safety aspects, often tend to get separated from other considerations with quality control staff sometimes developing their own procedures. Quality issues too should be fully integrated into a single task procedure. Aspects which may need to be covered include:

- raw material quality
- component quality
- process parameters
- product appearance
- product specifications
- hygiene standards.

There is often scope for considerable losses to be incurred if the employee carrying out the task is insufficiently aware of these quality considerations and the steps he needs to take to avoid or overcome possible problems.

Equipment damage prevention

Equipment used in carrying out a task, whether process equipment, tools or vehicles may be damaged if the task is not carried out correctly. A vehicle engine could be wrecked by a failure to check the levels of cooling fluids or lubricants, or a metal holding furnace could be blown apart if damp material is fed into it. Such incidents would obviously also have safety risks associated with them.

Environment and waste

The opening of the wrong valve, or the right valve at the wrong time, could result in a major pollution incident as could the failure to operate a process within the correct parameters of temperature or pressure. Clearly such considerations need to be fully taken account of in developing task procedures as does the safe disposal of waste products.

Security

Raw materials, products or the tools or equipment used in carrying out a task may be attractive to thieves and, where relevant, these aspects need to be addressed in task procedures. Some processes or products may be commercially sensitive or there may be a need to keep the public out of certain areas because of a general risk of thefts or unspecified risks to their health or safety.

Although by no means essential from a health and safety viewpoint, there is no reason why a written task procedure should not be subject to the full rigours of the BS 5750 approach. This is likely to have the additional benefit that the procedure is reviewed and, where appropriate, revised on a regular basis.

Figure 12.1 contains a specimen of a written task procedure that includes many of the features referred to in this sub-section.

Writing task procedures

The process of writing task procedures may be divided into five stages:

- research
- field work
- drafting
- revision
- issue

Research

It is important that those involved in writing the task procedure understand the task, are aware of its context within the overall activities of the organisation and know something of the environment in which it is carried out. A good starting point is to examine any procedures which may already exist, even if these are acknowledged to be out of date. Other documents may also provide useful information. These might include:

- engineering drawings
- manufacturers' manuals
- quality control documents
- training resource material
- assessments
  - general risks
  - COSHH
  - noise
  - manual handling
  - DSE
  - PPE needs
- safety rules
- accident records
- environmental standards
- waste requirements.

Fieldwork

An effective way of conducting fieldwork is for an insider and an outsider to work together. The insider knows his way around and what is normal and

TASK Preparation of base solution **DATE** 3 February 1995 **AUTHOR** K. Jones

STANDARD PPE REQUIRED Protective overalls
Safety footwear
Hearing protection

STAGES/STEPS	KEY POINTS	SAFETY POINTS
STAGE 1 MATERIAL PREPARATION		
1.1 Ensure mixing vessel is clean	This avoids product contamination	
1.2 Check surrounding area		Remove tripping hazards, clear up spillage
1.3 Check sufficient base constituent bags are available	See Product Specification document	
1.4 Check waste receptacle is in position	Replacements available from stores	
1.5 Lift each bag onto the edge of the vessel		
1.6 Slit the bag with a knife and tip the powder into the vessel	Only use the standard knife	Lift carefully, bending the knees Cut away from your body
1.7 Fold the empty bag and place in the receptacle		Particle dust mask may be worn but not essential
1.8 Continue until the correct quantity is in the vessel	See Product Specification document	Do not leave bags on the floor
1.9 Open the water valve (turn anti-clockwise)	Blue valve above vessel to right	
1.10 Add 1000 litres of water and close the valve (turn clockwise)	Check digital flow meter on water line	
STAGE 2 MIXING		
2.1 Switch mixer isolator to 'ON'	Isolator on the wall to the left	
2.2 Press impeller 'START' button	Labelled button on control panel	NEVER ADD MATERIALS WHEN THE IMPELLER IS MOVING
2.3 Mix for the designated period	See Product Specification document	
2.4 Press impeller 'STOP' button	Labelled button on control panel	
2.5 Check solution is clear	Mix for 5 more minutes if cloudy	

Fig 12.1 *Task procedure*

abnormal. The outsider can question the status quo more easily and can compare it with his experiences elsewhere. Observation is a very important part of the process, especially in identifying risks with which those carrying out the task may have become too familiar. It is almost impossible to prepare a task procedure without having thoroughly observed the task being performed. Extensive notes will need to be taken, use of a clip-board being essential.

In some cases it may not be possible to see part of the task or it may be over so quickly that it is difficult to comprehend. Here, as in other parts of the process, it is important to talk to the person carrying out the task. Operatives should be invited to describe their activities or explain why they perform certain actions. Questions should preferably be open-ended ie asking what, why or how?

Most people are only too willing to talk about their work to others who are interested and sometimes their comments can be very revealing, particularly in relation to what happens during breakdowns or other abnormal conditions.

Drafting

A first draft of the task procedure should be made very soon after the field work has finished. It may be possible to draft the procedure in stages whilst still carrying out the field work. If too long a time is left before drafting then not only will some of the detail have been forgotten but there may be difficulty in understanding the notes which were made.

A good rule is to draft a procedure that can be understood by a newcomer, since it is the newcomer and the person training him who are going to make the greatest use of the procedure. Many procedures written by experienced operators miss out important detail because the writer is so familiar with the task – more reason for involving an outsider in the process.

One danger to avoid is of making the procedure instantly obsolete by quoting details that change regularly. Some product formulations or process parameters may vary frequently and here it will be necessary to make a cross reference from the overall procedure to the documentation in which these variables are contained.

The next part of the process is to get the draft procedure reviewed by others who can make a contribution. Managers will need to be involved in this, but the comments of those who are more likely to make direct use of the procedure – supervisors, team leaders, trainers, operators – may well be much more important.

Constructive criticism should be invited – people are often reluctant to criticise for fear of hurting the author's feelings. Reviewers should be encouraged to realise that by offering criticism they can help the author to do a better job and help themselves by having a better procedure available.

Revision

Those drafting procedures should not be discouraged by the comments received, whether oral or written. A draft returned with extensive annotations shows that someone has read it thoroughly and has been interested enough to offer suggestions for improvement. For anything other than simple matters it is best to review comments face to face. This ensures the comments are fully understood and provides the opportunity to test out revisions of the text. It may also be necessary to go out and do some additional field work or research.

Once the revised version has been prepared it will probably need to be circulated again for comments and these may be sufficient to justify a further circuit of the 'revision' loop. Frustrating though this may be for the author, consolation can be drawn from the fact that people are still interested and that a better procedure should result.

Issue

When the final version is ready it will need to be issued and the organisation's own document control procedures may well have something to say about the way this happens. The procedure will be another wasted piece of paper if it is not effectively communicated to those with a direct interest in it, particularly those carrying out the task. There may well be some training implications arising from the new procedure.

EQUIPMENT ISOLATION

There are many circumstances where, in order to ensure a safe system of work, it is necessary to isolate equipment so as to separate it from sources of energy or to prevent the flow of materials. Some of the general principles of isolation are discussed below, although electrical isolation is dealt with more specifically in chapter 20.

Against what dangers?

Electricity is just one source of power and isolation may be necessary to prevent danger being created by all possible sources of power eg mechanical, hydraulic or pneumatic. These power sources could be responsible for creating dangers of a mechanical nature eg from moving machinery or they could provide the motive force to release materials which could cause dangers in other ways, for example:

- hazardous substances
- pressurised liquids or gases
- very hot (or very cold) materials
- water, creating a drowning risk.

Other types of isolation may also be required – for example the closing of valves to prevent the flow of liquids or gases in particular directions.

If equipment is not effectively isolated then there is always the possibility of it being unintentionally activated:

- by the operator, accidentally
- by another person, unaware of the implications or even maliciously
- by some other means eg automatic and semi-automatic processes, remote relays, magic eye devices etc.

Switching off

Simply switching off the source of power may be an acceptable level of protection in circumstances where the risk is relatively low and there is little or no risk of unintentional activation. This might be the case where the operator is going to remain at the controls and there is no possibility of the equipment being activated by remote means. Ultimately it is for the employer to decide for what tasks and in what circumstances switching off a power source provides a sufficiently safe system of work.

Good isolation practice

Secure isolation

The isolation must be secure. There must be no foreseeable risk of mechanical or electrical failure (the possibility of short circuits normally renders control circuits unsuitable for secure isolation). Additionally there must be no foreseeable risk of the isolation being reversed by human hand – whether deliberately or unintentionally.

The posting of caution notices in positions where power sources have been isolated or valves closed is certainly beneficial; however notices alone provide nothing like the security of physical locks.

Locks

Each lock used for isolation purposes should have a single unique key – the use of duplicate or master keys is a recipe for disaster. The physical configuration of the item to be isolated (eg a valve) may necessitate the lock being applied using a chain or strap.

Where more than one person is involved in the work it is good practice for each to apply his own lock to the isolation using a multi-lock hasp which can only be removed when the last person has removed his own lock. If an individual is clearly responsible for a group of people carrying out a task then it is acceptable for him to apply an isolation lock on behalf of his group, although he must then accept responsibility for ensuring that they are all clear of any danger areas before the lock is removed.

The removal of isolation locks by burning them off is a practice that should be strongly discouraged. It may however be permitted in certain circumstances but only by a decision at a relatively high level and after stringent checks have been made to ensure that nobody will be in danger.

Where the source of power is electrical then isolation by fuse removal or plug withdrawal is likely to be acceptable providing the precautions described on page 470 are taken.

Proving the isolation

In the case of purely electrical work, isolations should be proved at relevant points by the use of a suitable test instrument (see page 470). In principle isolations of other types should also be proved in practice, eg by attempting to start the isolated equipment. However, there are some practical considerations which must be taken into account. For example there may be no means of actually proving the isolation or the act of proving may be potentially damaging to the equipment, eg by pressurising it against a resistance. In such cases there is really no alternative to simply double checking the isolations to ensure that they have been made effectively *and at the correct point*. It may be preferable to use a different person to make this second check.

Where the isolation is proved this must always be done on the assumption that the equipment *may* not have been correctly isolated. No one should be in a position of potential danger should the equipment unexpectedly operate during proving and special precautions may be necessary to prevent or contain any possible release of hazardous materials should the isolation prove ineffective.

Permits to work

The process of isolation may be so complex or the need to check that it has been carried out satisfactorily may be so great that the use of a Permit to Work is necessary. This subject will be dealt with in the next main section of this chapter.

Who should isolate?

The person carrying out the isolation must obviously be competent for the purpose. Factors to take into account in assessing this competence are:

- knowledge and experience of the type of equipment
- awareness of the hazards involved
- knowledge of the location of the equipment and its isolation points
- physical capability to carry out the isolation
- knowledge of safe methods of proving the isolation
- a positive attitude to health and safety.

Knowledge and skill of both a general and a specific type are clearly necessary and even a very experienced person may not be immediately competent to carry out isolations in an unfamiliar environment. Many organisations require employees to satisfy someone in authority as to their competence before allowing them to make isolations.

PERMITS TO WORK

A permit to work system requires formal written permission to be given before work can start on certain tasks. It provides a rigorous means of ensuring that an assessment is made of the risks present in the task and that the necessary precautions are identified and implemented. It has additional benefits in encouraging communication between the parties involved in the task.

Why use a permit?

There are a number of reasons why it may be necessary to introduce the formality of the permit to work approach.

The scale of the risks

The risk of death or serious injury involved in the work to those carrying it out or to others may be particularly high. Examples of such situations are given below.

Unusual risks are present

The risks may be of a relatively unusual type, outside the normal knowledge and experience of those carrying out the work, eg from the presence of radioactive sources.

Complex isolations are necessary

There may be the need to isolate a variety of power sources, services or supplies, possibly at a number of different positions.

Special precautions are required

Atmospheric testing may be necessary to check for the presence of toxic or flammable gases or vapours or the absence of oxygen. Use of specialist equipment such as harnesses, crawling boards or alarm devices may be required.

Unfamiliarity of personnel

Those carrying out the work may be unfamiliar with the work environment and the risks associated with it. This is particularly likely to be the case for contractors who may need a detailed briefing before starting work. Such persons may not themselves be capable of taking the necessary precautions eg making isolations. (On the management of contractors see chapter 15.)

Where permits may be required

The types of situations where permits to work may be required include:

- electrical work, especially at higher voltages
- work on larger, more dangerous machines
- work on overhead cranes and their tracks
- entry into confined spaces
  eg vessels, pits, silos
- presence of flammable gases, liquid or dusts
  (possible risk of ignition by hot work, electrical or electrostatic sources)
- presence of toxic substances
- possible uncontrolled release of substances
  eg process materials, water, steam, gases
- pressure testing
- work involving ionising radiation or lasers
- situations of difficult or dangerous access
- excavation work
- major construction or demolition activities
- lone working in hazardous environments.

Designating permit situations

Employees and contractors alike need to be made aware of those types of work for which permits to work are required. In the case of contractors, some organisations exercise very close control over access to their sites (possibly using a type of permit) to ensure that these situations are highlighted before work starts.

There are a variety of means available for designating in advance those situations where permits are necessary. These include:

- inclusion in written task procedures
- references in safety handbooks
- signs and notices
  — on relevant equipment or in relevant areas
- information, instruction and training.

However, even in well regulated workplaces, it will not be possible to designate in advance every situation where a permit is necessary. All employees, but particularly those in management positions, should be alert to the possibility of new tasks emerging with the type of characteristics that justify the use of a permit to work.

Where permits are issued on a regular basis for identical or similar types of work it is desirable for the permit issuer to be provided with a checklist of standard isolations and other precautions that are required in that situation to avoid the possibility of important safeguards being omitted.

The issue and cancellation sequence

The sequence described below is a fairly typical one. Although slight variations may be possible, some of the steps must be taken in the order described, otherwise the integrity of the system may be prejudiced.

Request for issue

This may be made by the person directly in charge of the work although in some cases (eg where the work is to be done by contractors) it may be made by the person arranging the work.

Clearance is obtained

Some systems require clearance to be given by the person in charge of the area where the work is to be done. This is particularly common where maintenance staff issue permits and clearance has to be obtained from production. Such a step ensures that production are aware that the work is taking place and avoids the premature isolation of services required for production activities.

Isolations made/precautions taken

All the necessary isolations are carried out and other precautions taken (eg atmospheric tests are made, visual inspections carried out and necessary equipment obtained). Although the permit issuer may not carry out all these tasks personally *he is responsible* for ensuring that they have been carried out satisfactorily. All of these precautions must be taken before the permit is issued.

The permit issuer completes the permit form

Details of isolations made and all other precautions which have been taken are entered on the form. The permit form should contain details of relevant restrictions such as the type of work permitted, the equipment to be worked on or the areas where the work may be carried out. Warnings may also be given about other work taking place in the area. The permit should not yet be formally issued.

The permit holder signs

The person in charge of the work (the permit holder) signs all copies of the permit. In so doing he acknowledges that he and those working for him will abide by the terms of the permit, particularly in abiding by any restrictions imposed by the permit.

The permit is issued

The permit issuer signs all copies of the permit to confirm not only that all the precautions described have been taken, but also that he is satisfied that the permit holder understands and will abide by any restrictions on the work. If he is not confident in the permit holder's understanding or attitude then he should not sign.

Normally there are at least two copies of the permit – one being given to the holder and one retained by the issuer. Some systems also require a copy to be put on display and/or one to be given to the person in charge of the area where work is taking place.

Termination of work

When the work is complete or must terminate for other reasons *all* copies of the permit are brought together again. (The situation where all are not available is considered later.)

The permit holder cancels

The permit holder signs all copies of the permit to cancel them. In so doing he confirms that his work has terminated and that all the people and equipment under his control are clear of the area. He may also be required to confirm that he has left the area in a safe condition. It is possible that in some situations the permit holder may not be the same person to whom the permit was originally issued. (This possibility is considered later.)

An authorised person cancels

The permit issuer (or another authorised person) signs all copies of the permit to permit removal of isolations and the cessation of other precautions where relevant. Isolations must not be removed before the permit is cancelled, otherwise someone may enter a danger area in the belief that the permit is still in force.

Normal activity resumes

Once isolations etc have been removed and a check has been made of the general safety of the area or equipment, normal activities may be allowed to resume. This may be covered by a specific section of the permit form or may be communicated informally.

Design of the permit to work form

The design of the permit to work form will obviously need to match the needs of the organisation using it. There are a number of objectives which must be borne in mind when designing a form. A sample form is illustrated in figure 12.2.

Logical sequence

The form layout should encourage users to go through the various stages of the issue and cancellation process in a logical and safe sequence.

Clarification of responsibilities

Permit users should be absolutely clear on the responsibilities they are accepting as they sign at each stage of the permit.

Guidance on precautions

The permit issuer should be given guidance on what services may need to be isolated and what other risks may exist, necessitating other precautions. However, they should also be encouraged to think about other additional precautions which may be required.

The permit issuer

The qualities required of the permit issuer (and the person cancelling the permit) are similar to those of persons carrying out isolations. Before authorising employees to issue permits, organisations should ensure that they have the following capabilities:

- an appreciation of the reasons for permits to work
- a clear understanding of the permit issue and cancellation sequence
- knowledge and experience of their area of authorisation
  - — equipment and risks within it
  - — permit to work applications
  - — relevant isolation points
  - — other relevant precautions
- capability to carry out or supervise isolations
- knowledge of safe methods of proving isolations
- an awareness of problems which might arise and how to overcome them (see below)
- a positive attitude to health and safety.

Authorisation to issue and cancel permits should normally be in writing and should be given only when the individual's knowledge and practical capabilities have been fully assessed to be at a satisfactory level. Such an assessment might typically be made by a panel drawn from:

- senior engineering or production staff

PERMIT TO WORK Serial No. 00001.

DETAILS OF PROPOSED WORK		CLEARANCE	
Intended start date	Intended start time	Expected duration	Clearance is given for this work to take place Signature Position Date
CHECKLIST	SERVICE	ISOLATIONS MADE	
ELECTRICITY			
STEAM			
WATER			
GAS			
COMPRESSED AIR			
HYDRAULICS			
OTHER SERVICES			
CHECKLIST		OTHER PRECAUTIONS	
CONFINED SPACE RADIATION FLAMMABLES ACCESS TOXIC SUBSTANCES PPE LACK OF OXYGEN OTHER			
RESTRICTIONS ON WORK			
ACCEPTANCE	ISSUE	WORK TERMINATION	CANCELLATION
I accept this permit to carry out the work described. I understand the restrictions described above and all persons under my control will abide by them.	I confirm the isolations and other precautions described above have been carried out. I am satisfied the permit holder understands the restrictions on the work. THE PERMIT IS ISSUED	The work described above has terminated. All persons and equipment under my control are clear of the area which has been left in a safe condition.	THIS PERMIT IS CANCELLED. All isolations may be removed. Other precautions may cease. Normal activities may then resume.
SIGNATURE	SIGNATURE	SIGNATURE	SIGNATURE
PERMIT HOLDER	PERMIT ISSUER	PERMIT HOLDER	AUTHORISED PERSON
Date Time	Date Time	Date Time	Date Time

Figure 12.2 *Permit to work*

- specialist engineers
- health and safety specialists.

An authorisation to issue permits should never be simply something that 'comes with the territory'. A fresh assessment and authorisation is likely to be necessary even when an experienced permit issuer moves to an unfamiliar area.

Common mistakes and problems

There is plenty of scope for mistakes to be made and for problems to arise during the operation of a permit to work system. Since permits are generally used as a means of ensuring safety in high risk situations it is important that mistakes are avoided and that problems are dealt with in a systematic and responsible way.

Not considering all the risks

The permit issuer must give sufficient thought to the full range of risks that may be present in the work and the precautions necessary. The use of well-designed permit forms and standard lists of isolations and other precautions will help overcome this problem.

Issuing outside authorised areas

The permit issuer should never yield to pressure to issue a permit for an area or an activity outside the scope of his own authorisation. He may not have the specific knowledge necessary to identify all of the relevant risks and precautions in an unfamiliar situation. Even when operating within his authorisation the permit issuer should be aware of his own possible fallibility or lack of experience and should seek a second opinion in cases of doubt.

Failing to check precautions

Although the permit issuer may not carry out all of the isolations or take all of the other precautions necessary, he is responsible for ensuring that they are satisfactory. To sign a permit form without checking not only negates a large part of the value of the permit to work system but would also be a major omission in respect of the issuer's legal responsibilities.

Problems with the permit holder

There may be situations where the person in charge of the work (the potential permit holder) does not demonstrate a sufficient understanding of the restrictions and requirements being placed on him by the permit. The permit issuer may also not be satisfied about his overall attitude to health and safety and the permit system in particular. Clarity both in writing the permit and in explaining its contents will assist but if the permit issuer is not satisfied then he must not issue the permit. In the case of contractors it may be necessary to send for someone else with a higher level of understanding or a greater sense of responsibility.

Where monitoring of work being carried out under a permit demonstrates that a permit holder (or someone working for him) is working outside the terms of the permit or is abusing the permit system in other ways then very positive action must be taken.

In the case of employees this should be a disciplinary matter whereas for contractors either a ban from site or at least from permit work should be seriously considered. (Similarly a permit issuer operating irresponsibly should expect serious disciplinary action against him.)

Transferring permits

Some permit work may last for a considerable period and require more than one person in charge. It is important that the permit holder at a particular point in time is clearly identified. Some permit forms incorporate sections enabling this to be done. If this is not the case then four columns could be ruled on the reverse side of the holder's copy of the permit and entitled Date, Time, Holder's Name, Signature. This then enables the permit issuer to identify at a glance who is the person in charge at that time. He will need to be satisfied that all the possible permit holders have the necessary understanding and attitude.

Inadequate cancellation

A failure to collect in all of the copies of the permit form for cancellation or the absence of a signature on the cancellation section of the form could result in workers entering danger areas in the belief that the permit is still in force. *All* copies of the permit must be *signed* as cancelled before isolations and other precautions are removed.

Missing permits or permit holders

Even in well regulated organisations important documents go missing and a copy of a permit may disappear. If it can still not be found after appropriate checks then all parties concerned need to be advised of this. During the cancellation sequence a note should be made on the remaining copies that the missing copy has been lost. Depending upon the circumstances there may be a need to support this by the use of notices (either on noticeboards or on relevant equipment) in case someone is still in possession of the lost copy.

There is also the possibility of the permit holder disappearing before the permit can be cancelled. This is particularly likely to be a problem with contractors. Firstly stringent checks should be made to confirm the holder is not still at work or even present on site. Then attempts should be made to recall the holder – an inconvenient recall will often act as a major deterrent to future offences.

If the holder cannot be recalled then the issuer has little option but to cancel the permit without him, making relevant notes on the holder's section. He should normally refer such cases to his immediate superior before proceeding in this way.

Where the holder's copy of the permit has also disappeared, subsequent efforts should be made to obtain its return for cancellation or at least its destruction. Back-charging contractors for the administrative time involved in all of this additional work is another means of avoiding further offences.

Testing

Sometimes a permit to work may need to be removed for a short time to allow testing or similar work to take place. Where this is a common feature, permit forms can be designed to take account of testing needs and some companies use separate sanction to test forms. Where this is not the case then cancellation and re-issue is the most watertight way of proceeding. However, particularly if the issuer is able to physically supervise the work area during testing, it may be acceptable for him to temporarily collect in all the copies of the permit and check personally that the work area is clear.

Whatever the problems that arise, both the permit issuer and the permit holder (and their respective employers) have clear legal and moral duties to maintain the integrity of the permit system. It is important that employers regularly monitor the system's effectiveness. The breakdown of a permit system in 1992 resulted in a major brewing company being fined £75,000 and their technical engineering manager (the permit issuer) £5,000. More importantly it also led to the death of an employee of an electrical contractor.

TRAINING AND AUTHORISATION

Training is essential if employees are to adopt safe systems of work, a need that is underpinned by legal requirements.

Legal obligations

S2(2)(c) of HSWA states that an employer's duty includes:

> the provision of such information, instruction, training and supervision as is necessary to ensure, so far as is reasonably practicable, the health and safety at work of his employees.

This duty to provide 'information, instruction, training and supervision' is repeated in many sets of regulations. Regulation 11 of the Management Regulations also requires employers to:

- take into account employees' capabilities as regards health and safety

- provide employees with adequate health and safety training.

There are also many other specific training requirements contained in a variety of acts and regulations, some of which are referred to in chapter 7.

Training needs and objectives

Training needs analysis

Organisations must first identify what their health and safety training needs are. Some may be specific to health and safety but in other cases these aspects will be fully integrated into training in task procedures generally. Consequently health and safety training needs are best established as part of an overall training needs analysis.

Health and safety training needs can be categorised under a number of headings including:

- general induction
- general procedures
- task related
- general skills
- changes (equipment, technology, systems of work, legislation)
- managers and supervisors
- health and safety specialists.

Defining objectives

The objectives of the proposed training activity need to be established and these are increasingly being specified in terms of competencies and performance criteria. Trainees need to be able to demonstrate that they can carry out the task safely and competently rather than that they have attended a training event.

In some cases the competencies and performance criteria are set out in legal or quasi-legal documents such as the ACOP for the training of fork lift truck drivers.[1] However, increasingly they are forming part of the framework of NVQs. Some NVQs provide very clear guidance on the health and safety related competencies to be demonstrated and performance criteria to be met whilst others are rather vague and woolly.

In these cases and for those tasks and situations not covered by the NVQ system, the training organisation will need to set down its own performance criteria which must be met to demonstrate competence in the task or activity.

Training methods

The training methods adopted will to a large extent depend upon the objectives of the training activity. Classroom-based training is more likely to be suitable for developing the trainee's knowledge whilst an actual or simulated workplace is likely to be required to develop skills. In either case the training should be planned systematically to achieve the intended learning outcome.

Care is also necessary in the selection of who is to carry out the training. A knowledgeable and capable performer does not necessarily make a good trainer. Training is more likely to be effective where there has been investment in training the trainers.

Classroom-based training

Most classroom-based training involves some form of lecturing but this can be made more effective by the use of relevant visual aids (slides, videos etc). The introduction of group discussions, syndicate exercises and similar methods of involving participants can also aid the learning process.

Workplace training

Whilst workplace-based training is principally concerned with developing skills, the trainee must first understand the context in which those skills are to be performed. This includes gaining an awareness of any health and safety risks present and the precautions that need to be taken. Most types of training should then proceed through the following cycle.

The trainer should describe how the task (or the stage of the task) is to be carried in a simple logical sequence. He should then demonstrate the task fully, describing what he is doing as he carries out each step and highlighting any key points. The process should then be repeated but this time with the trainee first describing the task and then carrying it out whilst giving a running commentary. The trainer should provide constant feedback, correcting verbal errors and preventing physical errors being made, particularly if these could result in injury, damage or loss.

Once the trainer is satisfied that the trainee understands the task and can carry it out safely, he can allow the trainee to improve his skills by practising under supervision. Again, feedback should be given on sub-standard performance and errors corrected or prevented as far as possible.

Assessment and authorisation

A formal assessment will often need to be made in order to demonstrate that the trainee has achieved a satisfactory level of competence. Such assessments are usually necessary to meet the demands of the NVQ system but this may also be required by in-house standards. Some legislation and ACOPs also require the assessment of competence eg in the case of fork lift truck drivers.[1] The assessor (who could be the trainer or a third party) may have to satisfy specified qualification requirements.

The form of the assessment will depend upon the nature of the competence to be assessed. Knowledge will usually be assessed through written tests or orally whilst skill will be assessed through practical demonstration. In some cases the assessment will need to follow a pre-determined format. It may also be necessary to assess the trainee's capabilities in dealing with abnormal situations as well as normal conditions.

Those employees assessed as competent may then be formally authorised to carry out the relevant task. For some tasks (eg power press setting or abrasive wheel mounting) formal authorisation or appointment by the employer is a legal requirement. This is also the case for fork lift truck driving and the issuing of licences to drive other types of plant vehicles is strongly recommended. The importance of authorising employees to issue permits to work or carry out isolations was stressed earlier in the chapter.

Records

Records should be kept of all training activity, whether in the classroom or the workplace, and of assessments of trainee competence. The maintenance of training records is sometimes a specific statutory requirement but, in any case, the existence of adequate records provides proof that the many legal obligations to carry out training have been met. Good records will also be important in the effective analysis of future training needs.

PURCHASING AND ENGINEERING CONTROLS

A satisfactory level of control must be exercised in respect of the health and safety aspects of the purchase of equipment, materials and services by an organisation or of changes to its existing equipment and processes.

Hazardous substances

Chapter 23 deals with employers' obligations under the Control of Substances Hazardous to Health (COSHH) Regulations 1994 to carry out assessments of the risks involved in the use of hazardous substances and the control measures necessary for their continued safe use.

An important link in this process is that between the purchasing staff who place the orders for potentially hazardous substances, those requesting the order and those responsible for carrying out COSHH assessments. Whilst purchasing staff are not expected to have the detailed knowledge of health and safety specialists they should have a general awareness of hazardous substances and of the types of product about which they may need to seek more information.

Whenever a new product is to be purchased there are several questions which need to be asked.

What is the substance?

In many cases products use trade names and enquiries need to be made to establish what their chemical constituents are. Under s6 of HSWA suppliers are legally required to provide information about the contents of their products, although not necessarily the precise formulation.

What are the hazards associated with it?

The manufacturer's or supplier's data sheets are likely to provide this information on these although there may be the need for recourse to other sources of information as described on page 548.

What is its intended use?

This information has to be obtained from those placing the order.

Is a COSHH assessment necessary?

This judgement will probably be made by the individual responsible for making COSHH assessments once the three previous questions have been answered. The product may simply be replacing a similar one but from a different supplier or it could be a safer alternative to a substance at present in use. On the other hand it may be an extremely hazardous substance, totally unsuitable for its intended purpose.

In smaller organisations it may be perfectly acceptable for these questions to be posed on an ad hoc, informal basis. However, in larger organisations, especially those using significant quantities of hazardous substances, there may be the need to introduce much more formal procedures. Such procedures are likely to contain the following elements:

- purchase requisitions for substances must state their intended end use
- requests for data sheets are automatically made by the purchasing staff for all new products
- this information is passed promptly to the staff responsible for carrying out COSHH assessments
- the COSHH assessors have the right to veto the use of new products pending their making the assessment.

Personal protective equipment (PPE)

PPE is examined in greater detail in chapter 25. There are three main aspects of PPE to be considered:

- its technical performance
  — usually the domain of the health and safety specialist

- its comfort
  — very much the subjective judgement of the user
- its cost (unit price and usage)
  — generally the realm of the purchasing staff.

All three aspects need to be fully taken into account, particularly where changes to the status quo are being proposed. Again smaller organisations may be able to deal with such situations on an ad hoc basis but larger ones may need more formalised arrangements.

Consideration of proposed changes to PPE is frequently on the agendas of health and safety committees and a few organisations have set up formal sub-committees to deal with such matters. It is advisable to tread warily, especially on matters of comfort, and a field trial is often the best step to take before embarking on major changes, even though it may prove impossible to please all of the people, all of the time.

Contractors' services

The whole topic of the use of contractors is dealt with in chapter 15 but the purchasing department can act as an important first screen in identifying those contractors with adequate capabilities to provide the services in question. Aspects particularly worth including at a first screening are:

Insurance

Does the contractor have an adequate level of insurance in relevant areas, eg public or professional liability?

Competent staff

Do the contractor's staff have relevant trade or professional qualifications and experience in the type of work involved? What sources of specialist health and safety advice does he have available? Does he possess relevant licences?

Method statement

Does the planned method of carrying out the project constitute a safe system of work?

Health and safety documentation

Can the contractor demonstrate that he is managing health and safety effectively through providing a health and safety policy or relevant risk assessments?

It is not sufficient to state in the 'Terms and Conditions' section of order forms that this information is required. The relevant documents must be provided by the contractor, checked by the client and, where necessary, reviewed with the contractor. This is particularly important in respect of the method statement and there is of course a need to ensure that the work is actually being carried out safely – an aspect which is covered more fully in chapter 15.

Changes to equipment and processes

Where new equipment is to be purchased or changes are to be made to existing equipment and processes there are many health and safety aspects which may need to be considered. It is far better to identify these at an early stage so that potential problems can be engineered out or that appropriate performance standards can be included in the project specification. (Many other chapters may provide guidance on specific subjects.)

Where the equipment to be purchased is relatively simple or the project is relatively small there is likely to be need only for contact to be made by the purchaser of the equipment or the sponsor of the project with other employees or outside sources able to provide necessary advice. These may on many occasions be health and safety specialists but the nature of the equipment or the process may be such that others with knowledge and experience in the relevant field are better equipped to advise, eg electrical engineers, ventilation engineers, chemical engineers etc.

Major projects are likely to require contributions from a larger number of people providing a wider range of expertise in order to identify and eliminate health and safety problems. Their contributions will generally be made within a much more formalised framework and over a much longer period of time. Some or all of the following stages could be required.

Identify potential contributors

The co-ordinator will need to identify the types of health and safety related expertise likely to be required for the project and where this is available. As well as in-house employees there may be the need to involve outsiders – specialist consultants, the HSE or the Fire Authority. However, all parties will not necessarily need to be involved at all stages of the project.

Review of concept and design

This may take place over several meetings and could involve the use of specialised techniques such as Hazard and Operability Studies (HAZOP). At this stage it is particularly important that all relevant legal obligations are identified together with other important documents such as Approved Codes of Practice (ACOPs), HSE Guidance and related British and International Standards.

Specification

A specification must be drawn up for the necessary equipment or installations that reflects (and preferably refers directly to) relevant legislation and standards. These may be included in specifications for the materials to be used, the design to be adopted, the method of construction or erection or even the performance standards (eg noise may not exceed a specified level, measured in a specified way at a specified position).

Monitoring

The degree of monitoring that can be carried out during the manufacture, assembly or installation of necessary equipment will be influenced by many factors, including where the work is taking place and over what timescale. Nevertheless, if new health and safety problems or failures to conform to specifications can be identified at this stage they will be easier to rectify than later in the project.

Commissioning

An inspection should take place both immediately before the equipment or installation is commissioned and soon after commissioning has commenced. As well as checking that all aspects of the specification (particularly any performance standards) have been met, such inspections provide the opportunity to identify those types of problems which may not have been readily apparent at the design stage.

Preparatory work in identifying and, where necessary, documenting safe systems of work should have already taken place but this cannot really be finalised until the commissioning stage has been reached. Operator training will also need to be closely linked to this.

Early use

Further reviews should be carried out during the early use of the equipment to identify other problems which may have emerged such as:

- failure to perform correctly
- unexpected failures of components
- difficulties in non-routine tasks eg maintenance
- hazards not previously identified or not rectified
- unsafe systems of work
- need for better documentation of systems of work
- training needs.

Note

1. *Rider operated lift trucks – operator training: approved code of practice and supplementary guidance* – HSE 1988 (COP 26).

13

Occupational Health and Hygiene

OCCUPATIONAL HEALTH OUTLINED

Occupational health can be defined as the process of keeping the work environment fit for the workers and keeping the workers fit for the work environment.

Several specific aspects of occupational health are dealt with in other chapters:

- ergonomics, manual handling and VDUs – chapter 19
- noise, vibration and radiation – chapter 22
- hazardous substances – chapter 23
- asbestos and lead – chapter 24
- stress, smoking, drinking and drugs – chapter 27.

Many of the sets of regulations dealing with these subjects include requirements to assess the extent of the risks to health and the adequacy of control measures. There is also often an additional requirement to screen the health of employees exposed to these risks.

Assessing the risk

In some situations all that is required is an investigation into the nature of the risk followed by observations and enquiries in the workplace to determine its extent. Then a simple judgement can be made as to whether a significant risk does exist in practice and, if so, whether the existing control measures are adequate to control it. However, in other cases the situation will not be so clear cut, requiring more specialist knowledge and possibly the carrying out of detailed tests or surveys. This is generally known as occupational hygiene and is dealt with later.

Screening employees

In addition to the need to assess the extent of the risks present in the workplace, there may be a need to monitor or measure the impact of these risks on the employees exposed to them.

As well as carrying out such screening on a regular basis after exposure has started, there is often a need to check on employees' health before they are placed in their jobs:

- to establish a base line level of health against which future comparisons can be made
- to ensure that individuals who may be particularly susceptible to certain risks are not put into situations where these risks are present.

Whilst some simple screening work can be carried out by non-specialists, screening in higher-risk situations will be carried out as part of an occupational health service.

Controlling the risk

Various methods of controlling occupational health risks are considered in greater detail in the chapters of the book devoted to those risks or in chapter 25 dealing with personal protective equipment. Control measures can generally be grouped under four main headings:

Elimination or substitution

Fundamental questions need to be asked about the source of the risk, for example:

- Do the hazardous substances have to be used?
- Can a quieter method of carrying out the task be found?
- Is it necessary to work such long shifts?
- Do shifts have to rotate so frequently?

Equipment-based controls

It may be possible to introduce equipment that significantly reduces or even eliminates the risks present in the activity, for example:

- mechanical handling equipment to reduce manual handling
- local exhaust ventilation to remove dust or fume
- use of noise reduction measures or acoustic booths
- physical screens to attenuate radiation

Procedural controls

Systems of work can be developed which reduce the possibility of the risk affecting the employee, for example:

- the employee works in the least noisy position

- employees are encouraged to take breaks from repetitive or stressful tasks
- assistance is provided for more difficult manual handling operations.

People-based controls

Although the general principle of controlling risks at, or close to, source is to be preferred, the introduction of people-based controls may be the only effective way of dealing with some risks. Into this category would come:

- the wearing of personal protective equipment
- periodic exercises to reduce the risks from manual handling, vibratory tools, repetitive movement etc
- training, eg in manual handling techniques, correct position at keyboards
- training in techniques for dealing with stress.

PROVISION OF AN OCCUPATIONAL HEALTH SERVICE

The nature of the occupational health service and the means of providing it will be very dependent upon the size of the organisation and the types of risks existing within it. An occupational health service is not simply an overhead cost to be borne stoically or avoided if possible. When planned properly and managed effectively, even a fairly large occupational health service may more than pay for itself through the reductions it achieves in the cost of employee absence and in the amount of damages paid to employees.

Occupational health provision is likely to involve a combination of in-house and external services, drawn from the following:

The personnel section

In smaller organisations some occupational health work can be carried out by those responsible for the personnel function. This might include some pre-placement screening (of new and transferred employees) and dealing with health-related problems as they arise, probably with the assistance of specialists outside the organisation.

Health and safety specialists

Some organisations employ staff solely to carry out occupational hygiene work or combine the role with that of an environmental specialist. However, often such work falls within the remit of the overall health and safety specialist. General health and safety qualifications such as the NEBOSH certificate and diploma require a significant level of knowledge and expertise in the field of occupational health. Whilst personnel staff may retain a role in carrying out pre-placement screening work, the health and safety specialist should be seeking out and dealing with health-related problems in the workplace.

Occupational health nurses

Larger organisations or those with greater occupational health risks may employ one or more occupational health nurses but it is important that their role is properly defined. This means providing a proactive occupational health service rather than pursuing a purely reactive role as a provider of plasters and panadol.

Administratively, occupational health nurses are usually placed within the personnel department or a health and safety department if one exists, although professionally their work may be under the scrutiny of a medical practitioner. Occupational health nurses should preferably hold a relevant specialist qualification such as the Occupational Health Nursing Certificate or Diploma.

Medical practitioners

Doctors specialising in occupational health may be employed on a full-time or, more commonly, a part-time basis. As is the case for nurses, it is important that they have had training in occupational health and that their occupational health role is properly defined. Retaining the services of a general practitioner with no specialist training on an ill-defined basis is likely to prove a poor investment. Some doctors will have obtained formal qualifications in occupational medicine (approved by the Faculty of Occupational Medicine) whilst others may have acquired relevant training via attendance at short courses.

Other specialists

There are many other specialists who may be able to contribute to an occupational health service and whose services may be employed full-time or part-time or engaged on a consultancy basis. These include:

- Occupational hygienists
- Health physicists
- Audiometricians
- Toxicologists
- Ergonomists
- Psychologists
- Chiropodists
- Physiotherapists
- Dentists.

Such services are likely to be needed regularly only by larger organisations although even quite small employers may require them on a short-term basis to deal with specific problems.

Group occupational health services

There are a number of organisations, particularly in the major conurbations, who provide occupational health services to a wide range of employers. In this way the small or medium employer can obtain the services of a highly professional team but at a lower cost than employing them in-house.

Such services may be provided on a fee-paying basis or through an annually negotiated contract or subscription. As with any external service it is important that the employer ensures at the contracting stage that he is really getting what he wants for his money. The services may be provided at the employer's own premises or alternatively his employees may need to visit the premises where relevant facilities are located.

Shared services

Occupational health services may be shared between neighbouring employers or between the different locations of the same employer. Large firms with their own in-house services may make them available to smaller neighbours. As with group services, contractual arrangements need to be clearly defined.

Employment Medical Advisory Service (EMAS)

EMAS is a team of specialist doctors and nurses providing occupational health advice and operating within the HSE. They are particularly likely to be helpful in advising on health aspects of job placement or rehabilitation or in investigating unusual occupational health problems. Their role only includes providing routine health screening services in specialised areas such as lead, asbestos or radiation.[1]

Other agencies

There are many agencies and organisations which may be able to provide advice or practical help within their own specialist fields. These include:

- Careers Services (in respect of school leavers).
- The Disablement Advisory Service (part of the Employment Department).
- NHS specialist centres (in relation to work placement of those with special health problems).
- Health education services.
- Local authority environmental health departments (on food hygiene).

OCCUPATIONAL HEALTH SCREENING

Pre-placement screening

There are a number of reasons for screening the health of new employees or of employees being transferred into a new job:

- to avoid putting the employee's own health and safety at risk
- to avoid endangering other people's health and safety
- to identify those with high absence potential
- to identify those whose work performance may be poor
- to establish health baselines, against which future comparisons may be made
- to identify existing occupational health problems.

These reasons need to be balanced against the importance of not discriminating unnecessarily on grounds of health. An employee or potential employee who is unsuitable for a particular type of work may be perfectly acceptable for a different activity and employers should avoid placing blanket exclusions on people with certain types of health problems or disabilities.

Apart from the good health and safety reasons for identifying actual or potential health problems it is very much in the employer's interests to identify those who represent a poor risk in terms of absence or performance or those with health conditions (eg occupational deafness) arising out of previous employment but which could result in damages claims against him at a later date.

Some specific health problems

There are many specific health problems that may make individuals unsuitable for certain types of work. Some of these are listed in Figure 13.1, although this list is far from comprehensive. Employers will need to identify those particular health problems that are relevant to their types of activities. Situations of doubt should be fully discussed with an occupational health specialist – if one is not available in-house then EMAS are likely to be able to help.

Regular screening

The reasons for carrying out regular screening of the workforce are similar to those for carrying out pre-placement screening. In some cases the employer may wish to keep a general check on the health of the workforce, in others the need will arise because of the possible effects of the work environment on the employee's health or because of the possible impact in the workplace of an employee's deteriorating health. Many of these specific situations should have been identified through the general process of risk assessment or through other assessments carried out under specific legislation.

The health conditions or disabilities below may make employees unsuitable for the types of activities listed. Short term conditions eg pregnancy, will only require temporary restrictions. In cases of doubt seek specialist advice.

Condition	Activities
POOR EYESIGHT	Driving – vehicles, cranes, other mobile equipment Other work where good vision is important
COLOUR BLINDNESS	Electrical work Colour printing
DEAFNESS	Work where audible warnings are important
EPILEPSY, FITS, BLACKOUTS etc	Work at heights or near water or unfenced holes Work with machinery Working alone for extended periods
VERTIGO	Work at heights or near deep holes
LACK OF STRENGTH	Heavy physical work
HEART PROBLEMS	Driving Heavy physical work Working alone for extended periods
ASTHMA	Many agricultural or animal handling activities Work in dusty environments Exposure to certain chemicals
DERMATITIS	Exposure to certain chemicals (acids, alkalis, solvents, oxidising agents, sensitizers) eg in hairdressing, cleaning Food handling
INFECTIOUS DISEASES	Food handling Work involving close contact with others
PREGNANCY	Exposure to ionising radiation Work with lead and certain other chemicals

Figure 13.1 *Possible health problems*

If regular screening is necessary for health and safety reasons then it will have to be compulsory for all employees engaged in the particular activity. The introduction retrospectively of what is in effect a condition of employment is fraught with potential difficulties and there would certainly need to be full consultation with employees on the matter.

Thought also needs to be given to the frequency of screening. Whilst in some situations the same frequency might apply to all employees there will be others where it may vary to take account of the different ages or previous health condition of the employees concerned. Eyesight, for example, will generally deteriorate more rapidly for the older worker.

Some employers have introduced random screening tests although these are generally restricted to situations where abuse of alcohol or drugs could seriously affect work performance, particularly in respect of public transport undertakings.

Screening methods

Screening might involve the use of questionnaires, the administration of relevant tests or the conducting of physical examinations. Some simple screening work can be carried out by staff who have received relevant training but have no formal qualifications. However, the more complex or invasive types of screening are likely to require the use of qualified technicians, nurses or medical practitioners.

Pre-placement questionnaires

Most pre-placement screening takes place through the administration of a health questionnaire by an occupational health specialist (if one exists) or a member of the personnel section. The questions asked should be relevant to the job for which the candidate is being considered. It is better for the interviewer to ask the questions rather than the candidate complete a form, since unfamiliar terms can be clarified and the relevance of the questions to the job explained if necessary.

Candidates who conceal relevant health conditions could render themselves liable to dismissal if this is discovered at a later stage and they may also be in breach of s7 of HSWA.[2] Employers have a duty to keep health information confidential, which may mean that health questionnaires are kept separate from the main application form.

Any doubts or concerns about a candidate's health should be reviewed confidentially with any in-house occupational health specialists. If it is necessary to consult with the candidate's own doctor or to refer him to an external specialist then his written consent should be obtained. However, a failure to give consent is likely to mean that the candidate's application is rejected.

General physical examination

The checking of the general physical condition of potential employees is a common part of the pre-placement screening and some employers continue to carry this out as part of a regular screening programme. Measurement of weight, height, pulse rate and blood pressure are frequently involved and in some cases this is accompanied by a detailed physical examination by a medical practitioner. Such an examination is likely to be of particular relevance where heavy, physical work is to be carried out to ensure that employees have sufficient strength and mobility to carry out their work safely.

Some employers also offer mammograms, cervical smear tests and other services as part of their health screening or health promotion programmes.

Vision testing

A number of vision screening instruments are available, some designed particularly for use in the occupational sector. These are capable of being operated by nurses or trained technicians. As well as testing visual acuity at near and long distances they can test other vision capabilities, some of which may be of great importance for specific activities, for example:

- *colour vision* – electrical work, colour printing
- *peripheral vision* – most types of driving
- *depth perception* – crane driving

The screening procedures will need to take account of whether corrective spectacles are normally worn in carrying out the work activity. Vision screening may also be carried out for those working at display screens but users as defined in the regulations governing such work may require more detailed eye and eyesight tests (see chapter 19).

Lung function testing

Lung function testing instruments require employees to blow hard and for as long as they can into the mouthpiece of the instrument which then produces a trace on a graph or a digital display. The two most important values measured are the Forced Vital Capacity (FVC) – effectively the capacity of the lungs – and the Forced Expiratory Volume in one second (FEV1) – a measure of the elasticity of the lungs.

These are then compared to average values for people of the same age, sex and size. Interpretation of the results requires specific training. Simpler instruments are also available to measure peak respiratory flow. Regular calibration and maintenance is important for these and other screening instruments.

Audiometry

Audiometric testing is usually carried out in conjunction with questioning as to medical history and previous exposure to noise (occupationally and socially) together with a visual examination of the ears. Most testing now uses semi-automatic audiometers which are less prone to bias or error than manual instruments.

Such testing normally takes place within a soundproof booth although the practicalities of work environments may prevent the noise levels in the booths being as low as those sometimes recommended. Special noise-reducing headsets may be used instead of or as well as acoustic booths.

Again interpretation of test results requires specialist training and the test instrument must be regularly calibrated and maintained. One aspect which must also be taken into account is the short-term temporary loss of hearing caused by recent exposure to noise. This is known as the short-term threshold shift.

Consequently too much credence should not be given to a single, poor, audiometric result which could be due to a variety of factors – short-term threshold shift, faulty calibration, incorrect operation of the test, deliberate cheating etc. Adverse results should be followed up by a further more carefully controlled test, possibly using the devices built into some instruments to prevent cheating.

Blood tests

Blood tests have been used for many years to monitor employees' absorption of industrial pollutants such as lead. Advancing analytical techniques now make testing for a wide range of other substances relatively cheap. As well as alcohol levels, the longer-term effects of alcohol consumption can be measured as can other general health indicators eg cholesterol levels.

Urine tests

Analysis of urine can also provide an indication of substances present in the body's metabolism. Such tests are generally less reliable than blood tests and there is a greater risk of contamination during the taking of the sample. The aluminium smelting industry uses urine tests to monitor the effects of fluorides present in their processes on employees.

Specific health questionnaires

Some aspects of employee health may best be screened by periodic questioning about relevant conditions or symptoms. Questionnaires could be completed by employees themselves or administered by someone with occupational health knowledge. Clearly the questionnaires must be properly designed to identify the condition or symptom concerned.

Questionnaires may relate to areas such as skin condition (dermatitis, or the marks or growths which could be the early indicators of skin cancer) or lung condition (difficulties in breathing at certain times or problems when using certain substances). Where questionnaires elicit adverse responses these will need to be followed up by referring the employee to someone with the necessary specialist health knowledge.

Specific physical examinations

Employees may also be regularly examined for specific occupational health problems. Some examinations may need to be carried out by occupational health specialists but others (eg simple skin inspection for dermatitis or ulceration) may be made by responsible employees such as supervisors, providing they have received relevant training. Again adverse results need to be systematically referred for further investigation.

Action after screening

The need for screening to be followed up by re-testing or referral or for further investigation has already been identified and the view expressed that *necessary* screening should be compulsory for relevant employees. This begs the question of what might ultimately happen to employees who have problems identified by the screening process.

It is relatively simple not to recruit people who do not meet the required standards but dealing with existing employees is more difficult. Screening programmes will not be acceptable to employees if those with unfavourable results are dismissed or suffer in other ways. Any compulsory screening programme should be accompanied by a procedure which describes the mechanism for dealing with those employees screened out. Firstly every effort should be made to find alternative work compatible with the employee's condition – either on a temporary or permanent basis.

Termination should only be a consideration in a minority of situations and it should not take place until:

- there have been efforts made to find alternative work
- the employee's fitness to work has been fully investigated in conjunction with relevant health specialists
- the matter has been fully discussed with the employee.

Anyone contemplating dismissing an employee on ill-health (or any other) grounds should be familiar with the law on unfair dismissal.[3]

The results of individual medical screening are confidential to the employees concerned but this does not preclude the correlation of screening results overall, providing this is done in an anonymous way. The calculation of total or group averages, the banding of results or even the recording of passes and fails may be appropriate, depending upon the circumstances. These can be used to provide comparisons, indicating whether or not control measures are working and they may give an illustration to employees of how their individual screening is contributing to the presentation of an overall picture.

OTHER OCCUPATIONAL HEALTH SERVICES

Rehabilitation

The costs of absence

Absence from work due to injury or ill health results in huge costs to employers. Although only a relatively small part of that absence is due to workplace accidents or the work environment, reducing the overall costs of absence is very much the role of the occupational health service. This may involve personnel specialists or occupational health professionals or preferably the two parties working in tandem.

A proactive approach

Employers should lay down a clear policy that formal contact will be made with an employee once he has been absent for a specified period of time or been absent on a given number of occasions within a defined period. This contact could be at the employee's home or at the workplace and could involve a personnel specialist (often a welfare officer) and/or an occupational health specialist (a nurse or doctor).

As well as expressing genuine social concern for the prolonged or repeated absences the contact should include an exploration of the following:

- the nature of the illness or injury
- the likely duration of the absence
- specific problems preventing return or regular attendance
- factors which might speed up return or prevent future absence
- whether the employee will be able to carry out the job safely and effectively
- whether the condition could recur.

Simple notes should normally be made recording the employee's responses on these points. A plan can then be formulated whereby the employer provides assistance to the employee in returning to work at the earliest opportunity or in remaining at work.

Providing assistance

Employees may be absent for long periods because of lengthy delays in obtaining appointments with relevant specialists. Money spent by employers in arranging private consultations with such specialists will often be more than amply repaid by a much earlier return to work. There are many ways in which the employer can provide assistance including:

- providing or arranging specialist advice
- providing or arranging treatment to accelerate the return to work
- providing treatment after the return to work

- providing advice and counselling after the return to work
- modifying the demands of the job to take account of reduced capabilities (on a temporary or a permanent basis)
- finding alternative work (either temporarily or permanently).

Such efforts should pay dividends with reduced absence costs and appreciative employees. For the small minority of employees who do not wish to return to work or who indulge in persistent absenteeism there will at least be plenty of evidence of the efforts that the employer has made to deal with their condition or alleged condition.

Treatment services

Injury and illness

The first-aid treatment of injury and illness is dealt with in some detail in chapter 16.

However, many employers choose to go beyond their immediate obligations for first-aid and provide a much more comprehensive service. This has obvious benefits to employees and will often also benefit the employer in keeping people at work who otherwise would be absent.

Such services should be provided by competent and properly authorised staff and there will be a need to liaise with others providing medical care to employees (such as general practitioners or hospital specialists). Services provided may include the changing of dressings or the supply of medication. Authorisation by a medical practitioner is particularly important here. They may extend to cover the type of specialist services detailed below.

Specialist services

Some occupational health services provide specialist types of treatment either as a general benefit to employees (and indirectly to their employers) or to meet specific needs of their business. Some of the more common sevices are:

- *Physiotherapy* Remedial treatment can be provided for musculoskeletal injuries, whether originating in the workplace or elsewhere. The physiotherapist may also be able to contribute in a preventative function, either by providing guidance to individual employees on how to avoid injury or by helping to improve the design of workplace tasks.

- *Chiropody* Some activities cause special problems for the feet through long periods spent standing, the use of certain types of footwear or the potential for transmission of foot infections. The chiropodist too will be able to assist in the prevention of foot problems as well as in the provision of treatment.

- *Inoculations etc* Where there are significant amounts of foreign travel by employees, time-consuming trips to obtain necessary inoculations can be avoided or reduced by providing these in-house, if there is an appropriate degree of medical supervision.

Such services can usefully be augmented by the provision of health-related advice and other support for those embarking on foreign travel, for example:
- control of existing health problems
- use of medication (possibly providing extra supplies)
- matters relating to drinking water
- food hygiene issues
- jet lag, travel fatigue etc
- equipment to treat minor illnesses or injuries
- advice on diseases prevalent in the area to be visited.

Health education and promotion

The promotion of a healthy lifestyle among employees is another important role of the occupational health service. Several areas where such promotional activity is particularly relevant – stress (including the effects of shift working), smoking, alcohol and drugs – are reviewed in chapter 27. Other such areas include:

Diet and exercise

'Healthy eating' and the taking of regular and sensible exercise can be promoted via the occupational health service. Exercise activities could be directly organised by occupational health staff and an occupational health centre could provide an impartial (and correctly calibrated) weight monitoring service.

Health counselling

There are many other topics on which health-related counselling may be necessary, ranging from life-threatening or infectious diseases to matters of personal hygiene. Generally confidentiality should be the norm, although there may be occasions where the importance of confidentiality is outweighed by the potential health and safety risks, particularly those to other people.

Such conflicts of interest are best addressed within the framework of formal policies and procedures but, even if such documents do not exist, confidentiality should not be breached without the intention to breach being fully discussed with the person concerned. Any breach should only be to the extent necessary for safeguarding health and safety.

Occupational health advice

The dissemination of information and training on specific health matters taking place within the workplace can be greatly supported by occupational health staff. They can help to educate employees about the risks to their health within the workplace and about the importance and correct use of appropriate control measures, such as personal protective equipment, barrier creams etc.

OCCUPATIONAL HYGIENE

The role of the hygienist

Occupational hygiene can be described as the recognition, evaluation and control of risks to health at work. Larger employers may employ full-time occupational hygienists whilst in smaller organisations the role is often included in the duties of a single health and safety specialist or the services of consultants are engaged when necessary. The occupational hygienist will need to co-ordinate his activities closely with those of other health and safety specialists, particularly occupational health professionals. Whilst the occupational health service is primarily concerned with screening the workers, the occupational hygienist is more involved with the environment within which those employees work.

A sound understanding of scientific principles is a prerequisite for the occupational hygienist who may then acquire specific training by attendance at short courses relating to particular aspects of the subject, eg atmospheric pollutants, noise, radiation. Comprehensive degree courses in occupational hygiene are offered by some universities whilst the British Examining and Registration Board in Occupational Hygiene (BERBOH) award a Diploma of Professional Competence. Advice on training is available from professional bodies such as the British Occupational Hygiene Society and the Institute of Occupational Hygienists as well as the HSE.[4]

Recognition

The range of occupational health hazards present in the workplace may include any or all of those listed on page 276. The hygienist should have a broad appreciation of all these types of hazards and the types of circumstances in which they may present a significant risk to the workforce.

Preferably his expertise should be applied in a proactive way so that potentially unhealthy work activities are not embarked on, or are started only when all necessary control measures are in place. However, the realities of most work situations mean that the hygienist is often operating in a reactive way and responding to:

- *observations* – his own and those of others
- *complaints* – from employees and others
- *cases of ill health* – reported directly by employees or found by health specialists.

In many cases the hygienist's expertise will enable him to conclude that no significant risk is present or that the risk is so obvious that an appropriate control measure must be introduced. Some situations though will require a much more detailed evaluation or assessment to be made to determine the extent of the risk.

Evaluation

The evaluation of health risks forms the main part of the hygienist's role and involves a number of different aspects.

Observations and enquiries

There are several types of occupational hygiene work where it is impossible to carry out any sampling of the work environment, for example where problems relate to ergonomic aspects of the task or involve stress. Even where it is anticipated that sampling may be necessary, some preliminary work will be necessary to determine the most appropriate sampling equipment and sampling method.

Detailed observation of tasks being carried out may reveal new facets of health risks which have not previously been appreciated, or simply identify that the task is not being carried out correctly. Observations should normally be accompanied by enquiries. Those carrying out the task should be asked whether the conditions being observed are normal or otherwise. Some examples of points to be observed and questions to be asked in relation to hazardous substances are contained on pages 549–50.

In some situations it may be required to devise questionnaires in order to evaluate the effects of particular occupational health problems. These need to be designed with care to ensure that the appropriate information is obtained and the possibility of bias eliminated or reduced. It is always wise to test out the questionnaire on a small sample of people before using it on a widescale basis.

Enquiries may also be required to find out more about substances or equipment in use. Reference may be necessary to manufacturer's or supplier's data, HSE Guidance, specialist books etc in order to determine the nature of the health risk, its possible extent and the methods available or recommended for its control.

Selection of sampling equipment

There are many types of sampling equipment available for the hygienist to use. The choice will be dependent on many factors (besides the equipment's availability) including:

- the nature of the substance or physical phenomenon to be sampled
- the chosen sampling method

- the duration of the work activity
- availability of power sources
- portability
- availability of related analytical method
- desired accuracy.

Some common types of sampling equipment are:

- *airborne contaminants* (see chapter 23)
  - chemical indicator tubes
  - direct reading instruments
  - sampling pumps and filter heads
- *noise (see chapter 22)*
  - sound level meters } Both types can be used
  - dosemeters } with continuous recording devices
- *thermometers* (various types)
- *lightmeters.*

Advice on sampling equipment and methods is provided by HSE publications in their *Guidance Notes – Environmental Hygiene* (EH) and *Methods for the Determination of Hazardous Substances* (MDHS) series.

Selection of sampling method

In many cases the sampling will attempt to measure the employee's personal exposure to the hazard by sampling air from his breathing zone or measuring the noise in the vicinity of his ears. However, in some instances this will be impracticable either because of the nature of the sampling equipment or because the employees most at risk can not easily be identified.

In such circumstances it may be better to sample the source of the problem (whether an airborne contaminant or noise) or the general levels prevailing in the workplace. If these levels are higher than any individual worker will experience but are still inside the relevant standards then it may safely be assumed that the worker's exposure will be comfortably within those standards.

There are several factors to be taken into account when deciding the time period over which sampling is to take place. The time period may be pre-determined by the standard against which comparison is to be made, it may relate to the sampling equipment being used or it may be dependent on the duration of the work activity being sampled.

If it is not possible to sample over the full cycle of the work activity then a sample which includes the period of worst exposure can be used to demonstrate that this period, and therefore the activity as a whole, are within the relevant standards. In many cases it will be desirable to monitor the workplace during sampling, and record any unusual incidents, potential sources of high exposure or possible instances of sample contamination.

Employees should be kept informed both about the purpose of the survey and its results. Their co-operation may be essential in conducting the survey and may also be important in making changes to control measures. They should be invited to comment on whether the sampled conditions are representative and may be able to assist in explaining apparently abnormal results.

Analysis

The analytical methods involved will depend upon the pollutant being sampled and the sampling equipment being used – some sampling equipment does not require further analysis. HSE's *Methods for the Determination of Hazardous Substances* series provides plenty of specific advice on analytical methods. Analysis may be by:

- gravimetric methods (weighing)
- chemical analysis
- microscopic counting (especially for asbestos fibres).

In some cases the sampling head may divide the sample into separate components (eg gaseous and particulate pollutants or respirable and non-respirable particulate sizes) and each of these will require separate analysis.

Interpretation

This is the final stage of the evaluation process. The hygienist needs to decide firstly whether the sampling results are accurate and truly representative of the normal working conditions. The results may be so high or so low or so inconsistent that he is forced to conclude that something has gone wrong in the sampling or analytical procedures. Any records of observations made during sampling or of comments made by employees may be of particular value in explaining abnormalities in sampling results.

If the results are accepted as being accurate and representative, they must be compared with established standards – in the UK, usually those published by the HSE. For airborne contaminants there are two important types of standards:

- *Maximum Exposure Limit (MEL)* An MEL is the maximum concentration of an airborne substance, averaged over a reference period, to which employees may be exposed by inhalation under any circumstances. (These are specified, together with the appropriate reference period, in Schedule 1 of the COSHH Regulations.)

- *Occupational Exposure Standard (OES)* An OES is the concentration of an airborne substance, averaged over a reference period, at which, according to current knowledge, there is no evidence that it is likely to be injurious to employees if they are exposed by inhalation, day after day, to that concentration. (These are published by the HSE each year in their *Guidance Note EH40*.)

Some of the standards set by the HSE in relation to noise, vibration and radiation are reviewed in chapter 22.

Control

The occupational hygienist should be very much involved in the selection of appropriate control measures for health hazards and in the subsequent monitoring of their effectiveness. On pages 277–8 control measures were categorised into four groupings:

- elimination or substitution
- equipment-based controls
- procedural controls
- people-based controls.

Consideration of what is the most suitable method of control should always start at the top of the list and only where that option is not 'reasonably practicable' should the next one be considered. (This presupposes that the standard required by the law is what is reasonably practicable.) People-based controls are often cheap and easy to initiate but may be much more difficult to sustain in the long run. Investment in equipment-based controls (or elimination or substitution) may prove to be more cost-effective eventually.

Regulation 4 of the Management Regulations requires control measures to be effectively monitored and many other specific sets of regulations contain similar requirements. Those imposed by the COSHH Regulations are reviewed in detail in chapter 23. Use of the evaluation methodology described earlier may be necesssary to ensure continuing compliance as well as an initial investigatory tool.

Notes

1. *Introduction to EMAS* – HSE (HSE 5)
2. Such persons may also be guilty of an offence under the Theft Act 1968
3. See, for example, chapter 7 in Lewis, P. *Practical Employment Law*, Oxford: Blackwell 1992
4. The British Occupational Hygiene Society and the Institute of Occupational Hygienists can both be contacted at Georgian House, Great Northern Road, Derby, DE1 1LT.

14

Employee Involvement

INTRODUCTION

Health and safety today is very much concerned with safety awareness, safety behaviour and safety culture. These three elements are the keys to the management of injury and loss prevention and are vital to any organisation's long-term plans for health and safety. This chapter aims to present a number of methods by which employees can be practically involved in health and safety at work as a means of securing improvement in an organisation's overall health and safety performance. The ABC approach is adopted for this purpose, emphasising the importance of safety awareness, safety behaviour and safety culture. A safety infrastructure, discussed later, provides the foundation for this approach. The starting point, where trades unions are recognised, can be found in the legal framework for employee involvement – the Safety Representatives and Safety Committees Regulations 1977.[1]

As noted in chapter 3, these provide for recognised, independent trades unions to appoint safety representatives. The guidance notes accompanying the regulations suggest some criteria for determining the number of safety representatives to be appointed, no number being specified in the regulations themselves or in the ACOP. A recognised trades union is one which is recognised by an employer for negotiating purposes; an independent one has a certificate of independence from the statutory certification officer.[2] Where a union is not recognised by an employer, there is nothing to prevent that employer establishing his own system of safety representation.

Once an employer has been notified in writing that a person has been appointed as a safety representative in respect of a particular group of employees, the safety representative has a right to be consulted about the making and maintaining of health and safety arrangements and about checking the effectiveness of those arrangements.[3] In addition, a number of specific requirements are the subject of consultation as a result of the requirements in the Management Regulations. (See chapter 3 for details of these and the statutory functions of safety representatives.) The safety representative is given legal duties under HSWA in his capacity as an employee but is not subject to any further legal duties because of his position as a safety representative.

Safety representatives are entitled to inspect the workplace and to inspect relevant documents. They have an entitlement to receive information from employers if it is information necessary for the performance of their functions, but certain information is excluded (again, see chapter 3). Safety representatives are also entitled to receive information from inspectors. Important statutory rights are given to safety representatives in respect of paid time off for their duties and for training associated with those duties.

If requested in writing by two or more safety representatives, an employer must establish a safety committee. The composition of the committee is for the employer to determine, but only after consultation with safety and other representatives of a recognised trades union. The statutory purpose of the committee is to keep health and safety measures under review.

It can be seen that the SRSC Regulations provide an important framework for employee involvement in health and safety. This will be on a statutory basis where there is a recognised, independent trades union, or on a voluntary basis where, in the absence of such a union, the employer chooses to adopt the SRSC model.

A strong business case can be made out for involving employees in health and safety at work, but time and organisational resources will need to be invested. Above and beyond this managerial commitment to resources is a less tangible factor; that of a positive managerial and organisational health and safety culture. The first golden rule of successful employee involvement in health and safety is that commitment to a safety culture is required from the most senior management in the organisation. The second golden rule is that there must be exemplary senior management involvement.

While support and leadership must come from the top, employee involvement is likely to generate real change only if it begins at the bottom. Adoption of a corporate safety culture will therefore only be effective if driven by those at the bottom of the organisational hierarchy. Senior management must follow the golden rules of commitment and involvement. However, their involvement must be limited to placing health and safety programmes in the hands of their middle managers and employees and adopting a 'hands off' approach to the management of the programmes. Success is more likely to be achieved when the employees lead their own programmes than when dictated by senior managers. The most powerful way to bring about a positive change in individual attitudes and behaviour is to provide employees with new or additional roles, responsibilities and relationships which will become a part of their job.

The broad aim of any employee involvement project in health and safety at work must always be the prevention of ill health, the prevention of injuries and the prevention of loss to company operations, plant, equipment and products. In turn some of the objectives which can support this broad aim are:

- to raise the level of individual's knowledge of and commitment to safety
- to draw upon the safety expertise and experience of employees
- to encourage teamwork on health and safety issues at the workplace
- to develop or improve workplace safety standards
- to develop a proactive safety approach at the workplace.

Suggested methods and systems to meet these objectives are examined later. In the next section, a brief outline is given of the ABC approach to employee involvement in safety.

THE ABC APPROACH TO EMPLOYEE INVOLVEMENT

This approach focuses upon safety attitudes, safety behaviour and safety culture. It aims to show how areas can be developed through employee involvement projects. Each area is now defined and ideal standards identified.

Safety attitudes

Safety attitudes can be seen as the employee's state of mind or readiness to react in an individual manner when confronted with a given health or safety situation. Safety attitudes will also govern the way an individual employee will approach a dangerous situation. Very importantly, this readiness to react in a given way to health and safety situations could have a positive or negative effect on other employees' safety attitudes and indeed have a bearing on the outcome of the situation. Safety attitudes are made up of individual beliefs on health and safety issues connected to the employee's workplace and work group. These beliefs are in turn conditioned by an employee's safety knowledge, awareness and experience. If employees have extensive and correct knowledge on a given safety matter then this will greatly contribute to a positive safety attitude. When employees are made aware of why something is unsafe and informed about the correct safety procedure or safe system, this again will contribute towards a positive safety attitude. Finally if an employee's safety experiences on a given plant or work area have been generally positive, then this can only reinforce a more positive safety attitude.

Usually attempts are made to change safety attitudes by:

1. giving employees more safety information and knowledge; and
2. suggesting that positive safety attitudes could bring rewards and that negative safety attitudes could result in disciplinary action.

In health and safety work situations the knowledge approach can be implemented by:

- telling employees why certain practices are unsafe
- explaining what the injury consequences and costs of unsafe practices could be
- outlining safe systems of work
- giving training in safe working procedures.

The reward/disciplinary approach can be given effect by:

- taking formal disciplinary action against employees
- pointing out that inspectors could prosecute the offender
- making promotional prospects dependent on positive safety attitudes
- rewarding positive safety attitudes.

While the knowledge method can change safety attitudes for the better and result in positive long-term safety behaviour, reward/disciplinary methods will in most cases only result in temporarily improved safety behaviour. This is because it may not necessarily lead to an extensive change in employees' safety attitudes.

Safety behaviour

Safety behaviour is of course linked very closely to safety attitude. Safety behaviour can be seen as how an employee will react in terms of health and safety in a given situation, while safety attitude can be said to be how an employee thinks and feels in the given situation. The safety behaviour of an employee can be easily observed as he/she goes about their day to day tasks and jobs. For example, does he/she always wear safety glasses when required, do they work in trapping positions? An evaluation, however subjective, can easily be made of an employee's safety behaviour on a particular plant, task or job after a few months. However, it is much more difficult to evaluate an employee's safety attitude. It can usually be done only after considerable questioning and discussion and could require the use of a questionnaire. Consultants who specialise in attitude measurement may need to be used.

Another complex issue affecting an individual employee's approach to safety is the behaviour of the work-group. People will tend to conform to the general safety behaviour of the work group or groups to which they belong regardless of their own safety attitude. This interface between individual and group safety behaviour is at times referred to as a safety atmosphere or group safety culture. It comprises individual safety attitudes and manifests itself as a set of individual and group safety actions (ie safety behaviour) which has been shaped by the group. For example, an individual may elect to wear a safety harness when working alone as their own attitude to safety is to take precautions when possible. However, in a group situation, the same individual may discard the harness, as the attitude of the group is to take a risk rather than take the trouble to wear a harness. In such cases, the group has a negative effect on the individual's attitude to safety, perhaps due to peer pressure. Of course, the opposite may prevail and a positive-thinking group may influence negative thinking.

Safety culture

Safety culture is much more difficult to define. It can operate at three levels. There can be a company safety culture, a management safety culture and a workforce safety culture. Ideally, all three would come together and reinforce each other. Unfortunately, this is not always the case. Some of the key factors likely to encourage cultural harmony are:

- true senior managerial support both in spirit and resources
- a high profile for health and safety in every corner of the organisation
- time resource allocation given on a frequent basis to health and safety by management and employees
- a philosophy that work will be undertaken safely at all times and not just when production allows
- a clear visible commitment by middle managers and supervisors to the day by day, hour by hour practice of safety as a normal accepted part of their duties
- a belief that most serious injuries and all industrial ill health can be prevented
- a belief that the quality of working life of all employees and contractors is dependent on each individual's approach to safety
- a competent safety-trained workforce at all levels
- an effective health and safety communications infrastructure
- a workforce who want to work safely in addition to fulfilling other job satisfaction conditions.

This list is by no means exhaustive but it does begin to show the complex features that go to make up an organisation's safety culture.

Having described the ABC approach to employee involvement in safety, it is now necessary to consider the establishment and functioning of an organisation's safety infrastructure.

SAFETY INFRASTRUCTURE

Meaning and objectives

Whatever the size of an organisation, it must have some form of safety infrastructure. This is a collection of routine systems of communication and services. They are established to support and back-up health and safety operations throughout the various sections and departments of the organisation. Like any communications system they can and do relate to each other. Computers are ideal in many areas to assist in running a number of these routine systems and services. Other elements of a safety infrastructure include:

- safety meetings
- hazard reporting books
- accident investigations
- annual safety plans.

Appendix A presents a further picture of the elements of a safety infrastructure. Trade union safety representatives can assist in establishing such an infrastructure, which should have three main objectives:

1. to support a safety culture;
2. to facilitate the expansion of employee safety involvement into more complex fields; and
3. to act as a training base for (1) and (2).

Benefits include both the development of key elements of an organisation's health and safety culture together with the individual employee's development in the health and safety field. The list presented here should not be taken as exhaustive: rather, it indicates where experience in many companies demonstrates the greatest advantages are to be gained from employee involvement. Some of the aspects of safety infrastructure detailed later will show that limitation of damage to plant, equipment, products and materials can also be a benefit of such involvement.

Safety meetings

These are usually monthly and should include *equal* numbers (if not more) of employees to management. A large part of the meeting should concentrate on proactive safety issues, such as safety improvement plans, special initiative projects, audit planning, topics for future safety briefings and the study of injury trends with the view to drafting prevention measures. Employees should be encouraged by management to speak freely and openly at meetings so that an accurate picture of health and safety problems can be obtained and meaningful preventive measures developed. Minutes will normally be taken of meetings and it is a worthwhile exercise to make a summary of the action points that have been agreed at the meeting. This summary should clearly state the action agreed to be undertaken, who is responsible for undertaking it and the expected or required completion time. As well as being good management practice it will provide a method for showing employees that progress is being made.

As certain safety action items come off the list and others are added for progressing in a prioritised and agreed manner, the safety meeting should become a forum to review and monitor progress made in terms of safety and not just a forum for talking about general safety issues. A similar prioritising method can be employed on other reported health and safety issues tabled at the safety meeting. For example, outstanding safety items from audits, inspections and hazard report books could be considered. Every member of a safety committee should be encouraged to take at least one small item away from the meeting and make some progress on it for reporting back to the next committee. This can be further developed to involve employees in safety by allocating small projects to experienced employees and giving them time to develop them. An example would be the evaluation of the best gloves to be purchased for a given workplace operation. The project would need to examine hand protection, ease of use and cost considerations taking into account the comments of other workplace employees on the operation under study. When this sort of employee involvement in safety gains momentum, employees can be encouraged and coached to make presentations to the safety committee on more complex safety projects.

In some organisations, any employee can report to safety meetings on accidents within their own area of work. This assumes that they are given time to investigate accidents which they must be if they are a safety representative within the meaning of the Safety Representatives and Safety Committee Regulations. (See chapter 3) It is not suggested that the manager's duty to investigate accidents should be given to employees. Rather it is suggested that an additional set of eyes will be of help and may arrive at a cause more quickly with the possibility of practical, preventive measures.

It is suggested that employees' involvement can be on a union basis or on a non-union basis. It is the commitment of the individual employee which is vital. Effective safety meetings with employee involvement can be a major step in developing full scale employee involvement in safety throughout the organisation, and indeed can be the backbone of an organisation's safety culture. This, perhaps, is one reason why the legal framework is built upon the foundations of individual safety representatives and safety committee members.

Hazard reporting books

Most organisations run a system for written reporting of hazards. The design of the reporting format usually allows an action completed note to be placed next to the reported item. Employee involvement can be twofold in this area. Firstly the management must endeavour to encourage the wide use of a hazard reporting system by all employees. This can be done by open management techniques and by providing an adequate supply of properly designed books. It is most important that the management's acknowledgement and actioning of entries is carried out in a professional and prompt manner. However open and professional the system, there will still be reluctance by some employees to make entries in the book. This is where involving employees can again be of value. A small number of key personnel can be trained to fulfil the role of safety counsellor. The counsellor may be approached by a reluctant employee who would make an oral report on a given hazard, unsafe condition or operation. The counsellor would then enter the issue into the hazard report book under their own name. Properly organised and managed hazard reporting systems can be a valuable low cost contribution to accident and loss prevention.

Hazard telephones

As an alternative to a hazard report book or to complement it, a dedicated telephone number could be used. This should be fitted with an answerphone to record messages left by employees who have noted workplace hazards, regardless of the time. A system of this type has the added advantage of allowing employees to report hazards anonymously. This may result in many more hazardous items being reported than would be the case if hazard report books were used.

Experience shows that telephone systems are heavily used when first installed but that use tails off after ten to twelve months. They can, however, be given a fresh impetus from time to time using special projects which publicise the facility and employ an ever-changing safety theme. Employee involvement can be encouraged by giving key employees the role of extracting the information from the tapes for entry under their own names in the hazard report books. This can be done on a rota basis.

Near-miss reporting

This is a single sheet reporting system, an example of which is presented in appendix B of this chapter. The purpose of the report sheet is to facilitate the reporting of near miss incidents which do not cause injury or major plant or product damage. The aim is to use a standard format for such reporting.

EXAMPLE

An example of a near-miss would be a large timber box – unserviceable and to be scrapped – falling off a fork lift truck into a walkway. Nobody is injured and no damage is done to plant or products; only the scrap box is damaged. This incident should be reported and investigated. Next time the box may have a load in it and someone may be on the walkway.

Near-miss reporting is growing in application in the UK. However there is a reluctance by some employees to report this type of incident in writing and some confusion in people's minds as to what constitutes a near-miss. This is another example of a situation where involvement of selected employees is valuable. Again a small number of key employees could be trained to act as counsellors on near-miss reporting for their local workplace area. These employees would have access to report forms and could report information supplied by their workplace colleagues or derived from their own observations and investigations. Counsellors would need to understand the managerial reporting route for near-miss reports and be responsible for speedy communication of each near-miss incident to the relevant manager or engineer. They could encourage all other employees to report near-miss incidents.

Accident investigation

Most organisations carry out some form of investigation of the more serious type of accidents, ie those resulting in injury. Usually this is in three stages:

- on-site investigation of the accident
- local post-accident investigation
- formal managerial investigation.

In most organisations the last stage takes place only in cases of major injury accidents ie lost-time injury and injuries leading to hospitalisation. Site accident investigation and follow-up is one of the roles performed by union safety representatives where they exist. However, in many small and medium-sized companies there are no such representatives. It is suggested that accident investigation is a valuable area in which to develop the involvement of employees as by far the best people to study and investigate a serious accident, with the objective of preventing a similar accident happening again, are trained employees from the workplace. The training required for this site role and for formal participation at later stages can be very brief and simple. Any experienced safety professional could describe the key points in a half-day training session. These would cover interview techniques, documentation, site locations, report writing and the simple techniques of accident prevention. Trained employees in key locations could carry out their site investigations in conjunction with a management investigation, or separate from and in addition to a management investigation, depending on the circumstances. The insight and viewpoint of these employees can be important in reaching recommendations for the prevention of similar accidents. Where a more formal managerial investigation is required, possibly with a report setting out recommendations, again there is great value in securing employee involvement. Finally, by involving certain employees in an examination of accident records, practical prevention measures might be agreed. (On the subject of accident investigation more generally, see chapter 9.)

Annual safety plan

All organisations should have an annual safety plan to maintain the momentum on health and safety and to keep presenting a fresh approach to health and safety problems. An annual safety plan is not to be confused with a company safety policy or indeed with a company safety mission statement. The annual safety plan could be a one-page document listing a handful of health and safety topics applicable to the particular department or company. It should specify aims and objectives, and their means of achievement. The formation of annual safety plans provides an ideal opportunity for employee involvement in safety. A small team of managers and employees could first agree the subjects to be included in a particular plan. Subjects should be selected on the basis of accident trends and areas of serious injury potential. Following discussion, joint agreement could then be reached on aims, objectives and methods. Employees will be able to see that their safety concerns are similar to those of the company. They will also see their safety suggestions and ideas incorporated as part of the annual safety plan. Involved employees could in fact own specific elements of the plan and be charged with their achievement. This would require managerial agreement for time off and would need to have a 'hands off' management facilitator to assist where necessary. Some typical examples of annual safety plans are presented in appendix C.

Emergency plans

Every organisation needs to have an emergency plan. While the drawing up of emergency plans is a commonsense measure, it is also a requirement under regulation 7 of the Management Regulations. This regulation requires an employer to establish procedures to be followed in the event of serious and imminent danger to persons at work.

Emergency procedures provide another opportunity for involving employees in health and safety. Employees can be involved at the drafting stage and when the plan is subject to review. The employees selected should be given some training in emergency procedure formulation and danger evaluation. When it comes to making all employees aware of emergency plans and any changes therein, key employees can be vital in assisting management. This applies in particular to evacuation from the workplace. Employees who really know the workplace can be much better than management at explaining emergency procedures. All emergency plans should be practised from time to time, most especially fire drills. Key employees can be valuable additional eyes and ears for management in observing the effectiveness of emergency plans when they are put to the test by realistic practice. A review of emergency plans can take place on the basis of such observations.

EVALUATION OF THE SAFETY CULTURE: THE EMPLOYEE SURVEY

This section of the chapter links the basic methods of employee involvement in safety detailed previously to the more specific proposals in the next section. The central concern is to evaluate an organisation's safety culture by the use of an employee safety survey.

The objective of this section is to present a simple but effective vehicle with which to:

- develop employee involvement in safety
- draw on employees' health and safety expertise and experience
- establish a base level against which to measure future improvements.

These key objectives could in turn be linked to a number of specific objectives, for example:

- the development of a safety training programme
- the improvement of cooperation on health and safety
- the collation of information on unsafe conditions, operations and procedures
- the implementation of safety briefings
- to develop safety initiative projects

The safety survey itself will obtain the best results if it is split into three parts. Part one could be made up of health and safety knowledge questions of a yes/no and multi-choice type. Questions could be designed to allow personnel to demonstrate awareness by being asked to substantiate some of the points raised. The second part should attempt to establish attitudes to safety by asking leading questions on important health and safety topics. Finally the interviewee should be allowed to speak freely on the health and safety issues which cause them most concern. This should broaden the scope of employee involvement in safety by encouraging more specific proposals. The survey would seek to clarify the following two basic questions about an organisation's health and safety:

- where are we at present?
- where do we want to be in the future?

It may be that management and most employees feel they know the answers to these, but without a survey, nobody can really be sure. In addition to answering these two questions, the findings of the survey can be very useful in establishing whether or not genuine and long term improvements have taken place in health and safety standards. This is easy to establish by conducting another survey two or three years after the initial one.

Each organisation will have its own views on what the objectives of a survey should be. A few key objectives are listed here by way of example:

- to reduce accidents and losses
- to establish, by a test, current levels of safety knowledge and safety awareness
- to evaluate individuals' health and safety attitudes towards their own workplace operations.

The questionnaire itself will need to be in a printed format for ease of use. Appendix D presents some sample questions. It should be noted here that surveys of this sort tend to be accurate only if respondents are permitted to answer anonymously. A representative sample of employees should take part in the survey. This should include managers, supervisors, administrative and clerical staff, labourers, operatives, craftsmen and tradesmen.

A number of employees could be selected and trained to act as interviewers. Questions will be answered more fully and honestly if posed by peers, especially when a guarantee of anonymity is given. Similarly, for managers and supervisors the interviewer could be a manager involved in the project. The tasks of processing the data, drafting a report and drawing conclusions and recommendations could be given to the employee involvement team.

A safety survey is a time-consuming and expensive way to assess an organisation's safety culture but it is the most realistic and practical way of establishing the full picture. There is only one other method which could be used: a quantifiable safety audit. This method is likely to be more costly and tends to concentrate on evaluating managerial safety policy and systems rather than on employee safety attitudes and awareness.

The information gathered from a survey will evaluate employees':

- awareness of safety;
- attitudes to safety;
- competency in safety tasks; and
- safety behaviour patterns.

A considerable number of improvements, initiatives, projects and development programmes may be derived from the information processed. In particular the information may point to the use of a number of the methods and systems that are proposed in the next section.

PROPOSALS FOR EMPLOYEE SAFETY INVOLVEMENT

A few of the dozens of employee safety involvement schemes found in both the UK and the US are presented in the following pages as proposals for trial, modification and adaptation. The schemes have shown how involvement of employees can lead to major health and safety dividends. The techiques are in the main very simple, however, the skill of obtaining the necessary commitment at all levels is difficult. The approach centres on the human aspects of health and safety rather than the 'hardware' (plant and equipment) aspects. This is not to say that moving machinery does not need to be guarded or lifting tackle does not need checking, but human factors need to be closely examined and developed. The complex safety interface between an individual employee, their work group, and the management of the organisation is the focus that is rapidly developing. The following are free-standing ideas for involving employees in safety. They are put forward as practical schemes using the ABC approach outlined earlier – attitudes, behaviour and culture.

Safety attitudes

Standard safety manuals

These are local plant safety documents (possibly in a small booklet form), covering the critical and hazardous operations in a workshop or warehouse or part of plant. These manuals should not cover the well known generic safety issues but should deal with local safety operations in a best practice format. As a standard manual on a given local safety topic, they could then be used for safety training purposes ensuring that work groups are employing best practices on each shift and/or in each area.

Without doubt, the best standard safety manuals are those drafted by employees who have had some training in the preparation of manuals. Of course the final checking of these drafts and issue for use would be a managerial responsibility.

Standard safety training

Where selected employees are involved with the production of standard safety manuals, they can also be deployed as on-the-job trainers.

Group management safety surgeries

These are a method whereby given groups of employees can put forward their worries, concerns and problems about health and safety within their area, directly to a senior manager. This scheme is best applied in solving particular safety problems in a given area at a given time. For example, hand injuries may have risen dramatically or excessive plant damage could be taking place in certain areas. The group or groups would be scheduled to attend the senior manager's office at a set time with a set time limit. Ideally each group should be between six and ten employees. The senior manager would outline the current health and safety problem and ask for comments. The group should be briefed by a middle manager or supervisor as to the objective of the surgery before the group attends the senior manager's office.

One-to-one interviews

This simple technique attempts to improve an individual's attitude to safety. It should be used if an employee is having more accidents than the rest of the workgroup, or is not conforming to standard safety procedures or practices. A selected and specially trained safety employee should carry out a one-to-one interview with the 'problem' employee. No managers or union representatives should be present. The discussion at the interview should be confidential except for any jointly-agreed recommendations for improvement. This sort of interview should be distinguished from that carried out under a disciplinary procedure.

At the interview the employee interviewer should concentrate on attempting to discover the causes of the problem with the objective of jointly resolving it with the interviewee. Causes may include lack of safety training, peer group pressure, personal problems, a poor attitude to safety or lack of safety awareness.

Safety briefings

This is a safety technique where small groups of employees are given a short – five or ten minute – talk or briefing on relevant health or safety topics. Some examples of these topics are:

- use of eye protection
- scaffolding safety check
- what is COSHH?
- dangers of confined spaces
- VDU health and safety.

After the safety briefing, the group can ask questions and should be given a sheet or card containing the key points of the topic. A written or oral validation should then be obtained to ensure that the group has understood the content and importance of the briefing. The safety briefing leader, who could be a non-managerial employee, might draw up a list of topics for the safety briefings and prepare the text for management approval prior to implementation.

Contractor safety inductions

When organisations need to bring in contractors there will be a need for safety induction training. Indeed the Management Regulations (in regulation 10) call for appropriate instructions and comprehensible information to be given by employers to employees who are not from the host company, regarding risks to a person's health and safety. In addition to this legal requirement it is sound accident prevention management to ensure that contractors' employees are given safety induction training before commencing work on site. Again this provides an opportunity for the involvement of employees of the host organisation.

Safety behaviour

Computer communication

This requires sufficient computer capacity to store and send safety communications in the form of brief text messages, notices and memos etc. It is in fact the operation of an electronic mailing system. Under coded entry control, any employee authorised to use the system could, after a short training session, send a safety communication to or from any or all other stations. The main communication formats are:

- *warning notifications*, eg road works at main gate for ten days – roof work taking place on the main office
- *general safety notices*, eg manual handling training course next Monday at 10.00 am main office – safety manual on risk assessment now available see safety manager
- *safety messages* (to specific people), eg do we have any safety helmet chin straps on site? – can the brakes on forklift 47 be checked?
- *safety memos* (to specific people) putting forward health and safety ideas, suggestions, improvements, modification requests etc – these memos would be answered using the same open system

A system of this type requires a few basic rules, such as no personal comments and the use of good manners, but after that it should be totally open for any employee to access for text reading. Some senior managers will have reservations about this type of health and safety system. However in the few organisations where it is employed, the system is rarely if ever abused.

Safety grandparents

This is an extremely effective but uncomplicated method for employee safety involvement and can be applied to new starters or employees transferring between departments or units. Selected and coached employees should be encouraged to accept the formal task of 'keeping an eye' on another employee in terms of health and safety for a few weeks until the new employee finds their feet. Safety grandparent or 'safety mentor' duties would include, for example, pointing out local hazards, familiarisation with plant layout and access and egress, awareness of the plant emergency procedure and being brought up to speed on safe working procedures and safety briefing topics. The safety grandparent is simply a point of contact for new starters. It is an especially vital proposal for any new young employee. This method will positively influence the new starter's safety behaviour at a crucial time, and as a spin off will have a bearing on the safety grandparent's safety behaviour. Often, grandparents, knowing that their own performance is being closely observed by their charges, will improve their own safety behaviour.

Safe working procedures

These are in place in many organisations today. If an employee is involved in the drafting of safe working procedures it will have a beneficial effect on safety behaviour. When an employee accepted by his peer group has contributed to the draft and made sure the workgroup has been given a chance to make a contribution, the draft is likely to be practical and have the acceptance of the workgroup. It is therefore much more likely to be followed in practice. When implemented, those employees who are committed to the procedure will try to ensure other employees know and follow 'their' safe working procedures.

Task observation

Unless participating employees are properly trained in the techniques of workplace safety observation and counselling, this type of scheme can be difficult to establish. Moreover, instead of positively shaping safety behaviour it can have a damaging safety effect at both group and individual level. The scheme requires specially trained observers, at times direct management involvement, to carry out periodic monitoring of the health and safety issues on given work tasks. Positive and negative records are made on a standard score card. Names of people and times are not recorded. Some of the observational headings are presented in the following list as examples but the scoring detail and layout of record cards or sheets are designed by the observers in pre-start training sessions.

- *Changes of practice when under observation*, eg individual or team take up new positions – change work method – stop or leave work area – adjust work area housekeeping
- *Injury position at workplace*, eg individual or team in a trapping position – in a fall position – in a struck-by position – working in slip or trip conditions – in contact with hazardous substances
- *Personal protective equipment*, eg are the individuals wearing the PPE necessary for the task? Are adjustments made to PPE?
- *Safe working procedures*, eg are they available? – are they known and understood? – are they being followed? – are short cuts taken?
- *Operation of tools and equipment*, eg are they correct for the task? – are they used correctly? – are they in a safe condition?

The task observation would normally be only a few minutes, but may be up to some 30 minutes' duration and could cover a total task or a major component of it. It could centre on a single person or group of people.

The most important element of all in task observation is the counselling role of the employee carrying out the observation. After each observation they would discuss with the individual or group under observation how they have scored and why. This may need to be done in conjunction with management. Observers would also point out, for example, that the correct eye protection was not worn and discussion would focus upon whether the employee knew the correct eye protection to wear. If so, did they have the correct eye protection available for wear, if not why not and so on. Unsafe behaviour remains confidential between the individual and the observer provided that dangerous acts are corrected.

Hazard tags

This is a low cost, simple but effective method of challenging the effects of actions caused by poor safety behaviour. The method will have an indirect effect on safety behaviour of employees, but will mainly be directed at the safety behaviour of first line managers. The hazard tag, similar in design to a scaffolding inspection tag, is fastened on to an unsafe piece of equipment, unsafe part of plant, unsafe tools, unsafe machine or even an unsafe vehicle. It can be removed only after rectification of the unsafe condition.

Hazard tags will act as a reminder to management that corrective action is necesary, but, more importantly, it will give warning to all employees of an unsafe situation. The proposal to involve employees in safety is that the hazard tags could be placed during, say, a safety inspection, by a specially selected and trained employee.

Managers should be given notice immediately after the tag has been placed and the reason for it should be explained. Removal of the tag should be undertaken by the employee 'tagger' on request from management and be subject to a rectification inspection. Under proper control conditions, this is a tool which can have an immediate and positive influence on the safety behaviour of first line managers. After the initial rash of hazard tags when the system starts, considerable changes in behaviour take place. All employees start to observe the NIMBY principle, ie 'Not in my back yard'. Safety improvements take place to ensure the hazard tags do not appear in an individual manager's areas of responsibility. Tags are a more effective way to sort out safety hardware problems than the managerial blitz that can follow serious or fatal accidents. An example of a hazard tag is shown in Appendix E.

Safety culture

Safety discussion workgroups

As the name implies these are workgroups or teams which are permitted to get together from time to time to discuss safety issues associated with the particular workgroup or local workplace. There should not usually be an agenda or record of the proceedings and managers should not attend. Each person attending should be allowed to have an opportunity to express his/her views. The golden rules should be: no personal comments; individual confidentiality;

adopt a problem solving approach; stick to health and safety issues only. A time limit should be set for each discussion.

To give shape to the discussion and to act as a communication channel for management, an employee leader would be required. This leader ideally should come from the workgroup. A leader's role in these discussions should be to prioritise discussion time, to ensure focus on problems, to bring about a consensus, to make sure each person has a say and to stop obstructive discussions which prevent conclusions being reached. This type of scheme, tailored to organisational requirements, is an important foundation for establishing a workgroup safety culture.

Areas of safety responsibility

These are commonly used systems for building up a positive safety culture in local areas or plant. The system revolves around a trained employee becoming the owner of a clearly defined area in terms of the safety of hardware and systems. The safety employee(s) must be allocated time to undertake the duties linked to this responsibility. It should be understood that the area owner is not responsible for the health and safety supervision of employees within the area. That responsibility must remain with first line managers. The area owner's task is to inspect and monitor safety hardware eg guards in place, housekeeping, fire extinguisher availability, ventilation systems working etc. In addition the area owner should also check and monitor safe systems of work and the use and availability of safe working procedures. Results of these observations should be subject to a formal report to management for action or rectification by the relevant manager or engineer. Owners could be given limited authorisation, on simple items, to approach the relevant manager or engineer to discuss and agree rectification of the particular health or safety problem.

Outstanding actions could be listed. Other sources could supply further information which then can be added to a central list. Injury report forms, near miss reports, shift reports and weekly production reports are but a few examples of other systems which can be useful sources of health and safety action requirements. Some organisations refer to these listings as blacklists. They are a central register of outstanding health and safety action points, providing the base data for improvement and rectification planning and programming. Computer recording of these blacklist points with planned action, date of rectification and the person responsible for action can provide an excellent managerial control tool. Blacklist programming registers could be an agenda item at safety meetings permitting all interested parties to be aware of progress and keeping the action required centrally focused.

Audits and inspections

Audits and inspections are a high-profile area (see chapter 11). Employee involvement raises its profile even higher. In many cases employees will be quicker than management in identifying unsafe situations and operations in their own area. Therefore including employees in safety audits and safety inspections as members of an accident prevention team can be a major contribution to development of a positive safety culture.

Safety monitoring of contractors

The point here is to consider how the safety culture of contractors working in a host organisation can be evaluated in a realistic and practical manner. Safety monitoring is seen as an effective starting point. It will demonstrate the safety culture of the host organisation because in carrying out safety monitoring, the host's safety standards will become clear.

Contractors found by the host company's monitors to be carrying out unsafe operations should be provided with professional training to familiarise them with safe procedures, methods and systems. The employees of the host company can be involved as monitors.

Training courses

Training courses on health and safety, presented by trained and selected employees with the backing of management facilitators, have a surprisingly positive effect on safety culture at all levels. In addition, where organisations have safety representatives there will be a statutory requirement for time off for training.

Voluntary safety inspectors

Safety inspectors could be on a fixed retainer, for say two years, with duties laid down in a written but informal job description. The prime objective of this type of scheme would be to bring into play the sometimes unused experience of selected employees and involve them in the task of supervising health and safety in a given work area. The main role of a safety inspector would be the examination, via inspection and observation, of plant, procedures and people to a periodic remit detailed by management. The major objective would be to improve health and safety by the formulation and progressing of action plans and programmes. Given careful selection and training, the employee safety inspector could be a more cost-effective option than many external consultants. Selection of employee safety inspectors should be by formal interview, having first possibly advertised the post(s) in-house. As a pre-requisite of in-house health and safety training, consideration should be given to encouraging the safety inspector to attend a nationally-approved training course with the view to obtaining a relevant health and safety qualification.

Safety-related projects

There are two further closely-linked ideas for involving employees in health and safety. The first is running seminars on health and safety topics. With substantial employee involvement these can and do enhance safety culture. The second is mixed discussion groups ie managerial and non-managerial employees. These debate a company safety problem in small sub-groups, having first received a project brief from senior management. The findings of the sub-groups are then presented to the full seminar with further questions and points being put forward by the full seminar membership. At this point the sub-groups would take their project away for implementation and development over a period of months. The project should be led by an employee, closely supported by a management facilitator. At the end of the project a report should be drafted by the project leader and management facilitator.

SAFETY CULTURE COST CONSIDERATIONS

Accident costings presented by an HSE publication in 1993 show that the cost of operational accidents of all types is ten times more than recovered insurance costs.[4] The important lesson here is that until senior management are able to communicate a positive and real safety conviction throughout the organisation, then other managerial levels, and in turn employees, will only pay lip service to the objective of improved safety. The most powerful way to change this situation is to give people new health and safety roles, safety responsibilities and safety relationships. This creates a situation from which develops new and improved safety behaviour.

Safety culture is a difficult concept to grasp. A definition is presented in figure 14.1.

> "The safety culture of an organisation is the product of individual and group values, attitudes, competences and patterns of behaviour that determine the commitment to, and the style and proficiency of, an organisation's health and safety programmes. Organisations with a positive safety culture are characterised by communications founded on mutual trust, by shared perceptions of the importance of safety, and by confidence in the efficacy of preventive measures."
>
> Health and Safety Commission Third report: *Organising for Safety*, ACSNI Study Group on Human Factors, HMSO.

Figure 14.1 *A definition of safety culture*

Positive and decisive safety leadership at all levels is the only realistic approach to the development of a safety culture. What management does is far more critical than what it says. Employees will only become motivated towards active involvement in health and safety when the whole organisation is positively motivated towards that goal. Once that involvement is achieved, a significant improvement in health and safety can be expected.

APPENDIX A Basic company safety infrastructure

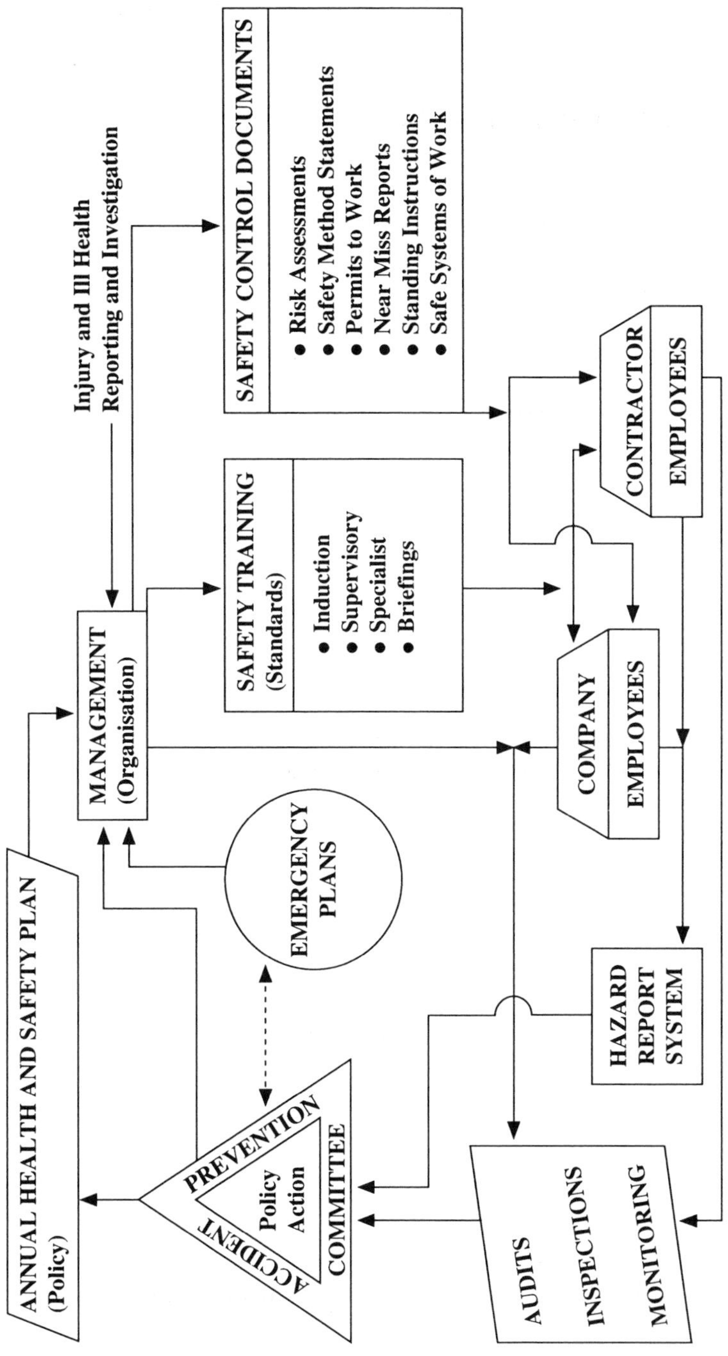

ANNUAL HEALTH AND SAFETY PLAN
(Policy)

MANAGEMENT
(Organisation)

Injury and Ill Health
Reporting and Investigation

ACCIDENT PREVENTION COMMITTEE
Policy
Action

EMERGENCY
PLANS

SAFETY TRAINING
(Standards)

- Induction
- Supervisory
- Specialist
- Briefings

SAFETY CONTROL DOCUMENTS

- Risk Assessments
- Safety Method Statements
- Permits to Work
- Near Miss Reports
- Standing Instructions
- Safe Systems of Work

AUDITS
INSPECTIONS
MONITORING

HAZARD
REPORT
SYSTEM

COMPANY
EMPLOYEES

CONTRACTOR
EMPLOYEES

APPENDIX B Near-miss report

<table>
<tr><td colspan="3">PLANT OR DEPARTMENT</td></tr>
<tr><td colspan="3">1. INCIDENT DETAILS
Date of incident ....
Time of incident ....
Location of incident ....</td></tr>
<tr><td colspan="3">2. INCIDENT REPORT</td></tr>
<tr><td colspan="3">3. IMPROVEMENT SUGGESTION(S)</td></tr>
<tr><td colspan="2">4. MANAGEMENT ACTION PLAN</td><td>Action by:-</td></tr>
<tr><td>5. SUBMITTED BY
Date:-</td><td colspan="2">6. MANAGER'S SIGNATURE
Date:-</td></tr>
</table>

APPENDIX C Annual health and safety operating plan

PLANT/DEPARTMENT		YEAR		
Ref	Subject	**Start Date**	**Completion Date**	**Facilitator Responsibility**
W.S.1	**GUARDING** – All machine guarding to be inspected (repaired if necessary) and fastening bolts fitted. All doors and access points to be fitted with interlock systems.	JANUARY	(end) MARCH	Plant Engineer
W.S.2	**LOG BOOKS** – Supervisors, Leaders, Managers and Engineers to be briefed in the use of Safety Monitoring Log Books.	FEBRUARY	(end) MARCH	Plant Manager
W.S.3	**FIRE POINTS** – To be re-painted, re-numbered (and where necessary repaired) then re-equipped with new powder extinguishers.	APRIL	(end) JUNE	Electrical Engineer
W.S.4	**WORK ON ROOFS AND AT HEIGHTS** – Safety Control System to be reviewed and a new Permit System to be devised and put into operation.	JULY	(end) DECEMBER	Safety Officer
W.S.5	**COSHH REVIEW** – Hazardous substances' data sheets and assessments to be reviewed and updated.	MARCH	(end) AUGUST	Planning Supervisor
W.S.6	**RISK ASSESSMENTS** – Current method of risk assessment control to be evaluated and recommendations for improvement/ development drawn up.	AUGUST	(end) OCTOBER	Safety Committee (via Chairman)

APPENDIX D Sample questions for employee safety survey

SAFETY AWARENESS

- What is your opinion of safety housekeeping?
- Have you ever carried out work regardless of safety?
- What is your opinion about company safety in general?
- Do you report hazards to your supervisor?
- What is your opinion of managers' attitudes towards safety?
- Are you aware of a safe working procedure?

SAFETY KNOWLEDGE

- Can you describe the local emergency procedure?
- In what situations must eye protection be used?
- Do you know any safety standards for using scaffolds?
- Can you think of the key safety points of immobilisation?
- Do you understand the basics of risk assessment?
- Are you familiar with company manual handling practices?

SAFETY STANDARDS

- Do you know of any dangerous or hazardous practices, conditions or situations within the company? (LIST BRIEFLY)
- Have you got any recommendations or ideas to improve health or safety in your area? (LIST)

APPENDIX E Hazard tag system

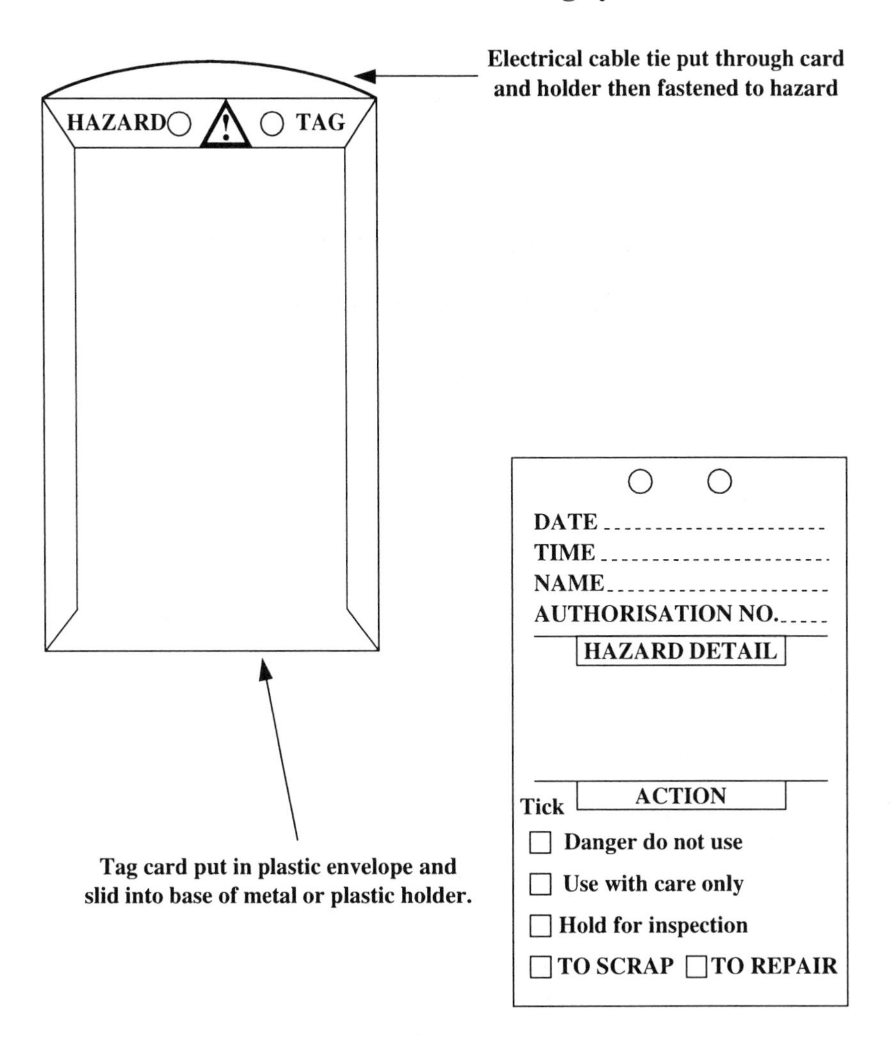

Notes

1. SRSC Regulations, SI 1977/500. Issued under HSWA.
2. On this see Lewis, P., *Practical Employment Law*, Blackwell, 1992, pp227–9 and 249–50.
3. HSWA, s2(6).
4. *The Costs of Accidents at Work*, HSE, 1993.

15

Selection and Management of Contractors

INTRODUCTION

Virtually every workplace plays host to contractors at one time or another. Recent years have seen an increasing trend towards contracting out services which were previously provided 'in-house' with the result that contractors have a permanent presence in many work locations.

This chapter firstly deals with legislation of relevance to the use of contractors and in particular to the host/contractor relationship. It then examines how contractors should be selected, the importance of communicating effectively with contracting companies and their employees (especially prior to the commencement of contract work) and how they should be managed once work has started. Specific aspects of the use of contractors in construction and demolition work are dealt with in chapter 26.

The benefits of effective health and safety management were examined in chapter 6 and certainly apply to the management of contractors. It must be remembered though that the range of contracted services is extremely wide, as illustrated in Figure 15.1. What might be a very effective and necessary system for selecting and managing contractors engaged in major construction or engineering projects would be unnecessarily bureaucratic and inefficient if applied to the servicing of a photocopier or the provision of a temporary telephonist. However, even such relatively low risk activities are not without their health and safety considerations and parts of the chapter will still be of relevance to them.

RELEVANT LEGISLATION

This review of legislation concentrates on the duties of the host organisation towards the contracting company and its employees and, to a lesser extent, the duties of the contractors towards the host organisation and its employees. However, it must not be overlooked that where contractors work together, each

1. **MAJOR CONSTRUCTION AND ENGINEERING PROJECTS**
   Such projects are likely to last over a significant period of time and might involve several contractors providing a variety of specialist services, possibly involving a further layer of sub-contractors.

2. **MINOR CONSTRUCTION AND ENGINEERING PROJECTS**
   These are likely to be short-term activities and will often involve only a single contractor.

3. **ROUTINE MAINTENANCE AND REPAIR**
   Many types of equipment will require routine maintenance or repair. Such services may be provided by the equipment manufacturers or by other organisations approved by them. The equipment may vary from major plant (eg cranes or ventilation systems) through vehicles and other mobile equipment to office equipment such as photocopiers or fax machines.

4. **PLANT HIRE**
   Equipment such as mobile cranes, fork lift trucks, mobile access platforms, scaffolding or temporary heating systems may be hired, often involving the use of the hire company's operators or their erection or installation staff.

5. **SERVICE SUPPLIERS**
   Many contractors maintain a permanent or semi-permanent presence on their clients' sites providing such services as cleaning, catering, security, transport or grounds maintenance.

6. **PROFESSIONAL SERVICES**
   Companies frequently engage the services of professional specialists such as architects, design engineers, inspecting engineers, quantity surveyors, accountants or medical practitioners. Health and safety consultants or training specialists could be included in this category.

7. **TEMPORARY STAFF**
   Many contracting companies provide temporary staff to work alongside the host company's employees on a short or a long term basis. Such staff might be engaged in manual work or clerical activities or provide specialist skills eg contract nurses or draughtsmen.

Figure 15.1 *The range of contracted services*

contractor and his employees will have duties towards each of the others. The principal legislation in this area is:

- The Health and Safety at Work etc Act 1974.
- The Management of Health and Safety at Work Regulations 1992.
- The Construction (Design and Management) Regulations 1994.

The Health and Safety at Work etc Act 1974

Section 3 of HSWA sets down 'General duties of employers and self-employed to persons other than their employees'. Sub-section (1) states:

> It shall be the duty of every employer to conduct his undertaking in such a way as to ensure, so far as is reasonably practicable, that persons not in his employment who may be affected thereby are not thereby exposed to risks to their health or safety.

Sub-section (2) places a similar duty on the self-employed.

Contractors are just one example of "persons not in his employment" that the host employer must take into account. Visitors, neighbours, customers, passers-by, members of the emergency services and even trespassers must also be considered. However, section 3 applies equally to the contractor (whether an employer or self-employed person) who has his own duties towards the host organisation and its employees and towards other contractors.

A major precedent in the application of this section was established following the Court of Appeal ruling in the case of *R v Swan Hunter Shipbuilders Limited.*[1] The case resulted from a major fire on board a warship HMS *Glasgow* which was under construction within one of Swan Hunter's shipyards on Tyneside.

A firm of contractors (Telemeter Installations Limited) was installing equipment on the ship. In the course of their work they were using oxy-acetylene equipment which they left in a poorly ventilated area on board. Overnight there was a leak of oxygen from the equipment which created an oxygen-enriched atmosphere throughout a large area of the ship. Early the following morning a source of ignition created an intense fire in which eight men died.

Swan Hunter had well-established safety instructions in respect of oxy-acetylene equipment which were described in their safety manual but this information was not routinely passed on to contractors or sub-contractors. The Court concluded that an employer cannot ignore the actions of non-employees and how those actions affect the safe system of work. If the provision of a safe system depends upon non-employees being adequately trained and instructed, then HSWA requires such training and instruction to be provided.[2]

The principles developed by the Court of Appeal in the Swan Hunter case were followed by the Crown Court in the Case of *R v Derby City Council, Bardon Contractors and City Plant Hire*. The case resulted from the death of a young woman on the top deck of a bus which was struck by the bucket of an excavator as it swung to load a lorry.

It was established that engineers from the City Council had made several visits to the site but had taken no action on safety infringements. As a result the City Council (the clients) were found guilty under section 3(1) of HSWA and fined £15,000, the same amount as the plant hire company. The main contractor was fined only £500.

The Management of Health and Safety at Work Regulations 1992

The duties of employers to others have been further expanded by the Management Regulations. Under regulation 3 both employers and the self-employed must carry out risk assessments which must include risks to the health and safety of persons not in their employment. Hosts must assess the risks to contractors and vice versa. The subject of risk assessment is dealt with in some detail in chapter 8.

Once the measures necessary to deal with these risks have been identified through the risk assessment process, regulation 4 of the Regulations requires that:

1. Every employer shall make and give effect to such arrangements as are appropriate, having regard to the nature of his activities and the size of his undertaking, for the effective planning, organisation, control, monitoring and review of the preventive and protective measures.
2. Where the employer employs five or more employees, he shall record the arrangements referred to in paragraph (1).

ie Employers must effectively implement those precautions and procedures described in this chapter that are "appropriate" to their circumstances.

Several other specific requirements of the Management Regulations are of particular relevance to contractors.

Co-operation and co-ordination (regulation 9)

Where two or more employers share a workplace (whether on a temporary or a permanent basis), this regulation places a three-fold duty on each employer. (For this purpose self-employed persons are treated as employers.)

- *Co-operation* Each employer must co-operate with the other employers so far as is necessary to enable them to comply with their legal duties.
- *Co-ordination* All reasonable steps must be taken by each employer (taking into account the nature of his activities) to co-ordinate his health and safety measures with those of the other employers. The ACOP accompanying the Regulations highlights the role of the employer who controls the site in establishing site-wide arrangements and informing new employers accordingly. If no employer is in overall control, there may be a need for those present to agree on the appointment of a site health and safety co-ordinator.

- *Information* Employers must "take all reasonable steps" to inform other employers concerned of any risks to their employees' health or safety. Such information might relate to unusual risks of which the other employers may be unaware or to high risk activities being carried out eg the use of explosives, the spraying of highly flammable liquids or the removal of asbestos insulation.

Persons working in host employers' or self-employed persons' undertakings (regulation 10)

To some extent the requirements of this regulation overlap with those of regulation 9. However, it gives host employers (including self-employed) greater duties both in respect of employers who have employees working in the host's undertaking *and* the visiting employees themselves. Under this regulation, a visiting self employed person is treated as both an employer and an employee.

Paragraph (1) of the regulation requires the host employer to provide the visiting employer with "comprehensible information" on:

- the risks to his employees' health and safety arising out of or in connection with the host's undertaking
- measures taken by the host employer to comply with his legal obligations to the visiting employees.

Taking the Swan Hunter case as an illustration, this regulation would mean that the host ship builder would have to spell out to visiting contractors the risks of leaving oxy-acetylene equipment in confined spaces and identify areas where such equipment may safely be stored.

Many hosts have become so familiar with risks associated with their activities that they over-estimate the awareness of others. Such risks (with which outsiders may be unfamiliar) should be identified through the risk assessment process and then communicated to contracting employers using the means described later in this chapter.

The onus is then on the visiting employer to pass this information on to his or her employees but the regulation recognises that this may not always be the case. Paragraph (3) places an *additional* duty upon the host employer to *ensure* that any visiting employee "is provided with appropriate instructions and comprehensible information regarding any risks" to his health and safety arising out of the host's undertaking. Under paragraph (4) of the regulation this requirement includes information relating to emergency evacuation arrangements.

Whether the host chooses to provide such instructions and information himself, or simply ensures that the employer of the visiting employees has provided it, is for the host to decide.

Whilst regulation 9 relates to shared workplaces, ie major construction sites or situations where contractors have a significant presence, regulation 10 can apply even where the contracting employer is not present. It is of particular relevance in safeguarding individual contractors' employees working on clients' premises, eg service engineers, cleaners or temporary contract staff working under the host's control.

Capabilities and training (regulation 11)

Paragraph (1) of this regulation states that "Every employer shall, in entrusting tasks to his employees, take into account their capabilities as regards health and safety". Paragraph (2) identifies situations where the employer must ensure employees are provided with adequate health and safety training and augments the basic training requirement contained in HSWA and specific requirements in various sets of regulations.

In the case of contractors' employees this is clearly the duty of their employer but nevertheless the host employer may well want to ensure that they have received an adequate level of training before they embark upon work on his premises.

Temporary workers (regulation 13)

This regulation provides additional safeguards to ensure that temporary workers receive health and safety-related information. Both those employed directly under a fixed-term contract of employment and those workers supplied by an 'employment business' (eg an agency) must be provided by their host with information on:

- any special occupational qualifications or skills required to be held to carry out their work safely
- any statutory health surveillance to which they will be subject.

Similarly employment businesses providing temporary employees must also be provided by the host with information on any special qualifications or skills required and additionally information about any special features of the jobs to be filled which may affect the health and safety of those employees. Such features might include the work having to be done at heights or requiring an adequate level of physical strength.

Construction (Design and Management) Regulations 1994

This important new set of regulations creates many further specific obligations for the host organisation (the client) and the contractors they are dealing with in respect of significant "construction work" (the term is defined in the regulations). The detailed requirements of the regulations are reviewed in chapter 26.

SELECTION OF CONTRACTORS

The need for selection control

The health and safety standards of contractors are just one consideration within the selection process alongside the capability to perform the task in hand, price, quality of work, promptness of service etc. However, as the previous part of the chapter demonstrated, there is a legal obligation to consider health and safety standards carefully and the costs of not doing so may also be very significant. The degree of formality required in the selection process will be influenced by four factors – the task to be performed or the service provided, the nature of the host organisation, the nature of the contractor and previous experience.

The task or service

As can be seen from Figure 15.1, the range of contracted services is extremely wide. A far greater degree of care will need to be taken in selecting a contractor to carry out a major building project or to supply a mobile crane for a particularly tricky operation than in selecting one to replace a broken window or to provide a temporary clerk.

The host organisation

In a small organisation an owner or senior manager, having selected a contractor, will be able to exercise personal and almost constant supervision over that contractor's work. However, in a large organisation those selecting the contractor may never actually see them in action. A more formalised selection procedure (preferably one involving those who will have to manage the project) should ensure that the right contractor is chosen and thus reduce the need for future management effort.

Whatever the size of the host organisation, the hazards present within its activities should also play a major part in determining the extent of the selection process. If the host is storing or using large quantities of highly flammable liquids they will need to be careful about the type of contractors invited onto the premises.

The contractor

Whilst it is not always the case that the health and safety standards of large contractors will be better than those of small contractors, there are nevertheless some contractors who have earned a favourable reputation. There are also contractors who have achieved such a dominant position in the market for certain services that the client will find it extremely difficult to select an alternative, even if he wishes. In this latter situation the client may well be forced to place reliance upon his management of the contractor in order to ensure satisfactory health and safety standards.

Experience

Experience of previous health and safety problems associated with the carrying out of a particular task may highlight the need for a greater degree of care in selecting the contractor to perform it the next time. However, experience will also be utilised in respect of previous performance standards of known contractors. In smaller organisations such experience may be used informally to determine which contractors are invited to carry out work, whereas in larger organisations this experience may be applied in a more formal sense with only 'approved' contractors being invited to tender. This concept of approved status will be developed later in the chapter.

The selection process

The degree of formality necessary in the process of selecting contractors will be influenced by the factors outlined above. The flowchart in figure 15.2 sets out the stages which might be included in a formal selection process. A less formal approach may be more appropriate in lower risk situations but the same basic selection principles should be applied.

The concept of 'approved' status contractors may allow some stages of the process to be leapfrogged, particularly where routine services of a similar nature are provided. Even formal procedures may need to contain a method by which authorised members of management are permitted to by-pass certain stages in cases of emergency, although again it must be stressed that the basic principles of selection should still be adopted. Figure 15.2 shows how the selection process continues into communication with and management of contractors. The stages of the selection process are described in more detail below.

Identify the work required

The scope of the work to be carried out by the contractor needs to be clearly identified. Often this will be a relatively straightforward task but sometimes extremely detailed specifications may need to be drawn up in respect of technical or health and safety requirements.

Specify the selection criteria

There are many health and safety aspects to take into account in selecting contractors for particular projects. Does the contractor possess an adequate level of insurance for the task, does he have staff who are competent in this type of work, how does he plan to approach the tasks and is he aware of relevant risks and how to overcome them? What is his overall standard of management of health and safety issues and what sort of a track record does he have?

These types of selection criteria are outlined in figure 15.3 and described in detail later. Contractors who have already satisfied their clients in general terms

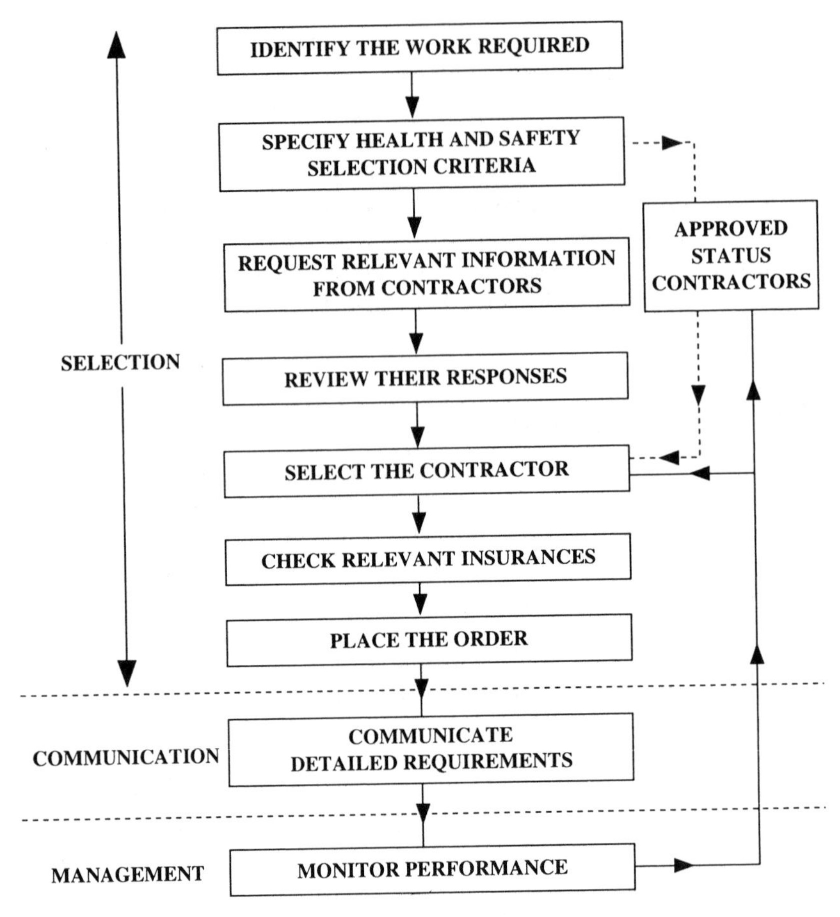

Figure 15.2 *Contractors: the selection and management process*

in respect of their health and safety standards may be accorded 'approved' status and therefore not have to resubmit the same information in respect of every piece of work for which they tender. However, even approved contractors may still need to be asked how they propose to tackle a specific task, particularly if it is outside the normal range of services they provide to a client or if it involves a high level of risk.

Request relevant information

Contractors who are to be asked to tender for work or who are seeking approved status should be asked to provide relevant information. This could be chosen from the 'menu' contained in figure 15.3 and might be split into two sections – one providing basic health and safety information, the other relating specifically to the work to be performed. At this stage it is important that the client

BASIC HEALTH AND SAFETY INFORMATION

- INSURANCES
  - – liability to client
  - – public liability
  - – professional indemnity
  - – motor vehicles
  - – employer's liability

- HEALTH AND SAFETY MANAGEMENT
  - – health and safety policy
  - – risk assessments (general and specific)
  - – availability of specialist assistance
  - – monitoring and supervision

- PREVIOUS RECORD
  - – prosecutions by enforcing authorities
  - – correspondence with enforcing authorities
  - – accidents reported under RIDDOR

WORK-RELATED INFORMATION

- COMPETENT STAFF
  - – trade and professional qualifications
  - – specialist skills

- EVIDENCE OF CAPABILITIES
  - – relevant licences
  - – work of similar type
  - – availability of sites to visit
  - – equipment possessed

- PROPOSED SYSTEMS OF WORK
  - – method statement
  - – relevant risk assessments
  - – costings for key health and safety items

Figure 15.3 *Contractor selection criteria*

spells out to the contractor the health and safety standards he is expected to meet and site-related risks that he is likely to encounter. The supply of a contractor's handbook (see later in the chapter) would be particularly relevant.

Review contractors' responses

Contractors' responses to the requests for information must then be reviewed by someone competent to do so. It is surprising how often such responses are filed away or just given a cursory glance by purchasing staff. It may be necessary to seek further clarification of the information supplied: the responses of some contractors may be totally inadequate, rendering them unsuitable for selection.

Selection

As stated earlier, there are many factors other than health and safety which need to be taken into account when selecting contractors. However, a minimum standard of health and safety performance must be established and only those contractors who satisfy the basic criteria should be allowed to remain within the selection process. Even when this minimum threshold has been passed, health and safety should still actively be considered in evaluating the relative merits of different contractors. Selection of contractors by price alone is unlikely to meet the client's legal obligations.

Check insurances

In many contracted operations there is a significant time lag between the submission of information by tendering contractors and the final selection of the successful bid. Clients would be wise to seek confirmation that relevant insurances are still in place. Similarly insurances held by approved contractors are likely to be renewable on an annual basis and clients may wish to receive confirmation that the renewal has been effected. The use of computer-generated standard letters at or around renewal time is recommended.

Placing the order

Many order forms contain standard health and safety-related clauses, often drawing attention to standards applied by the client or expected of the contractor. Whilst these may provide useful reminders to the contractor, they in no way relieve the client of the legal obligations described earlier. It may be appropriate to include specific health and safety references within the detail of the order, eg directly relating to information previously supplied by the contractor or to HSE Codes of Practice or Guidance. Even then, the client is still under a duty to effectively communicate with and manage the contractor.

Contractor selection criteria

Figure 15.3 spells out the criteria which might be used to assess the health and safety capabilities of contractors. These are divided into those which might be required of any contractor and those which relate to a particular task or category of work.

Insurances

The client needs to be satisfied that the contractor possesses an appropriate level and type of insurance.

- *Liability to client* The contractor could cause damage to the client's buildings, equipment or employees. Such damage could result in major disruption to the client's activities. There have been many instances where contractors have been involved in major fire losses eg the Windsor Castle fire. However, very small contractors may not have the financial resources to insure against major losses and the client may need to arrange separate insurance or agree to bear the potential loss.

- *Public liability* The responsible client will wish to make sure that the contractor is adequately insured against injury to members of the public or damage to their property, especially if these members of the public are his customers or neighbours. Even if such injury or damage is solely down to negligence on the part of the contractor, the client is still likely to be dragged into the proceedings and could well suffer from any adverse publicity.

- *Professional liability* Where the contractor provides specialist professional services the client should ensure that there is adequate insurance available to protect him against the consequences of poor service or advice. This is particularly necessary in relation to the services of architects and design engineers but even health and safety consultants may need to be insured against such risks.

- *Motor vehicle* Hauliers and other contractors who use vehicles on the highway must possess adequate insurance to comply with legal requirements. The client may also wish to ensure that his load is adequately insured against damage or theft when being transported by road or other means.

- *Employer's liability* Apart from those organisations which are exempt, all employers must hold a satisfactory level of employer's liability insurance. Any contractor who does not comply with this basic legal requirement is not the sort of organisation with which the responsible client should wish to deal. The client's moral obligation to ensure this is even greater where the contractor is providing temporary staff to work alongside his own employees.

Health and safety management

Clients should also wish to ensure that the contractors that they are engaging manage health and safety issues in an effective manner. Key indicators of this are the following:

- *Health and safety policy* The contractor (unless employing fewer than five employees) is legally obliged to have a health and safety policy. The client is at liberty to request a copy of this policy and to determine whether its three main components (the policy statement, the responsibilities for its implementation and the related arrangements) provide for the effective management of health and safety in relation to his project. (Chapter 7 refers in greater detail to health and safety policies.)
- *Risk assessments* General risk assessments are required under the Management Regulations with other more specific assessments being necessary under other sets of regulations. Despite the period which has elapsed since these regulations became operative there are still many organisations (contractors and others) who have failed to comply with their statutory obligations. Whilst clients may not choose to debar contractors on the basis that they cannot supply suitable assessments, the availability of satisfactory assessments is another indicator of a contractor who places a priority on health and safety. Risk assessments may also have a specific relevance to the task to be carried out (see below).
- *Availability of specialist assistance* Regulation 6 of the Management Regulations requires every employer to appoint one or more competent persons to assist him in undertaking the measures he needs to take to comply with the law. Clients should make enquiries of contractors: who is to provide this assistance, what are their qualifications and to what extent will their services be available? Well-qualified individuals (whether based at a national office or engaged occasionally on a consultancy basis) may have little impact at local, site level.
- *Monitoring and supervision* The availability of supervision and the degree of health and safety monitoring relate closely to the above. The client needs to be satisfied that someone is responsible for supervising the work on his site and that that work will be monitored by the contractor on an ongoing basis.

Previous record

The client may also wish to find out more about the contractor's previous health and safety record, particularly in relation to the enforcing authorities (HSE or local authority). He may choose to ask about

- *Prosecutions* Convictions, acquittals or cases pending

- *Correspondence* Copies of correspondence with enforcing authorities over a previous period of time might be requested.
- *Accidents reported under RIDDOR* The client may be more concerned if the contractor is unaware of his RIDDOR obligations than if he supplies a list of relatively innocuous accidents.

Competent staff

Contractors should be able to provide details of relevant trade or professional qualifications held by those who are to be involved in the contract work. Many professional bodies are willing to verify such qualifications in cases of doubt. In other situations it may be important to ensure that contractors' employees have necessary specialist skills eg that fork lift truck drivers are competent to the ACOP standard or that users of cartridge-operated tools have the necessary certificate of competence.

Evidence of capabilities

Clients should also look to contractors for evidence that they are capable of carrying out the task in question. Some work may require the contractor to hold a specific licence, eg for the removal of asbestos. Contractors may be asked for details of similar work they have carried out elsewhere and whether reference can be made independently to their clients. Even better evidence could be provided by visiting other sites where the contractor is working, particularly if such visits can be made unannounced. Visits to workshops or work bases may also be appropriate to ensure that the contractor has the necessary equipment and support organisation to provide a satisfactory service from a health and safety and other standpoints.

Proposed systems of work

For major projects or even for minor ones where there is a significant degree of risk the contractor should be asked about the systems of work he proposes to adopt. This could be done in a number of ways.

- *Method statement* The client might request a detailed safety method statement from the contractor which specifies how the health and safety aspects of the work are to be addressed, along with any technical and other aspects.
- *Relevant risk assessments* The contractor may have available relevant model or generic risk assessments relating to the type of work to be performed. These may satisfy the client that not only is the contractor fully aware of all the risks but he is also aware of the related precautions and has the means to implement them. Such documents may form the basis on which a more detailed method statement could be prepared if required.

- *Costing for key health and safety items* A slightly different approach may involve the client identifying key health and safety aspects associated with the work and insisting on the contractor stating the anticipated expenditure in each of these areas. This approach can force the contractor to more proactively consider these aspects, with the contract being unlikely to be awarded to those who make minimum or no provision in these areas. Typical areas where such costings may be requested are:

  — provision of scaffolding or other access equipment
  — personal protective equipment
  — fire precautions
  — training
  — supervision
  — welfare facilities.

Approved status

The concept of approved status for contractors has been referred to earlier. Contractors who have provided basic health and safety information to an acceptable standard and who have satisfactorily carried out work for the client may be accorded such status. This would eliminate the need for the contractor to resubmit the same information every time he wishes to tender for work. However, regular confirmation that insurances have been renewed may be required and the contractor may still be asked to supply health and safety information on how he proposes to tackle specific tasks.

Some client companies may wish to extend the approved concept to include other aspects of contractors' work such as technical capability, financial viability and price. Even where contractors have been approved, it is important that there is regular feedback from those charged with monitoring their performance. Where contract managers identify significant shortcomings in the contractor's standards they should be able to imperil the approved status and there should also be periodic reviews of the performance of approved contractors which might allow repetitive lesser transgressions to come to light. The frequency of such reviews may vary to take account of the changing nature of contract services supplied.

COMMUNICATIONS WITH THE CONTRACTOR

In this section four important means of communicating with contractors are reviewed. These are:

- the contract order
- contractors' handbooks

- pre-work meetings
- induction of contractors' employees.

The contract order

As stated earlier, the inclusion of standard health and safety clauses in order forms is of only limited value in communicating effectively with contractors. The information contained in such clauses is unlikely to percolate through to individual employees carrying out the work and does not relieve the client of his duty to *ensure* that visiting workers are informed about risks in the workplace. However, official orders (and schedules to them) can be used to draw important health and safety issues to the contracting company's attention. Such orders may be used in conjunction with contractors' handbooks. The types of issues which may be covered in contract orders are:

- reference to information previously supplied by the contractor
- relevant HSE codes of practice or guidance (or other standards)
- special features of the contract work
- relevant PPE standards
- responsibilities for supplying necessary services or equipment (see below)
- the use of permit to work systems (if relevant)
- sanctions which may be imposed in cases of non-compliance with standards.

The following are examples of the categories of services or equipment for which responsibility for provision may need to be identified within the contract order:

- access equipment
- personal protective equipment (PPE)
- first-aid
- canteen and other welfare facilities
- training
- advice and assistance
- supervision
- power supplies
- transport
- storage facilities
- waste disposal.

Contractors' handbooks

Contractors' handbooks are only likely to be published by larger client companies although smaller companies may still find it appropriate to produce a simple information sheet. In some cases such handbooks may be devoted entirely to health and safety but there is no reason why these topics cannot be combined with other information of relevance to contractors.

It is important to decide whether such handbooks are to be targeted primarily at the contracting company or at their individual employees as this will influence the style to be adopted. Some large companies produce separate handbooks for each category. In either case the handbook should aim to be 'user-friendly' with illustrations and diagrams where appropriate. The purpose of the handbook (or information sheet) should be to inform the contractor, not to serve as a backside-covering device for the client.

Information should be included on risks present within the client's workplace and on the precautions necessary to deal with such risks. Reference should also be made to procedures affecting contractors and services which they may require. Handbooks should not simply include a long list of 'don'ts'. The types of topics which may be covered are set out in figure 15.4.

Special or unusual risks present on the site should also be highlighted within the handbook. Examples of such risks are provided in figure 15.5.

Although contracting companies may be provided with a handbook at the time of placing an order (or even at tender stage) the contents of the document should also be reviewed during any pre-work meetings. Similarly, contractors' employees should be made familiar with topics of relevance to them as part of a structured induction programme. Handbooks which are simply issued, whether to companies or employees, are unlikely to be read.

Both contracting companies and their employees should be required to sign to confirm that they have received a handbook and that they will comply with the standards and requirements therein. The contract company may also be asked to sign to confirm that they have read the handbook, understand its contents and will pass on information of relevance to those employees affected. However, the client should not assume that the existence of such signatures provides a guarantee that this will be the case.

Pre-work meetings

This term is intended to cover informal briefings of the contractor by the client as well as more formal meetings. A meeting may be necessary because the contractor is working on the premises for the first time or it may be required to discuss how a particular contract task is to be approached. It provides an opportunity to pass on to the contractor's representative general information of the types listed in figure 15.4 or to highlight special or unusual risks as illustrated in figure 15.5.

Sections of the contractors' handbook which are particularly important to the contract activities can be reviewed in detail as can any of the client's procedures which are relevant. The use of a permit to work procedure to control contractors' activities (see chapter 12) is particularly advantageous in high risk situations as it forces contact between the client's representative and the contractor and, if properly used, should ensure that all relevant risks are considered

Contractors may need to be provided with information on the following topics, using means such as:

- CONTRACTORS' HANDBOOKS
- PRE-WORK MEETINGS
- INDUCTION OF CONTRACTORS' EMPLOYEES

1. **SITE LAYOUT AND ACCESS**
   Locations of:
   Main departments/buildings
   Major hazards
   Permitted storage areas
   Traffic routes
   Speed limits
   Parking areas

2. **SECURITY**
   Permitted entry routes
   Entry control arrangements
   Contractors' pass systems
   Material removal
   Vehicle searches
   Personal searches

3. **FACILITIES AVAILABLE**
   First-aid
   Canteens
   Eating areas
   Vending machines
   Toilets
   Washing facilities
   Showers
   Changing accommodation

4. **USE OF CLIENT RESOURCES**
   Electric power
   Compressed air
   Water
   Drains
   Waste disposal
   Cranes and lifting equipment
   Materials
   Vehicles
   Access equipment
   Workshop equipment
   Portable equipment
   PPE

5. **HEALTH AND SAFETY RISKS AND PRECAUTIONS**
   Major risks present
   Unusual risks (see figure 15.5)
   Related precautions

6. **PROCEDURES FOR FIRE AND OTHER EMERGENCIES**
   Raising the alarm
   Emergency telephone numbers
   Recognising alarms
   Evacuation routes
   Assembly points
   Roll-call arrangements

7. **STANDARDS AND PROCEDURES**
   PPE standards
   No smoking areas
   Equipment inspection requirements
   Prohibited items or activities
   Isolation procedures
   Permit to work procedures
   Hot work procedures
   Accident and incident investigation
   Health and safety inspections

8. **SANCTIONS FOR NON-COMPLIANCE**

Figure 15.4 *General information for contractors*

What constitutes a special or unusual risk to a contractor will depend on that contractor's level of knowledge or previous experience. The list below provides examples of risks which may specifically need to be pointed out to contractors.

TRANSPORT
The presence of rail traffic, fork lift trucks, cranes, and particularly quiet vehicles.

DANGEROUS ACCESS
Fragile roofs or other non-load bearing areas. Proximity of open water.

ELECTRICAL
High voltage equipment, earth-free areas, unprotected electrical conductors and strong electromagnetic fields.

SERVICES
Buried or overhead supplies of gas, electricity, water etc; high pressure water and compressed air.

HAZARDOUS SUBSTANCES
Chemicals, pathogens and asbestos.

NOISE
Hearing damage risks; sudden or unusual noises eg explosions.

RADIATION
Ionising radiations, lasers and ultra violet radiation.

FLAMMABLE OR EXPLOSIVE SUBSTANCES
Flammable or explosive gases, liquids or dusts.

METAL
The presence of moisture or pressurised containers in molten metal areas. Some hot metal can look the same as cold metal.

PEOPLE AND ANIMALS
The presence of people with disabilities, disturbed people or potentially dangerous animals.

Figure 15.5 *Information to be given to contractors about special or unusual risks*

and all necessary precautions taken. The meeting or briefing may be accompanied by a tour of the client's premises or at the very least a visit to the contractor's intended worksite.

Induction of contractors' employees

The host employer is under a specific duty to *ensure* that visiting employees of contractors are properly briefed about health and safety risks 'arising out of the host's undertaking'. In most cases the host prefers to perform this task himself rather than check that the contracting employer has done it. The extent of the induction necessary and the manner of its delivery will depend on many factors such as:

- the task to be performed or the services to be provided
- risks present on the site
- size of the site
- capabilities or experience of the contractor's employees
- number of contractor's employees
- frequency of use of contractors.

Content

The content of the induction should be chosen from the general information listed in figure 15.4 and any special or unusual risks present, as illustrated in figure 15.5. Whilst some host organisations develop standard induction packages which are used for all contractors' employees, there are also advantages in keeping some flexibility and only providing information which is of relevance to the contractor's activities or work location. It will be both tedious and inefficient (from both parties' viewpoints) to put a contractor carrying out a short-term, low risk activity in a low risk environment through a lengthy and comprehensive induction – unless that contractor is expected to be involved in higher risk work later.

Delivery

An informal oral briefing of the contractor's employee may be all that is required, although the representative of the host employer carrying it out may be wise to prepare an ad-hoc 'aide memoire' to ensure that no important aspects are overlooked. A more structured induction briefing may involve providing the contractor's employee with a handbook or information sheet and highlighting relevant sections of the document. Some larger organisations, particularly those with higher risk sites, have chosen to invest in contractors' videos or computer-based learning packages in order to ensure that a standardised and well-illustrated programme is delivered.

Whichever induction method is chosen, it is important that the visiting contractors are given the opportunity to ask questions. Consequently the person delivering the induction should be in a position to answer such questions or to be able to make prompt contact with someone who can. A tour of the premises or a visit to the worksite and other relevant locations may also be necessary.

Testing and screening

The risks present on the host employer's site may be such that it is necessary to ensure that the contractor's employees have fully understood the information they have been provided with. Their knowledge and understanding may be tested orally (either formally or informally) or the host may choose to administer a written test. Computer-based packages can be made interactive in order to test the understanding of users. In devising tests, simple wording of questions is recommended: the use of multiple choice formats may be desirable.

It may also be necessary at this stage to check whether the contractors' employees actually do possess the qualifications, capabilities or experience that has been claimed for them or which is an essential pre-requisite of the work they are to carry out. Inspection of certificates of competence or informal questioning may be appropriate depending on the circumstances. If the contractor's employees are to be exposed to health risks the host employer may also wish to include them in his health screening arrangements as he would his own employees (see chapter 13).

Records

For higher-risk situations it would be prudent for the host employer to keep records of inductions carried out for contractors' employees. Such records should include the names of employees, the content of the induction, the name of the person carrying out the induction and the results of any tests or screening.

Attendance at an induction could be linked to a contractor's pass system so that only those who have attended an induction (and passed any relevant tests) can gain access to the host employer's premises or the higher risk parts of those premises. Where the host employer wishes to carry out periodic refresher training for contractors' employees he can arrange for such passes to have an appropriate expiry date.

MANAGING CONTRACTORS

Responsibility for liaison

A suitable member of the client's staff must be made responsible for liaising with the contractors on a day-to-day basis. There are advantages if this is the person who has been involved in the pre-work meetings and who has carried out the inductions of the contractors' employees. However, organisational constraints (particularly in large client companies) may mean that these roles are carried out by different people. As well as dealing with queries and requests for assistance from the contractors, the client's representative should be responsible for monitoring the contractor's work from health and safety and other standpoints. He would also be responsible for issuing or arranging the issue of permits to work where these become necessary (see below).

Similarly, a suitable member of the contractor's own staff must be responsible for supervising their work. In the case of small, low-risk operations this supervising role may well be carried out by someone located elsewhere. However, such supervisors should still have a sufficient awareness of what is happening on site (through site visits or regular contact with their staff) to be able to carry out their duties effectively. If the client is not satisfied with the level of supervision at local level he should raise this as an issue with the contracting company. Concerns about the competence or effectiveness of the contractor's supervision should also be raised immediately. For higher risk activities or contracts of lengthy duration, supervision levels may well have been specified before the order was placed.

Access control

There are a number of reasons why the host employer may wish to control the access and egress of contractors' employees to and from his premises. The importance of contractors receiving a suitable induction programme was stressed earlier and there may well be security problems in allowing open access. Contractors' employees will also have to be accounted for in the event of a fire evacuation or other emergency. Several options are available as to how such access is controlled, the choice being dependent on the level of risk, the amount and nature of contractor activity and the layout of the premises.

Signing-in systems

Contractors (or visitors) are required to sign in at the entrance to the premises, usually at a reception desk or a gatehouse. A simple signing-in book is often used but there are also various proprietory systems where, in signing-in, as well as leaving a record of his details, the contractor also prepares a temporary pass which he can wear whilst on the premises. A simple induction can be carried out by the receptionist or security officer or the contractor may be passed on to another member of staff who will carry out this task. Health and safety information may be printed on the temporary pass or provided in the form of a simple information sheet. However, such details are seldom read by the contractor without some encouragement from the host's representative. Another weakness of such systems is that signing-out is seldom as rigorously policed as signing-in, with the result that the host is never really sure who is still on his premises – a major problem in the event of an emergency evacuation.

Pass-card systems

On satisfactory completion of an induction programme, contractors' employees are issued with a pre-printed pass-card. This should certainly contain their personal details and signature and may also include a photograph. Generally such passes should be renewed periodically, typically annually. Whilst these types of passes show that the bearer is authorised to be on the premises and has received a health and safety induction they do not in themselves provide an indication of who is on the site at any given time. Some companies achieve this by arranging for the pass-cards to be placed on a clearly visible 'IN' board on the contractor's arrival. The contractor must then leave by the same route to collect his card or place it on a similar 'OUT' board. A signing-in system may also be used in conjunction with pass-cards but a lack of discipline in signing-out again may be a weakness.

Electronic systems

Such systems are based on cards similar to those described above but with the contractor's details encoded into a magnetic strip within the card. The card is then 'swiped' through a control box on entering and leaving the premises. Whilst the set up costs for such systems are greater, there are many advantages, particularly on a large site on which many contractors operate. Access can be controlled by turnstiles or other physical barriers which open only on the use of a valid card. Cards can be revalidated or cancelled instantaneously by the host company. Problems associated with multiple entrances and exits can be eliminated if the control boxes are linked. The software can also be programmed to give an immediate print out of who is on the premises at any given time (extremely useful in emergency evacuations) or to provide details of the times at which individual contractors arrived or left.

Permits to work

The topic of permits to work is dealt with in some detail in chapter 12. Permits provide a very useful tool in controlling the activities of contractors, particularly in higher risk situations. Some companies are so concerned about the potential dangers associated with contractors that they require all work by contractors to be under a permit to work. Others have developed systems whereby contractors' work is controlled by a simpler contractor's access permit with the more detailed permit to work reserved for their high risk activities.

As stated earlier, the principal advantage of the permit to work approach is that it forces the client's representative and the contractor's representative into contact with each other and ensures that risks are fully considered and appropriate precautions taken before work starts. However, if permit systems are not correctly used they lose much of their value in ensuring high standards of health and safety and become more of a bureaucratic exercise. Chapter 12 described the types of problems which may occur. Use of permits to work to control contractors involved in low risk activities is generally inappropriate.

Monitoring

The importance of the client appointing a representative to liaise with each contractor was stressed earlier as was that representative's role in monitoring health and safety standards. However, all other members of the host employer's workforce (particularly managers and supervisors) have a duty to monitor the activities of contractors and, where necessary, to take appropriate action. Such action may involve making direct contact with the contractors (particularly in situations of imminent danger) or reporting shortcomings to the person liaising with the contractor (or otherwise through the chain of command). Human nature dictates that employees are often more diligent in identifying health and safety failings in respect of contractors than in respect of their own colleagues.

The activities of contractors should also be included in routine health and safety inspections and audits carried out by the host employer. Conformance with PPE standards, suitability of access equipment and compliance with permit to work systems or even their own method statements are all aspects which can be given particular attention during inspections and audits. The host may also wish to be kept informed of monitoring work carried out by others eg the contracting company's own health and safety inspections or visits to the contractor by inspectors from the HSE or local authority.

In most cases it will be sufficient to require the contractor to take corrective action in respect of his health and safety shortcomings. However, in the case of major failings, repetitive problems or a failure by the contractor to take necessary remedial action the host will need to take stronger action. This might involve reading the riot act to the contractor's staff locally, calling in more senior staff from elsewhere or even dispensing with the contractor's services, although this last course may present practical difficulties. Certainly the client should not expect the HSE to carry out any dirty work on his behalf – the responsibility is his. The use of work completion reports is particularly useful in ensuring that contractors who do not conform to standards are not used again. This subject is dealt with later.

Accident and incident investigation

The host employer will certainly wish to know about any significant accidents or incidents involving contractors working on his premises, since all such occurrences have the potential to also involve his own employees or other contractors. Whether the host requires the contractor to follow the local investigation procedure or accepts copies of reports made under the contractor's procedure will be a matter for negotiation between the parties. However, the host can never be sure whether all accidents and incidents are actually being reported to him by the contractor. The periodic checking of accident books together with close monitoring of the contractor's activities should help ensure a good standard of compliance.

Accidents or incidents resulting in investigations by HSE or local authority inspectors will be of particular interest to the host employer. Often during such investigations the effectiveness of his own systems for controlling contractors will come under the inspector's scrutiny. As with monitoring activities, the client should ensure that suitable action is taken by the contractor after investigations, or be prepared to take appropriate action himself. Accident investigating and reporting is dealt with in greater detail in chapter 9.

Review meetings

Meetings to review the performance of contractors should be held periodically throughout the duration of long contracts or at the time of renewal of service contracts. Compliance with health and safety standards may well be just one of a number of issues to be addressed, as is the case during the initial selection of contractors. Feedback from the client's representative in dealing with the contractor will be particularly important in this process as will the contents of work completion reports where these are used (see below).

During such reviews the client will have to decide whether or not to continue to use the contractor's services. In the majority of cases the relationship is likely to continue. Consequently the contractor's representatives should be involved in the review process in order to determine how the parties can more effectively work together to ensure high standards in health and safety and other aspects of the work. This will be particularly important in long contracts where there will need to be planning for the health and safety implications of different phases of major projects.

Work completion reports

Such reports are likely to be of particular value to large client organisations for whom significant quantities of contract work are being carried out under the control of a variety of local co-ordinators. They help the client to build up an overall picture of the quality of each contractor's work in respect of health and safety and other relevant criteria.

Reports would normally be completed by the representative of the client company appointed to liaise with the contractor. They would take account of such features as the methods of work used by the contractor, the quality of their management of health and safety, any significant incidents or problems that occurred and the condition of the work site on completion of the work. Figure 15.6 contains an example of a work completion report form which can be

CONTRACTOR'S WORK COMPLETION REPORT		
Contract company		
Nature of work	Order no.	
Report made by	Date	
		Comments
1 METHODS OF WORK 1.1 Suitability of systems of work 1.2 Conformance with method statement 1.3 Compliance with permit to work system 1.4 Conformance with PPE standards 1.5 Housekeeping standards 2 CONTRACTOR'S MANAGEMENT OF HEALTH AND SAFETY 2.1 Availability of employees for induction 2.2 Quality of supervision 2.3 Capabilities of employees 2.4 Response when health and safety action required 3 SIGNIFICANT INCIDENTS OR PROBLEMS 3.1 Injuries (contractor's employees or others) 3.2 Damage or near miss incidents 3.3 Adverse comments by HSE inspectors 3.4 Environmental incidents or problems 3.5 Security incidents or problems 4 COMPLETION INSPECTION OF WORKPLACE 4.1 Workplace left clean and tidy 4.2 Guards replaced 4.3 Any other residual risks? 4.4 Borrowed equipment returned in good condition		
5 OTHER COMMENTS (positive or negative)		
6 HEALTH AND SAFETY RATING (award from 0 – 10 marks) 0 = should never be used again 10 = excellent performance		

Figure 15.6 *Contractor's work completion report*

adapted to meet local needs. Whilst smaller client organisations may not see the need to use such a form they should nevertheless apply the same principles in evaluating the performance of their contractors.

Notes

1. [1981] IRLR 403. Law report abbreviations are explained on p16.
2. If the provision of a safe system of work for the employer's employees depends on the training of non- employees this will be required by s2. Where training is needed for the safety of the non-employees themselves it will be required by s3.

Part III

Practical Health and Safety Issues

16

Workplace Health, Safety and Welfare

INTRODUCTION

The law relevant to the workplace itself – dealing with matters such as the fabric of the workplace, the working environment and welfare facilities – is found in the Workplace (Health, Safety and Welfare) Regulations 1992 (the Workplace Regulations).[1] These are general in that they apply to various types of workplace, although with exceptions (see below). They are accompanied by an ACOP and guidance notes.

In addition to the Workplace Regulations, this chapter deals with the statutory provisions governing first-aid. These are found in the Health and Safety (First-Aid) Regulations 1981. Finally, it deals briefly with three other, unrelated areas: safety signs; violence to staff; and pressure systems.

The Workplace Regulations seek to implement much of the EU's directive on Workplace Health and Safety,[2] although some areas of UK law (eg first-aid) are thought to meet the directive's requirements already, and in such cases there are to be no changes. Implementation is also being achieved through the Fire Precautions (Places of Work) Regulations (see chapter 21).[3] Health and safety on temporary construction sites is covered by the Construction (Design and Management) Regulations 1994 (see chapter 15).[4]

'Workplace' is defined to mean any premises (except domestic premises) which are made available to any person as a place of work.[5] It includes "any place within the premises to which such person has access while at work" and other places (eg corridors and staircases) used for access, egress and the provision of facilities in connection with the workplace, but excluding public roads. "Premises" includes any place and can be an outdoor place.[6] Workplace also includes the common parts of shared buildings and private roads and paths on industrial estates and in business parks. The regulations cover both 'permanent' and temporary workplaces: the latter include sites used infrequently or for short periods and structures, such as fairs, which occupy a site for a short period. Some of the regulations are modified in their application to temporary workplaces: provisions for sanitary conveniences, washing facilities, drinking water, accommodation for clothing, changing facilities and facilities for rest and eating meals apply only "so far as is reasonably practicable".[7]

The regulations apply to all workplaces except:

- *means of transport* – ships, boats, hovercraft, trains, aircraft and road vehicles are excluded. However, the provisions relating to falls and falling objects apply to aircraft, trains and road vehicles when stationary in a workplace, although not when on a public road. Means of transport which are not operational (eg a stationary boat or train used as a restaurant) would not be excluded from the regulations.
- *extractive industries* – mines, quarries and other mineral extraction, where this is the only activity, are excluded, as is work preparatory to this activity. Separate legislation applies.[8]
- *construction sites* – that is, where the only activities are "building operations" or "works of engineering construction".[9] Where construction is taking place within a workplace and the area is fenced off from normal operations, this is a construction site and is excluded. Otherwise, both these regulations and construction regulations apply.
- *farming and forestry* – outdoor workplaces which are away from the main buildings are excluded except that the provisions for sanitary conveniences, washing facilities and drinking water apply so far as is reasonably practicable.[10]

The regulations place duties upon employers in respect of workplaces under their control where any of their employees work. Duties are also placed on controllers of premises as regards matters over which they have control, with an exception for the self-employed in respect of the work of themselves or of any partner.

LEGISLATION COVERING SPECIFIC TYPES OF WORKPLACE

Factories Act 1961[11]

Until repealed and replaced by regulations under HSWA, parts of the Factories Act 1961 remain operative. In general, the person or company in immediate control of the premises is responsible for ensuring compliance with the Act and the many sets of regulations issued under it.

A factory is a place where people are employed in manual labour in any process for or incidental to the making, repairing, altering, cleaning, adapting for sale or demolition of any article.[12] With a number of exceptions (for example government, local authorities) this must be done for gain or for trading purposes. An employer will have rights of access and control. Manual labour means that the work is done primarily or substantially with the hands. A number of specific processes are also defined as factory processes so that they are covered by the Act; these include shipbuilding or repair or breaking up.

A person intending to use premises as a factory must notify the inspector at least four weeks in advance.[13] Every factory must have a general register – HMSO Form F31 suffices – which includes, among other things, details of the work carried on, and has attached to it evidence of various tests and examinations as required by statute.[14] In prescribed circumstances a list of homeworkers must be provided to the local authority and inspector.[15]

Where premises are occupied by separate factories, the owner is responsible for the common parts. Otherwise, the occupier is liable.[16] Where a particular individual fails to fulfil a statutory duty imposed upon them and the employer has taken all reasonable steps, the employer will not be liable.

Offices, Shops and Railway Premises Act 1963

Most of the Act's provisions are similar to those of the Factories Act 1961 including the requirement to register the business premises with the relevant enforcing authority.[17] Usually this will mean notifying the local authority. However, some offices and shops, such as those in places covered by the Factories Act need to be notified to the HSE. With a few exceptions, the Act applies wherever people are employed. It applies generally to all offices and shops, and to most railway buildings near the permanent way. The Act covers offices and shops which are part of buildings used for other purposes as well as associated areas such as stairs, passages and toilets.

Office premises means a building or part of a building the sole or principal use of which is an office or for 'office purposes'. The latter includes administration, clerical work, handling money and telephone and telegraph operating. Ticket offices, travel agents and betting offices are all within the coverage of the Act. Shop premises include retail shops, wholesale premises or warehouses, catering establishments open to the public and fuel storage premises. The definition includes hairdressers, repair shops, dry cleaning shops, launderettes, retail sales by auction – wherever retail trade or business, repairs or treatment are carried on in the public domain.

Generally, the occupier is responsible for complying with the Act, although some responsibilities are given specifically to employers (such as the duty to notify the inspectorate of the commencement of business). If the premises are used by more than one employer the owner becomes responsible for health and safety in the common parts. The defence against alleged breach of duty under the Act is that all due diligence was used.[18]

Mines and quarries

The principal Act regulating the health and safety and welfare of those employed in the extractive industries is the Mines and Quarries Act 1954: the provisions of the Act have been substantially extended by regulations made under section 141. Also of note is the Mines and Quarries (Tips) Act 1969 and the associated Mines and Quarries (Tip) Regulations 1971: the 1969 Act resulted from the recommendations of the enquiry into the Aberfan disaster in 1966.

The 1954 Act is unusual in so far as it imposes specific duties on owners of mines and quarries to make provision for, and ensure that, the workplace is managed and worked in accordance with the requirements of the Act and relevant orders and regulations. The Health and Safety at Work etc Act 1974 also applies. Part II of the 1954 Act, in the case of mines, and part IV, in the case of quarries, provide for the formal appointment of managers and specify their duties, powers and responsibilities. The duties under the 1954 Act and related legislation are extensive and complex and a detailed analysis is not attempted here.

Progress has been made on the gradual process of modification and amendment of health and safety legislation in relation to mines and quarries so as to bring it under the umbrella of the Health and Safety at Work etc Act 1974 and related regulations and approved codes of practice: this will continue. It should be noted that whilst the provisions of the Electricity at Work Regulations 1989, the Noise at Work Regulations 1989 and the Control of Substances Hazardous to Health Regulations 1994 apply to both mines and quarries, Regulations 6 to 12 of the Control of Substances Hazardous to Health Regulations are specifically not applied to work carried out below ground at a mine.

Unlike the Factories Act and Offices, Shops and Railway Premises Act, the relevant provisions of the mines and quarries legislation have not been repealed in favour of the Workplace Regulations. Therefore, the mines and quarries legislation remains in force for workplaces within its scope and the Workplace Regulations do not apply, even after 31 December 1995.

Agriculture

Agriculture, including horticulture, forestry and associated industries, is subject to the Health and Safety at Work etc Act 1974 and regulations thereunder including COSHH, Noise at Work and Electricity at Work (see chapters 20, 22 and 23). Indoor workplaces, and those outdoor workplaces near to the main buildings, are covered by the Workplace Regulations. However, as noted above (p350), outside workplaces, away from main buildings are excluded or are subject to the regulations in modified form.

The construction industry

Health, safety and welfare on construction sites is provided for principally by separate legislation: as noted, construction sites are excluded from the Workplace Regulations 1992. Health and safety on construction sites is dealt with in detail in chapter 26, with transport and material handling aspects considered in chapter 18. Here the aim is solely to consider the legal provisions which are equivalent to those laid down for other workplaces in the Workplace Regulations. In the main, these are found in the Construction (Working Places) Regulations 1966.

The regulations require suitable and sufficiently safe means of access to and egress from any place of work and the Construction (General Provisions) Regulations provide specifically for means of exit in case of flooding. The workplace must be kept safe. Scaffolding work is subject to detailed regulation, as is work on sloping roofs and ladderwork. There are detailed requirements relating to working platforms, gangways and runs.

The Construction (Health and Welfare) Regulations 1966 provide for shelter for employees and accommodation and drying facilities for clothing, including protective clothing. There must be dry facilities for taking meals and boiling water. Drinking water and facilities for hot food should be available. Accommodation must be kept clean and orderly and not used for storing plant or materials. The regulations also provide for washing and sanitary facilities. Some of the requirements differ according to the number of employees and the duration of the work. Adequate and suitable protective clothing is required for anyone who has to work outside in snow, rain etc.

THE WORKPLACE REGULATIONS

General requirements

By virtue of regulation 4, every employer has a duty to ensure that a workplace under his control, and where any of his employees works, complies with the requirements of the regulations. Any person, who in connection with the running of a business or undertaking is in control of a workplace also has a duty to ensure such compliance as regards matters under his control. This affects liability for the common parts of shared buildings, such as stairs and landings. Everyone deemed a factory occupier by the Factories Act 1961 must comply. There is no such duty placed upon the self-employed in respect of their own work or the work of any partner.

Managers will need to ensure that standards in 'new workplaces' – those coming into use after 31 December 1992 – meet the requirements of the Workplace Regulations. The same is true of workplaces where changes were started and completed after that date. Existing workplaces – those already in existence on or before 31 December 1992 – should be upgraded by the beginning of 1996. Particular attention may be necessary in the case of workplaces not previously within the ambit of specific, workplace legislation (eg schools, hospitals and hotels) although these were already covered by the general requirements of HSWA.

The remainder of this section sets out the requirements of the Workplace Regulations, beginning with those that are of a general nature.

- *Maintenance* (regulation 5) – The workplace and related equipment, devices and systems must be maintained in an efficient state, efficient working order and good repair. Maintenance includes cleaning and "efficient" means efficient in terms of health, safety and welfare rather than productivity or economy. The equipment, devices and systems concerned are mechanical ventilation systems and any other equipment where a fault is liable to result in a breach of the regulations.

Where appropriate, there should be a suitable *system* of maintenance. Such a system involves:

- regular maintenance at suitable intervals
- remedying dangerous defects, with access prevented in the meantime
- properly carried out maintenance and remedial work and
- keeping a suitable record.

Examples of equipment and devices where such a system would be "appropriate" include: emergency lighting, fixed equipment used for window cleaning, devices to limit the opening of windows, escalators and mechanical ventilation systems.

- *Cleaning and waste materials* (regulation 9) – The workplace and its furniture, furnishings and fittings must be kept sufficiently clean. (The ACOP states that standards will depend on the type of workplace: floors and indoor traffic routes should be cleaned at least once a week.) Surfaces of internal floors, walls and ceilings must be capable of being kept sufficiently clean. So far as is reasonably practicable, waste materials must not be allowed to accumulate except in suitable receptables.

Workplace safety

- *Floors and traffic routes* (regulation 12) – These must be of suitable construction and surface, have no hole or slope nor be uneven or slippery, and have drainage where necessary. (Transport is dealt with more generally in chapter 18.) "Suitable" means suitable for the purpose for which it is used. Staircases should have guards and hand rails. So far as is reasonably practicable, floors and traffic routes shall be kept free from obstructions and articles or substances which may cause slips, trips or falls.

- *Falls or falling objects* (regulation 13) – So far as is reasonably practicable, physical safeguards must be provided to give protection against: falls from heights; falling into dangerous substances; and falling objects. (On safety in respect of ladders, scaffolding, access platforms etc see chapter 26.) Other measures, such as restrictions on access, must be taken when physical safeguards are not reasonably practicable. The application of regulation 13, unlike the other substantive workplace regulations, extends to means of transport, but only when the transport is stationary inside a workplace, and in the case of a licensed vehicle only when it is not on a public road.

- *Glazed doors and partitions* (regulation 14) – Where necessary for reasons of health or safety, these must be of safe material or protected against breakage, and must be appropriately marked or otherwise apparent.
- *Windows, skylights and ventilators* (regulations 15 and 16) – These must be designed for safe opening, closing and adjusting, with provision for the safe cleaning of windows.
- *Traffic routes* (regulation 17) – These should be organised for pedestrians and vehicles to circulate in a safe manner. They should be in suitable positions, and be of sufficient number and size, with sufficient separation of vehicles and pedestrians. (On transport, see chapter 18.) The general requirement for suitability applies also to existing workplaces, but only so far as is reasonably practicable. A "traffic route" is for pedestrians, vehicles or both, and includes stairs, fixed ladders, doorways, gateways, loading bays and ramps.
- *Doors and gates* (regulation 18) – These must be of suitable construction and should incorporate various design features.
- *Escalators and moving walkways* (regulation 19) – These should function safely, have any necessary safety devices and incorporate at least one emergency stop.

Workplace health: the working environment

- *Ventilation* (regulation 6) – There must be sufficient fresh or purified air and a breakdown warning device if "necessary for reasons of health and safety". The requirement is for "effective and suitable" provision in every enclosed workplace, with certain exceptions.
- *Temperature in indoor workplaces* (regulation 7) – This should be reasonable during working hours, with non injurious/offensive methods of heating/cooling used and sufficient thermometers provided. (The ACOP specifies a minimum of 16°C, reduced to 13°C if "severe physical effort involved".) These temperatures do not apply where lower temperatures are required by law for food safety reasons.
- *Lighting* (regulation 8) – This must be suitable and sufficient and natural so far as reasonably practicable. Emergency lighting should be provided where special risks exist if artificial lighting fails. Lights should also be placed at places where there are particular risks, such as pedestrian crossing points on traffic routes.

- *Room dimensions* (regulation 10) – There should be sufficient floor area, height and unoccupied space. (The ACOP requires 11 cubic metres per person.) Space more than 3m high should be ignored when making the calculation, and in determining the number of people to work in a room account needs to be taken of the space occupied by furniture and equipment. There are exceptions for workplaces where space is "necessarily limited", such as kiosks and cabs.
- *Workstations and seating* (regulation 11) – Workstations must be arranged so as to be suitable for the people and the work. There are some specific requirements for outdoor workstations. A suitable seat must be provided where work can be done sitting, together with a suitable footrest where necessary. (These and related issues are considered in more detail in chapter 19.)

Welfare facilities

As noted, temporary workplaces need to comply only as far as is reasonably practicable.

- *Sanitary conveniences* (regulation 20) – "suitable and sufficient" sanitary conveniences must be provided at readily accessible places. The minimum numbers of facilities are set out in tables in the ACOP.[19] Sanitary conveniences will not be "suitable" unless they are adequately ventilated and lit, kept in clean condition and provided separately for men and women. Sanitary conveniences may still be "suitable" even if not separately provided if each convenience is in a separate room the door of which can be secured from the inside.
- *Washing facilities* (regulation 21) – There are detailed requirements in the ACOP. Showers (or baths) should be provided for particularly strenuous or dirty work.
- *Drinking water* (regulation 22) – This should normally be provided by water jets or through taps/drinking vessels. Use of enclosed refillable containers (refilled daily) is acceptable where there is no mains supply.
- *Clothing storage* (regulation 23) – At least pegs or hooks should be provided for outdoor clothing and special work clothing. Lockable lockers or other secure storage must be provided if changing facilities are required.
- *Facilities for changing clothing* (regulation 24) – These will be needed if special clothing is required for work and the person cannot change in another room.
- *Facilities for rest and to eat meals* (regulation 25) – Suitable and sufficient rest facilities must be provided at readily accessible places. In new workplaces facilities must be separate (if necessary for health and safety).

Facilities for eating meals must be provided if there is a contamination problem. Suitable arrangements must be made to protect non-smokers from discomfort from tobacco smoke. (The ACOP indicates separate areas, rooms or prohibition.) Rest facilities must be provided for pregnant women or nursing mothers.

FIRST-AID

Legal requirements

Most legal requirements in relation to first-aid are contained in the Health and Safety (First-Aid) Regulations 1981 supported by an Approved Code of Practice (ACOP) and Guidance Notes. All of these are brought together in the HSE booklet COP42.[20]

Meaning of 'first-aid' (regulation 2)

First-aid is defined in the Regulations as meaning:

- In cases where a person will need help from a medical practitioner or nurse, treatment for the purpose of preserving life and minimising the consequences of injury and illness until such help is obtained.
- Treatment of minor injuries which would otherwise receive no treatment or which do not need treatment by a medical practitioner or nurse.

Duty of employer to make provision for first-aid

Under regulation 3 an employer must provide, or ensure that there are provided:

- such equipment and facilities as are 'adequate and appropriate'
- such number of suitable persons as is 'adequate and appropriate'

to enable first-aid to be administered to his employees if they are injured or become ill at work.

A 'suitable person' may be a trained first-aider (supported for temporary absences by an appointed person) or in low risk situations could simply be an appointed person (see p360).

What is 'adequate and appropriate' is dealt with in the ACOP and guidance notes (see below).

Information

Regulation 4 requires the employer to "inform his employees of the arrangements that have been made in connection with the provision of first-aid, including the location of equipment, facilities and personnel".

The ACOP states that employees should be given this information during their initial induction, with fresh information being given each time they move to a new place of work. It requires at least one notice to be posted in a conspicuous position giving the locations of the equipment and facilities and the name(s) and location(s) of the first-aider(s) or appointed person(s).

The notice should comply with the Safety Signs Regulations 1980 and should be in English with any other language commonly used displayed alongside.

Self-employed persons (regulation 5)

The self-employed must provide, or ensure that there is provided, adequate and appropriate first-aid equipment for their own first-aid.

Application/non application (regulation 7)

The regulations DO NOT APPLY:

- where the Diving Operations at Work Regulations 1981 apply (these contain specific requirements relating to diving first-aid)
- where the Merchant Shipping (Medical Scales) (Fishing Vessels) Regulations 1974 apply
- where the Merchant Shipping (Medical Scales) Regulations 1974 apply
- on vessels which are registered outside the United Kingdom
- to a mine of coal, stratified ironstone, shale or fireclay
- in respect of the armed forces of the Crown and any force to which any provisions of the Visiting Forces Act 1952 applies.

The regulations DO APPLY to:

- mines not subject to the exclusion above (with the mine manager holding the employer's duties)
- offshore employment not subject to the exclusions above.

First-aid equipment and facilities

The ACOP and the guidance notes leave much to the employer's judgement as to what is 'adequate and appropriate' for his workplace.

Factors to take into account

The guidance notes refer to the following factors which must be taken into account in deciding the level of provision:

- the type of work being carried out and its level of hazard
- availability of other places of treatment (are there hospitals nearby?)
- work away from the employer's premises.

Other factors to be considered might also include

- agreements to provide equipment and facilities for the employees of other employers (such agreements should be in writing with a copy kept by each employer)
- the presence of others who might need first-aid treatment, eg school pupils, shop customers etc (there is no obligation under these regulations to provide first-aid for such people).

Equipment and facilities to be provided

The ACOP states very clearly that first-aid boxes and kits should contain a sufficient quantity of suitable first-aid materials **and nothing else**. The treatment of minor illnesses with tablets or medicines is outside the scope of the regulations.

The minimum contents of boxes and kits is specified in the guidance notes accompanying the regulations, as is a range of supplementary equipment (eg blankets) which may be appropriate, depending on the circumstances.

Higher risk workplaces should be provided with a suitably equipped and staffed first-aid room. It is left to the employer to decide when this is needed. However, where a first-aid room is provided, the ACOP sets out detailed requirements.

First-aiders and appointed persons

Employers must provide 'such number of suitable persons as is adequate and appropriate' to render first-aid, although for temporary absences or in low risk situations an 'appointed person' may be substituted.

Suitable persons

A 'suitable person' is defined in the ACOP as a first-aider with a current first-aid certificate issued by an organisation approved by the HSE. Practising medical practitioners and practising nurses who are registered on Part 1, 2 or 7 of the Single Professional Register are also regarded as first-aiders.

Adequate and appropriate

The ACOP steers away from relating the number of first-aiders to the numbers of employees at work. It identifies several factors to be taken into account:

- the distribution of employees
- the nature of the work
- the size and location of the workplace
- shift working
- the distance from outside medical services
- the numbers of employees.

The appointed person

An appointed person is appointed by the employer to take charge of the situation (eg to call an ambulance) if a serious injury or illness occurs in the absence of a first-aider. If trained to do so, the appointed person can also give emergency first-aid treatment.

In most circumstances the 'appointed person' provides cover for the absence 'in temporary and exceptional circumstances' of a first-aider. Foreseeable absences such as planned annual leave do not fall into this category.

However in workplaces where, because of their nature and location, there are no specific hazards and the number of employees is small, an appointed person may be provided instead of a first-aider. In such cases the appointed person is responsible for the first-aid equipment and facilities.

First-aider training

In order to gain HSE approval organisations must satisfy several criteria: the detail of relevant guidance is contained in COP 42. Specific or additional training may be required if special risks exist.

Emergency first-aid training

The guidance notes also allow for shorter emergency first-aid courses to be given at the workplace by occupational health staff or in short courses by HSE approved organisations. Such courses are intended to cater for 'appointed persons', employees working in small groups away from their employer's establishment or where specific hazards exist.

Records

A written record should be kept of the dates on which first-aiders obtained their certificates (including any additional or specific hazard training) together with details of refresher training. Records must also be kept in a suitable place (eg with the first-aid equipment) of all cases treated.

OTHER ASPECTS OF WORKPLACE SAFETY

Safety signs

The Safety Signs Regulations 1980

The 1980 regulations state that safety signs must comply with the statutory requirements. However, they do not require safety signs to be erected. The requirements apply to all workplaces except coal mines. Signs on internal roads must conform to the Road Traffic Acts.

There are four types of signs:

- *prohibition* – circular with red border and red bar across a black symbol on a white background
- *warning* – triangular, with black border and black symbol on yellow background
- *mandatory* – circular with a blue background; symbols in white giving specific instructions
- *safe condition* – square or oblong, green background with white symbols.

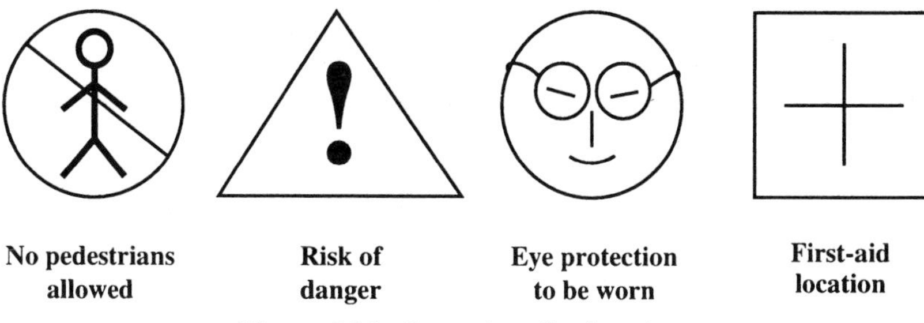

Figure 16.1 *Examples of safety signs*

The Health and Safety (Safety Signs and Signals Regulations 1995

These will replace the 1980 Regulations and will give effect to a 1992 European Union directive on safety signs.[21] Employers will now be required to use safety signs whenever there are hazards to health and safety which cannot be avoided or adequately controlled by other means. The term "safety sign" is extended to include other means of communication, such as:

- hand signals
- verbal signals
- acoustic signals, eg fire alarms
- coding, eg of pipework
- marking, eg of traffic routes and
- illuminated signs.

The number of conventional signs has been increased, particularly with regard to the identification of fire-fighting equipment. Most of the safety signs in use in the UK are thought to meet the new EU requirements and it is not expected that there will be need for significant change. The new regulations do not apply to signs used in connection with transport or with the marking of dangerous substances. Currently-permitted fire safety signs are likely to be allowed to remain in use for a transitional period, provisionally until 24 December 1998.

Violence to staff

In a number of occupational areas, risk assessments may point to the need for measures to protect staff against violence, eg by patients in the health services or shoplifters in the retail sector. In retailing, there were over 12,000 incidents of violence to staff during the year 1993-4, including three fatalities.[22] Almost half of the incidents arose when staff confronted shoplifters. Banks and building societies face similar problems and violence is also an issue for local authorities and government departments, especially where staff deal with the public.[23]

Violence in this context is principally physical, but the concept can also be said to include verbal abuse and threatening behaviour. The psychological effects of violence may develop into anxiety or depression.

Staff exposed to this type of risk should be trained to identify potentially violent situations so that they can handle them effectively. There may also be a need for physical safeguards such as protective glass screens or for the control of access within buildings. Panic buttons, personal alarm devices, portable telephones or radios may be appropriate.

Violence or the threat of it involves 'serious and imminent danger' to employees and therefore 'appropriate procedures' must be established by the employer to deal with such situations. Such procedures are required by regulation 7 of the Management Regulations. Chapter 10 deals with emergency procedures in more detail.

Support services for those who have experienced violence or serious threats could also be needed. Whilst managers or occupational health services may deal with such cases in the first instance there may be a need for specially trained counsellors.

Violence prevention programmes should include a system for the reporting of violence, specifying who was involved, a full description of what happened and the reasons. The report should also specify where and when the incident occurred and which protective systems failed. Such reports should be signed by the victim and the victim's immediate supervisor. Use could be made of established accident and incident investigation and reporting procedures (see chapter 9).

However, even if the organisation has well-developed policies and procedures to combat violence, the individual employee should also do everything possible to ensure his or her own safety. This could include identifying any risks and considering how they might be reduced or removed. The guidance below is given by the Suzy Lamplugh Trust:[24]

- *Do not ignore warnings* eg feelings of uneasiness; or being scared.
- *Be alert*: wear clothing which allows movement; keep some money, keys and emergency numbers for credit cards separate from your bag or wallet.
- *Avoid putting yourself at risk*: when away from the office leave written details of your movements. Walk facing the traffic on the street-side of the pavement; if necessary take taxis after dark.
- *Take action if in real danger*: if in risk areas, such as car parks, carry a shriek alarm and use it next to the attacker's ear; then move away quickly.

The above points could be set out in any guidance or training provided by an employer. From April 1996 employers will have to notify to the enforcing authorities acts of violence against persons at work. (On notification generally, see pp194–9).

Pressure systems and transportable gas containers

The Pressure Systems and Transportable Gas Containers Regulations 1989[25] were introduced in stages and eventually became fully operational in 1994. They extend many of the safety requirements previously applying to steam and compressed air plant to include other types of pressurised system. Their intention is to prevent the risk of serious injury from stored energy as a result of a failure within the pressure system.

Pressure systems are those containing or liable to contain a 'relevant fluid' ie steam, gases under pressure or fluids artificially under pressure which become gases on release to the atmosphere. The regulations do not apply to hydraulic systems, systems containing traces of dissolved gas or to liquids in storage tanks which exert a static pressure. Systems in which gases exert a pressure of 0.5 bar or below are also excluded. Some vessels with only small amounts of stored energy are exempt from parts of the regulations.

A 'pressure system' includes pressure vessels of rigid construction, pipework, pipelines and associated protective devices. The term 'transportable gas container' in effect means a gas cylinder. Full definitions of the terms used in the regulations are contained in regulation 2.

Most of the duties in respect of pressure systems are placed upon 'users' in the case of fixed (installed) systems or 'owners' in the case of mobile systems. However, the owner is allowed to assume the user's obligations under contract in respect of leased installed systems – often the case for bulk oxygen or LPG systems. The 'user' is defined as the employer or self-employed person having control of the operation. Designers, manufacturers, importers and suppliers of both pressure systems and gas cylinders have duties under regulations 4 and 5 as do those involved in installation, modification or repair work. Installers have additional duties under regulation 6.

A very important part of the regulations is the requirement for the drawing up of a written scheme of examination for each pressure system (regulation 8) and then for the carrying out of such an examination in accordance with the written scheme (regulation 9). Both of these activities must be carried out by a competent person. Although a definition of competent person is not provided within the regulations, guidance is given within the ACOP *Safety of Pressure Systems*.[26] This subject is reviewed in greater detail on p136 above.

The competent person may be an individual or an organisation and a different person may be involved in drawing up the written scheme of examination from the one carrying it out. For those employers who do not have a suitable resource available in-house the services of a competent person are likely to have to be provided by the suppliers of the pressure system, an insurance company engineering inspection section or by specialist independent engineers.

Pressure systems must not be operated unless safe operating limits have been established (regulation 7) nor if the examination by the competent person reveals a situation of imminent danger (regulation 10). In the latter circumstances the competent person must within 14 days send a written report to the enforcing authority.

Adequate and suitable instructions must be provided for the safe operation of pressure systems and for the action to be taken in the event of an emergency. The system must be operated in accordance with the instructions (regulation 11). Systems must be properly maintained in good repair (regulation 12) and various records must be kept in respect of each system (regulation 13).

Where a vessel has a permanent outlet to the atmosphere (or a space at atmospheric pressure), precautions must be taken to prevent the obstruction of this outlet if the obstruction would cause a build up of pressure within the vessel (regulations 14 and 15).

Transportable gas containers are covered specifically by regulations 16 to 22. Requirements for design standards, approval and certification are contained in regulation 16. Regulation 17 requires checks to be made before the filling of containers to establish that they are marked as having been examined within the appropriate period and that they are suitable for the particular fluid. Any other appropriate safety checks must also be carried out. After filling there must also be a check to ensure that the container is within its safe operating limits and is not overfilled. Any excess fluid must be removed in a safe manner. Non-refillable containers must not be refilled. Further details of the types of checks required are contained in the relevant ACOP.[27]

The requirements for transportable gas containers to be examined at appropriate intervals by a competent person and then marked to show the date of examination is imposed by regulation 18. Modifications to containers, repair work and re-rating of containers are covered by regulations 19 to 21 whilst regulation 22 details the records which must be kept.

Regulation 23 allows for due diligence defences to be made against charges under the regulations and regulation 24 gives the HSE the power to grant exemptions. Many exceptions to the regulations are listed in Schedule 2. Full details of these are contained in the two ACOPs associated with the regulations and in the comprehensive booklet of guidance.[28]

THE WORKPLACE AND CIVIL LIABILITY

Negligence

As indicated in chapter 4, the common law duty of care embraces a duty to provide safe premises. The duty of care is defined by weighing the risks – likelihood and seriousness of injury – against the costs and other inconvenience of taking precautions. Thus, an employer can be fulfilling his duty of care even where the workplace gives rise to significant risks if the only precautions available are extremely costly and disruptive (eg turning off the current on an electric railway to allow minor maintenance – *Hawes v Railway Executive*). However, in most practical situations alternative precautions will be available which can reduce the risk without entailing disproportionate costs.

The incurring of substantial costs will not be required where the foreseeable risk is slight, such as the risk of slipping on a wet and oily floor. This will be so even if the actual injury suffered, a broken leg, is more serious, because such an injury was not reasonably foreseeable. The risk is from slipping, and a broken leg is unlikely. Closing part of the factory would be disproportionate to the risk of slipping (*Latimer v AEC*). Where the risk is of imminent serious injury or death, the expenditure required to discharge the duty of care may be considerable. An employee working on a narrow ledge at the top of a dock, with the risk of serious injury or death if he falls, will therefore need protection against falling even if the precautions result in the employer incurring substantial costs (*Bath v British Transport Commission*).

Breach of statutory duty

As noted in chapter 4, another basis for civil action, in respect of workplace health and safety is breach of statutory duty. In *Ebbs v James Whitson and Co Ltd*, the court reserved its opinion on the point of whether there could be an action for damages based upon breach of the Factories Act requirement for adequate ventilation. However, subsequently there have been successful actions on the basis of this (*Nicholson v Atlas Steel Ltd*) and other Factories Act provisions. The position as regards welfare requirements such as washing facilities is less clear, but an action based on breach of the Factories Act provision for suitable accommodation for clothing succeeded in *McCarthy v Daily Mirror*. A distinction might be possible between welfare (ie comfort and convenience) and health and safety provisions, but where the former affect the latter (such as absence of washing facilities leading to dermatitis – *Reid v Westfield Paper Co Ltd*) there should be a cause of action. However, the opposite view was expressed in *Clifford v Challen and Son Ltd.*

Nothing in the Workplace Regulations expressly excludes their use in civil actions, unlike, for example, the Management Regulations, so the presumption may be that they can be so used. There are, in addition, duties under the Occupiers' Liability Act 1957 which require the occupier to ensure the premises are reasonably safe.

An action for damages for breach of statutory duty in respect of an unsafe workplace will succeed only if the plaintiff can show that he is one of a class of persons whom the statutory provision seeks to protect (eg an employee). The injury must be of a kind that the statutory provision aims to prevent, and the employer must be in breach of the duty. Thus, the requirement to provide a work platform at least 34 inches wide if a person is liable to fall 6 feet 6 inches (two metres) is not activated when the person is liable to fall only 6 feet (*Chipchase v British Titan Products Ltd*). The breach of duty must be shown to have caused the injury.

As with negligence cases, what is reasonable is a question of fact and degree in each case.

Notes

1. SI 1992/3004
2. 89/654/EEC
3. As yet, still in draft form
4. SI 1994/3140
5. Workplace Regulations, regulation 2(1). On the meaning of work, see chapter 3.
6. HSWA, s53
7. Regulation 3(2)
8. See particularly, Mines and Quarries Act 1954
9. Factories Act 1961, s176
10. Regulation 3(4)
11. See *Redgrave, Fife and Machin: Health and Safety*, Hendy J & Ford, M, Butterworths, for comprehensive coverage of the subject.
12. Factories Act 1961 (FA), s175
13. FA, s137
14. FA, s140
15. FA, s133
16. FA, s155
17. Offices, Shops and Railway Premises Act 1963 (OSRPA), s49
18. Offices, Shops and Railway Premises Act 1963, s67
19. Paras 201-5
20. Health and Safety (First-Aid) Regulations, SI 1981/917, ACOP and guidance notes
21. EC directive 92/58/EEC replaced EC 77/576/EEC, on which the 1980 regulations were based.
22. *Retail Crime Costs*, 1993–4, London: British Retail Consortium, 1995. For HSE guidance see: *Preventing Violence to Retail Staff*, HS(G)133, HSE, 1995.
23. See the HSE guidance booklet: *Prevention of Violence to Staff in Banks and Building Societies*, HS(G)100, HSE, 1994
24. *The Guide to Safer Living*, London: The Suzy Lamplugh Trust
25. SI 1989/2169
26. *Safety of Pressure Systems: Approved Code of Practice*, HSE 1990 (COP37)
27. *Safety of Transportable Gas Containers: Approved Code of Practice*, HSE 1990 (COP38)
28. See notes 26 and 27 and: *A Guide to the Pressure Systems and Transportable Gas Containers Regulations 1989*, HSE 1990 (HS(R)30).

17

Machinery Safety

INTRODUCTION

From the time of the Industrial Revolution, workers have been seriously injured by moving machinery of all types, shapes and sizes. In many cases the knowledge to prevent these injuries is available and can be applied with little cost. When examining the safety of moving machinery some general principles need to be considered and some terms defined.

General principles

First, it is clear that whatever additional or contributing factors are involved, the most common moving machinery accidents arise from human error and poor judgement. Second, there is a widely-held belief that as the danger can be clearly seen and observed, little or no preventive action is necessary. In many industries there is a reluctance to ensure that plant is safe when the safety procedure takes as long as the actual work activity itself. Similarly there is a reluctance to fully replace guards and other elements of preventive equipment if this means delaying the start-up of operations.

It is a widely-known and well-tested principle that the separation of humans from danger is fundamental in the safe operation of moving machinery.

Definition of terms

Definition of some of the key terms is attempted below.

1. *Moving machinery* – all industrial, commercial and testing apparatus, equipment, plant or tools employing power to effect movement to any or all activating units or parts. This definition includes equipment and plant which can move due to gravity forces, latent energy and the effect of product or stock movement.

2. *Hazard* – a situation or condition that may result in injury to people or damage to plant and equipment.
3. *Danger* – a situation or condition in which injury to people is reasonably foreseeable.
4. *Risk* – a combination of the probability of injury from moving machinery hazards or danger and the serious consequences of that injury.
5. *A guard* – a generic term for any physical body or body-part restraining barrier, that prevents access, or effects the maximum reduction of access, to any hazardous or dangerous moving machinery. This includes access to products being manufactured and waste being produced by moving machinery if such products or waste can cause physical injury to the body.
6. *Safety device* – purpose-designed and engineered safety guarding equipment (other than a guard). These are usually electrical, electronic or mechanical systems (though they can be a combination of all three) which eliminate or greatly reduce the risk of injury from the associated dangerous or hazardous moving machinery.
7. *Safeguards* – a term applied to the combined deployment of guards, safety devices and safe systems of work for the safety control of dangerous or hazardous moving machinery.
8. *Danger zone* – an area within and/or around a moving machine where there is a risk of serious injury.

THE LEGAL CONTEXT

The legal aspects of moving machinery safety are best understood if considered under the following divisions:

- *Safe manufacture* – this includes design and installation of plant and equipment; and
- *Safe operation* – this includes maintenance and modification of plant and equipment.

Design and manufacture

Whilst not a legal document, perhaps one of the best publications on the subject of moving machinery safety is British Standard 5304 entitled *Safety of Machinery*.[1] The HSE has participated in drawing up the standard and commends its application to those with relevant duties under HSWA. British Standard 5304 contains information and data for both manufacturers of plant and equipment with moving machinery and for the operators and managers of that plant and equipment. The legal publication superseding BS 5304 is the Supply of Machinery (Safety) Regulations 1992.[2] The supply regulations specifically address the *design and manufacture* of machinery which is defined as:

- An assembly of linked parts or components, at least one of which moves, with the appropriate actuators, control and power circuits, joined together for a specific application, in particular for the processing, treatment, moving or packaging of a material.
- An assembly of machines which, in order to achieve the same end, is arranged and controlled so that it functions as an integral whole.
- Interchangeable equipment modifying the function of a machine which is supplied for the purpose of being assembled with a machine (or a series of different machines or with a tractor) by the operator himself in so far as this equipment is not a spare part or a tool.

The Supply Regulations have their origins in the EU. In Great Britain the HSE will have powers under HSWA to enforce them, although there are transitional arrangements until 1997. The penalty for supplying non-compliant machinery (or for marking the CE logo onto machinery which does not comply with the essential health and safety requirements) or for machinery which is simply not safe is:

- a fine of up to £5,000; or
- imprisonment of up to three months.

However, the regulations do provide for a defence of due diligence (see chapter 3).

The Supply Regulations cover many well-known safety aspects associated with engineering, design and manufacture which are aimed at the prevention of moving machinery accidents and injury. Whilst these regulations are specifically designed to ensure the safe manufacture, supply and sale of machinery, certain features also have a direct bearing on safety in use. An appreciation of the content of the Regulations and a summary of safety in respect of moving machinery is provided in figure 17.1.

Designers and manufacturers must undertake a risk assessment of moving parts of machinery so that they are built, supplied to end users and installed in such a way as to avoid hazards. Where hazards do exist, moving parts must be fitted with guards or protective devices in such a way as to prevent contact with people which could lead to accidents and injury. The choice between guards and other protective devices must be selected by the manufacturer according to the types of risk identified. The regulations should be applied in conjunction with s6 of HSWA.

Safe operation

The Provision and Use of Work Equipment Regulations 1992 (PUWE) are now

Characteristics of guards and safety protection devices

General requirement
- must not be easy to bypass or render non-operational
- must be located at a safe distance from the danger zone
- must be designed to enable essential work and maintenance to be carried out by restricting access ie only to the area where work has to be done

Special requirements for guards/fixed guards
- must be fixed by a system that can only be removed with tools
- whenever possible must be unable to remain in place without their fixings

Moveable guards
- must remain fixed to the machinery when open
- must be adjustable only with the use of a tool
- absence or failure of one of their components must prevent start up

Adjustable guards restricting access
- must be strictly necessary
- must be readily adjustable without the use of tools

Other related safety issues
- precautions must be taken to prevent risks from ejected objects (eg products of work and waste etc)
- adjustment, lubrication and maintenance points must be located outside of danger zones
- all machinery must be fitted with means to immobilise it from all energy sources

Figure 17.1 *Guarding characteristics*

the principal legislation covering the guarding of dangerous parts of machinery.[3] Regulation 11 is of particular importance and is reproduced in full on p371. Regulation 11(1) lays down a clear requirement to prevent access to dangerous moving machinery or to stop movement before anyone enters the danger area. This is almost in the form of an absolute requirement. Regulation 11(2), however can be interpreted as saying that if it is not practicable to guard or provide guarding devices, then protective appliances, provision of information, instruction training and supervision can be employed instead.

In practice lower levels of safety measures (ie instruction, training and supervision) should be applicable only in the special cases of live testing, necessary close inspection or production setting and adjustments. They should not be employed for normal production running of moving machinery in plant, or equipment in workshops. Nothing less than guards or guarding devices will effectively safeguard against possible disabling injury. The principle of safeguarding all moving machinery with physical or electrical protection (except for necessary live-work tasks) is the only really appropriate safety measure.

The regulations exclude some working faces of machine or machine tools where people have to be in close proximity to operate effectively. This exclusion is perhaps the only one where the lower level of the safety hierarchy should be applied full-time. It can be argued that the exemption for necessary live-work tasks is a different situation again. To gain access for live work an interlocked gate or panel control system should be installed to permit authorised entry only by trained and validated personnel. Professional management must continue to employ the principle of separation of people from dangerous equipment. In general, instruction, training and supervision should not be used as a substitute to guarding or interlocking. Moving machinery should be guarded with close or area guards or guarding devices, and where necessary there should be engineered safe and controlled access sytems.

Regulation 11

(1) Every employer shall ensure that measures are taken in accordance with paragraph (2) which are effective –

(a) to prevent access to any dangerous part of machinery or to any rotating stock-bar; or

(b) to stop the movement of any dangerous part of machinery or rotating stock-bar before any part of a person enters a danger zone.

(2) The measures required by paragraph (1) shall consist of –

(a) the provision of fixed guards enclosing every dangerous part or rotating stock-bar where and to the extent that it is practicable to do so, but where or to the extent that it is not, then

(b) the provision of other guards or protection devices where and to the extent that it is practicable to do so, but where or to the extent that it is not, then

(c) the provision of jigs, holders, push-sticks or similar protection appliances used in conjunction with the machinery where and to the extent that it is practicable to do so, but where or to the extent that it is not, then

(d) the provision of information, instruction, training and supervision.

(3) All guards and protection devices provided under sub-paragraphs (a) or (b) of paragraph (2) shall –

(a) be suitable for the purpose for which they are provided;

(b) be of good construction, sound material and adequate strength;

(c) be maintained in an efficient state, in efficient working order and in good repair;

(d) not give rise to any increased risk to health or safety;

(e) not be easily bypassed or disabled;

(f) be situated at sufficient distance from the danger zone;

(g) not unduly restrict the view of the operating cycle of the machinery, where such a view is necessary;

(h) be so constructed or adapted that they allow operations necessary to fit or replace parts and for maintenance work, restricting access so that it is allowed only to the area where the work is to be carried out, if possible, without having to dismantle the guard or protection device.

4) All protection appliances provided under sub-paragraph (c) of paragraph (2) shall comply with sub-paragraphs (a) to (d) and (g) of paragraph (3).

(5) In this regulation –

"danger zone" means any zone in or around machinery in which a person is exposed to a risk to health or safety from contact with a dangerous part of machinery or a rotating stock-bar;

"stock bar" means any part of a stock-bar which projects beyond the head-stock of a lathe.

Any part of a person's body which comes into contact with unguarded moving machinery will usually result in some sort of injury, however minor. Therefore the outcome of a formal risk assessment under the Management Regulations is likely to be the need for preventive measures. Organisations have little choice in practice but to fully guard or fit tamper-proof guarding devices rather than the less effective options such as signs, open fencing, hand rails and instruction. This is especially true as time goes by because low-cost and non-technical engineering solutions become available. Indeed there is little reason or excuse for serious injury resulting from unguarded moving machinery.

Health and Safety at Work Etc Act 1974

HSWA contains three sections particularly relevant to moving machinery:

- *Section 2.2* "(a) The provision and maintenance of plant and systems of work that are, so far as is reasonably practicable, safe and without risks to health."
- *Section 7* "It shall be the duty of every employee while at work – (a) to take reasonable care for the health and safety of himself and of other persons who may be affected by his acts or omissions at work."
- *Section 8* "No person shall intentionally or recklessly interfere with or misuse anything provided in the interests of health, safety or welfare in pursuance of any of the relevant statutory provisions."

Though not covered in detail in this chapter, portable machinery and tools, including works transport vehicles and mobile plant, should have all the guarding principles applied to dangerous moving machinery. Examples are: winch nip points on mobile cranes, power take off units on tractors, mast slides on forklift trucks, drive drums on portable conveyor units and mixer paddles on pug mills. Portable power tools include grinders, drills, disc cutters, circular saws and pipe cutting/threading machines. There are many others. While these moving machinery hazards may be known to the unit operators they may not be known to other people in the area in which the machinery is being used.

HAZARDS OF MOVING MACHINERY

To undertake an effective risk assessment of moving machinery, those engaged in protection and prevention duties should understand the relationship between injury causation and the way in which injuries occur. All injuries connected with moving machinery are derived from human contact with dangerous moving parts (or elements being worked by moving machinery) which at the time of contact will not be guarded or will have had the guard removed to allow access. Contact arising from the failure of protective devices comes about because the protective devices are inoperable, disconnected, removed or bypassed.

In addition to coming into direct contact with moving machinery, other contact with moving machinery can be as a result of the impact of, entrapment by, or entanglement with plant or equipment. This includes by-products or waste material associated with the machinery operation, or contact with materials, product or waste being ejected from moving machinery.

Movement of machinery can be classified under four headings:

- rotary movement
- sliding movements
- reciprocating movements or
- a combination of the above.

These are likely to produce eight categories of injury. They are illustrated in figure 17.2 which also gives examples of the type of moving machinery motions.

As can be seen, the safeguarding of moving machinery is a task which needs both technical and practical understanding.

When considering the application of safeguarding the most important issue is the human factor. One of the key human factors which must always be considered is the old and now largely unaccepted principle of 'safe by position'. As operatives nearly always need to be in close proximity to moving machinery at some time or another, it is almost impossible to guarantee that moving machinery is safe by virtue of its position. Therefore safeguards must always be provided and used even in remote and out-of-the-way locations. The issue of whether risks were reasonably foreseeable plays an important part in legal cases.

TYPES OF SAFEGUARDS

Moving machinery is of diverse design today, especially when microprocessor-controlled robotic applications are taken into account. Those who have design and operational responsibilities for the safeguarding of moving machinery should consult BS 5304: Safety of Machinery, the Supply Regulations and the HSE guidance document on Industrial Robot Safety.[4] These provide more than adequate information on the engineering aspects of the subject. In considering types of safeguarding the main issue, after risk assessment, is to examine which guards or guarding devices should be employed. It could be that a combination is required. The first group is physical guards. Later, guarding devices are listed.

Physical guards

Physical guards are of four main types:

- fixed guards
- interlocking guards
- adjustable guards or
- automatic guards.

Fixed guards

A fixed guard has no moving parts and is kept fixed in position by fastenings that can be removed only with the use of a specific tool. The design must prevent access to the danger zone.

Perimeter/area guarding
One of the main fixed guard designs is perimeter guarding sometimes referred to as area guarding. An example of a stylised design for perimeter guarding is shown in figure 17.3.

- *A straightforward fall into or strike against or contact* with moving machinery can result in a cutting and laceration injury from equipment such as band saws, circular saws, abrasive wheels, fan blades and gear cog teeth.
- *Contact by being drawn or pulled into an in-running nip* will result in crushing or severence injuries from two contra-rotating parts such as product rolls, intermeshing gear cogs, feed rollers, worm gearing and rack and pinion gearing.
- *Contact which develops into entanglement* can be the result of loose clothing, hair or flesh being caught up in single or multi-rotating equipment such as drill chucks, milling heads, grooved shafts, keyways or joining boltheads on shaft additions, also fan blades or mixer tools. This type of contact can cause multiple injuries such as crushing, lacerations and severence. In addition full body entanglement can be the result of the classic body pulling, in-running nip on conveyor belts and pullies and chain and cog drives.
- *Contact by being crushed* usually results from traverse movements of a machine against parts of the body and some fixed point such as a wall or bench or floor. These types of accidents can arise from travelling tables, lathes, counterweight blocks, gravity take ups, power presses, drop forge heads, traverse ram pushers and hydraulic doors.
- *Contact with two elements* of moving machinery can cause body parts to be torn or sheared off. This can also happen between a moving machinery part and the product. Examples of this type of injury are those caused by guillotines, fan blades, spoked wheels, connecting rods and linkages, oscillating tables, transfer and turnover equipment, press shears and hydraulic devices running in tight clearances.
- *Contact with non-cutting edges* of rotating machinery, surfaces of conveyor belts, flywheels, lips or rims of rotating drums, drive shafts and moving ropes will all result in friction burns and abrasions. When contact time is extensive then friction injuries can become serious lacerations and even severence injuries.
- *Contact with punching or hole making* machinery can result in puncture and possible penetration of the skin. This type of injury is caused by flying objects from machinery (eg staples, fastening pins and nails along with ejected production material or waste). Hole-making type machinery such as sewing machines, drilling or punching heads, cartridge tools and flying swarf can also cause such injuries.
- *Impact to the body* with puncturing is caused by contact with moving machinery or parts of moving machinery which rotate, traverse, lift or oscillate. Being sruck by this type of equipment can cause bruising, broken bones and body and organ damage. Examples are robotic machinery, moving tables, product ram-pushers, overhead runway equipment and hydraulic arms.

Figure 17.2 *Moving machinery injury types*

Perimeter guarding, to function effectively, should comprise the following key design features:

- Be constructed of material suitable to withstand normal activity within the work conditions applied to it.
- Be of sufficient safe distance from the danger zone of the moving machinery it is intended to guard.
- If meshed, be of sufficient mesh size to prevent accidental access of body parts into the moving machinery it is intended to guard (see figure 17.4).
- Be designed to be released from its fixing point by a tool of some description.
- Where applicable have a 0.25 metre gap at its base to allow sweeping up to take place and passing of heavy portable engineering items in to and out of the machinery area.
- If required, be designed and constructed to include an access gate or gates which should be interlocked to prevent unauthorised entry when hazardous or dangerous moving machinery is in operation.
- Be of sufficient height and be devoid of possible foot or hand holds to prevent easy access to or to present a temptation to access. Two metres has traditionally been the minimum height as per BS 5304. However 1.40 metres allows inspection viewing over the top of guards (and manual passing over of light equipment) whilst ensuring that only a purposeful act of entry can occur.

Fence type barriers (1.00 metre post and rail) should not be considered as adequate perimeter guarding as they can easily be entered by climbing over or through.

Enclosed guarding

The next type of fixed guard to consider is an enclosed guard. This type of guard is common and can be in the form of an enclosed box which fits closely over the danger zones. Another form of enclosed guard is a horse-shoe guard which fits closely over the hazardous part of a rotating shaft while the smooth ends of the shaft are allowed to run from the horse-shoe unguarded but possibly covered. If the smooth shaft is a hazard, then a loose sleeve guard can be employed which is bolted together after the two halves have been placed around the shaft. The material for this type of guard is light-weight and fireproof. It rides loosely on the shaft and provides a high level of protection where there are fast-moving shafts. A horse-shoe guard is ideal for fitting in difficult, short-drive-shaft equipment where a bolted coupling or gearing on the shaft is a danger.

Perhaps the most common application of a fixed guard is to prevent injury from an in-running nip point. These are often seen for example on chainwheels or belt drives and can be oval or pear shaped in appearance. Conveyor belt drives and free rotating belt drums also have in-running nip situations which require a standard design application of fixed guards. It is important to remember that if any form of belt or chain or similar drive can run in reverse: then both the upper and lower in-running nip points need guarding. Should the wheels of those drives have spokes or holes, then these again must be guarded. It is necessary to examine all angles of approach when considering closed fixed guarding, not just the obvious protection faces. Access to simple in-running nip points can be prevented by the use of nip bars or nip plates.

Larger enclosed guards can be designed to fit over the full machine covering all of the moving parts instead of just enclosing one or two dangerous moving parts.

Distance guarding

Distance guarding is equipment whereby the body or body parts are kept a safe distance from hazardous moving machinery while allowing the operators to feed or work products or materials. A tunnel guard usually bolted into position is designed to a size which prevents fingers, hands or arms entering the danger point but allows safe feeding and safe removal of products, materials or waste. Another variation of a tunnel guard is a shaped access guard, again allowing feeding or removal but in three steps in accordance with safety 'reach distances'. Firstly it stops upper arm entry, then hand entry, then finally it prevents finger entry but allows the operator considerable movement towards the danger point.

Interlocking guards

An interlocking guard is a guard with movable elements which are interlinked with the control systems of a machine. The interlock system is connected so that the guard cannot be removed or entered while the machine is in operation. Interlocking systems can be electrical, mechanical, hydraulic, pneumatic or a combination of these. This principle of interlocking can be applied to perimeter guarding employing a gate, and to the other types of guarding detailed in this chapter. A variation of an interlocking guard is a control guard which, when the guard is open or removed, prevents the operation of a machine.

Adjustable guards

Adjustable guards are associated mainly with bench and portable power tools. This type of guard is fixed into position but has moving manual or self adjusting elements to allow adjustment to the size of tool heads or the materials to be processed. Perhaps the best example of this is the top guard of a circular saw bench machine, where part of the guard is adjustable to the timber size being cut, and remains fixed in position while the saw is in operation.

Automatic guards

This guard is operated automatically by being moved into position by the machine itself. A guillotine cutting machine is a good example of the use of automatic guards. A variation of this is a guard linked to the continuous operation of a machine which gently but firmly pushes the operator away from the danger point (eg a power press down stroke) but allows safe feeding and recovery when on the up stroke. This is known as the 'push away' or 'sweep away' guard.

As a basic safety principle all guarding should be opened, removed or adjusted only by the use of a specific tool. This means it becomes a purposeful act to remove or adjust. Gaps in all types of guarding must be permitted only after what is known as an 'Anthropometric Consideration'. This is the assessment of openings in guards and whether any body parts can enter the gap, eg fingers (see figure 17.4).

A major consideration in respect of all types of guards is their material of construction. The final selection of material for guards will generally depend on a number of operational factors. These include, for example, strength to withstand operational production forces, minor operator errors (eg overhead cranes and forklift trucks etc), environmental conditions, corrosive liquids, dust and fumes along with possible damage from maintenance operations. The weight of the guard, important for removal and replacement, will be a function of the type of material used. Guards which have vision and observational requirements will very much dictate the sorts of material to be employed in construction. Flame and heat resistant properties will be important in some cases. The need to frequently adjust guards by hand is the final operational factor to be considered in the selection of material for guards. The most common types of material used are:

- solid sheet or plate metals
- metal or carbon fibre rods for cage guarding
- perforated or expanded metals or plastics for mesh guarding
- special laminated, reinforced or toughened glass
- plastic sheeting, polycarbonate or PVC sheeting.

A combination of these might be used.

Guarding devices

Another aspect of safeguarding moving machinery is that of protective guarding devices. This term should not be confused with safety devices for guards which should be seen as elements of guarding equipment. Protective guarding devices are mainly stand-alone guarding systems. Three of the main types are:

- trip out devices
- photoelectronic beams
- pressure sensitive mats.

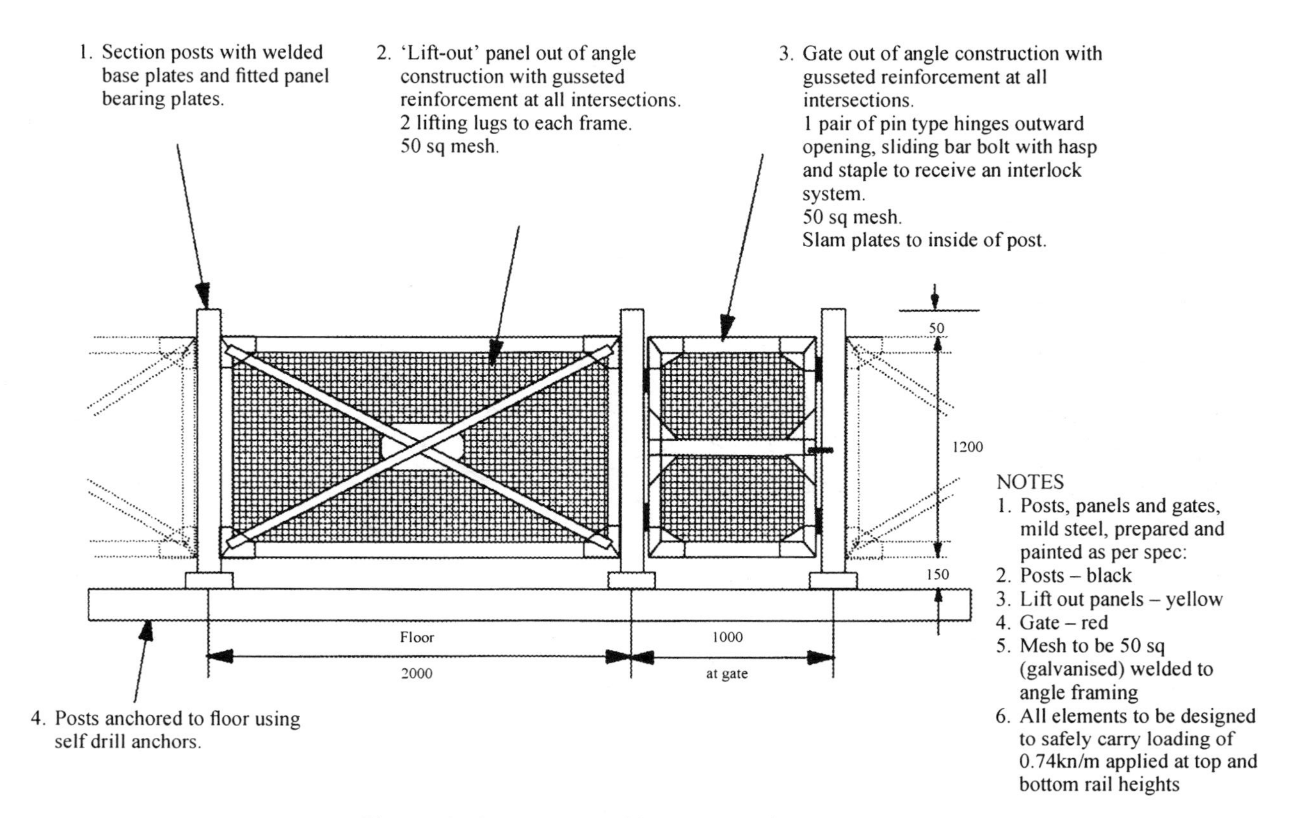
1. Section posts with welded base plates and fitted panel bearing plates.
2. 'Lift-out' panel out of angle construction with gusseted reinforcement at all intersections. 2 lifting lugs to each frame. 50 sq mesh.
3. Gate out of angle construction with gusseted reinforcement at all intersections. 1 pair of pin type hinges outward opening, sliding bar bolt with hasp and staple to receive an interlock system. 50 sq mesh. Slam plates to inside of post.
50
1200
150
Floor
2000
1000
at gate
4. Posts anchored to floor using self drill anchors.
NOTES
1. Posts, panels and gates, mild steel, prepared and painted as per spec:
2. Posts – black
3. Lift out panels – yellow
4. Gate – red
5. Mesh to be 50 sq (galvanised) welded to angle framing
6. All elements to be designed to safely carry loading of 0.74kn/m applied at top and bottom rail heights

Figure 17.3 *A suggested layout of area guarding*

REACHING IN AND THROUGH ELONGATED OPENINGS WITH PARALLEL SIDES

Safety distances are given on page 382 where
a is the smaller dimension of the aperture
b is the safety distance from danger point

Fingertip	Finger		Hand to ball of thumb	Arm to arm-pit
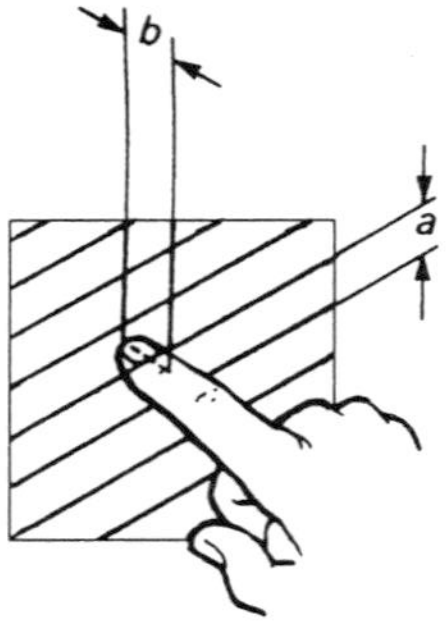	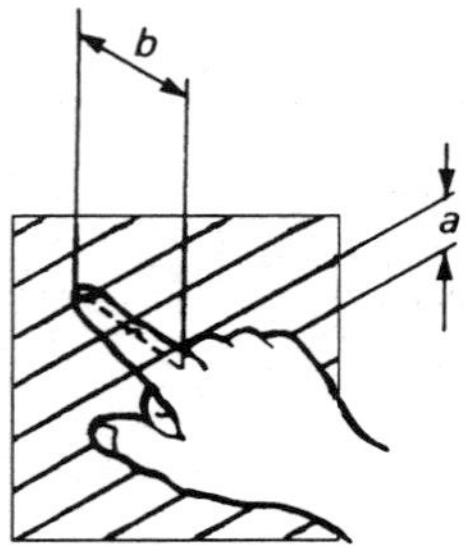		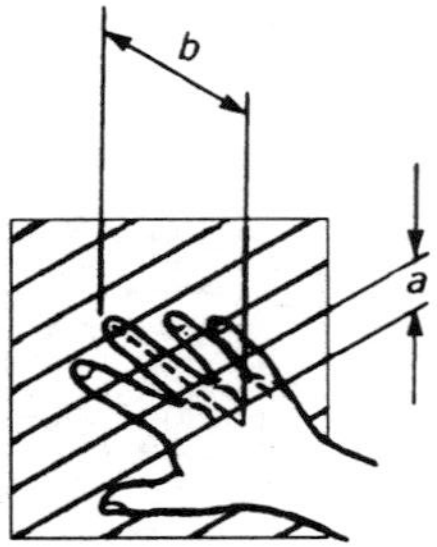	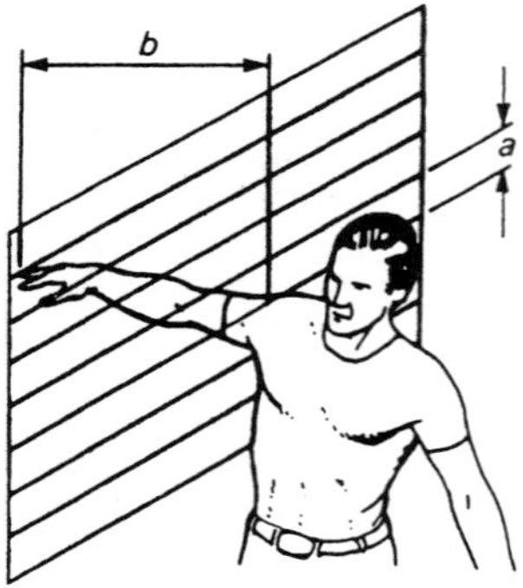
$4 < a \leqslant 8$	$8 < a \leqslant 12$	$12 < a \leqslant 20$	$20 < a \leqslant 30$	$30 < a < 150$ max.
$b \geqslant 15$	$b \geqslant 80$	$b \geqslant 120$	$b \geqslant 200$	$b \geqslant 850$
All dimensions are in millimetres.				

Figure 17.4 *Values of* a *and* b

REACHING IN AND THROUGH SQUARE OR CIRCULAR APERTURES

Safety distances are given on page 382 where
a is the aperture diameter or length of side
b is the safety distance from danger point

Fingertip	Finger		Hand to ball of thumb	Arm to arm-pit
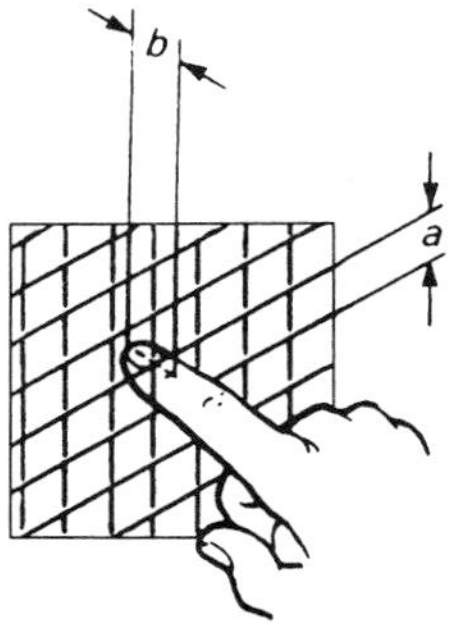	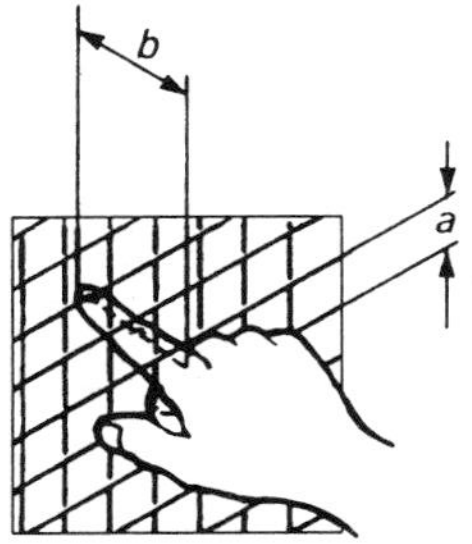		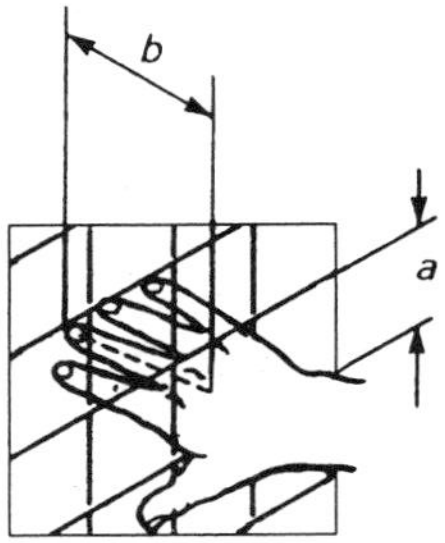	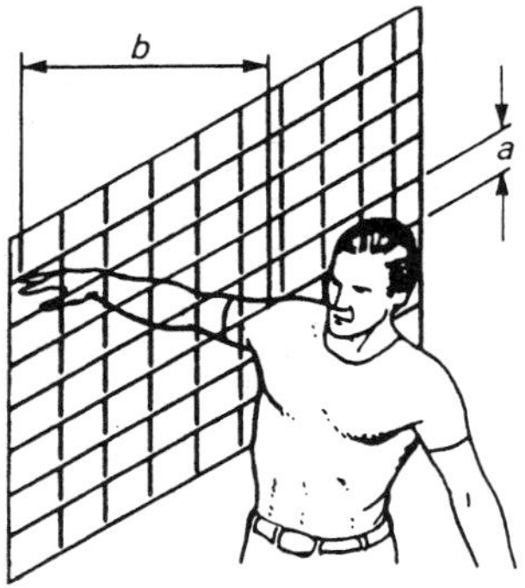
$4 < a \leqslant 8$	$8 < a \leqslant 12$	$12 < a \leqslant 25$	$25 < a \leqslant 40$	$40 < a < 150$ max.
$b \geqslant 15$	$b \geqslant 80$	$b \geqslant 120$	$b \geqslant 200$	$b \geqslant 850$
All dimensions are in millimetres.				

Figure 17.4 *(Continued)*

Source: Extracts from BS EN 294: 1992 are reproduced with the permission of BSI. Complete editions of the standards can be obtained by post from BSI Customer Services, 389 Chiswick High Road, London W4 4AL.
Appendix A BS 5304:1988 has been used to illustrate the safety issues of guarding. It should be noted however that it has been superseded by BS EN 294:1994. Ensure you work to latest, current standards.

Trip devices

Trip out guarding devices come in many forms but their function is always the same – to stop the moving machinery when anyone enters the danger zone within which injury could take place. A typical example is a simple trip bar at a specified distance from the danger point which when touched or contacted by any part of the body (usually chest or waist) will shut down the moving parts of the machine in a safe manner. Trip outs can also be in the form of front and rear bumper boards fitted to automatic (underfloor wire) guided vehicles. Contact with the legs of people will stop the machine. The most common type of trip out seen in industry is a telescopic trip probe fixed to radial arm drilling machines (sometimes called a whisker guard) which will shut down the drill if anything caught up in the drill tool strikes the telescopic trip probe.

Light beams

Photoelectric beams or light curtains are increasingly used throughout industry and commerce as protective guarding. The danger zone of a plant or machine with moving parts can be surrounded by single light beams or where necessary for increased safety, multi beams or even light curtain beams. When the beam or beams are broken by people approaching the danger zone the moving machinery is brought to a stop. This system is ideal for constant access or forklift access to the danger zone instead of having to open gates or remove perimeter area guarding. In particular, a light beam system is ideal for large automatic or manual rail-borne transfer car guarding when trip out bumpers are difficult to maintain due to damage from loading/unloading equipment or machines.

Pressure mat guards

Safeguarding of machinery danger zones can be controlled by the use of pressure sensitive mats placed under the moving part(s). The pressure mat is linked to the machine control system and until people stand clear of the mat the machine cannot be operated. Anyone stepping on to the mat during operational running of the machine will cause the machine to stop.

PRINCIPLES AND PRACTICE OF IMMOBILISATION

General introduction

The final section of this chapter relates to control of the safety of major parts of plant in terms of access to perimeter guarded moving machinery and equipment. However these basic principles can be applied (with certain modifications of scale) to any plant or equipment with moving machinery of whatever type or size. It will examine how people can safely enter these area-guarded systems in a controlled manner. More importantly, it will also examine controls which must be applied to allow safe but necessary entry when the plant is live (ie operating or capable of operation). This presents, in broad terms, the principles of a safe system of work.

The principles of immobilisation

The danger

The danger lies in inadvertent movement of machinery, plant or equipment while people are located on or near the moving machinery or are within guarded areas. When this potential danger exists it becomes necessary to IMMOBILISE machinery, plant or equipment so as to minimise the risk of accidents. These accidents can result in serious injury which can in some cases be fatal.

Any part of moving machinery, plant or equipment which, when in operation presents any form of danger, should generally be fully immobilised before work commences. Where it is not reasonably practicable to fully immobilise machinery there must at least be a part immobilisation procedure.

There are only two conditions under which moving machinery, plant and equipment may be safely worked on or near (ie within the danger zone) and these are:

- *Full immobilisation* (totally safe state, *all* machinery incapable of movement)
- *Part immobilisation* (in this state, *some* machinery is still capable of movement).

Definitions of immobilisation and isolation

Immobilisation is the act of rendering fully incapable of movement any machinery liable to cause injury to people or damage to plant and product. The movement may be due to the machine's own motive power source, its stored energy (such as air hydraulics or compression pressure) or to natural forces (such as flow, wind and *gravity*). Movement could also be due to product energy transmission.

There are two major forces at work that need to be safely controlled. They are:

- machinery forces
- natural forces.

It is necessary to distinguish immobilisation and isolation. Isolation is mainly an electrical term which means the separation of electrical power. Isolation can also be applied in engineering terms to the disconnection of mechanical power such as hydraulics and pressurised systems. Thus the distinction between the terms is as follows:

- *Immobilisation*
  — to render motionless
  — to put out of operation
  — to make safe
- *Isolation*
  — to place in detached situation
  — to separate from the power source
  — to electrically and/or mechanically disconnect the power supply.

From the above definitions it can be seen that immobilisation (along with any pre-immobilisation requirements) is the more complete and comprehensive act. Therefore it is a safer condition (in terms of movement) than isolation. So while isolation *may* render *some* machinery motionless, full immobilisation *will* in every case render *all* required machinery, plant and equipment motionless.

Immobilisation is a system of work designed to ensure machinery, plant and equipment are totally safe from unplanned movement, while isolation can only separate and disconnect electrical and mechanical power supplies and pressurised systems. Certain engineering work tasks and production tasks will require both immobilisation and isolation to ensure total safety. This is because in some tasks a number of risks are present.

Elements of immobilisation

Any safe system of work for moving machinery, plant and equipment which seeks to effectively immobilise against unplanned movements must contain three basic elements or steps. These are:

- *A procedure* – this is an established *sequence of events* to safely carry out an immobilisation. It should be written in the form of full or part immobilisation procedure for major areas or sections of plant. For straightforward local immobilisation a written safety practice should be detailed but not necessarily in the form of a safe working procedure. Part immobilisation should be detailed in a part immobilisation procedure.

- *A mechanism* – this will include the *electrical* or *mechanical* control *devices* whereby immobilisation is achieved. It could utilise circuit breakers, switches, valves, retarders, chocks, blocking plates, stops, mechanical anchors etc.

- *Security* – involves the method *whereby the means of* the immobilisation is *guaranteed.* One method which can be employed is a locked personal danger board. Another method is to hang a personal danger board (see Appendix to this chapter) over the means of immobilisation though this can be less secure. As a back up to personal danger boards, suite danger boards (again see Appendix) could be employed and locked on or over to secure the means of immobilisation control devices.

Without an immobilisation procedure, a control mechanism and a method of achieving security, a totally safe immobilisation has not been carried out. These requirements should apply to electrical and mechanical engineering and production staff. In addition they should apply to all contractors' staff.

Types of immobilisation

- *Major plant immobilisation* – this type of immobilisation is associated mainly with complex machinery and equipment and is usually surrounded by perimeter guarding which would include interlocked entry gates. It is the most complete form of immobilisation. The control points of this type of immobilisation can either be an electric switch or a valve. Major plant immobilisations should be fully detailed in an Immobilisation Procedure document. They may require experienced electrical and mechanical staff to apply and secure the various elements of the immobilisation.
- *Local immobilisation* – this is associated mainly with single units of machinery or small units of plant and equipment. The control points for this type of immobilisation are usually local electrical switches or breakers and/or local valves near to the machine or plant. Procedural instructions could be detailed on notices or signs near to the local immobilisation control mechanism. This type of immobilisation should be applied only by staff who have been trained and authorised to undertake local immobilisation.

It may be necessary to add associated procedures to immobilisation. These could include:

- confined spaces control
- exhaust fume control
- fluid (water/chemicals) control
- gas safety control
- overhead crane and electrical conductors safety
- pressurised systems control.

Where these additional issues are identified, special safety requirements should be encompassed in the pre-immobilisation procedure.

Part immobilisation

It will at times be necessary to leave some machinery, plant or equipment live. 'Live' in this context means still operational and therefore capable of movement. This is to facilitate certain tasks to be undertaken that can only practicably be carried out when plant is in an operational state. Examples of such tasks are: testing, setting limits, inspection, minor adjustments and product removal.

The practice of *part immobilisation* can be applied under two broad conditions which are:

- exemption immobilisation or
- operational immobilisation.

Exemption immobilisation is the act of making live under controlled conditions (and therefore capable of movement) parts of plant or equipment. The majority of the plant will still remain in a fully immobilised state.

Operational immobilisation is the act of entry into plant areas or onto live machinery or equipment while it is predominantly in an operational state and therefore capable of movement. Some equipment may still need to be immobilised. This will depend on the tasks to be undertaken. Operational immobilisation should only be utilised when full or exemption immobilisation is not reasonably practicable.

Four important safety principles apply to part immobilisations:

- As few as practicable elements of plant and equipment should be left in an operational mode.
- All work in areas under any form of part immobilisation must be safely supervised. The supervisor should safely supervise all people and be at the point of work full-time.
- Only the people required to carry out the task should be present at the part immobilisation. (This should include supervision and could include any safety lookout men required.)
- It is always safer to plan and undertake work with an immobilisation which is as near to a *FULL* immobilisation as possible, with only *the minimum of equipment capable of movement*.

Before any part immobilisations are placed, any personnel who may be working in the area under a full immobilisation must withdraw and remove their personal danger boards. The full immobilisation may then be removed to allow the part immobilisation to be applied.

Care must be taken to ensure a drive or element of equipment is not made live which may form part of an adjoining immobilisation. Careful consideration needs to be given when applying part immobilisations because of the possible interactions between equipment in any adjoining areas. Buffer zones and overlapping areas may need to be considered when preparing written procedures.

Part immobilisations controlled from switching rooms or control stations or similar should employ a part immobilisation certificate as proof of validity. Suite notification boards should be employed on all control points. Part immobilisations should be subject to written part immobilisation procedures. Where an authorised part immobilisation document is not available, a hand-written document could be prepared by an authorised person in control.

Components of immobilisation

When applying any immobilisation a number of component stages must be worked through and checked to ensure the complete, and therefore safe, state of immobilisation. The five major categories which must be examined when devising or applying immobilisations are set out below:

- *Product immobilisation* – product movement protection both upstream and downstream, such as product energy forces, movements or slippage causing impact, knock-on effect, product deflection or simple product section over-run.
- *Latent energy immobilisation* – protection against, for example, tensioning devices, compressed spring forces, falling equipment due to vibration and unplanned release from pressurised systems – (eg high pressure water sprays).
- *Gravity force immobilisation* – ensure protection from entrapment by components of plant or equipment which are held in position by hydraulic or pneumatic forces or mechanical latching and fastening devices – any possible movement of plant or equipment (due to depletion, exhausting or leakage) even though *unpowered* must be secured.
- *Special immobilisation* – to ensure full protection it will be necessary to consider special requirements for immobilisation. Examples are:
  - piped services (gas, steam, water chemicals)
  - hydraulics
  - lubricants
  - pneumatics
  - exhaust systems
  - heat sources
  - ionising radiations
  - lasers
  - open electrical conductors
  - constituent elements of pressurised systems.

The hazards of open water and liquid flows may need to be considered.

- *Motive power immobilisation* – protection against main motive power drives – eg electrical power drives, fluid power, pneumatic power, steam power, combustion engine power.

The first four of the above components can be termed pre-immobilisation requirements (ie they come prior to the main motive power immobilisation).

Depending on the type of immobilisation being carried out, permission should be obtained from operational staff in charge before proceeding with any of the stages of immobilisation. This is an important part of pre-immobilisation: therefore it is vital to build in a communication check at this stage of the procedure.

Safeguarding against further injury

When work is to be done *within* electrical power systems or pressurised systems or electronics control equipment, vessels and pipelines or within moving machinery, it is important to take into account the following:

- Immobilisation does not mean that all electrical power supplies have been isolated. Although machinery movement is prevented, electrical equipment often still has live connections within it and live cabling running within the immobilised area.
- Electrical equipment and cabling must not be interfered with or disturbed in any way without first securing an electrical isolation on the applicable section of equipment.
- Immobilisation does not mean that all pressurised systems or their power supplies have been isolated. Although machinery movement is prevented, pressurised systems and equipment often still have live pressure within them and the piplines running through the immobilised area.
- Pressurised equipment (eg hydraulic and pneumatic systems, lubrication and water systems, steam systems and their associated pressure vessels, tanks and pipelines) must not be interfered with or disturbed in any way without first securing a mechanical isolation. This may also require system de-pressurisation on the applicable section of equipment.

Visual indication signals

This chapter has so far examined moving machinery safety and provided an outline of the methodology to be used to safely immobilise major plant and machinery systems.

A practical safety device employed by a number of organisations which gives visual meaning, warning and indication to the practice of immobilisation is worth consideration. This is a visual entry signal lamp cluster. The unit consists of a three coloured signal lamp housed in a vertical cluster one above the other. A red lamp would be on top followed by a yellow, centre, and a green lamp at the bottom. These signal lamp units are located at entry doors to perimeter guarding or placed adjacent to opening or removable guards. Signal lamp units would be electronically linked to immobilisation control systems. This facilitates the following visual indication signals through the illumination of one lamp at a time depending on machinery condition, for example:

- red lamp on indicates plant or equipment running, no access (danger)
- yellow lamp on indicates plant or equipment part immobilised, limited access only (caution)
- green lamp on indicates plant or equipment immobilised, full access (safe condition).

APPENDIX

1. *Personal danger board* – a plastic-type board 150mm x 75mm with the employee's name and identity number engraved on both sides. It should also contain in large letters the words "Danger Do Not Operate" or similar, and could be equipped with a unique padlock chained to the board. This personal danger board would be employed for safely locking off immobilisations of all types.
2. *Suite danger board* – similar to the above but usually 200mm x 100mm for locking off and safely securing sets of immobilisation equipment forming part of a full immobilisation suite. Instead of a name it would show the identity of the unit of immobilisation it is to secure. A padlock could be attached for extra security.

Notes

1. *Safety of Machinery*, BS 5304, British Standards Institute, 1988
2. Supply of Machinery (Safety) Regulations, SI 1992/3073
3. Provisions and Use of Work Equipment Regulations, SI 1992/2932
4. *Industrial Robot Safety*, HSE Guidance Note HS/G43.

18

Workplace Transport

This chapter does not propose to deal with any aspects of sea or air transport. The main body of the chapter will cover:

- road transport
- railway operations.

It will also deal briefly with the associated in-works vehicles, plant, mobile machinery and mobile equipment which assist in general product and material movement, its storage, handling and lifting.

INTRODUCTION

Overview

In the UK about 20% of workplace fatalities are associated with transport or support operations. This means that over 100 people die each year as a result of works transport-related accidents. About one third of these deaths occur during building and construction work. There is also a high incidence of accidents involving serious injury, eg major spine damage, severed limbs and body damage. Serious crushing and impact accidents are the main causes: very few transport accidents involving employees result in minor injury. The causes of most transport accidents are well-known and understood, the prevention measures are simple and in the main inexpensive.

HSWA is specific in its application to works transport and associated operations. In s2(1) it states:

"It shall be the duty of every employer to make arrangements for ensuring, so far as reasonably practicable, safety and the absence of risk to health in connection with the use, handling, storage and *transport* of articles and substances."

(emphasis added)

In addition to causing injuries, transport accidents cause damage to plant, infrastructure and vehicles. In a four-month study by HSE, *The Costs of*

Accidents at Work, it was found that a transport company with a fleet of 145 tanker and refrigerated vehicles had uninsured losses eight times higher than the sum recovered under its insurance policies.[1] The study recorded a total of 296 accidents, albeit with no personal injuries, at a total cost of £49,000. This sum was equivalent to some 37 per cent of annual profits. The picture throughout industry generally is that considerable costs can arise from lack of effective managerial control of works transport.

There is a general perception within management that works transport is simply about lorries and cars, and as everyone understands traffic issues, works transport is not seen as important enough to command the necessary slice of managerial resources. Yet workplace transport is one of the few health and safety subjects which most, if not all employees, come into proximity with on almost a daily basis.

As a first step in prevention a company transport safety policy should be considered. This must be seen to be enforced and supported by senior management and will need to fully address the following five broad areas:

- vehicle design and construction standards
- driver and operators' standards
- infrastructure standards
- maintenance standards
- safe systems of work.

When planning to implement a works transport safety policy one of the major problems to be overcome is that of double standards. This problem applies mainly to road transport vehicles. Some drivers who have operated quite safely on the public highway drive into a works or factory or construction site and behave in a totally different manner. They will operate at high speeds, cut corners, reverse over main roads, jump red lights and even go around roundabouts the wrong way.

The term transport is employed to mean the carrying or movement of goods or materials from one place to another. The term traffic is employed to mean the transportation flow of goods, materials or persons on and throughout works traffic routes. The various types of works transport and materials handling vehicles, machines, equipment and systems can be classified under four broad headings, ie:

- *Road transport* – eg cars, vans, pickups, lorries, tankers, trucks and articulated vehicles.
- *Rail vehicles and equipment* – eg locomotives, wagons, rail cranes, track machines and special transport equipment (rail cars).
- *Mobile plant* – eg mobile cranes, fork lift trucks, man lift platforms, shovels and bulldozers etc.
- *Off-highway vehicles* – eg (site) dumpers, dump trucks, pallet and self loading trucks and tractor/trailer units.

It is important to remember that health and safety issues connected with transport are predominantly the responsibility of management. Workplace transport and material movements should have clear lines of operational accountability in terms of health, safety and environmental control.

Legal aspects

HSWA and Management Regulations

In addition to s2 of HSWA, the Management Regulations cover works transport through the general requirement to carry out a risk assessment. The ACOP, in relation to regulation 3, specifically mentions machines, methods of work and layout of premises. All three are important aspects of works transport.

The Workplace Regulations 1992

The Workplace Regulations contain a number of references to the problematic interface of people and certain types of works transport. The regulations employ the term traffic route for this interface. The ACOP, in relation to regulation 12, mentions that traffic routes should be kept free of obstructions which may present a hazard or impede access. Further detail applicable to works transport can be found in the ACOP in respect to regulation 13. Information is provided in particular on stacking and racking, and loading and unloading vehicles. While there is no direct reference to rail and road transport in regulation 13 it does provide general guidance on loading and unloading of vehicles and rolling stock. The guidance makes reference to the sheeting of loaded lorries. Guidance is also given on fall prevention from the top of vehicles, especially tankers.

Regulation 17 – organisation of traffic routes – contains a number of references to works transport. For example, it states that it should be organised in such a way to allow pedestrians and vehicles to circulate in a safe manner. Where vehicles and pedestrians use the same traffic route, sufficient separation should be ensured. The ACOP, in respect of regulation 17, suggests the consideration of one-way systems and the protection of people from vehicle exhaust fumes. Attention to sensible speeds and speed retarders is also given in the ACOP, and the guidance covers the problem of reversing vehicles. The ACOP takes up on the critical issue of pedestrians and vehicle route crossings, suggesting that at particularly dangerous points the design of crossing points and methods needs consideration.

Provision and Use of Work Equipment Regulations 1992

The ACOP, in respect of regulation 2, gives guidance by listing examples of work equipment including dumper trucks, cranes, road tankers, tractors and lift trucks. These are all forms of workplace transport and handling equipment which are covered by the regulations. The regulations exclude private cars.

The PUWE Regulations can therefore bring to bear a number of major safety considerations linked to workplace transport operations. In the main these are:

- suitability of work equipment (regulation 5)
- maintenance (regulation 6)
- specific risks (regulation 7)
- information and instruction (regulation 8)
- training (regulation 9).

Factories Act 1961

When considering the operation of overhead and mobile cranes in a transport system, the Factories Act 1961 should be considered in terms of sections 26 (chains, ropes and lifting tackle) and 27 (cranes and other lifting machines). Any railway operations on site will come under the Locomotives and Wagons (Used on Lines and Sidings) Regulations.

Although not applicable to works transport, many of the provisions in the Road Transport and Traffic Acts can provide useful guidance. This legislation is applicable to company transport using public roads and the same comparable safety standards should operate in the workplace. In particular the Highway Code, although not legislation, should be considered as a basic standard.

ROAD TRANSPORT

Introduction

The HSE has produced a useful guidance note, *Road Transport in Factories and Similar Workplaces*.[2] The message of the note is that there are legal requirements and standards applicable to transport operations in public places, and that nothing less should apply within a workplace.

Vehicle design and construction standards

Flat-back lorries, articulated lorries and tipper trucks form much of the works traffic flowing in and out of and around factories and work sites. This type of transport varies in design, size, weight, carrying capacity, number of wheels and operational purpose. However, many of the safety design and construction standards have a common base: the vehicle must be fit for its purpose. The major considerations are as follows:

- The vehicle design must include stability of operation under all working conditions. It must be especially designed for sideways stability.
- Cab windscreens and windows must be designed to give safe, all-round

vision for the driver, and be fitted with effective windscreen wipers and screen washers.

- Ergonomic design should be applied to the driver's seating, the design of vehicle controls and the relationship between them. Where additional people are to be carried, provision should be made for the necessary seating.
- Safety belts should be fitted for drivers and any passengers.
- A safe means of access and egress must be a feature of the design of cabs and other locations on the vehicle where drivers or operators need to work. Non-slip equipment should be employed.
- The design must provide for an effective, audible warning device.
- The design must provide front and rear vehicle lighting and night vision lighting plus rear reflectors and any necessary side lights.
- High visibility marking in day glow type paints along with low light reflective panels should be considered.
- For extra wide or long vehicles consideration should be given to a red flashing lamp on the front. This must not interfere with the driver's vision.
- All vehicles should have automatic audible and visual warning devices activated by a reverse mechanism.
- Driver operated amber flashing lamps should be fitted for vehicles that are required to operate inside buildings and plant.
- All vehicles should be fitted with a speedometer.
- All vehicles should be fitted with direction indicators.
- All vehicles should be fitted with brake stop lamps.
- In hazardous situations consideration should be given to the fitting of automatic shut down sensor bars at the rear of vehicles with restart controls placed at the vehicle rear.
- All wheels must have mudguards or stone guard flaps and incorporate anti-spray protection.
- Clear unique identification letters and numbers of large print should be shown on vehicle sides, rear and front.
- Rear view mirrors need to be considered on most vehicles.
- Under-run protection bars on front and rear should be considered and where necessary be fitted to sides of vehicles.
- Driving controls must be designed and positioned so that they can be satisfactorily operated without impairing safe and proper control of the vehicle. Controls should also be clearly marked and identified.
- Fuel tank filler caps should be fitted with a no-loss retaining device.
- All vehicles must be fitted with parking brakes.
- Plates inside cabs must indicate vehicle width, length, height and turning circle plus carrying capacities and axle loadings.
- Notices inside cabs should indicate pre-start safety check procedures and key safety points applicable to the vehicle, especially in the case of tankers.
- All vehicles should be equipped with fire extinguishers.

Driver and operator standards

Selection of drivers

Employees who are required to drive on the public road must possess a current driving licence appropriate for the classes of vehicles they may be expected to drive. There is no such specific legal requirement for drivers of internal transport vehicles. However the HSE recommend that similar standards of training, testing and certification should be applied. This principle is now accepted as managerial practice in line with the requirement for safe systems of work as laid down in HSWA.

Drivers and operators of transport classified as Loaded Goods Vehicles (formerly Heavy Goods Vehicles) will, when driving into a works site or factory, have the necessary LGV licence for the class of vehicle being driven. For operators of works transport vehicles classified under the LGV grouping, a simple and effective managerial policy is to ensure that internal transport operators meet the same standards as outside operators. Loaded Goods Vehicle licence requirements come under the following categories:

- *C* – a goods vehicle exceeding 3.5 tonnes inclusive, drawing a single axle trailer not exceeding 5 tonnes or other trailer types of up to 0.75 tonnes.
- *C* + *E* – a combination vehicle in category C drawing a single axle trailer exceeding 5 tonnes or other trailer type exceeding 0.75 tonnes.

Cars, vans and pick-ups of all types up to a carrying capacity of 3.5 tonnes will require the driver to possess a motor vehicle licence category B for operation outside works. Again the same safety standard should apply to works vehicles falling within this grouping.

Passenger carrying vehicle licences fall into four categories: D, D1, D + E and D1 + E.

Drivers operating goods vehicles classified as carrying dangerous goods must hold a full vocational training certificate in addition to an appropriate licence. In the main, these vehicles are road tankers or vehicles carrying tank containers. Workplace licence standards should be fully considered where passenger carrying and dangerous goods carrying vehicles operate only on work sites.

The possession of a valid current driving licence is not in itself an indication that the holder is capable of taking control of any vehicle. It is, however, a useful indication of a minimum level of basic driving skill when selecting suitable staff for workplace transport. A person not in possession of at least a motor vehicle driving licence should not normally be permitted or trained to operate site or works transport; nor should anyone who is not physically fit. Management should insist on a full medical examination before employees are permitted to drive or operate works transport, and in particular eye sight and hearing should be tested.

The selection of drivers and operators is probably one of the most important aspects of works transport operations and in addition to having the necessary skills and experience, a person must be assessed on their attitude towards safety. Low priorities given to employee skills and attitudes are the main causes of works transport accidents and injuries. HSE experience shows that most accidents can be prevented by improved selection and by placing more importance on the development of safe drivers and operators. Skills with vehicles and mobile plant can be obtained by recruiting people with the necessary experience, but are best developed by hands-on training supervised by senior staff who are themselves competent and experienced with the particular vehicle or machine. These trainers should be specially trained in teaching and training techniques in the context of health and safety at work.

Training of drivers

A training policy for drivers should have objectives and aims, and laid-down standards and methods, so that these policy issues can be monitored, reviewed and updated. The framework for such training should conform broadly to the following list of key points.

- There should be a written company policy document with an overall objective supporting aims and standards.
- The transport training plan should have a managerial owner responsible for overseeing its strategic operation.
- The training must be formal and under the control of an approved instructor.
- There should be driver and operator manuals (or for simple vehicles pocket cards) containing essential and practical operational and safety information and instruction, along with pre-start up and fault-finding checks.
- The training must be designed in a modular manner and tailored to the classification and type of vehicle or mobile machine.
- Drivers and operators should be trained only for the vehicles and machines they need to use in their job.
- All drivers and operators must be assessed by examination and practical testing.
- Assessment should be undertaken by a licensing authority, usually by a person who is not an instructor. In the case of small companies this may need to be an outside body.
- Drivers or operators who achieve the required standard in training and are successful in examinations and testing should be issued with an in-work driving licence or operator's licence. This should be limited to the workplace transport for which they have been trained and tested.

Monitoring of driver standards

Age limits are applied to drivers of vehicles on the public highway: these should apply in a factory or on a works site. Minimum age limits are usually 18 or 21 years.

A system of monitoring driving standards needs to be in place. Monitoring should be included in the duties of security and/or safety staff and/or supervisory management. In particular a system should be alert to and monitor:

- driving without due care and attention;
- driving too fast
- driving too close to other vehicles, fixed structures or people
- driving through red lights and ignoring traffic signs and markings
- turning without indicating
- driving with insecure loads
- use of a vehicle for purposes other than that for which it is designed.

A code of discipline needs to be drawn up to make sure that the effort put into monitoring is effective and that the message of safe traffic operations is understood by both visiting drivers and company drivers. After appropriate warning phases the code should provide management with powers to prevent named visiting drivers (or indeed haulier or delivery companies) from access to the workplace. Similarly, there should be a procedure to take appropriate action against employees who consistently fail to follow company safety standards when driving or operating within the workplace.

Traffic routes

The safety interface between pedestrians and traffic is a crucial consideration when reviewing traffic systems. Regulation 17(1) of the Workplace Regulations states: "Every workplace shall be organised in such a way that pedestrians and vehicles can circulate in a safe manner".

Pedestrian traffic

Clearly defined and marked routes should be provided for people going about their business at work. In particular, safe and separated access to and egress from the workplace gate, car parks and any bus or train service should be provided. Where it is necessary for pedestrians to cross main traffic routes, zebra crossings could be considered. Consideration could be given to pelican crossings at busy or hazardous crossing points. The use of simple, meshed hand-rails will assist in separating people from dangerous vehicles or routes. In buildings where transport operates, doorways are a point of risk. Separate pedestrian pass doors should be provided to allow people to walk from building to building on a route separated from vehicles. Again meshed hand-rails at each side of the pass door will channel people into the safe pedestrian route. Where design of existing buildings and plant does not permit a pedestrian route to have safety clearances from transport movements, then a raised pedestrian walkway could assist in segregation.

Parking areas

Traffic accidents may be caused by vehicles being parked unsafely, causing obstruction and reduced visibility. Parking areas should have entrance and exit routes and not free-for-all access and egress. Sufficient slip road shoulder should be provided to allow all vehicles using parking areas to enter safely in relationship to following and oncoming traffic. Security lighting and level hard standing for private cars will encourage employees to leave cars in parks rather than in unsafe locations around the site. Careful consideration should be given to the provision of locations or parks for outgoing transport to carry out load checking and preparation eg sheeting, load securing, load levelling and adjustment. Allowing these operations to be carried out on works roads can cause a hazard to both transport staff and others.

Vehicle traffic routes

Wherever practicable, vehicle traffic routes should be constructed of concrete or tarmac and be built with the sub-structures designed to take the axle loadings and operations of vehicles using them. Traffic routes of various types should fit within the following design parameters:

- Clear segregation from pedestrians.
- Segregation of hazardous transport ie oversized vehicles, rail traffic and dangerous load carrying vehicles.
- Avoid steep gradients where forklift trucks operate.
- Junctions should be kept to a minimum and clear right of way indicated.
- Roadway width should consider the widest vehicles passing each other with a safe margin; if this is not practicable wheel spill-over strips or passing laybys should be considered.
- Bends in roadways should consider the longest vehicle and side throw of any rear tracking or steering units; in addition vision on bends and clear warning is vital.
- Drainage of works roadways is important, even simple soak-aways; flooding on roads, and ice in winter can be very dangerous.
- Ensure hazardous substances in pipe lines, storage tanks or silos are adequately protected by crash barriers or similar to prevent them being struck by vehicles (design new plants and pipe line locations to be clear from traffic routes).
- Engineer road/rail crossings to high standards in terms of location, visibility and the engineering of crossing pad construction; check rails should be an engineering feature on busy crossings.
- Where traffic routes have to pass close to or run inside buildings and structures, design must avoid load bearing structures and brick built constructions; should this not be practicable then vulnerable structures and buildings must have crash barriers, or heavier protection if necessary.

- Bridges of all types, especially service line bridges, must be at a height that prevents them being struck by fixed parts of the highest vehicles or normal loads, and be protected by goal post bunting poles to ensure high lift mobile plant cranes, man lift platforms etc. do not collide when jibs, cradles and booms are in the semi-raised position.
- Where possible traffic routes should carry normal road markings eg
  - lane markings
  - give way and stop markings
  - no waiting/parking markings
  - controlled crossing markings
  - obstruction, keep clear hatching and markings
  - emergency vehicle access.

To allow the safe and free movement of works transport, the design, layout and construction of traffic routes should be examined periodically and reviewed to see if they are still compatible with the dimensions, weights and types of vehicles using them. In addition, current traffic flows – both normal and peak – should be reviewed and related to the traffic route design.

Speed limits

It is important to install practicable and effective maximum speed limits. In many cases, outside works roads can stand a maximum such as that of built up areas ie speed limits of 30 mph. Inside buildings, a 10 or 15 mph maximum limit would be more appropriate. It is always possible that in very hazardous situations a 5 mph maximum may be necessary. Above all, maximum speed limits must be displayed throughout the workplace, especially at entrances. Correctly designed anti-speed road ramps should be considered in areas or at locations where speed limit action is critical. Monitoring of speed limit compliance along with effective action against persistent offenders is a valuable aid in ensuring that the message and policy of driving within speed limits is high on management's agenda.

Pedal cyclists

Pedal cyclists are very much a minority issue in most workplace traffic systems. However they can be vulnerable to serious injury. Where possible, segregation on special cycle tracks can reduce accidents. The provision of reflective clothing along with projecting reflectors and heavy duty front and rear lamps will assist in making cyclists visible to other workplace traffic users. If cyclists have to use road/rail crossings, a warning sign should instruct them to dismount and walk-over rail tracks. Rail tracks can cause slipping and falling accidents especially in wet weather.

Transport maintenance standards

The maintenance standards of workplace road transport operations cover two main areas:

1. maintenance of the road transport infrastructure
2. maintenance of road transport vehicles and mobile plant.

Maintenance of infrastructure

- *Upkeep* Infrastructure maintenance will include repair of traffic routes. In particular, this covers the upkeep of segregating hand rail barriers, crash barriers and raised kerbstones. It is important to have safety inspections of these items, with reports being sent to the engineer responsible for repairs. Infrastructure includes traffic and safety signs and ground surface markings, along with various items of road furniture. All these require repair from time to time but signs in particular must be kept clean so that they can be seen. Where traffic lights and electrical warning systems form part of a road transport infrastructure it is vital to have a system of speedy reporting to the known repair authorities.

  Cleaning and repair of roadway and traffic routes, including lighting, is another important area of safety maintenance. Traffic vision in poor light or night conditions can be a major contributing factor to road transport accidents. Steam or similar vapour or mists emanating from pipe lines near roads can be a danger because it reduces visibility.

- *Obstruction removal* A system should exist for ensuring traffic routes are kept clear of obstructions. A known reporting authority should be available for speedy removal of obstructions from main transport routes. This same authority could be responsible for ensuring, where applicable, that certain vital transport routes are kept free of snow in winter and gritted to prevent traffic accidents caused by icy conditions. A simple but not often undertaken maintenance task is that of keeping surface water drains and drainage channels free from obstruction so that they are able to carry away rain, melting snow and in certain cases spillages.

- *Roadwork repairs* Risks are associated with the repair of traffic routes. The use of stop/go boards or temporary traffic lights, cones and bunting should always be considered. The employment of a competent flagman with high visibility clothing and effective marshalling battons could be considered as a minimum control measure where there are risks of injury. Roadwork repairs can and do result in fatal accidents. In the workplace, a standard safe working procedure should be developed to control the various categories of repairs and maintenance.

Maintenance of vehicles

The HSE considers that the well established maintenance standards that apply to transport on the public highway should be applied within workplaces.

- *Inspections and testing* Routine inspections as part of a planned preventive maintenance programme should be carried out by competent and trained maintenance staff at fixed time or mileage intervals. Manufacturers' servicing manuals and schedules can form the basis of maintenance programmes. Repairs to road transport vehicles and mobile plant should be undertaken only by skilled and trained staff who have the necessary qualifications to make sure vehicles and plant are fit for their purpose when returned to service. In particular, critical mechanisms such as brakes and steering must be subject to maintenance by competent staff. No vehicle or plant should be permitted to return into service unless key safety features are in place and in good working order eg audible warning devices and direction indicators. It is important that provision is made for maintenance staff to thoroughly examine vehicles and plant, especially the underside structures and mechanisms.

  Facilities and equipment should be provided to fully test and check critical equipment and parts of vehicles and mobile plant. Maintenance staff must be empowered to withdraw from operation any vehicles or plant they consider dangerous.

- *Maintenance records* The maintenance unit should be responsible for keeping and updating vehicle and mobile plant records. One file or log per vehicle would be effective. These maintenance records should set out:

  - dates of engineering inspection
  - details of any faults or damage found
  - dates of maintenance and repair work
  - details of testing and examination
  - details of modifications and alterations, with dates and reasons and names and signatures of people carrying out the work
  - dates and faults reported by named drivers or operators
  - records of parts replaced or renewed
  - records of consumables used on each vehicle.

- *Defect reporting* An integral element of the safe operation of works road transport and mobile plant is driver/operator defect reporting. This can be divided into two categories:

  1. faults developed in operation
  2. faults discovered during pre-operation safety checks.

  Faults of any type developing or observed by the driver or operator should be recorded on a fault report tear-out sheet. Details should be entered and the sheet should be handed to the reporting authority. In the case of faults affecting the operational ability or safety of a vehicle or mobile plant, it must be returned to the maintenance unit immediately, if safe to do so. Vehicles or mobile plant which become dangerous in operation should be taken out of operation until they receive maintenance attention by qualified staff.

Pre-operational checks are important: some of these checks should be subject to a record.

Start-up check list

- fuel, coolant lubrication and hydraulic fluid levels
- battery – power level, secure and electrolyte level
- tyres – pressure, general condition, tread depth
- wheels free from rim damage and nuts secure
- steering – safe and effective
- brakes – safe and effective, including parking brake
- lights – working and clean, including brake lights
- indicators – working and clean, including hazard warning
- audible alarms – working and effective
- visual warnings – working and clean
- windscreen – clean; wipers and washers working
- cab – seat belts effective, vibration and noise suppression, ventilation, heating and cooling systems working
- access – steps, hand-holds and rails effective and clean
- fire extinguisher – fitted and charged
- exhaust system – noise and volume of fume discharged within set limits
- mirrors – wing and rear view units clean and set correctly

Workshops and garages

Accidents involving vehicles and material handling equipment can and do take place in garage and workshop areas. Potential causes include:

- starting engines while vehicle is in gear
- fumes from battery acids causing burns and possible explosion
- uncovered repair pits
- axles propped on bricks or timber etc. (axle stands must be used)
- explosion and fire caused by hot repairs to fuel systems
- repaired tyres exploding during inflation (use protective cage)
- brakes not applied
- asbestos dust from brake pads
- failure of raised bodies of cabs or hydraulic lift systems (use props or latches before work)
- oil and grease spillage
- unguarded power-take-off units
- paint spray units (respiratory protection needed)
- exhaust fumes from running engines

- burns or shock from electrical short circuits
- defective vehicle lifts
- high pressure lubrication systems.

Some of the more important items could be displayed on notice panels inside cabs while the rest can be listed on plastic pocket cards issued to drivers, operators and maintenance staff.

- *Starter key control* One of the most critical safety control areas is the starter ignition key. Such keys should be unique to vehicles and mobile plant and not made common or wired to bypass the key control. When vehicles and mobile plant are at maintenance workshops in workshop parking areas, starter ignition keys must be secured by a control authority when they are not required for testing or moving the vehicles. Drivers and operators should also be directed to always immobilise by removing starter keys.
- *Fit for purpose* The engineering management of maintenance units should be accountable for ensuring that no vehicles or mobile plant are returned to tasks or duties if they are not fit for their purpose or designed for their proposed tasks or duties.

GENERAL SAFETY

Policy

- It should be a first principle built into managerial health and safety transport policies that drivers and operators are fully responsible for the safe operational condition of vehicles and mobile machines, and where applicable for the weight, height and overhang of loads.
- Where high-reaching mobile machines, high-sided vehicles or high attachments can come into contact with bridges, service pipe lines or overhead electrical cables etc, goal posts with bunting or a similar device slung between should be positioned on both sides of the traffic route.
- A company disciplinary procedure is important in the safe control of traffic operations in workplaces.
- Spot-check monitoring should be considered to maintain the safety standards of vehicles and mobile machines in terms of their fitness for purpose and driver/operator skills.
- Both works' and contractors' vehicles and mobile machines should be subject to a registration and licence system of control before they can operate; this should include hire vehicles and machines.
- All vehicles and mobile machines (including those of contractors) should carry a simple daily log book detailing their main operational records and any checking schedule, along with maintenance carried out.
- In addition to being medically fit, drivers and operators of in-works units

should be subject to appropriate mandatory age limits before being licensed to drive or operate.

- The need to reverse a transport unit should be eliminated or reduced to a minimum.
- Where conditions permit, consideration should be given to the installation of audible and visible driver activated no-go reversing signal lights which are automatically cancelled when the vehicle touches a loading bay or gantry end or a predetermined floor pad.
- Licensed drivers and operators should be approved to enter hazardous areas of work only if they have a permit to work or know the safe working procedure.
- Consideration must be given to the Control of Substances Hazardous to Health Regulations when transporting or handling toxic, dangerous or poisonous chemicals, fluids, materials or substances.
- Consideration should be given to the flammable and explosive nature of loads.
- Before transport or mobile plant is operated, all guards and guard panels must be in position and secure.
- Where combustion engine powered vehicles or moving machinery have to constantly work in buildings alongside people, consideration should be given to exhaust ventilation or catalytic converters.
- Effective emergency procedures should be in place to cope with major transport hazards and incidents.

Design focus

- As a basic safety requirement all vehicles and mobile machines capable of travelling at 5 mph or more should be equipped with effective speedometers.
- Passengers should not normally be carried unless cab or vehicle seating is provided – passengers must never be permitted by drivers to travel on flat-back vehicles.
- Safe means of access and egress must be designed into load-carrying vehicles where access for loading or load inspection is necessary.
- Where used, pallets should be designed to BS2629 or a similar EU standard, and be inspected on a regular basis by a nominated competent person.
- Tachographs should be considered for high-risk, in-works transport operations.
- Levers and buttons which control travel or lifting/lowering mechanisms must be designed to return to a neutral stop when hands or feet are removed.

Drivers and operators

- Training and operational instructions must ensure that drivers and operators understand and practice safe operation on rough terrain, inclines and steep gradients and when traversing across slopes. The danger lies in the overturning or rolling of vehicles and mobile machines and the attendant risk of crushing of drivers and operators.

- Forward and rear vision, including the rear view mirror, should not be obstructed by loads.
- All drivers and operators must be trained and instructed in the dangers of running or manoeuvring too close to incorrectly battered (sloped) or unsupported excavation edges of temporary roadways, or through material stock grounds.
- Extra wide or long loads on vehicles must display marker boards and lights at night: the standards should be those applicable on the public highway.
- All vehicles and mobile machines should be provided with practical driving and operating instructions and guidance: all drivers and operators should be trained and assessed on these.
- The parking of vehicles and mobile machines must allow a two metre clearance all round, and seven metres at the rear, in order to avoid trapping people.
- Engines should always be shut down and keys removed when drivers or operators leave their units unattended; it would also be good practice to lock cabs.
- Drivers and operators should be trained in load distribution security and compatibility.
- Push starting of vehicles and mobile machines is dangerous; responsible maintenance staff should oversee tow starting using a rigid draw-bar designed for the purpose.
- Drivers or operators unfit to undertake duties through alcohol, drugs or illness must not be permitted to drive or operate any type of transport, materials movement machines or mobile plant.
- Consideration should be given to staff being trained to act as banksmen for reversing vehicles where required.
- Drivers' and operators' working hours should be strictly controlled and detailed in a works transport policy.
- Cranes and lift trucks and shovels etc must not be used for moving rail or road transport vehicles.

Load considerations

- Axle weights and configurations are critical design factors in relation to traffic routes and road damage, including the effect on load-bearing structures such as bridges and weighbridge platforms.
- All load-carrying vehicles should have instructions displayed in the cabs in respect of loading capacities and load distribution.
- Where vehicles are designed to carry more than current national standards, they should carry clear plating details – consideration should be given to

total loading and axle loads imposed on transport infrastructures.

- When vehicles are being loaded by crane or lift truck, drivers must always be out of cabs at a safe distance from loading operations and within the vision of loading staff.
- Loading or unloading of vehicles by magnets will cause costly damage or injuries if the power fails and should be subject to a safe working procedure.
- Loading (including the securing of loads) and unloading must not be carried out where it obstructs normal traffic flows.
- Where pallets are not employed, soft wood timber packing should be available to ensure forks, lifting chains and similar equipment can be safely removed and damage prevented during loading and unloading operations.
- Where staff must board semi-trailers, hard standing pads should be available to safely stand trailer landing legs.
- Skip-carrying vehicles should not be driven until skips are fully located and lifting arms and chains secured.

SAFETY OF PARTICULAR TYPES OF VEHICLES

Articulated tractor unit

An articulated tractor is a power unit for the hauling of articulated trailers of all types (including tank trailers). It employs a fifth wheel towing system. It can have various axle and steering axle configurations, including up to four axles. Safe propping and latching of hinged cab units must be carried out when under repair. For hazardous loading and carrying operations, consideration should be given to cab protection hoods.

Where fitted, elevating fifth wheels must be designed to give adequate and safe clearance of semi-trailer legs and be fit for purpose. For rough terrain operations they should be fitted with a safety locking device in the event of lift suspension power failure. Before reverse coupling onto a semi-trailer, drivers must check that semi-trailer brakes are set, landing legs are sound and the route to and rear of the semi-trailer area are both clear. After coupling up to a semi-trailer, drivers should check that the coupling operation is latched and secured before driving away.

Dumpers

A dumper is usually a rear steer, two axle front drive machine with a driving position that is mostly open. It is a general purpose mechanised wheelbarrow found mainly on building and construction sites, and has a carrying capacity of up to 7.5 tonnes. Dumpers over this limit tend to become dump-trucks.

Dumpers should be fitted with running lights and indicators (many manufacturers do not provide lights). No dumper should operate at night without lights. The problem of thumb injury from starting handle 'kick-back' on small dumpers is well known – instruction to drivers and clear instruction labels on engines is important. Electric start should be a design requirement when hiring or purchasing dumpers. Carrying capacity and centre of gravity labels should be affixed to dumpers.

The Construction (General Provisions) Regulations (regulation 37) (see chapter 26) require vehicles tipping into an excavation to employ adequate measures (eg stop-blocks) to prevent the dumper over-running the excavation edge. If other people have to ride on dumpers a safe seat and position must be provided. It is vital that drivers are trained and licensed to drive especially if it is necessary to use dumpers for towing – a current category B licence should be only a basic requirement. When it is necessary for dumpers to operate on works roadways, a rear view mirror should be considered. Dumpers should be fitted with roll over bars and fall protection hoods. When dumpers operate in off-road locations, consideration should be given to high visibility markings and a flashing light mounted on a mast.

All non-standard loads should be secured by ropes or chains and one hand driving (while the other holds the load) should be discouraged by training. Loads that restrict driver forward vision should not be permitted. Dumpers must only operate with dumper bodies in the latched down position. This applies whether they are loaded or empty.

Dump trucks

A dump truck is an off-highway tipper of heavy duty construction, with two or three axles. Some have articulated coupling between cab and dump body. Most are fitted with a driver's cab forward of the tipping body. At the larger end of the range some units can have a carrying capacity in excess of 100 tonnes. A management policy decision should be taken as to whether or not dump trucks are to operate on works roadways or be confined to closed traffic routes or to closed site operations. Drivers must be of sufficient skill to ensure that when tipping, all wheels are on flat and solid ground and where loads are lodged in place, there is not sudden forward or rear movement of the truck so dislodging the load. Dump trucks must not travel when the tipper body is in the raised position.

Because of the nature of the loads carried by most dump trucks, an effective, stepped access and egress with hand-holds is necessary for drivers. Auto reverse warning devices are vital on dump trucks, especially the larger ones. When dump trucks operate on works roads, all the normal construction regulations and standards should apply ie lights, indicators, mudguards, front bumpers etc.

Drivers of articulated dumpers should be trained to ensure they do not operate when anyone is within the vehicle's trapping zones. Rearward obstruction and people sensor devices should be fitted to dump trucks to prevent the classic reversing accidents.

Earth-moving machines

Earth-moving machines include both tracked and tyred machines such as bulldozers, scrapers, graders, compactors and excavators. Most of these machines can and do move and handle raw materials, wastes, process scrap and consumables. Many are equipped with secondary mechanical tools and devices to assist and supplement the main design requirement. Some are designed solely for special operations.

Engine noise levels experienced by both operators and adjacent staff must be considered and brought within legal requirements when operating machines of this type. A safety method statement or safe working procedure should be drawn-up as a safe system of work prior to use of earth moving machines, especially when on lone site tasks. The high level of skill required to operate earth-moving machinery must be proved by a licence to operate, and nationally accepted standards should be applied.

Special safe systems of work must be in place before earth-moving machines work near water or on tips, especially toxic tips. Before removal of surface takes place below 200 mm, an engineering check should be carried out to locate possible underground services. The design and maintenance of access and egress to cabs is vital on this type of machinery, especially on tracked machines. Operators must ensure buckets, blades and arms are placed in a fully lowered position when the machine is at rest. It is important to prevent wheel spin on tyred excavators when they are operating in jacked-up fashion.

Lift trucks

Lift trucks are normally fitted with a two fork unit. Some have more and some versions have one. The latter are usually referred to as pin trucks. Lift trucks come in a combination of carrying capacities and design configurations eg high lift, side lift, rough terrain and articulated. Most are rear wheel steered and many of the larger machines are fitted with cabs.

Driver skills are critical to the safe operation of lift trucks. Such trucks should always be driven with forks in a fully lowered mode (approximately 200 mm above ground) with or without a load. Drivers of lift trucks should be instructed to cease operations when anyone comes within two metres of the machine. Lift trucks of any size or capacity should be fully equipped with a set of road lights and indicators. The driver's panels should include a permanent notice of the maximum height clearance of the machine.

Special safety instructions should be given to drivers when operating on rough or uneven ground as stability can be affected. Drivers should not drive lift trucks backwards except for load manoeuvring operations. Where it is not reasonably practicable to avoid the forward driving of lift trucks with partial load obstructions, a banksman must be provided to give guidance. When travelling lift trucks must always have the mast in the tilted back mode. Lift truck masts must be absolutely vertical (forks level) when entering pallets – when the forks have fully entered the pallet the mast can then be tilted back.

When stacking, the parking brake must be applied. When unloading or loading in situations which require lift trucks to board vehicles, ensure vehicle parking brakes are set, one set of wheels is chocked and the bridging ramp is secure and load bearing. Drivers of lift trucks should be trained to consider the rear steer end swing of the truck and articulated centre-pivoted steering axles, especially on slopes. Drivers should not leave trucks when loads are suspended on the forks.

Open driver positions on lift trucks must have an overhead shield/cover to give protection against possible load fall, and have roll over protection. All moving parts of lift trucks must be guarded where practicable, in order to prevent drivers from being trapped. Lift trucks should not be used to lift people unless equipped with man lift platforms designed for the purpose.

Loading shovels

These are tyred or tracked (JCB type) machines used for moving, feeding, stocking and loading raw materials, products or waste. Operators should be instructed to make clear signals to vehicle drivers when they are ready to load, and when loading is completed. Loading shovels must not undercut stocks of materials. Operators should be trained to ensure transport vehicles are not overfilled causing spillage, or loaded above maximum carrying capacity. Loading shovels in particular should be always fitted with automatic reversing warning devices, both audible and visible. Operators should be trained in the hazards of loading on and/or across inclines, slopes and gradients.

Passenger vehicles

These include workplace operating buses, coaches or mini-buses and passenger transit vans. Where applicable drivers should have a public service vehicle licence. Passenger carrying vehicles should not be used for carrying any other goods or materials when passengers are in transit. Consideration should be given to fitting mini-buses with passenger seat belts and front facing seats. A safe access/egress should be included in the design of passenger vehicles. Passenger carrying vehicles should have reinforced roof members for roll over protection.

Tankers

Tankers can be self-powered, rigid units of up to four axles, or constructed on semi-trailers up to three axles and towed by an articulated tractor. Other trailers of both rigid and steering drawbar types can be fitted with tanks usually towed by a two axled agricultural type tractor.

A risk assessment is essential for tanker operations – in some cases hazardous tankers may need special tanker routing through factories or works. A number of deaths have occurred because people entered tanker hatches and were overcome by fumes and/or lack of oxygen – hatches should be kept locked with keys in the control of drivers or authorised controllers. Confined space notices should be fitted to hatches and entry instructions placed on labels fitted to cabs. Tankers of all types must be fitted with pressure relief valves which should be checked by drivers and tested periodically as per legal requirements. Special emergency plans should exist for coping with incidents involving tankers carrying hazardous loads.

Tanker design must conform to legal requirements especially for separation of compartments and strength. Tankers must have earthing points when flammable or explosive loads are carried. Tanker units must be subject to tank examination and testing. Hazard warning signs and emergency information must be carried and displayed.

Drivers of tankers carrying hazardous loads must be properly trained and in possession of an appropriate certificate. Collision and roll over protection should be considered for some tankers. The design of hazardous tankers must prevent overfilling and fire/explosion risk. Some hazardous tankers should be parked in secure locations when not in transit. Driver records are necessary for some types of tankers.

Tractors and trailers

These include agricultural-type tyred tractors of either normal speed or fast tractor design. They tow mainly through a draw bar coupling, hauling trailers of various carrying configurations, and have steering or rigid draw bars. Drivers of tractors with trailers should have an agricultural category F licence or a LGV licence if they are to operate safely within a workplace. Tractors are not normally designed to be both load-bearing *and* capable of carrying loads. Tractors should always be limited to towing only *one* trailer and should not carry passengers unless extra seating is a design feature of the cab and passengers do not impede the driver.

Tractor draw bar hitch units must always include a designed for purpose draw bar pin chained to the unit for security from loss or damage. All trailers should be fitted with hand brakes and trailers with more than one axle, or with a carrying capacity of over 7.25 tonnes, should also have operational braking systems. Trailers equipped with central axles must be fitted with draw bar landing legs to permit safe tractor hitching and unhitching. Tractor power take off units must be close guarded to comply with legal requirements. Tractor and trailer units should be equipped with a full set of lights, indicators, reversing warning lights plus braking lights. Tractor cabs must be equipped with protection against roll over and falling objects.

Tipper lorries

These are usually a highway design vehicle of rigid construction with up to four axles. They can also be a tractor/trailer articulated unit of up to six axles. Under no circumstances should anyone enter or work in the space beneath raised tipper bodies until bodies are securely supported and chocked. Drivers should be of sufficient skill to ensure that when tipping, all wheels are on flat and solid ground and when loads are lodged in place, sudden forward or rear movement of the tipper does not occur, so dislodging the load.

Unless fitted with sloped rear decks, tipper vehicles should always operate with effective and safe rear tail gates or tail boards. Side boards fitted to tippers must not have loads higher than the top board and the side boards need to be fully maintained around all four sides of the tipper top to be effective. Tippers should not be driven when the tipper body is in the raised position especially near power lines or pressurised pipe lines.

Trailers (flat backs)

A semi-trailer is a unit of up to three axles towed by an articulated tractor unit. Other types of flat back trailers are rigid and steering draw bar types: rigid trailers are up to two axles and steering trailers up to three axles. The draw bar trailers are usually towed by an agricultural tractor.

NB Most of the items listed below would also apply to flat back lorries.

- Semi-trailers must have design for purpose landing/parking legs.
- Operational consideration should be given to the possible 'cut-in' of semi-trailer wheels on corners, which can also result in roadway damage.
- Double air brake lines must be fitted to all semi-trailers.
- Semi-(flat back) trailers must be equipped with headboards and no loading should be made above headboard height.
- Loads must be safely secured for travelling: this should be done by means of chains, nylon belts or ropes or trailer sheets.
- Trailers should be fitted with rear and side under-run prevention bars.
- Low loader trailers with multi-wheeled rear axles must have their wheels guarded (box covered or mudguards) to prevent people being trapped.

Power operated mobile work platforms

Mobile work platforms are usually of three main types; scissor lift platforms (self propelled), telescopic boom and articulated telescopic boom machines, both types self propelled. The final type is an articulated telescopic boom, mounted on a lorry or truck carrier. The following are important considerations:

- Ensure ground conditions are level and load bearing, and use outriggers where fitted.

- Where platforms are capable of slewing, checks must be made on clearances.
- Operators should wear safety harnesses when on platforms.
- The safe working load of platforms must never be exceeded.
- Checks must be carried out for possible overhead hazards, eg electric cables and overhead cranes etc.
- Operators must ensure people do not leave the platform when at working heights.
- Platforms should never be used in high winds.
- Ladders or scaffolding or any other means of extending the height of platforms should not be employed.
- Platforms should not be used to transfer materials from one level to another.
- Platforms should not be used to lift loads.
- Consideration should be given to the use of barriers, bunting and flashing lights where platforms are in operation in hazardous areas.
- Safety devices, especially emergency buttons, must be tested at the start of each work period.

Mobile cranes

Mobile cranes come in a number of basic types, sizes and configurations. There are self propelled and truck mounted cranes; both operate on pneumatic tyred wheels. Cranes can also be crawler tracked usually with a fully slewing superstructure. Crane jibs can be solid or lattice struts and telescopic or fixed. Most jibs will be luffing ie can jib in/out and most mobile cranes will be capable of slewing. Power is usually from a diesel engine via mechanical, electrical or hydraulic transmission and control systems. Important considerations are:

- There must always be a competent person in overall control of any lifting operation.
- A mobile crane should carry copies of relevant safety documents ie
  - — weekly inspection register (F91)
  - — six monthly standard lifting tackle inspections (F91) and test certificates (F97)
  - — fourteen monthly thorough examination certificate (F96)
  - — four yearly inspection and test certificate (F97)
- A pre-operation driver's check list should cover items such as
  - — ropes, sheaves, anchors, pins and hook block(s)
  - — hoist and luffing machinery, jib, telescopic gear and slewing mechanism
  - — safe load and radius indicators plus limit switches and warning devices
  - — outriggers, wheels and tyres
  - — steering, brakes, lights, indicators, wipers, mirrors and horn
- Ground conditions and structure must be examined before lifting operations start.

- Outriggers should always be fully extended utilising a level base.
- Cranes should be clear of overhead power lines, downshop power conductors and overhead cranes.
- Cranes should be clear of all types of rail tracks and slewing obstructions.
- Weights of loads must always be known or calculated by a competent person and communicated to the crane driver.
- Mobile cranes should only employ competent slingers who should also be trained to give standard operational signals.
- Weather conditions, eg fog, ice, snow and especially high winds should be considered before crane lifts are made.
- Lighting levels (natural or artificial) should always be considered.
- Mobile cranes must not lift loads over people; safety distances should be observed when operating near people.
- Cranes should be filled with load hook safety catches.
- Cranes should be fitted with flashing lights to operate when they are carrying out lifting duties.
- Drivers should not leave cabs with loads suspended on hooks.
- Mobile cranes should only travel with jibs in the fully parked position and hook block restrained.

RAILWAY OPERATIONS

Legal framework

As noted in chapter 3, risk assessments are required by the Management Regulations. Therefore any organisation with railway operations should subject the whole system to a risk assessment, enabling a safe system of work to be drawn up and reviewed perhaps every two to three years. There is also legislation specific to workplace railways – the Locomotive and Wagons (Used on Lines and Sidings) Regulations, 1906.[3] These cover a number of safety issues including some of the matters dealt with below.

Locomotive driver

- Loco drivers shall give audible warning on approaching a level crossing and when moving off as a general safety warning.
- Drivers must ensure locos carry lighting at night or in adverse weather conditions and if pushing wagons a lamp should be on the lead wagon or a person with a lamp should precede the lead wagon.
- Where a loco pushes more than one wagon a person (if it is safe and reasonably practicable) shall accompany or precede the lead wagon.
- No person shall ride on buffers of rail wagons or locos or ride in a position outside of the vehicle gauge.

- A signal or sign shall be displayed when wagons or locos are under repair.

Loco design

Design of locos and rail wagons must comply strictly to British Rail loading and travelling gauges, and be of adequate construction so as to ensure they are fit for purpose. Locomotives should be designed or modified to give effective access and egress to operators. Where overhead electric power supplies are involved, a lockable hatch device should prevent non-controlled access to the top of locos. Diesel powered locos should be checked to ensure the safety of operators when fuelling takes place. Some additional loco design and safety considerations are listed below.

- Cabin windows must be of adequate size and so disposed as to give maximum field of vision and to be shatter resistant.
- Outline, shape and positioning of engine covers must be such as to allow maximum space for windows and must interfere as little as possible with vision from the windows, front and rear.
- Doors to the cab should be sited to allow free and safe access from the four corners of the locomotive.
- There should be low level platforms at all four corners of the locomotive; these should be amply proportioned and have non-slip treads. Platforms should be so positioned relative to the frame and footplate as to allow a person to stand on them and be clear of all obstruction. Steps and cross galleries should permit a person to cross over the locomotive in safety without opening doors and passing through the cab, if practicable.
- Sufficient and properly positioned handholds are required especially for mounting and dismounting from the locomotive.
- An audible warning should be provided, with a distinctive note, easily operated from normal driving positions.
- Operating controls should be in duplicate on each side of the cab and so arranged as to be easily handled by the driver from the normal driving positions.
- There should be provision for carrying shunting poles in a safe and accessible position.
- There should be provision for carrying scotches when required.
- There should be efficient windscreen wipers of robust construction with window heating arrangements for severe weather conditons.
- Rectangular buffer beams should be fitted in order to protect people when riding on lower steps, and to limit the extent of any derailment.
- There should be lifting lugs and jacking pads at the four corners of the locomotive to assist in safe re-railing operations.
- Pneumatic or hydraulic buffers should be fitted to reduce intensity of shunting shocks.

- There should be visible warning including distinctive marking in bright day glow and reflective colours of prominent parts, and provision of suitable lights for identification at night, including flashing warning lights.
- There should be holders for portable fire extinguishers in accessible positions.

Railway wagons

Internal railway wagons should have park braking units fitted and a safe means of access should be considered. Rail wagons used internally should be uniquely numbered and marked up (on both sides) with maximum carrying capacities and wagon tare weight. Axle loading parameters of rolling stock is an important design consideration. This will be dictated by the axle bearing engineering design of rail tracks especially gantries, bridges, locations where tracks cross over electrical and service networks and any rail weighbridges. Ill-matched design will result in damage and possible structural failures and derailments.

Locomotive driver selection and training

Locomotive driver selection and training is a vital safety consideration and as such must be subject to a clear written company policy statement and manual. These documents would need to detail some of the following safety factors:

Driver selection

- Display a safety awareness of railway operations, hazards and risks.
- Medically and physically fit.
- Previous experience with railway operations.
- 21 years of age or over.
- Hold some sort of vehicle licence to demonstrate aptitude.
- Possess a positive safety attitude towards railway operations.
- Attain a competency standard by test.

Driver training

This should be subject to a training manual and should contain the following four elements:

- Awareness of the railway system and loco familiarisation (walking the operational areas and locos).
- Practical driver tuition (accompanied by a qualified driver).
- Study of relevant legislation and railway operations (a safety handbook should be provided).
- Loco driving and operational test plus a written examination (multiple choice questions).

General safety considerations

- Rail tracks shall be periodically inspected and maintained for traffic work-loads.
- No person shall cross a railtrack by crawling under wagons or walking between wagons.
- Vehicles, equipment and plant should not be permitted within three metres of railtracks.

MAIN CAUSES OF TRANSPORT ACCIDENTS

Areas where transport accidents are common are set out below.

Traffic entry into buildings

Road or rail traffic entry into buildings or bays etc can result in crushing injuries and people being struck by vehicles, especially reversing road vehicles or by rail wagons being propelled. To reduce the possibility of serious accidents in these situations a control system based on the following safe system of work should be considered.

1. *A warning sign* should be positioned in the bay with high visibility flashing lights activated by a central control switch. A number of warning signs and light units could be considered in long bays or where a single unit could be masked.

2. *A unique audible warning* should also be fitted (or a number of them) and be activated from the same central control switch.

3. *Signs outside the bay* to instruct incoming transport operators must ensure basic safety control. These could state no entry without activation of warning system.

4. *The central control switch (or button)* should be enclosed and located inside the loading or unloading bay on a 1.5 metre post to ensure the road transport driver or railway operator must come inside the bay. This means that they can check the prevailing safety conditions before bringing in the vehicle(s).

5. *When entry is prohibited on safety grounds*, authorised staff from the loading or unloading bay should place a preventive token or lock off on the enclosed switch preventing transport staff from activating the system and gaining unsafe and unauthorised entry to the bay.

6. *When entry is safe* the transport staff could place a preventive token, or lock off, to stop unauthorised cancellation of the warning system while transport movements are taking place.

7. *The preventive token* can take the shape of a personal danger board, tag-out board or personal identity card.

Falls from vehicles

People may fall from vehicles during loading, unloading, sheeting or load securing operations. It is inevitable that staff will from time to time have to work on top of loads and although it is not possible to ensure total safety in this work, the risk of injury may be reduced by reviewing the following points:

- provision of safe means of access to and egress from the vehicle (portable or fixed steps)
- instruction of all staff of the dangers involved
- the use, where appropriate, of suitable mechanical handling equipment which eliminates or reduces the need to board vehicles.

Loading

- No vehicle should, under any circumstances, be loaded beyond its rated capacity or beyond the legal limit of gross weight.
- Before loading is commenced, the floor of the lorry should be checked to ensure that it is safe to load. Points to look for are damaged or loose boards and tripping obstructions.
- Loads should be evenly distributed in line with vehicle design and axle loading.
- Loads should be secured or arranged so that they do not slide forward in the event of the driver having to brake suddenly, and will not slide off the lorry if it has to ascend steep gradients. Where loads of a particular type are carried regularly, it may be possible to provide racking or bolsters as aids to load stability.
- Tail board and side boards should be closed wherever possible. If over-hang cannot be avoided it should be kept to a minimum and the projecting part of the load should be suitably marked.
- No loading should take place near overhead electric cables.

Unloading

- Before ropes, tarpaulins etc are removed, steps should be taken to ensure that the load is stable and the removal of the securing ropes will not allow any part of the load to collapse and fall.
- Care should be taken to ensure that the removal of any part of the load does not cause the remainder to become displaced or unstable;
- Unloading should be carried out so as to maintain as far as possible a uniform spread of the load. Uneven unloading can result in the vehicle or trailer

becoming unstable, especially if it is an articulated or similar type of trailer which has been detached from the tractor power unit.
- Unloading near overhead electric cables should, like loading, be avoided. With tipping lorries, particular attention should be paid to the risk of the raised body of the vehicle fouling the cable.

Pocket cards with key safety points could be issued to appropriate staff. These could be backed up by poster signs, which can do much to make preventive measures a day to day feature of loading and unloading.

Road/rail crossings

Possibly one of the most hazardous workplace situations, which gives rise to a number of serious accidents, is the lack of control at road/rail crossings. Rail machines – transfer bogies or transport cars – may not always cross roads but a number of small road vehicles, such as dumpers and lift trucks, could cross railtracks on unmarked routes, and with serious results. It is worth conducting an audit of these official and unofficial crossing points on both railway tracks and rail runways as these locations are potentially very dangerous. Where road/rail crossings are required there must be clear signs and crossing points indicated by marker posts or barriers. For crossings with considerable traffic flows, interlinked traffic light signals should control both road and rail traffic. Any unofficial crossings should be removed and temporary crossings should be installed through a permit to work system which should be then removed when the task necessitating the crossing is completed.

Additional causes of transport accidents

Some of the most common transport accidents are listed here to assist in risk assessment.

- *Overturning* when tipping or operating on undercut or unsupported edges or floor support.
- *Driving too fast*, especially on bends in wet or icy conditions.
- *Turning right sharply* and without signals.
- *Manoeuvering and reversing* too close to people and plant.
- *Rolling over* when operating on gradients or lifting loads on unsupported ground.
- *Load spillage or shedding* due to overloading.
- *Vehicle component failure* caused by poor maintenance.
- *Operator errors* due to lack of training and testing.
- *Defective vehicle* or machine safety eg brakes, steering etc.
- *Entanglement* in dangerous moving machinery or equipment (power take off units and hydraulic lift units etc).

- *Collision* of road or rail traffic due to close parking of vehicles, close stacking of materials or close erection of temporary structures eg scaffolds.
- *Unsafe repairs* to rail tracks or rail wagons by maintenance teams (platelayers etc). This work should only take place by permit to work, which could include red flag/red light working and/or locked points.

To leave transport operations to some sort of natural control is a flawed safety policy. Competent management and supervision of transport operations is vital in any organisation.

Notes

1. *The Cost of Accidents at Work,* HSE, 1990
2. *Road Transport in Factories And Similar Workplaces*, HSE, GS9(R), 1992
3. The Locomotive and Wagons (Used on Lines and Sidings) Regulations, SR&O 1906/679.

19

Ergonomics, Manual Handling and VDUs

This chapter starts with an examination of the concept of ergonomics and how it relates to health and safety at work. Ergonomic issues in health and safety can be seen as being derived from the interaction between the worker, the workplace and the equipment, machinery and materials which are being used. Each of these interfaces is then explored. Finally, the chapter focuses upon two areas where there is now substantial and important legislation relating to the ergonomics of health and safety at work: first, manual handling, and second, the use of visual display units (VDUs), referred to in the regulations as display screen equipment.

THE CONCEPT OF ERGONOMICS

The meaning of the term

The Oxford dictionary defines ergonomics as study of the efficiency of persons in their working environment. The word is derived from the Greek, ergon, meaning work. This is perhaps too narrow, unless the word efficiency can be taken in a broad sense to include health and safety at work as well as output. Arguably, a better 'fit' between a worker and his or her equipment, work methods and workplace will result in higher output anyway, although it is likely that there will be a greater initial investment required to produce that better fit. Ergonomics means that differences in individuals' height, weight, physical strength and mental qualities need to be taken into account in the design of work.

Ergonomic problems

In its leaflet *Ergonomics at Work*, the HSE gives examples of ergonomic problems:[1]

- work surfaces which are uncomfortable to sit at because there is not enough clearance for the legs
- tools with handles which require a wide grip
- tools which require excessive force to be applied
- control panels with layouts which cause difficulty of use, for example by requiring the operator to bend or stretch, or by causing him or her to select the wrong control.

The problems are not restricted to equipment. For example, an office window which is difficult to open can easily cause a wrist sprain; an incorrect method of lifting can easily give rise to a back problem.

Ergonomic problems are likely to result in mistakes, accidents and near-misses, injuries and illnesses. These will arise in situations such as those where workers are unable to see important controls, unable to work comfortably, unable to cope with the amount of information presented at one time and are inattentive because there is too little to do. The HSE leaflet offers a checklist as a means of establishing whether equipment and systems are ergonomically satisfactory. It asks the worker various questions about the equipment and system being used:

- Does it suit your body size?
- Does it also suit all other users?
- Can you see and hear all you need to readily?
- Do you understand all the information that is presented?
- Do errors occur frequently, and is it easy to recover from them?
- Does the equipment or system cause discomfort if you use it for any length of time?
- Is it convenient to use?
- Is it easy to learn to use?
- Is it compatible with other systems in use?
- Could any of these aspects be improved?
- Do other users have similar reactions?

Ergonomic approaches

Ergonomic approaches towards achieving a better fit between worker and work environment usually involve:

- observing the job – which may extend to video-recording it
- identifying the components in the worker's behaviour
- defining safe behaviour for these components.

The last of the above will require safe working methods to be established in terms of specifics such as: body position; lifting; wrist-posture; selection and use of tools; and pushing and pulling. Employee involvement is often a feature of ergonomic approaches since co-operation is necessary to identify what is done and how it needs to be altered. This contrasts with other approaches, such as work study.

ERGONOMICS AND HEALTH AND SAFETY AT WORK

Principal sources of ergonomic risk

The principal sources of risk arising out of a mismatch between the worker and his or her work environment are outlined below. They may arise from defects in equipment, from poor methods of working or from deficiencies in the infrastructure of the workplace itself.

- *Bad posture*: this has the effect of exerting too much force upon the muscle and tissue of the body joints. It can be caused by poor seating arrangements and awkward reaching and stretching.
- *Excessive force*: here the worker is having to exert too much force in order to perform the job, resulting in muscle and tissue damage.
- *Repeated contact with objects*: known as mechanical stress, this is where a worker's body makes repeated contact with an object or surface, and can result in nerve damage. Repeated hand and arm work, such as keyboard work or packing, often affects the wrist and forearm.
- *Repeated exposure to vibration*: this may be through tools being used or through contact with a vibrating object. It can lead to white finger, involving a loss of sensation to the finger, and back injuries (see chapter 22).
- *Extreme temperatures*: these can be a further source of physical stress.

Repeated exposure to these is likely to result in injury. Manual handling and lifting, and highly repetitive activities such as keyboard work, are areas in which some of the above may be present. These are also areas which have their own specific legislation. This is described later in the chapter, although it should be noted that all risks will in any event be covered by the general provisions of HSWA, s2.

Ergonomic injury and illness

The term repetitive strain injury (RSI) has been used to describe the injuries arising from the above risks, but as can be seen, while repetition is the immediate cause of the problem the causal factors are varied and deeper. This has prompted the use of the term 'work-related upper limb disorder' (WRULD)[2] A WRULD is any one of a number of injuries and illnesses which occur when a worker is exposed to one or more of the risks already described. They include a number of specific conditions such as tendinitis, tennis elbow and white finger, as well as back injuries and other muscle problems.

The HSE refers to these as musculoskeletal disorders, which involve the muscles, tendons, joints and skeleton, particularly in the back, hands and arms. The symptoms include aches, pains, swelling and inflammation. It is estimated that in the UK 52 million working days are lost through back problems alone.[3]

In the United States, the term adopted is cumulative trauma disorder. This reflects the fact that the worker is not exposed to a specific event, but rather to a cumulative and gradually-developing phenomenon. A major consideration is that the injuries are internal but may not be visible externally. For example, carpal tunnel syndrome may produce numbness, tingling, burning and aching which may not always show up externally through bruising, swelling or abrasion. In the US, more than half of the workplace illnesses reported would be described in the UK as WRULDs. By 2000, they are expected to account for half of US medical expenditure. Even in 1984 they accounted for $27 billion in lost earnings and medical expenses.[4]

MANUAL HANDLING

The nature and scale of the problem

Over a quarter of all accidents reported each year are associated with manual handling. Most of these result in injuries requiring more than three days' absence from work, typically because of a sprain or strain, and often of the back.[5] The case law of health and safety at work includes many examples of manual handling accidents.

There is no general proposition that loads in excess of a specified weight are unsafe, but in *Fricker v Perry* the court accepted that, in the circumstances of that case, a weight of approximately 140 lb was the most that a man could be expected to lift. However, the worker himself may be regarded as the best judge of his own capability, so that the absence of a complaint to the employer or a request for assistance may be important (*Hall v McLaren*). Where it is normal for two people to carry out a lift, the unwillingness of one employee to wait for assistance from another may result in the defeat of that employee's claim (*Ashcroft v Harden*).

Where the work is clearly beyond the capability of one person, the employer is likely to be blameworthy if the work system is unsafe and he has not instructed the employee in safe handling techniques and told him to obtain assistance (*Ioannou v Fisher's Foils*). Where needed, assistance will have to be made available rather than assumed (*Hardaker v Huby*). It will be the worker's responsibility to inform his employer if already affected by an injury or weakness. The employer will then need to take account of this in fulfilling his duty of care.

It should be borne in mind that the above cases all pre-date the MHO Regulations. Future civil cases are likely to be strongly influenced by the Regulations, and especially by whether the employer has assessed the risk.

Weight itself may not be the overriding factor: for example, the load may be of an awkward size or shape, or may be an open container with dangerous contents capable of spilling. The analytical framework of the MHO regulations – task, load, environment, capability – is helpful to employers in determining an appropriate response to manual handling risks (see pp430–33).

Introduction to the Manual Handling Operations (MHO) Regulations 1992

The regulations repeal existing law on the subject, including s72 of the Factories Act 1961 which stated that an employee should not lift anything so heavy as to cause injury, but gave little guidance.[6] MHOs include both transporting a load and supporting it in a static position, as well as the intentional dropping or throwing of a load. The process must be carried out by hand or by bodily force, and this includes lifting, putting down, pushing, pulling, carrying or moving. A load will include persons and animals, but not an implement, tool or machine whilst in use since these are covered by the Provision and Use of Work Equipment Regulations. Injury, in the regulations, does not include injury caused by toxic or corrosive substances because this is within the scope of the COSHH Regulations. The MHO Regulations, which came into force at the beginning of January 1993, apply therefore to the transporting or supporting of a load by hand or by bodily force.[7] They have general application with the exception of normal ship-board activities on a merchant vessel.

Key stages of the MHO Regulations

The regulations transpose into British law the requirements of the European Union's Manual Handling of Loads directive of 1990.[8] They establish a hierarchy of requirements:

- avoid hazardous MHOs so far as is reasonably practicable
- make a suitable and sufficient assessment of any hazardous MHOs that remain
- reduce the risk of injury from those operations so far as is reasonably practicable.[9]

The flow chart in figure 19.1 sets out the steps employers need to take in ensuring that they comply with the regulations.

The philosophy of the regulations is that the safe handling of a load will depend upon a variety of factors which may be classified under five broad heads:

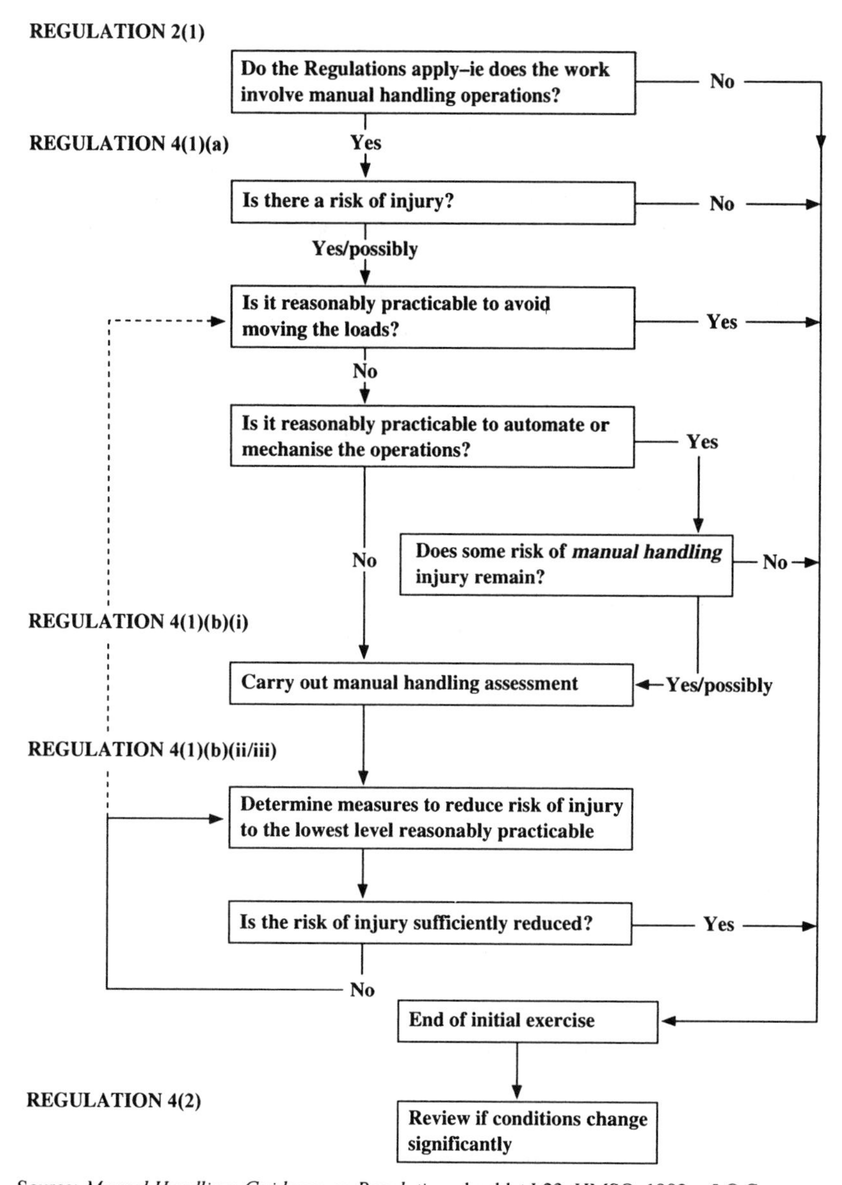

Source: *Manual Handling: Guidance on Regulations*, booklet L23, HMSO, 1992, p5 © Crown Copyright reproduced with the permission of the Controller of HMSO.

Figure 19.1 *Key stages of the MHO Regulations 1992*

1. the load
2. the task
3. the working environment
4. individual capability
5. other factors.

By virtue of regulation 4 and schedule 1 to the regulations, these factors, and specific questions relating to them, must be taken into account when making an assessment of risks. The guidance notes suggest that a similar approach should be taken to risk reduction, with the emphasis on ergonomics.[10]

The HSE also provides some numerical guidelines to indicate what weights of loads may safety be lifted or lowered to certain heights (eg shoulder height, elbow height) at arm's length and next to the body. This information is found in Appendix 1 to the Guidance and is intended as an initial filter which can help to identify those MHOs requiring further investigation. The guideline figures are not limits, but anything exceeding them by a factor of two or more should warrant careful scrutiny.

Duties are also placed upon employees: they must make full and proper use of any system of work provided for their use by their employer in order to reduce the risk of injury from MHOs to the lowest level reasonably practicable. Added to the duties laid down in the HSWA and in the Management Regulations, this means that as regards MHOs, employees are under a duty to:

- co-operate in risk assessments
- make proper use of safety equipment which is provided
- co-operate with their employer as regards health and safety training
- follow laid-down safe working procedures.

Avoidance of manual handling

As far as is reasonably practicable, MHOs should be avoided if there is a risk of injury. The numerical guidelines in appendix 1 to the HSE guidance can be of assistance in determining the likelihood of injury. Where there is a risk of injury, the first step should be to consider avoidance. It may be possible to eliminate manual handling eg by taking treatment to a patient rather than vice versa. Alternatively, the operation may be mechanised or automated eg by use of lifts and/or a powered wheelchair. If neither approach is "reasonably practicable" or if some manual handling with risk of injury still remains eg lifting the patient into a powered wheelchair, then a risk assessment must be made.[11]

Carrying out risk assessments under the MHO Regulations

The duty of the employer to avoid MHOs which involve risk, and to assess the risks in respect of those operations that remain, applies even if the employee is working away from the employer's premises. The degree of control exerted by the employer in such circumstances, will, however, be relevant in determining what is reasonably practicable. The test is satisfied, as in other areas of health and safety at work, by demonstrating that the cost of any further preventive steps would be grossly disproportionate to the further benefit accruing from their introduction.

The assessment must be 'suitable and sufficient'.[12] Suitable means that it must address the problem; sufficient means that it must address it in sufficient depth. Other points to be noted are as follows:

- The assessment must include routine work, varied work eg maintenance, work away from premises eg delivery men, installation crews.
- Assessor(s), who will probably be in-house, should
  - — be familiar with the regulations and associated guidance
  - — be aware of the nature of the handling operations (perhaps they should make a list of them)
  - — observe manual handling operations where necessary
  - — seek additional information (eg accident/ill health records, industry specific data)
  - — involve employees and their representatives
  - — seek outside assistance where necessary.
- Assessments must take into account
  - — task
  - — load
  - — working environment
  - — individual capability
- Assessments must be recorded unless they are simple and obvious or of low risk and short duration. There is no standard form for records but they should include the assessor's conclusions and recommendations (which must then be implemented, so far as reasonably practicable). A specimen recording form is shown in Figure 19.2.
- 'Generic' assessments of broadly similar operations are acceptable.
- A checklist is included in the HSE Guidance as an example but any form of checklist can be used to suit individual circumstances. A checklist is to be found on pp430–3.

The form of assessment

The duty to carry out an assessment lies with the employer. The assessment should be in writing unless it is very simple, and there should be a valid conclusion. Where there are more than five employees, the assessment should be part of the employer's health and safety policy. Generally, assessments can be done in-house. It is important to involve employees because they know the tasks which have to be performed. In assessing risks it is important to look at previous accident data.

MANUAL HANDLING ASSESSMENT

<table>
<tr><td>OCCUPATION</td><td>DEPARTMENT</td></tr>
<tr><td colspan="2">ACTIVITY/TASK</td></tr>
<tr><td colspan="2">DESCRIPTION

</td></tr>
<tr><td colspan="2">POTENTIAL RISKS

</td></tr>
<tr><td colspan="2">SCOPE FOR MECHANISATION? (OR ELIMINATION OF MANUAL HANDLING)</td></tr>
<tr><td>NUMBER</td><td>RECOMMENDATION</td></tr>
<tr><td></td><td></td></tr>
<tr><td colspan="2">OTHER RISK REDUCTION MEASURES</td></tr>
<tr><td>NUMBER</td><td>RECOMMENDATION</td></tr>
<tr><td></td><td></td></tr>
</table>

SIGNATURE(S)

DATE OF ASSESSMENT
DATE FOR REVIEW OF RECOMMENDATIONS

Figure 19.2 *Specimen form for recording MHO risk assessments*

Who should make the assessment?

This is a matter for the employer to decide. The person needs to have authority and to be familiar with the MHO Regulations.

Varied and/or peripatetic operations

These can cause some difficulty since it may not be possible to detail every task that is performed. Training may assume more importance here, as might the employee's own judgement about the risks. Peripatetic work may require liaison with the person in control of the premises or work system. It may be possible to conduct a generic assessment where there are similar MHOs in varying contexts.

Reducing the risk and providing information for employees

Risks must be reduced to the lowest level reasonably practicable. The risk assessment should be used as a basis for risk reduction. In some cases, the MHO may be a two person rather than a one person operation.

Although costs may be involved in risk reduction (as in the avoidance of manual handling by the introduction of mechanisation), the potential for savings through a reduction in lost-time injuries and other costs is very substantial. Investment in better manual handling may at the same time increase productivity, reduce costs and improve health and safety at work.

As well as taking "appropriate steps" to reduce the risk of injury, employers must provide employees with information about the weights of loads and the heaviest sides of irregular loads. This should be precise, where reasonably practicable. The checklist on pp430–3 gives details of the questions to be considered in an assessment of MHO risks, and of possible risk reduction measures. Further detail may be found in the HSE guidance to the regulations.

Key action points

The MHO Regulations can be approached as a four-stage process:

1. Preliminary appraisal of occupations
   — prepare a list of occupations
   — list MHOs associated with them
   — form a view of risks attached to these MHOs (Appendix 1 to the HSE Guidance may help)

2. Avoid MHOs where there is a risk eg via mechanisation or work redesign.

3. Assess those MHOs which cannot be avoided
   — identify the assessors: it is unlikely that health and safety specialists will have the time or detailed knowledge to carry out all the assessments, although they will play a key role in the process. Significant involvement of other staff will be essential.
   — train the assessors
   — carry out the assessments

4. Reduce risks
   — evaluate the assessments: what is reasonably practicable?
   — implement recommendations

The assessment should be used as a basis for implementing risk-reducing methods.

MANUAL HANDLING CHECKLIST

THE TASK

ASSESSMENT FACTORS	**REDUCING THE RISK**
Does the task involve:	*Risk can be reduced by:*
• The load being held or manipulated at a distance from the trunk?	Attention to task layout
• Unsatisfactory bodily movement or posture: – twisting? – stooping? – reaching? – sitting?	Using the body more effectively eg sliding or rolling the load Use of special seats
• Excessive movement: – distance (10m)? – height (perhaps requiring change of grip)? – pushing or pulling?	Use of team handling Use of protective equipment
• Risk of sudden movement?	The introduction of handling aids
• Frequent or prolonged effort? • Insufficient rest or recovery? • A rate of work imposed by a process?	Improved work routine

THE LOAD

ASSESSMENT FACTORS	REDUCING THE RISK
Is the load:	*Risk can be reduced by:*
• Heavy?	Having a lighter load
• Bulky (> 75cm)?	Having a smaller load
• Unwieldy?	Having a load which is easier to grasp:
• Difficult to grasp?	– handles – handgrips – indents
• Unstable?	– slings – carrying devices
• Hot or cold?	Using well-filled containers
	Ensuring that loads are clean
	Removing or reducing projections or roughness
	Use of containers
	Use of protective equipment

WORKING ENVIRONMENT

ASSESSMENT FACTORS	REDUCING THE RISK
Are there:	*Risk can be reduced by:*
• Space constraints?	Having adequate gangways, floorspace and headroom
• Uneven, slippery or unstable floors?	Good housekeeping
• Variations in work-surface level?	Well-maintained surfaces
• Extremes of temperature?	Prompt spillage clearance
• Extremes of humidity?	Gentle slopes or steps
• Movement of air?	Uniform bench height
• Poor lighting conditions?	Environmental control
• Movements of the workplace itself?	Relocating the work
– boat? – train? – vehicle?	Use of protective equipment
	Well-directed light

INDIVIDUAL CAPABILITY

ASSESSMENT FACTORS	REDUCING THE RISK
Does the job:	*Risk can be reduced by:*
• Require strength? • Require height? • Present hazards for pregnant employees?	Special considerations eg work restrictions – back trouble – hernia – temporary injury – pregnancy
• Present hazards for anyone with particular health problems?	Providing advice to seek help
• Require special knowledge?	Specifying a working procedure
• Require special training?	Provision of training – recognising hazards – dealing with unfamiliar loads – use of handling aids – protective equipment – safe work environment – housekeeping – recognising own limitations – good technique

DISPLAY SCREEN EQUIPMENT (VDUs)

Introduction

Objectives

The Display Screen Equipment (DSE) Regulations 1992 and associated guidance deal with what are commonly known as visual display units (VDUs).[13] The regulations aim to combat upper limb pains and discomfort, eye and eyesight effects and general fatigue and stress associated with work at display screen equipment. The HSE do not consider there are any radiation risks or special problems for pregnant women.

Workstations first put into service on or after 1 January 1993 must conform to specified standards immediately, while those in service before that date must conform by 31 December 1996, but in any case assessments must be carried out to identify any health and safety risks.

'Users' have the right to an eye and eyesight test and any spectacles (or contact lenses) *necessary* for their DSE work must be provided.

Definitions

Display screen equipment
Any alphanumeric or graphic display screen regardless of the display process involved.

Workstation
An assembly comprising display screen equipment (and any keyboard), any optional accessories, any disk drive, modem, printer, document holder, work chair, work desk, work surface etc and the immediate environment around.

User
An employee who habitually uses display screen equipment as a significant part of his normal work.

According to the HSE guidance, a person can generally be classified as a user if most or all of the following apply:

- the individual is dependent on DSE to do the job
- there is no discretion as to whether to use it
- particular DSE skills or significant DSE training is required
- there is normal use for prolonged spells of more than one hour
- there is prolonged use more or less daily
- fast transfer of information between the individual and the screen is an important requirement of the job
- high attention and concentration is required, for example, where the consequences of an error may be critical.

Detailed examples of who is likely to be a user are contained in the HSE Guidance.[14]

Operator
A self-employed person who habitually uses display screen equipment as a significant part of his normal work.

Exclusions

- Drivers' cabs or control cabs for vehicles or machinery.
- DSE on board a means of transport.
- DSE intended mainly for public use.
- Portable systems not in prolonged use.
- Calculators, cash registers or other equipment with small displays.
- Window typewriters.

Assessments (regulation 2(1) and (2))

An employer must perform a 'suitable and sufficient' analysis of those workstations:

- used for his purposes by users
- provided by him and used for his purposes by 'operators'

to assess the health and safety risks in consequence of that use.

To carry out an assessment, employers will need to:

- identify users (including 'temps'), operators and their workstations
- select assessors (familiar with the regulations)
- assess workstations covering organisation, job, workplace and individual factors (HSE Guidance gives a checklist)
- review the assessment if they suspect that it is no longer valid or there has been a significant change.

As a matter of guidance:

- assessment depth should match likely risk
- users and safety representatives should be involved
- employers should record and communicate findings
- employers should seek assistance if necessary.

An assessment checklist is included below on pp439–40.

Reducing risks

General (regulation 2(3))

Identified risks must be reduced to the lowest extent reasonably practicable.

The Schedule to the Regulations and the HSE Guidance on "Workstation Minimum Requirements" should be consulted. (The former is reproduced on pp441–5 below.)

Examples of steps which could be taken:

- Postural problems
  — reposition equipment or adjust chair
  — train the user
  — provide footrest or document holder
- Visual problems
  — reposition screen
  — clean screen
  — provide window blinds or modify lighting
- Fatigue and stress
  — vary the task, include other duties
  — give users more control over their task

Workstations (regulation 3)

Workstations which may reasonably foreseeably be used by a user or operator must meet the requirements of the Schedule by:

- 1 January 1993 – new workstations
- 31 December 1996 – existing workstations

In this context 'new' means first put into service on or after 1 January 1993 and 'existing' means in service on 31 December 1992. (The requirement must be relevant in respect of securing health, safety or welfare eg a document-holder will not be relevant if input is primarily from sources other than documents.)

Daily work routine (regulation 4)

Users' work should be planned for periodic interruptions by breaks or other activities. Such breaks need not be formal but should be included in working time. Short frequent breaks are recommended.

The HSE guidance notes state that breaks should be taken before the onset of fatigue and preferably away from the screen, and that where practicable, users should be given some discretion as to when to take their breaks.

Protection of users' eyes and eyesight (regulation 5)

Users have the right to an appropriate eye and eyesight test by a competent person if the user so requests:

- before becoming a user (as soon as practicable for current users)
- at regular intervals afterwards (take professional advice on the frequency, which may vary between individuals)
- on experiencing visual difficulties (related to DSE work).

Such tests must not be provided against the employee's will.

Eye *and* eyesight tests must be carried out by ophthalmic opticians (or medical practitioners) but employees may choose vision screening tests carried out by competent operators with screening instruments. 'Special corrective appliances' (spectacles or contact lenses) must be provided for users if 'normal' appliances cannot be used.

The *basic* costs must be met by the employer but employees must pay for extras eg designer frames, tinted or coated lenses. So called 'VDU spectacles' are not special corrective appliances.

Training for users (regulation 6)

Employers must provide users with adequate health and safety training in the use of workstations

- before they become users
- if workstations are substantially modified.

Training should cover:

- the causes of DSE related problems eg bad posture
- the user's role in recognising potential causes
- adjustment mechanisms, workstation layout, screen cleaning
- how to report problems
- break arrangements
- eye and eyesight tests
- the user's role in assessments.

The provision of information

Information has to be provided to the employer's own users, other users in the undertaking and to operators in the undertaking. The details are set out in Figure 19.3.

Employers Must Provide Information To On	Own Users	Other Users in Undertaking	Operators in Undertaking
DSE/workstation risks	Yes	Yes	Yes
Risk assessment and reduction measures	Yes	Yes	Yes
Breaks/activity changes	Yes	Yes	No
Eye and eyesight tests	Yes	No	No

Initial training	Yes	No	No
Training after modifications	Yes	Yes	No

Sources: DSE Regulations 2–7

Figure 19.3 *Provision of information under the DSE Regulations*

Key action points

- Identify who comes within the definition of user or operator. (A questionnaire asking the types of questions contained in section 1 of Figure 19.4 may be useful.)
- Select and train assessors.
- Carry out assessments (using a suitable checklist/ assessment form).
- Evaluate and implement recommendations.
- Make arrangements to offer eye and eyesight tests to users (and to provide necessary corrective appliances).
- Make arrangements for
  — training of users
  — information for users and operators.
- Identify who will check that new equipment is up to the standards required.
- Make sure existing equipment is up to standard by 31 December 1996.

The schedule to the regulations

The schedule is derived from the European Union directive which gave rise to the regulations.[15] It sets out the minimum requirements for workstations and is reproduced in full on pages 441–5.

USER'S NAME: .... DEPARTMENT: ....

ASSESSOR(S) SIGNATURE(S): .... DATE: ....

FACTOR	COMMENT
1. **WORK PATTERNS** 1.1 Most time spent per day at DSE 1.2 Average time per day at DSE 1.3 Number of days per week at DSE 1.4 Longest spell without break 1.5 Can breaks be taken 1.6 Alternative available to use of DSE 1.7 Concentration important 1.8 Fast information transfer important	
2. **INDIVIDUAL PROBLEMS** Significant experience of: 2.1 Back aches or pains 2.2 Aches or pains in arms or hands 2.3 Tired eyes or headaches 2.4 Problems with software used 2.5 Other problems	
3. **LIGHTING/ENVIRONMENT** 3.1 Artificial lighting adequate 3.2 Reflection or glare problems (artificial lighting) 3.3 Artificial lighting adjustable (if required) 3.4 Reflection or glare problems (daytime) 3.5 Blinds available (if necessary) 3.6 Noise hindering communication 3.7 Noise distracting or stressful 3.8 Temperature (summer and winter) 3.9 Ventilation (summer and winter)	
4. **SCREEN** 4.1 Stable image without flicker 4.2 Character contrast and colour 4.3 Brightness adjustable 4.4 Contrast adjustable 4.5 Swivels and tilts easily 4.6 Cleaning kit available	
5. **KEYBOARD** 5.1 Separate and tiltable 5.2 Sufficient space in front 5.3 Matt surface 5.4 Keys legible and adequately contrasted	
6. **DESK AND CHAIR** 6.1 Desk size adequate 6.2 Sufficient leg room 6.3 Desk surface low reflectance 6.4 Stable/adjustable document holder 6.5 Chair comfortable and stable 6.6 Chair height adjustable 6.7 Back adjustable (height and tilt) 6.8 Footrest available (if required)	

Figure 19.4 *DSE user assessment checklist*

OTHER COMMENTS

ACTION REQUIRED	RESPONSIBILITY	PROGRESS

DATE FOR NEXT ASSESSMENT

RECOMMENDATION FOLLOW-UP/ROUTINE REVIEW (Delete as appropriate)

Figure 19.4 (contd.)

SCHEDULE TO THE HEALTH AND SAFETY (DISPLAY SCREEN EQUIPMENT) REGULATIONS 1992

THE SCHEDULE

(Which sets out the minimum requirements for workstations which are contained in the Annex to Council Directive 90/270/EEC(a) on the minimum safety and health requirements for work with display screen equipment)

The Schedule

Extent to which employers must ensure that workstations meet the requirements laid down in this Schedule

1 An employer shall ensure that a workstation meets the requirements laid down in this Schedule to the extent that –

(a) those requirements relate to a component which is present in the workstation concerned;

(b) those requirements have effect with a view to securing the health, safety and welfare of persons at work; and

(c) the inherent characteristics of a given task make compliance with those requirements appropriate as respects the workstation concerned.

Equipment

2 (a) ***General comment***

The use as such of the equipment must not be a source of risk for operators or users.

(b) ***Display screen***

The characters on the screen shall be well-defined and clearly formed, of adequate size and with adequate spacing between the characters and lines.

The image on the screen should be stable, with no flickering or other forms of instability.

The brightness and the contrast between the char-

(a) OJ No. L156, 21.6.90, p14

The Schedule

acters and the background shall be easily adjustable by the operator or user, and also be easily adjustable to ambient conditions.

The screen must swivel and tilt easily and freely to suit the needs of the operator or user.

It shall be possible to use a separate base for the screen or an adjustable table.

The screen shall be free of reflective glare and reflections liable to cause discomfort to the operator or user.

(c) ***Keyboard***
The keyboard shall be tiltable and separate from the screen so as to allow the operator or user to find a comfortable working position avoiding fatigue in the arms or hands.

The space in front of the keyboard shall be sufficient to provide support for the hands and arms of the operator or user.

The keyboard shall have a matt surface to avoid reflective glare.

The arrangement of the keyboard and the characteristics of the keys shall be such as to facilitate the use of the keyboard.

The symbols on the keys shall be adequately contrasted and legible from the design working position.

(d) ***Work desk or work surface***
The work desk or work surface shall have a sufficiently large, low-reflectance surface and allow a flexible arrangement of the screen, keyboard, documents and related equipment.

The Schedule

The document holder shall be stable and adjustable and shall be positioned so as to minimise the need for uncomfortable head and eye movements.

There shall be adequate space for operators or users to find a comfortable position.

(e) ***Work chair***
The work chair shall be stable and allow the operator or user easy freedom of movement and a comfortable position.

The seat shall be adjustable in height.

The seat back shall be adjustable in both height and tilt.

A footrest shall be made available to any operator or user who wishes to have one.

Environment

3 (a) ***Space requirements***
The workstation shall be dimensioned and designed so as to provide sufficient space for the operator or user to change position and vary movements.

(b) ***Lighting***
Any room lighting or task lighting provided shall ensure satisfactory lighting conditions and an appropriate contrast between the screen and the background environment, taking into account the type of work and the vision requirements of the operator or user.

Possible disturbing glare and reflections on the screen or other equipment shall be prevented by co-ordinating workplace and workstation layout with the positioning and technical characteristics of the artificial light sources.

The Schedule

(c) ***Reflections and glare***
Workstations shall be so designed that sources of light, such as windows and other openings, transparent or translucid walls, and brightly coloured fixtures or walls cause no direct glare and no distracting reflections on the screen.

Windows shall be fitted with a suitable system of adjustable covering to attenuate the daylight that falls on the workstation.

(d) ***Noise***
Noise emitted by equipment belonging to any workstation shall be taken into account when a workstation is being equipped, with a view in particular to ensuring that attention is not distracted and speech is not disturbed.

(e) ***Heat***
Equipment belonging to any workstation shall not produce excess heat which could cause discomfort to operators or users.

(f) ***Radiation***
All radiation with the exception of the visible part of the electromagnetic spectrum shall be reduced to negligible levels from the point of view of the protection of operators' or users' health and safety.

(g) ***Humidity***
An adequate level of humidity shall be established and maintained.

Interface between computer and operator/user

4 In designing, selecting, commissioning and modifying software, and in designing tasks using display screen equipment, the employer shall take into account the following principles:

(a) software must be suitable for the task;

(b) software must be easy to use and, where appropriate, adaptable to the level of knowledge or experience of the operator or user; no quantitative or qualitative checking facility may be used without the knowledge of the operators or users;

The Schedule

(c) systems must provide feedback to operators or users on the performance of those systems;

(d) systems must display information in a format and at a pace which are adapted to operators or users;

(e) the principles of software ergonomics must be applied, in particular to human data processing.

Source: *Display Screen Equipment Work: Guidance on Regulations*, booklet L26, HMSO, 1992, pp29–31 © Crown Copyright reproduced with the permission of the Controller of HMSO.

Notes

1. *Ergonomics at Work*, HSE leaflet no. IND(G)90L, May 1992. A number of ergonomic case studies can be found in *A Pain in Your Workplace*, HSE, 1994
2. See HSE: *Work-Related Upper Limb Disorders*, Booklet HS(G)60, HSE Books, 1991
3. *Lighten the Load*, HSE booklet no. IND(G)109L, September 1991
4. Krause, Langston and Hidley, The Behaviour-Based Approach to Ergonomics, *The Safety and Health Practitioner*, April 1993, pp 32–7
5. See: *Manual Handling: Guidance on Regulations*, HSE, L23, London: HMSO, 1992, pp 1-4; also, *Getting To Grips With Manual Handling*, HSE booklet, IND(G)143L, 1993
6. MHO Regulations, Regulation 8 and Schedule 2
7. MHO Regulations, Regulation 2
8. European directive 90/269/EEC, OJ 1990, L156
9. For details of an early prosecution under the MHO Regulations see *HSC Newsletter* 99 (February 1995), p3. An engineering firm was find £8500 for not carrying out an assessment, not reducing the risk and not reporting the accident. The system of work involved heavy lifting which was eventually eliminated by use of a hydraulic lifting trolley costing a few hundred pounds.
10. Paragraphs 104–9, pp 20–1
11. On manual handling in the health services see: *Guidance on Manual Handling of Loads in the Health Services*, HSE booklet, ISBN 0 11 886354 1
12. Regulation 4(l)(b)(i)
13. *Display Screen Equipment Work: Guidance on Regulations*, HSE booklet L 26, ISBN 0 11 886331 2, HMSO, 1992. See also: *Working with VDUs*, HSE leaflet, IND(G)36(L)R2, 1994; *VDU Regulations: Do They Apply to Your Company?* HSE leaflet, 1994; and *VDUs: An Easy Guide to the Regulations*, HSE booklet, HS(G)90 ISBN 0 7176 0735 6, 1994.
14. para 12 (pp 7–9)
15. EC directive 90/270/EEC, OJ 1990, L156/14.

20

Electrical Safety

THE NATURE AND DANGERS OF ELECTRICITY

The nature of electricity

Electricity provides energy for heating, lighting, motive power and other purposes through the flow of electrons through conductors. Electrons are negatively charged particles which, together with protons and neutrons, make up atoms – the basic constituents of all matter. The flow of electrons through a conductor is known as a current. Just as water flows through a pipe because of the pressure behind it, the electric current flows due to differences in electrical 'pressure' or potential difference as it is often known. Differences in potential are measured in volts.

In some systems the current flows continuously in the same direction – direct current or DC. However the current may also constantly reverse its direction of flow – alternating current or AC. Most public electricity supplies are AC systems. The UK system reverses its direction of flow 50 times every second – it is said to have a frequency of 50 cycles per second or 50 Hertz (Hz).

Most substances will carry an electrical current if the electrical pressure or voltage forcing it through is large enough. They are said to conduct electricity and are known as conductors – the majority of metals are good conductors. Substances which do not conduct electricity well are known as insulators and they can be used to prevent electric currents moving in unwanted directions. However, if the voltage builds up across an insulator it may break down and conduct electricity. Air is normally a good insulator but it too can break down as can be seen during electrical arcing or lightning storms.

The level of resistance to electrical current flowing through a conductor is measured in ohms and the heating (or lighting) effect produced during the passage of current is the dissipation of the electrical energy in overcoming this resistance. There is a simple relationship between electrical pressure (voltage), current and resistance represented by Ohm's Law:

Voltage = Current x Resistance
or Volts (V) = Amps (I) x Ohms (R)

The greater the resistance to a given voltage, the lower the current that will flow.

The rate of consumption of electrical energy or power is measured in watts and has the relationship:

$$\text{Power} = \text{Voltage} \times \text{Current}$$
$$\text{or Watts (P)} = \text{Volts (V)} \times \text{Amps (I)}$$

Combining the two formulae produces the equation:

$$\text{Power} = \text{Current}^2 \times \text{Resistance}$$
$$\text{or Watts (P)} = \text{Amps (I)}^2 \times \text{Ohms (R)}$$

The dangers of electricity

In the Electricity at Work Regulations 1989 'danger' is defined as meaning risk of death or personal injury from:

- electric shock
- electric burn
- electric explosion or arcing
- fire or explosion initiated by electrical energy

associated with the generation, provision, transmission, transformation, rectification, conversion, conduction, distribution, control, storage, measurement or use of electrical energy.[1]

Electric shock

The passage of electric current through the body can produce several effects, one of which is the physical sensation of shock. Other effects can be muscular contractions, respiratory failure, fibrillation of the heart, cardiac arrest or injury from internal burns, any of which can be fatal. The nature and the severity of the injury will depend upon a number of factors

- the magnitude of the current
- the duration of the current flow
- the current path through the body
- the supply frequency (for AC).

The physical condition of the victim and his position and activity must also be taken into account. A relatively small shock could trigger off a fatal fall or a serious injury from machinery or substances being used at the time.

Table 20.1 links the magnitude of an alternating current in milli-amps (ma), the duration of the current flow and the likely physical effects on the human body.

Applying Ohm's Law, the current will be dependent upon the voltage involved and the resistance of the body:

$$I = \frac{V}{R}$$

The higher the voltage and the lower the resistance, the greater the danger.

The resistance of the body will depend upon the moisture and thickness of the skin making contact with the live conductor and the ease with which the current can flow to earth. Damp conditions or contact with other conductors (eg metal objects) will aid this flow, reducing the overall resistance and increasing the risk. The body's resistance has also been shown to decrease at higher voltages.

At 240 volts the resistance of the body is usually between 1000 and 2000 ohms. Taking a figure of 1500 ohms this produces a current of:

$$\frac{240}{1500} \text{ amps or } 160 \text{ ma.}$$

Table 20.1 *Physical effects of alternating current*

CURRENT (ma, RMS)	LENGTH OF TIME CURRENT FLOWS	LIKELY EFFECTS ON HUMAN BEINGS
0–1	Not critical	Region up to threshold of feeling. Undetected by person.
1–15	Not critical	Region up to threshold of cramp. Independent loosening of hands from objects no longer possible. Strong, partly painful, effects on muscles in fingers and arms.
15–30	Minutes	Cramplike pulling together of arms, breathing difficult. Limit of tolerance.
30–50	Seconds to minutes	Strong cramp effects, loss of consciousness due to restriction of breathing. Long times at upper end of range may lead to heart irregularities, even fibrillation.
50–500	Less than one heart period – 750 mS	No fibrillation. Strong shock effects.
	Greater than one heart period	Fibrillation. Loss of consciousness. Current marks.
Over 500	Less than one heart period	Fibrillation during vulnerable period (¼of cycle). Loss of consciousness. Current marks.
	Greater than one heart period.	Reversible stopping of the heart. Loss of consciousness. Burns.

Such a current is potentially fatal.

As already stated, the pathway taken by the electricity is also important. A current flow from hand to hand across the heart area is particularly dangerous but so also are routes involving the brain or other vital organs.

Electric burn

Where electric current passes through body tissues there will be a heating effect along the route which could cause significant burns. Whilst there are likely to be burn marks on the skin at the points of contact there may also be deep-seated burning within the body which is usually painful and very slow to heal. As the outer layer of the skin is burnt its resistance is reduced and the current will increase. At high frequencies (such as microwaves) direct contact with conductors may not be necessary for burns to occur. This type of risk is considered in more detail in chapter 22.

Electrical explosion or arcing

Electrical explosion or arcing whether caused accidentally such as an unintended flashover or deliberately (eg during electrical arc welding) involve considerable evolution of electrical energy. This energy is likely to cause localised melting of metal (it is the principle behind welding) and spattered metal or radiant heat generated could cause severe burns or fires. Ultra violet radiation may also be produced which could be damaging to the skin or create the painful condition known as arc-eye. (Radiation risks are covered in more detail in chapter 22.)

Fire or explosion

Fires and explosions of an electrical origin are another significant cause of danger. Some of the more common sources of fires are:

- Overheating of cables or other electrical equipment through overloading with currents above their design capacity.
- Current leakage due to inadequate or deteriorated insulation.
- Overheating of flammable materials placed too close to electrical equipment.
- Ignition caused by arcing or sparking.

Electrical equipment itself may explode or arc violently, as was considered earlier, but it may also act as a source of ignition of flammable vapours, gases, liquids or dust through electric sparks, arcs or high surface temperatures of equipment.

AN OUTLINE OF THE ELECTRICITY AT WORK REGULATIONS 1989

Some definitions

Regulation 2 defines several terms used in the regulations. Since it is not the intention in this chapter to cover the regulations applying only to mines (regulations 17 to 28), some terms which are of relevance only to mines have been excluded.

Circuit conductor

> *Circuit conductor* means any conductor in a system which is intended to carry electric current in normal conditions, or be energised in normal conditions, and includes a combined neutral and earth conductor, but does not include a conductor provided solely to perform a protective function by connection to earth or other reference point.

This definition, which is used only in regulations 8 and 9, includes phase and neutral conductors and also a combined neutral and earth conductor (CNE).

Conductor

> 'Conductor' means a conductor of electrical energy.
> Any material which is capable of conducting electricity is included, rather than just those which are intended to carry current. Consequently factors such as temperature need to be taken into account eg glass can conduct in its molten state but when solid is a good insulator.

Danger

> 'Danger' means risk of injury. (see pp447–9)

Electrical equipment

> *Electrical equipment* includes anything used, intended to be used or installed for use, to generate, provide, transmit, transform, rectify, convert, conduct, distribute, control, store, measure or use electrical energy.

All types of electrical equipment are included by this definition, as the HSE Guidance states, everything from a 400 KV overhead line to a battery-powered handlamp. The demarcations in the previous regulations between low, medium and high voltages have been removed. Low voltage or low power equipment may cause an explosion, even though it presents no shock or burn risk. The important test under these regulations is whether danger may arise.

System

> *System* means an electrical system in which all the electrical equipment is, or may

> be, electrically connected to a common source of electrical energy, and includes such source and such equipment.

This definition includes all the conductors and electrical equipment connected to the system, not just the normal current conductors. Anything readily capable of being made live by the system is considered to be part of it.

Some systems may therefore be quite large and cover areas over which several people have control. Such people only have duties (under regulation 3) in respect of matters which are within their control. Portable generating sets, transportable systems and systems on vehicles are all systems under this definition. Systems can range in size from a digital watch (with a battery power source) up to the national grid.

Application and duties

Application

These regulations were made under HSWA and therefore apply to everyone at work and to their workplaces. Consequently not just industrial plant is covered but a huge range of other workplaces including offices, shops, farms, construction sites and even (subject to some limitations) work in domestic property.

The definition of injury is not restricted to injury to people at work and therefore under the regulations dangers to customers and other members of the public must be taken into account, as well as those to employees, contractors, visitors etc.

Duties

Regulation 3 places duties upon employers, self-employed persons and employees (plus managers of mines and quarries). Paragraph (1) states:

> Except where otherwise expressly provided in these regulations, it shall be the duty of every employer and self-employed person to comply with the provisions of these regulations in so far as they relate to matters which are within his control.

Paragraph (2) states:

> It shall be the duty of every employee while at work –
>
> (a) to co-operate with his employer so far as is necessary to enable any duty placed on that employer by the provisions of those regulations to be complied with; and
> (b) to comply with the provisions of these regulations in so far as they relate to matters which are within his control.

Consequently both an employee carrying out electrical work in a domestic situation and his employer will have duties not just in respect of his own safety but also of residents in and visitors to the property – so far as they relate to matters within his control.

Standard of care

Some of the duties in the regulations are qualified by the term reasonably practicable. As previously noted (see chapter 3), this requires a comparison to be made between the level of risks on the one hand, and the costs of compliance on the other. The costs include physical difficulty, time and trouble as well as financial expense. If the risks are low and the costs of a particular precaution are high then the precaution need not be taken. However, an alternative precaution may still be reasonably practicable.

The ability of the duty holder to comply (whether in terms of finance or other capabilities) is not a factor to be taken into account and the onus is always on the duty holder to prove to a court that it was not reasonably practicable for him to take more precautions than he did.

Other requirements under the regulations are absolute and must be complied with, whatever the costs or other practicalities involved. However, regulation 29 does allow for a defence to be made in respect of offences under a number of regulations for the duty holder "to prove that he took all reasonable steps and exercised all due diligence to avoid the commission of that offence". (This due diligence defence is dealt with in more detail in chapter 3.)

Basic requirements

Regulation 4 contains some important basic requirements for electrical safety, many of which are expanded on in subsequent regulations. The regulation has four paragraphs:

> (1) All systems shall at all times be of such construction as to prevent, so far as is reasonably practicable, danger.
> (2) As may be necessary to prevent danger, all systems shall be maintained so as to prevent, so far as is reasonably practicable, such danger.
> (3) Every work activity, including operation, use and maintenance of a system and work near a system, shall be carried out in such a manner as not to give rise, so far as is reasonably practicable, to danger.
> (4) Any equipment provided under these regulations for the purpose of protecting persons at work on or near electrical equipment shall be suitable for the use for which it is provided, be maintained in a condition suitable for that use, and be properly used.

These four main requirements, each of which will be explained in more detail later in the chapter, are:

Construction

The system and the equipment comprising it must be designed and installed to take account of all likely or reasonably foreseeable conditions of application or use at all times throughout the life of the system.

Factors to be taken into account are:

- The manufacturer's rating of the equipment.
- Likely load or fault conditions.
- The need for suitable electrical protective devices.
- The fault level at the point of supply and the ability to handle likely fault conditions.
- Other contributions to fault levels.
- Environmental conditions.
- User's requirements.
- Commissioning, testing and maintenance activities.

Maintenance

This requirement is concerned with maintaining the system in a safe condition rather than the way that maintenance is carried out. Systems need be maintained only if danger would otherwise result. The frequency of maintenance and the type of maintenance required will very much be a matter for judgement by the duty holder. In exercising this judgement he must take account of guidance from manufacturers, the HSE and other sources as well as his own knowledge and experience. Physical inspection of equipment constitutes maintenance as well as testing and actual attention to the equipment.

Maintenance records provide both proof that the maintenance has been carried out and a means for management to monitor the compliance with and effectiveness of their maintenance standards.

Work activities

All work activities – whether operation, use and maintenance of a system or work near it – must be carried out in such a manner as not to give rise to danger, as far as reasonably practicable. Electrical work is covered more specifically in regulations 12 to 16 but regulation 4(3) provides a safety net for such work and also protects users of electrical equipment and others who might be endangered by it. Excavation work in the vicinity of power cables would be covered by this requirement as would work near overhead lines and work in some electrochemical processes.

Protective equipment

There is an absolute duty for equipment provided to protect persons at work on or near electrical equipment which is likely to be live, to be suitable for its use and to be maintained in a suitable condition. Such equipment might include:

- Insulated tools.
- Protective clothing.

- Insulating screening materials.

However, this absolute requirement is subject to the due diligence defence.

DESIGN AND INSTALLATION

Although regulation 4(1) creates a general duty in respect of the construction of electrical systems, regulations 5 to 12 and regulation 15 contain many other detailed requirements. Guidance is available from a variety of sources as to how these requirements can be complied with in practice.

Guidance sources

HSE guidance

The HSE have provided some excellent general guidance on the implications of the regulations in their *Memorandum of Guidance on the Electricity at Work Regulations 1989*.[2] As well as the guidance it gives in its own right the booklet lists many other sources of guidance.

IEE Wiring Regulations

The Institution of Electrical Engineers' Regulations for Electrical Installations (generally known as the IEE Wiring Regulations) are now in their 16th edition.[3] They have also gained British Standard recognition (BS7671:1992). Despite the use of the word regulations in their title, they do not have any formal legal status. However, they do provide detailed guidance on the design and installation of systems operating at up to 1000 volts AC or 1500 volts DC. Compliance with the IEE Wiring Regulations can be cited in court as evidence of having complied with the requirements of the Electricity at Work Regulations.

British Standard publications

Many other standards and codes of practice published by the British Standards Institution are of relevance to the safe construction of electrical systems.

Detailed requirements of the regulations

Strength and capability

Regulation 5 states that:

> No electrical equipment shall be put into use where its strength and capability may be exceeded in such a way as may give rise to danger.

"Strength and capability" relates to the equipment's ability to withstand the effects not only of its load current but also of transient overloads, fault current and pulses of current. The effects might be thermal, electro-magnetic or electro-chemical.

Ratings of equipment provided by manufacturers need to be compared with load currents and likely fault levels, taking into account how long the fault current might flow before the excess current protection comes into operation. Earthing conductors must be capable of surviving beyond fault clearance times. The IEE Wiring Regulations give detailed guidance on such calculations and various other factors which may need to be taken into account.

Adverse or hazardous environments

Regulation 6 states:

> Electrical equipment which may reasonably foreseeably be exposed to:
>
> (a) mechanical damage
> (b) the effects of the weather, natural hazards, temperature or pressure
> (c) the effects of wet, dirty, dusty or corrosive conditions or
> (d) any flammable or explosive substance, including dusts, vapours or gases, shall be of such construction or as necessary protected as to prevent, so far as is reasonably practicable, danger arising from such exposure.

- *Mechanical damage* The likelihood and extent of mechanical damage will vary considerably between working environments. Protection may need to be provided against vehicles, loads being moved, impacts from people or animals as well as long term effects such as vibration. Electrical cabling may need to be protected by conduit or trunking or by armouring or sheathing and switchgear of a sufficiently robust type should be selected or be protected in other ways.

- *Weather, natural hazards, temperature or pressure* The effects of the weather including ice, snow, wind, rain, lightning, solar radiation or temperature extremes need to be taken into account on both a short term basis and in relation to long term effects. Other natural hazards may also be present eg bird excrement, insect swarms etc. Variations in temperature or pressure may be due to natural phenomena or due to the operation of plant or other equipment.

  BS EN60529, through its Index of Protection (IP) system, classifies the degrees of protection provided by enclosures against the ingress of water and foreign bodies, while BS 6651 details precautions for the protection of structures against lightning.[4]

- *Wet, dirty, dusty or corrosive conditions* BS EN60529 is also of relevance in ensuring that electrical equipment is protected against dirt, dust and water. The possible corrosive effects of water and chemicals on all metallic parts of electrical equipment must be considered – conductors, operating parts and enclosures. Insulating materials and other materials may also be affected by solvents or other chemicals. In some cases pressurisation or purging of enclosures housing electrical equipment may be necessary to protect the equipment from its environment. Regular inspection and cleaning of equipment is also likely to be needed.

- *Flammable or explosive substances* Electrical equipment may provide a source of ignition for dusts, vapours or gases to cause fires or explosions. The likelihood of an explosive atmosphere developing can be assessed using the following zone classification system which is contained in BS 5345.[5]

  0 Explosive atmosphere continuously present or present for long periods
  1 Explosive atmosphere likely to occur during normal operation
  2 Explosive atmosphere not likely to occur in normal operation and, if it does occur, will only exist for a short time

Once this has been done, electrical equipment of that category, suitable for use in the dust, vapour or gas concerned should be selected from the range available. The British Approvals Service for Electrical Equipment in Flammable Atmospheres (BASEEFA) operate a certification system for such equipment.[6] Maintenance and repair of such equipment is a specialised task.

Insulation, protection and placing of conductors

This is covered by regulation 7 which states:

> All conductors in a system which may give rise to danger shall either –
>
> (a) be suitably covered with insulating material and as necessary protected so as to prevent, so far as reasonably practicable, danger; or
> (b) have such precautions taken in respect of them (including, where appropriate, their being suitably placed) as will prevent, so far as is reasonably practicable, danger.

The prime purpose of the regulation is to prevent danger from direct contact and therefore if none exists (eg with very low voltage equipment) no action is necessary. However, conductors normally will need to be insulated and in many cases they will also need some other form of protection, usually from mechanical damage.

This protection could come from the use of ducting or conduits, from an additional layer of insulation or from the use of armoured cable. Alternatively conductors with single insulation may be housed within enclosures where they are adequately protected against mechanical damage. In all cases some assessment needs to be made of the level of risk and the costs or practicalities of providing insulation and protection in order to decide what is reasonably practicable. Following published standards such as the IEE Wiring Regulations is desirable in this respect.

There will be circumstances where it is impossible to cover conductors with insulating material and other precautions become necessary. Examples of such situations are:

- overhead power lines
- downshop leads for overhead travelling cranes
- live rails or overhead wires for electric trains, trams etc
- electrolytic and electrothermal processes.

In some cases danger may largely be prevented by the position of the conductors eg detailed standards exist for the placing of power lines. However, additional precautions may still be necessary eg to prevent people climbing pylons to gain access to power lines. Appropriate precautions might include:

- placing of conductors
- physical measures to prevent access
- instructions, training, warning notices
- separation of conductors at different potential
- use of unearthed or isolated supplies
- use of insulated earth free areas
- segregation, screening or barriers
- safe systems of work
- protective clothing.

Earthing or other suitable precautions

Earthing is covered by regulation 8 which states:

> Precautions shall be taken, either by earthing or by other suitable means, to prevent danger arising when any conductor (other than a circuit conductor) which may reasonably foreseeably become charged as a result of either the use of a system, or a fault in a system, becomes so charged: and for the purposes of ensuring compliance with this regulation, a conductor shall be regarded as earthed when it is connected to the general mass of earth by conductors of sufficient strength and current carrying capability to discharge electrical energy to earth.

Conductors, such as metal casings, which are not intended to carry current may still become live under fault conditions. The purpose of this regulation is to prevent danger arising in such circumstances either from electric shock or from burns, fire, arcing or explosion. There are a number of techniques available for achieving this:

- *Earthing* This is the most common means of protection. Any exposed conductors which could become live are connected to something which will remain at earth potential, so that fault currents pass directly to earth.

  In the UK the electricity supply system is deliberately referenced to earth at substations or transformers so that if a fault does develop, whereby current is flowing to earth, fuses or automatic circuit breakers will detect this and operate to interrupt the supply.

Until this happens the earth circuit conductors must be capable of carrying the maximum fault current – if the earth fails then the exposed conductor will remain live and a source of danger. Consequently both the earthing and the current interruption times must meet certain standards, and these are published in the IEE Wiring Regulations and various British Standards. Electric arc welding may damage protective earthing conductors. Advice on this is given in an HSE publication.[7]

- *Double insulation* Double insulation is often used for protection of portable tools. The casing of the equipment itself is made of insulating material and any conductors inside are protected by a further layer of insulation. Thus providing the insulation remains in good condition there is little chance of an insulation failure and there is no need for any external metalwork to be earthed.[8] The insulation will need to be regularly inspected and maintained.

- *Reduced voltages* Again reduced voltage systems are often used for portable equipment, particularly on construction sites and in other situations where there is a high risk of mechanical damage and also of damp conditions. In the UK these systems operate at 110 volts (AC) with a 'Centre Tapped Earth' transformer ensuring that the maximum voltage to earth is 55v.

  While such systems provide a much greater degree of safety than with normal mains voltage, some adverse effects may still be experienced and it must not be assumed that 110 volt systems will be completely safe in all conditions.

- *Earth free areas* Earth free areas are mainly used in specialised situations such as for electrolytic processes or for electrical testing. If the area is insulated from earth then even if live metal is touched there is no path for the electricity to flow to earth.

  It is important that the earth free area is not prejudiced by the presence of structural parts of the building or metal pipework which is at earth potential. Such items will need to be effectively screened and working practices will need to ensure that other possible sources of earths (eg portable equipment connected to earthed supplies) are excluded. The possibility of accidental bridging between conductors at different potentials must also be considered and effective measures taken to prevent this.

- *Current limitation* If the system is designed so that fault currents cannot exceed 5 milli-amps then any shock experienced is unlikely to cause direct physical damage, although the risks of indirect dangers (eg the shock causing a fall) must also be considered.

Integrity of referenced conductors

The requirements of regulation 9 are that:

> If a circuit conductor is connected to earth or to any other reference point, nothing which might reasonably be expected to give rise to danger by breaking the electrical continuity or introducing high impedance shall be placed in that conductor unless suitable precautions are taken to prevent that danger.

In the UK electrical supply systems the neutral is a circuit conductor connected to earth. The purpose of the Regulation is to ensure that electrical continuity is never broken; thus conductors which should be at the same potential cannot reach different potential.

The use of proper joints or bolted links and even removable links or manually-operated knife switches is allowed. It is important that precautions are taken to ensure that these are not removed or operated in circumstances which could give rise to danger. Devices such as fuses, thyristors and transistors which could create an open circuit or high impedance cannot be introduced into referenced conductors.

Connections

Regulation 10 requires that:

> Where necessary to prevent danger, every joint or connection in a system shall be mechanically and electrically suitable for use.

As well as having suitable insulation and conductance (taking into account likely fault conditions), connections must have adequate mechanical protection and strength. The requirement applies both to temporary and permanent connections. Plugs and sockets must conform to appropriate standards as must connections between cables and equipment or within cables.

Means for protecting from excess of current

This is covered by regulation 11 which states that:

> Efficient means, suitably located, shall be provided for protecting from excess of current every part of a system as may be necessary to prevent danger.

Faults or overloads can always occur on electrical systems and protection must be provided against their effects. The type of protection will depend upon a number of factors:

- the nature of the circuits and equipment
- the fault level
- the environment
- whether the system is earthed or not.

Guidance on selection is given within the IEE Wiring Regulations, the choice resting between various types of fuse and circuit breaker.

It may well be impossible to eliminate danger completely in the event of a fault or overload – a potential danger could exist at the point of the fault for a finite time until the fuse or circuit breaker operates. The due diligence defence is available, but it would have to be shown that due diligence was exercised in the choice of equipment. Immediate interruption of current may be undesirable in certain circumstances eg where electromagnets are being used for lifting purposes.

Means for cutting off the supply and for isolation

The requirements of regulation 12 are:

> (1) Subject to paragraph (3), where necessary to prevent danger, suitable means (including where appropriate, methods of identifying circuits) shall be available for:
> (a) cutting off the supply of electrical energy to any electrical equipment; and
> (b) the isolation of any electrical equipment.
>
> (2) In paragraph (1) "isolation" means the disconnection and separation of the electrical equipment from every source of electrical energy in such a way that this disconnection and separation is secure.
> (3) Paragraph (1) shall not apply to electrical equipment which is itself a source of electrical energy but, in such a case as is necessary, precautions shall be taken to prevent, so far as is reasonably practicable, danger.

Firstly, means must be provided to switch off electrical supplies either directly or through 'stop' buttons in control circuits. In addition to this there must be a suitable means of isolation, so that the supply remains switched off and inadvertent reconnection is prevented. In some cases the functions of switching off and isolation will be carried out by separate pieces of equipment but in others they may be performed by the same equipment or even by a single action.

Means for cutting off the supply should:

- be capable of operating in all likely conditions
- be in a suitable location (related to risks, positions of people and speed of operation)
- be clearly marked (unless obvious).

Means of isolation should:

- be capable of positively establishing an effective isolation gap
- prevent unauthorised interference eg by locking off facilities
- be reasonably accessible
- be clearly marked (unless obvious).

Both switches and isolators should only be common to several items of equipment if this is appropriate. The IEE Wiring Regulations and various British Standards give guidance on the selection of switches and isolators.

Working space, access and lighting

These aspects are covered by Regulation 15 which states that:

For the purposes of enabling injury to be prevented, adequate working space, adequate means of access, and adequate lighting shall be provided at all electrical equipment on which or near which work is being done in circumstances which may give rise to danger.

- *Working space and means of access* Where there are dangerous exposed live conductors, dimensions should be adequate:
  — *to allow persons to pull back away from the conductors without hazard*
  — if necessary, to allow persons to pass one another with ease and without hazard.

  Some detailed space requirements from the previous regulations have been reproduced as guidance in respect of this regulation and are contained in the memorandum of guidance on the regulations.[2]

- *Lighting* The preference is firstly for natural light, then for permanent artificial lighting and finally for portable lamps or torches. The light should be adequate in both the horizontal and vertical planes. Detailed guidance on standards is provided in an HSE booklet.[9]

Note: Regulations 13, 14 and 16 are dealt with later in the chapter (see pp468–74).

MAINTENANCE

The principal requirement to maintain electrical equipment at high standard is contained in regulation 4(2) which states:

As may be necessary to prevent danger, all systems shall be maintained so as to prevent, so far as is reasonably practicable, such danger.

It sets the simple objective – to prevent danger. What type and level of maintenance is necessary to do this is not stated in the regulation and the aim here is to give guidance on this subject.

Guidance on maintenance

Health and Safety Executive

The HSE have published a booklet on the maintenance of portable and transportable electrical equipment.[10] This is supported by two free leaflets relating to the maintenance of portable equipment in relatively low risk environments.[11]

IEE Wiring Regulations

The IEE Wiring Regulations contain guidance on the inspection and testing of electrical installations including model inspection certificates.

British Standards

Several British Standards codes of practice are concerned with maintenance standards and methods although these mainly relate to switchgear, control gear, 'flameproof' equipment and other specialist areas.

Manufacturers' recommendations

Many manufacturers of electrical equipment will provide recommendations on maintenance and it would be foolish to ignore this, although experience of using the equipment may indicate that changes should be made to the frequency or method of maintenance (as could be the case in relation to guidance from other sources).

What needs to be maintained?

All electrical equipment needs to be maintained to some extent, unless it is not a reasonably foreseeable source of danger. Equipment operating at a low voltage, especially if it is battery powered (eg torches or calculators), is likely to come into this category unless its circumstances of use are unusual.

In recent years maintenance of portable equipment has sprung into prominence but the requirement is to maintain all equipment – fixed, portable and transportable. Portable and transportable equipment is more likely to be subject to damage and abuse than fixed installations but the latter must not be neglected in planning the maintenance regime.

Employees may bring their own electrical equipment to work either formally (eg tools used by a tradesman) or informally (kettles, radios, heaters, fans etc). If the employer decides to allow the use of this equipment then he must ensure that it is satisfactorily maintained and include it in any inspection and testing system. Both employers and employees have duties under these regulations relating "to matters within their control" and there will need to be co-operation and co-ordination between the two in respect of private electrical equipment.[12] It must be remembered that unsafe electrical equipment represents a danger not only to the user but also to other employees and outsiders.

What should be done?

Regulation 4(2) requires whatever maintenance "as may be necessary to prevent danger". Here the various sources of guidance become particularly relevant in deciding what is "reasonable practicable" in the light of the experience and knowledge of others.

HSE guidance[13] refers specifically to:

- user checks (visual)
- formal visual inspections

- combined inspections and tests.

A fourth category may be added to these:

- physical attention.

User checks

Users of more potentially dangerous types of equipment should be required to inspect the equipment for possible defects both before and during use. Such inspections could be made after very basic training and should be aimed at identifying the following:

- damaged cable sheaths
- damaged plugs – cracked casing or bent pins
- taped or other inadequate cable joints
- outer cable insulation not secured into plugs or equipment
- faulty or ineffective switches
- burn marks or discolouration
- damaged casing
- loose parts or screws
- wet or contaminated equipment
- loose or damaged sockets or switches.

Clearly the users must know where to report such faults and there must be arrangements for withdrawing the equipment from use and repairing it.

Formal visual inspections

User checks will normally need to be supported by a system of formal visual inspections. The users may not be sufficiently knowledgeable or conscientious to identify all faults and arrange for their rectification or it may not be reasonable to expect them to inspect certain parts of the equipment (eg inside the plug) every time they use it. Visual inspections are likely to need to look for the same types of defects as user checks but should also include the following:

- Opening plugs of portable equipment to check for:
  - — use of correctly rated fuse
  - — effective cord grip
  - — secure and correct cable terminations
- Inspection of fixed installations for:
  - — damaged or loose conduit, trunking or cabling
  - — missing, broken or inadequately secured covers
  - — loose or faulty joints
  - — loose earth connections
  - — moisture, corrosion, contamination
  - — burn marks or discolouration

— open or inadequately secured panel doors
— access to switches and isolators
— presence of temporary wiring.

As with user checks there must be arrangements for ensuring that faults are rectified and, where necessary, equipment is taken out of use.

Combined inspections and tests

There is a common misconception that the Electricity at Work Regulations require all portable electrical equipment to be tested annually. Whilst testing together with a thorough visual inspection may be the only means of detecting certain types of faults, not all equipment needs testing and the frequency of test will vary according to the type of equipment and circumstances of use.

In the case of portable equipment, inspection accompanying testing should include all of those aspects covered in user checks and formal visual inspections together with:

- checking of correct polarity
- checking the effective termination of cables and cores
- checking the equipment's suitability for its environment.

Testing would include earth continuity and insulation integrity and might involve use of a simple portable appliance tester (PAT) or a more sophisticated type of instrument.

The more long-term inspection and, where appropriate, testing of fixed electrical installations is a much more specialist area and is likely to require someone of considerable electrical competence. Advice on such matters is available in the IEE Wiring Regulations and relevant British Standards.

Physical attention

The rectification of faults identified during user checks, formal visual inspections and combined inspections and tests is very much part of the maintenance process. This work needs to be carried out competently, and using safe methods, by a person with the necessary knowledge, experience and skill.

However, there are various other aspects of the maintenance of electrical equipment which will require physical attention to be given to the equipment itself or its surroundings on a regular basis. Examples of such attention are:

- cleaning areas where dirt or dust are liable to accumulate eg ventilation openings or identification labels
- removal of corrosion
- renewal of surface coatings
- removal of weeds (or other potentially combustible material)
- periodic replacement of components with finite lifespans.

Who should do it?

Persons carrying out this type of maintenance work must have adequate technical knowledge and experience to

- know what they are looking for
- recognise it when they find it
- take appropriate action about it.

The need for this technical knowledge and experience (as required by regulation 16) is much greater if the maintenance work itself presents danger (see pp 473 and 474).

User checks

Users should need only very basic training in order to be able to identify defects in their electrical equipment. Oral instruction combined with some illustrations (pictures or samples) of the type of fault they are looking for will probably be all that is required.

Formal visual inspections

These require a greater degree of knowledge and experience and hence more formalised training. Nevertheless those appointed to carry out formal visual inspections will not generally need to have formal qualifications. They should be provided with a simple checklist of the types of fault they should look for and guidance as to what action to take when defective equipment is found. An appreciation of the limits of their own knowledge and experience is also important.

An employee (or employees) could be appointed specifically for this purpose. Alternatively these types of electrical inspections could form part of an overall safety inspection programme (see chapter 11).

Combined inspections and tests

Here a higher level of competence still is required, particuarly if test results require interpretation. It may be possible to train a non-electrical specialist to reach the necessary standards (especially if a simple pass/fail portable appliance tester is used) but it is more likely that someone with specialist electrical abilities will be required. This could be an employee or an external contractor.

Physical attention

The level of electrical competence required will depend on the type of activity carried out. Simple cleaning operations with no danger involved may require no electrical knowledge at all whereas other tasks might need considerable electrical capabilities and require extensive precautions.

How often?

User checks on the more potentially dangerous types of equipment should be made whenever the equipment is used but the frequency of other types of maintenance is a matter for judgement, based on an assessment of risk. Factors to take into account when assessing the risk include:

- the type of equipment
- is it hand held?
- manufacturers' recommendations
- its initial integrity and soundness
- age
- the working environment
- likelihood of mechanical damage
- frequency of use
- duty cycle
- foreseeable use
- use by the public
- modifications or repairs
- past experience.

HSE have given some general guidance on where user checks are necessary and on frequencies for formal visual inspections and combined inspections and tests but these may need to be increased if local circumstances (eg conditions of use) merit. Table 20.2 has been combined from three HSE publications.[14]

Maintenance records

Although the Electricity at Work Regulations do not contain any specific requirements for the keeping of maintenance records, there are a number of good reasons for keeping records.

Proof of compliance

The existence of records will be of considerable assistance in proving that maintenance has been carried out in accordance with the regulations if this is ever necessary in relation to criminal or civil proceedings.

Management monitoring tool

A record of maintenance is an essential tool for management to monitor whether maintenance work is in fact being carried out in accordance with its own standards. If it contains sufficient detail eg records of defects found or variations in test readings, then it can provide feedback to review whether or not the existing standards are set at the right level, especially in respect of inspection or test frequency. Breakdown records will also be useful in this respect.

Table 20.2 *Electrical inspections and tests*

Type of business	Type of equipment	User checks	Formal visual inspection	Combined inspection and electrical tests
Equipment hire	Most *	Yes	Before issue/ after return	Before issue
Construction	Most *	Yes	Before initial use – 1 month	3 months
Industrial use–	Most *	Yes	Before initial 3 months	6–12 months
Offices, hotels and other low-risk environments	Battery-operated: (less than 20 volts)	No	No	No
	Extra low voltage: (less than 50 volts AC) eg telephone equipment, low voltage desk lights	No	No	No
	Information technology: eg desk-top computers. VDU screens	No	Yes, 2–4 years	No if double insulated – other-wise up to 5 years
	Photocopiers, fax machines: NOT hand-held. Rarely moved	No	Yes, 2–4 years	No if double insulated – other-wise up to 5 years
	Double insulated equipment: NOT hand-held. Moved occasionally, eg fans, table lamps, slide projectors	No	Yes, 2–4 years	No
	Double insulated equipment: HAND-HELD eg some floor cleaners	Yes	Yes, 6 months – 1 year	No
	Earthed equipment (Class 1: eg electric kettles, some floor cleaners, some kitchen equipment and irons)	Yes	Yes, 6 months – 1 year	Yes, 1–2 years
	Cables (leads and plugs connected to the above) Extension leads (mains voltage)	Yes	Yes, 6 months – 4 years depending on and type of equipment it is connected to	Yes, 1–5 years depending on the type of equipment it is connected to

* Guidance should be adapted from other parts of the table for low-risk equipment eg battery-operated, extra low voltage or double-insulated.

NB: Experience of operating the maintenance system over a period of time, together with information on faults found, should be used to review the frequency of inspection.
It should also be used to review whether and how often equipment and associated leads and plugs should receive a combined inspection and test.

Inventory

Records provide an inventory of equipment, making it less likely that individual items can escape the inspection and/or testing system and making it easier to identify the introduction of employees' own equipment. The use of labels on inspected or tested equipment is also helpful in this respect.

Records do not have to be kept on paper but they may be put into a computer system. This could then be used to generate a list of equipment requiring inspection or test. Some test instruments can be downloaded directly into a computer database. Those with a healthy distrust of computers may still want to retain a hard copy.

SAFE WORKING PRACTICES

There is an overriding duty contained in Regulation 4(3) which states that:

> Every work activity, including operation, use and maintenance of a system and work near a system, shall be carried out in such a manner as not to give rise, so far as is reasonably practicable, to danger.

Regulations 13, 14 and 16 are of particular importance in relation to work methods whilst some note must also be taken of working space, access and lighting (Regulation 15) which was covered earlier (see pp460 and 461).

Whether to work dead or live

The first decision to be made in relation to work on electrical systems is whether this work is to be done dead or live. Regulation 14 spells out the way this choice must be made:

> No person shall be engaged in any work activity on or so near any live conductor (other than one suitably covered with insulating material so as to prevent danger) that danger may arise unless
>
> (a) it is unreasonable in all the circumstances for it to be dead; and
> (b) it is reasonable in all the circumstances for him to be at work on or near it while it is live; and
> (c) suitable precautions (including where necessary the provision of suitable protective equipment) are taken to prevent injury.

Unreasonable for it to be dead

The first consideration is that it must be unreasonable (as opposed to inconvenient) for the work to be done dead and factors to be taken into account in making this judgement are:

- Is it practicable to work dead? eg do conductors need to be live for testing

- Will hazards be created elsewhere? eg for other users or for continuous processes
- What economic considerations are there?

Apart from the need for some testing work to be done live, the HSE Guidance on this regulation suggests other situations where it may be unreasonable to work dead:

- the electrical supply industry, eg live cable jointing
- some types of work on electric railways (see p77 for a relevant legal case)
- work on telephone networks.

In some cases it may only be unreasonable to work dead for a small part of a task.

The need for live working can be eliminated or the risks reduced by thought and planning at the design stage. Provision of alternative power in-feeds and suitable distribution systems should allow sections of the system to be isolated. Power circuits should also be separated or segregated from logic and control circuits.

Reasonable for it to be live

This is the second criterion to be satisfied. Whatever the practical or economic arguments against working dead, the safety of all persons in working live is a major consideration. Those not directly involved in electrical work need to be considered eg a crane driver under an overhead line. The main questions to be asked are:

- What is the level of risk in working live?
- What is the effectiveness of the precautions available?

Only if the risk can be reduced by available precautions to an acceptably low level can the work proceed. Precautions necessary in live working will be considered later (see pp471–3).

Precautions on equipment made dead

Working dead should always be the norm: it should not usually be unreasonable to work dead. Regulation 13 describes the precautions which must be taken in working dead.

> Adequate precautions shall be taken to prevent electrical equipment, which has been made dead in order to prevent danger while work is carried out on or near that equipment, from becoming electrically charged during that work if danger may thereby arise.

While there may be a few low risk activities where switching off the supply is adequate to prevent danger, for most types of electrical work it is necessary to isolate in such a way that the equipment cannot be re-energised. The HSE booklet *Electricity at Work, Safe Working Practices*[15] provides detailed guidance on the precautions necessary. Some or all of the following steps are likely to be required.

Identification

The circuit to be worked on should be identified. It must never be assumed that identification labelling is correct and the equipment or circuit will still need to be proved to be dead at a later stage.

Disconnection and isolation

The difference between switching off and isolation was discussed earlier (p460), although in practice these functions may be performed by the same piece of equipment. There may be a need to isolate the equipment from different potential sources of supply. What is essential is that the isolation is secure. There have been many tragic accidents where power has been inadvertently restored due to inadequate isolation arrangements. Methods of isolation will vary dependent upon the circumstances but the following are the most common:

- *Isolation switches* The isolation should be put to the 'off' position and preferably locked there using a lock with a unique key which is kept in a secure place. If several people are involved in the work then each should apply their own lock using a multi-lock hasp which can only be removed when everyone has taken away their own lock. Where isolators cannot be locked off, at the very least a 'caution' notice should be placed on the switch.
- *Fuse removal* A fuse (or fuses) may be removed as a sole means of isolation or as an additional precaution where the isolator switch cannot be locked off. It is essential that reinsertion of the fuse (or another one) is prevented. This might involve taking the fuse away (to a secure place), locking the fuse box, use of a 'caution' notice or label or even taping over the fuseholder.
- *Plug withdrawal* Isolation may be achieved simply by removing a plug from its socket but precautions must be taken to prevent its reinsertion. Again this could involve the use of a 'caution' notice or label or the application of tape in suitable positions. Clearly such precautions are unnecessary if the plug is to remain within the immediate control of the person carrying out the work.

Notices and barriers

As well as the type of 'caution' notice already mentioned, it may be necessary to post 'danger' notices adjacent to the work in order to identify equipment which is still live. In some circumstances the use of physical barriers to prevent inadvertent contact may be required.

Proving dead

The point of work and accessible parts around it should be proved dead before work commences. This test should always be carried out assuming that the point is still live – there have been many instances of injury being caused where isolators have been wrongly labelled and no test has been carried out or an unsafe test method has been used. The test device must itself be proved both immediately before and after testing.

Earthing

The application of earthing devices at the point of work may also be necessary in circumstances such as those where:

- high voltage apparatus is involved
- equipment is linked to stored electrical energy sources (eg capacitors)
- re-energisation is still a risk, eg from the starting of a generator under someone else's control.

Such earthing devices must be properly designed for their purpose and must only be applied once the equipment has been proved dead.

Permits to work

Various other precautions may be necessary in high voltage and other high risk situations. Such circumstances often involve the use of a permit-to-work system which should ensure that all the appropriate precautions have been taken before work commences. The operation of permit-to-work systems is covered in chapter 12.

Precautions when working live

If the first two criteria for working live are satisfied, ie it is unreasonable to work dead *and* it is reasonable to work live, then the third part of regulation 14 comes into operation:

> suitable precautions (including where necessary the provision of suitable protective equipment) must be taken to prevent injury.

What constitutes "suitable precautions" will very much depend upon the circumstances and the level of risk involved. HSE Guidance refers to a variety of precautions which may be appropriate:

Competent staff

Regulation 16 (see pp473 and 474) goes into more detail about the need for competence in electrical work.

This is of particular importance to live work.

Adequate information

Staff already competent in electrical work will still need to be provided with adequate information in order to work safely. They will need to know where live conductors are, what other risks exist, what precautions have already been taken and any other precautions they are expected to take.

Suitable tools

Insulated tools, protective clothing and other specialist equipment are likely to be required for live work. A variety of British Standards exist in relation to tools, equipment and clothing, some of which include guidance on examination and testing. Under Regulation 4(4) there is an absolute duty to maintain such items in a suitable condition.

Barriers or screens

The provision of suitable barriers or screens (whether permanent or temporary) is an important precaution in preventing electric shock or short circuits. In some cases physical barriers or screens (possibly insulated) may be all that is required to prevent people of relatively low electrical competence being endangered by nearby live conductors. Again these must be maintained in a suitable condition.

Instruments and test probes

Suitable instruments and test probes may need to be used in order to identify what is live and what is dead or in live testing work. As previously stated, the testing should always assume the item to be tested is still live and the test device must be proved both immediately before and after testing.

Accompaniment

Accompaniment is no longer an automatic requirement for live working but may be relevant depending upon the circumstances. The accompanying person should be able to "substantially contribute towards the implementation of safe working practice". Possible examples of such a contribution are:

- assisting in the planning of work methods
- preventing confusion between live and dead conductors
- keeping other people clear of the area
- isolating the supply in an emergency
- providing prompt first-aid treatment in an emergency.

However too much emphasis must not be placed on the accompanying person's emergency role. Precautions should be such that such an emergency is extremely unlikely to arise.

Designated test areas

Routine live test work should be restricted to areas where suitable precautions can be taken eg earth-free areas with isolated power supplies.

Effective control

Only those whose presence is necessary and who are competent should be allowed into areas where there is danger from live conductors. The level of precautions will depend upon the environment and the level of risk but could include:

- lockable enclosures
- temporary barriers
- warning notices
- sentries.

In small areas the presence of the person doing the work may be all that is needed to exercise effective control.

Competent persons

The need for electrical competence is covered in regulation 16 which states:

> No person shall be engaged in any work activity where technical knowledge or experience is necessary to prevent danger or, where appropriate, injury unless he possesses such knowledge or experience, or is under such degree of supervision as may be appropriate having regard to the nature of the work.

This requirement relates to all types of work associated with electrical equipment where danger (of an electrical type) may arise (directly or indirectly). It could include:

- working live
- working on dead equipment
- the quality of installation work
- the quality of maintenance work
- the use of electrical equipment
- casual exposure to electrical risks.

The extent of technical knowledge or experience necessary to prevent danger in each of these circumstances will vary considerably. Aspects to consider are:

- knowledge of electricity
- experience of electrical work
- understanding of the system to be worked on
- practical experience of that type of system
- understanding of the hazards which might arise
- understanding of the precautions to be taken
- ability to recognise whether it is safe for work to continue.

At the lower end of the risk scale all that might be required is a knowledge that the individual cannot enter certain areas, must obey instructions in notices and must not continue to use defective equipment (which must be reported). However for higher risk situations more detailed controls are required. Electricians under training should be allowed to carry out more complex and risky tasks as they demonstrate that their knowledge has advanced sufficiently and as their experience increases. Even experienced electricians may be incompetent in a new environment until they acquire knowledge and experience of the type of equipment present there, the hazards which might arise and the precautions which should be taken. Care must be taken in the use of multi-skilled maintenance employees. A contractual obligation to be prepared to carry out a wide range of maintenance tasks does not necessarily mean that the individual concerned is competent to carry out all electrical work safely. Restrictions are particularly necessary in relation to work on or near live equipment and many organisations allow this only to be done by individuals who have formally been authorised for this type of work, with such authorisations often being in writing.

Those with insufficient knowledge or experience may be put under the supervision of others who do have the necessary qualities. Such supervision does not require constant attendance – the degree of supervision must relate to the level of danger and the circumstances of the work. It is the responsibility of the duty holder (normally the employer) to satisfy himself that those carrying out electrical work have sufficient knowledge and experience for the demands of the activity or are under an adequate degree of supervision.

Specialist guidance

The HSE publishes much guidance relating to safe working practices in specialist areas of electrical risk. Some of these documents also contain guidance on design and installation standards or on maintenance in these specialist applications. Several of the more important sources of guidance are listed in the reference section.[16]

Notes

1. The Electricity at Work Regulations, SI 1989/635, regulation 2(1)
2. *Memorandum of Guidance on the Electricity at Work Regulations* – HSE 1989 (HS(R)25)
   (This publication also contains the regulations themselves, apart from those only applicable to mines.)
3. *Institution of Electrical Engineers' Regulations for Electrical Installations*, 16th edition (BS 7671:1992)
4. BS EN 60529:1992 *Specification for 'Degrees of protection provided by enclosures'* (IP code)
   BS 6651:1992 *Code of practice for protection of structures against lightning*
5. BS 5345 (various parts) *Code of practice for selection, installation and maintenance of electrical apparatus for use in potentially explosive atmos-*

pheres (other than mining applications or explosives processing and manufacture)

6. *BASEEFA list 1993; certified and approved explosion protected electrical equipment*, 8th edition 1993 (published by HSE Books)
7. *Electrical safety in arc welding* – HSE 1994 (HS(G)118)
8. BS 2754:1976 *Construction of electrical equipment for protection against electric shock*
9. *Lighting at Work* – HSE 1987 (HS(G)38)
10. *Maintaining portable and transportable electrical equipment* – HSE 1994 (HS(G)107)
11. *Maintaining portable electrical equipment in offices* – HSE 1994 (IND(G)160

    Maintaining portable electrical equipment in hotels – HSE 1994 (IND(G)164)
12. The Electricity at Work Regulations 1989, regulation 3
13. See note 10, above
14. See notes 10 and 11, above
15. *Electricity at Work, Safe Working Practices* – HSE 1993 (HS(G)85)
16. See note 7 above. The following HSE publications also relate to specialist areas of electrical risk:

COP 34	The use of electricity in mines, Electricity at Work Regulations 1989, Approved Code of Practice *1989*
COP 35	*The use of electricity in quarries, Electricity at Work Regulations 1989, Approved Code of Practice* 1989
HS(G)47	*Avoiding danger from underground services* (1989)

Guidance Notes

GS6(Rev)	*Avoidance of danger from overhead electric lines* (1991)
GS34(Rev)	*Electrical safety in schools* (1990)
GS34	*Electrical safety in departments of electrical engineering* (1986)
GS38(Rev)	*Electrical test equipment for use by electricians* (1991)
GS47	*Safety of electrical distribution systems on factory premises* (1991)
GS50	*Electrical safety at places of entertainment* (1991)
PM29(Rev)	*Electrical hazards from steam/water pressure cleaners* (1988)
PM32	*Safe use of portable electrical apparatus* (1990)
PM38	*Selection and use of electric handlamps* (1992)
PM53	*Emergency private generation: electrical safety* (1985)

Leaflet IND(G)102L – Electrical safety for entertainers (1991)

The following HSE publications are out of print but are understood to be in the process of revision.

HS(G)13 *Electrical testing*

Guidance Notes

GS24	*Electricity on construction sites*
PM37	*Electrical installations in motor vehicle repair*

21

Fire Safety

This chapter is concerned with the prevention and control of fires. The approach to be taken comprises three stages:

- the acquisition by management of the necessary knowledge about fire safety
- the taking of various fire safety precautions and
- the development by management of an effective policy in relation to fire safety.

This approach is set out diagrammatically in figure 21.1.

The starting point is a brief introduction which includes definitions of the main terms used in fire safety. The legal framework is then described before the first of the key stages is addressed – fire safety knowledge.

INTRODUCTION

Terminology

It is appropriate at the outset to look briefly at some of the terms employed in fire safety.

- *Fire* Fire is the combustion (an exothermic reaction) of a fuel source in an oxygen environment after ignition from a heat source, resulting in light and heat from burning. Fire is simply a generic term given to a mass of burning matter.

- *Flame* A flame is a visual result of fire, ie gaseous or vapourous matter undergoing combustion.

- *Explosion* There are many types of explosion, but here the concern is with those where fire is also present. These are known as red explosions. An incident not involving fire, such as a pressure vessel explosion, is known as a black explosion. In the main, red explosions are in fact extremely fast fires.

Figure 21.1 *Fire safety – an overview*

- *Fire safety* Fire safety is a general term for the prevention and control of fires and explosions.
- *Preventive measures* Preventive measures centre around the removal of one or more of the three elements making up the 'triangle of fire' (see p490) ie the fuel, heat (the ignition source) and oxygen, although in the case of oxygen this is possible only in specialised circumstances, for normally if you remove oxygen you have to remove people. Remove one element from the triangle and fire cannot occur.
- *Control measures* Control measures are concerned with the extinguishing of fire and the limitation of its effects. Such measures fall into two main categories – active and passive.
- *Active control measures* These include:
  — fire detection and protection systems
  — fire appliances

- fire fighting teams.

- *Passive control measures* These include:
  — fire compartmentalisation
  — protection of structural steelwork
  — use of fire-resistant materials.

The importance of fire safety

The arguments in favour of having an effective fire safety policy are quite clear.

- Fire kills and maims people.
- Fire destroys or damages property.
- It reduces companies' productivity and profits.
- Failure to prevent or control fires will also produce additional costs in the form of increased insurance premiums.
- The law requires employers to have fire safety knowledge and to put fire safety procedures in place. Failure to comply could result in prosecution and in some cases the closure of the premises.

The size of the problem nationally, including domestic fires, is approximately 1,000 lives lost, 15,000 serious injuries and a property fire damage bill of around £1,000 million per year. The property damage bill is for direct damage only; if consequential loss is added the figure is more than doubled.

Examples include:

- Textile factory, Birmingham – £0.5 million.
- Electronics factory, London – £0.75 million.
- Offices, Essex – £0.94 million.
- Chemical plant, Wilton – £2.6 million.
- Hotel, Hull – £12 million.

In one recent incident a chemical plant undergoing maintenance and cleaning was suddenly engulfed by a fire almost explosive in nature. A resultant fireball ripped though the plant and through a nearby control room. Two people in the control room were killed instantly while two further employees died later. In addition a female office worker was overcome by smoke when trapped in an adjacent office block. She died later in hospital.

An effective fire safety plan

Any effective fire safety prevention and control plan must have a number of key elements.

- It must have visible corporate management commitment and support.
- It must be a combined effort involving management and employees (and

unions where present), together with any internal specialists and outside experts (eg insurance technicians, fire brigade staff etc).

- It should set out to identify and quantify potential exposure to fire risks and recommend effective measures to eliminate or minimise them.
- It must have clearly defined and attainable aims and objectives and be flexible enough to be altered to meet changing circumstances.
- It must be reviewed regularly.

LEGAL FRAMEWORK

The main UK fire legislation is The Fire Precautions Act 1971. It has a general application, including factories, offices and railway premises. Its cornerstone is the need to obtain from the local fire authority a fire certificate as evidence of conformity to the requirements of the Act. Fire certificates have to cover four fire protection measures which must be in place at the time of the inspection by the fire authority. These are:

- escape routes
- fire safety signs
- fire resistant construction requirements and
- fire equipment eg alarm call points, extinguishers and hose reels.

For industrial buildings and offices a fire certificate is required under the following circumstances:

1. where more than 20 persons are employed at one time
2. in respect of floors above or below ground level where ten or more persons are employed at one time and
3. where there is storage or use of explosive or highly flammable materials.

The Fire Certificates (Special Premises) Regulations apply to certain industrial premises where there is a high risk of fire or explosion from highly flammable material or dangerous chemicals.[1] The fire precautions required for the special premises certificate are the same as those under the 1971 Act itself but the special premises regulations are enforced by the HSE and not by the local fire authority.

The Building Regulations 1991 apply to all new buildings and alterations to existing buildings in terms of the design standards necessary to provide adequate means of escape in the event of fire.[2] Measures to restrict the spread of fire over surfaces and limit the spread of fire within a building are contained in the regulations as are requirements to prevent fire spreading from one building to another. While in the main the regulations apply to domestic and dwelling buildings they also apply to offices and retail and wholesale shops.

To fulfil the fire safety requirements of the EU Framework and Workplace

Directives a new and fundamentally different piece of fire safety legislation is set to become the central focus of industrial and commercial fire safety.[3] This is the Fire Precautions (Places of Work) Regulations 1995 henceforth, the FP(PW) Regulations.[4] The regulations, currently only in draft form, will apply to most buildings where one or more persons are employed. However premises already certificated under the FP Act 1971 will be exempt from the FP(PW) Regulations. Those certificated under FA 1961 and the OSRPA 1963 will not be exempt. The key to the FP(PW) Regulations is the requirement for a risk assessment to be made in the workplace. Where five or more persons work, the risk assessment must be made in writing. Arising from the risk assessment, an emergency plan must be produced covering the following points:

- action to be taken by persons in the event of fire
- procedures to be followed by staff
- arrangements for calling the Fire Brigade and informing them of any special risks.

Other important requirements of the draft regulations are:

- means of escape
- means of fighting fires
- means of detecting fire and giving warning
- maintenance and testing of fire equipment
- information, instruction and training
- keeping of records
- assistance for disabled people
- fire prevention housekeeping.

A practical employers' guide to the regulations is available containing:

- an action checklist
- risk assessment worksheets
- inspection checklists
- examples of fire precautions record sheets.

This proactive approach to fire safety will not permit employers in future to rely simply on the fact that a fire certificate is available. Risk assessments on fire safety will need to be made, recorded and reviewed. Finally before the end of the 1990s it is likely that the Fire Precautions Act 1971 itself will be reviewed and updated to take account of some of the new thinking on fire safety.

FIRE SAFETY KNOWLEDGE

As noted, acquiring the necessary knowledge can be seen as the first stage in a three-stage approach to fire safety. That knowledge can be described under a number of headings:

- risk assessment
- fire classification
- causes and sources
- spread considerations.

Risk assessment

Regulation 3 of the Management Regulations clearly applies equally to fire safety as it does to safety more generally. As a first stage in fulfilling this requirement a generic risk assessment of a factory, production plant, warehouse or other unit should be undertaken. The assessment should:

- identify possible hazard areas
- evaluate the risks to people, property, products and productivity
- lead to the development of an effective prevention and control strategy.

These three risk assessment stages could be the subject of a reference manual which includes plans and drawings of works, plants and buildings. Review of this central document on a regular basis by discussion and inspection is important. Critical examination by senior management of risks to people, property and productivity should be undertaken on an annual basis. It is for management to decide whether outside agencies (such as fire insurance surveyors, fire prevention officers, or factory inspectors, all of whom will also be carrying out fire prevention audits of the premises) should be directly involved. Their visits could be treated as a separate matter.

Throughout the risk assessment stage it will be necessary to undertake a number of visits to the assessment area, the purpose of which will be to carry out fire safety audits and inspections and to identify on-site hazards and risks.

Figure 21.2 is intended to assist in identifying and evaluating fire risks in a given area. It can be tailored to the specific circumstances of each organisation. A plan of this nature can form the keystone of a programme of regular audits, inspections and monitoring of fire and explosion hazards.

Most fires have one root cause – people. The failure of people may cause a fire. In addition, their reaction to a fire may lead to a small fire becoming a major fire. In general, these human failures are due to:

- lack of training or instruction in written procedures
- failure to follow procedural instructions
- inadequate oral instruction or direction
- panic, leading to defective judgement
- human error.

Fire classification

A detailed classification system can be found in British Standard 4547. In the

MAIN IGNITION CAUSES		**SOURCE SUBSTANCES OF COMBUSTION**
(1) Faults or misuse of electrical equipment	Any	(a) Waste materials and rubbish
		(b) Packaging and wrapping materials
(2) Burning and spark generating equipment	numbered	
		(c) Building and structural components eg wood and fibre boards etc
(3) Oil, gas and electrical heating, boiler and drying equipment	*cause*	(d) Personal and work clothing
(4) Smoking, matches, cigarettes etc	can	(e) Electrical insulations and circuit boards
(5) Uncontrolled rubbish burning		
(6) Hot products or waste products	**IGNITE**	(f) Office fittings and decoration, wallpaper, carpets, curtains and chairs
(7) Friction generation – bearings, conveyors, drive belts etc	any	(g) Gases, vapour, mists or dusts
(8) Static electricity	lettered	(h) Flammable liquids, paraffin, petrol, oils and greases etc
(9) Spontaneous chemical ignition		(i) Conveyor belts and drive belts
	substance	
(10) Arson and deliberate ignition		(j) Solid fuel stock, coal, coke etc

REASONS FOR MAJOR SPREAD OF FIRE

(i) Delayed discovery of fire
(ii) Missing, damaged or uncharged fire fighting equipment
(iii) Lack of fire separation and compartmentalisation
(iv) Lack of effective fixed fire protection systems
(v) Leaks or spillage of oils, greases, hydraulic fluid and fuels
(vi) Sub-standard levels of housekeeping
(vii) Restricted or difficult access for fire fighters and fire engines
(viii) Unmanned or unmonitored areas
(ix) Overstocking or overstoring of products, raw materials or engineering equipment
(x) Lack of suitably trained staff available to deal with fire in the vital first few minutes following ignition

Figure 21.2 *Risk elements of industrial fires*

main, fires are grouped according to the nature of fuel. BS 4547 is used as a standard format for communication and training. Its main purpose is to help in deciding upon the suitability of the types of extinguishing agents available for fire fighting.

- *Class A – ordinary combustibles* Fires which involve solid materials (carbonaceous and organic) forming glowing embers; examples are cloth, coal, wood, paper, rubbish and shavings.
- *Class B – flammable liquids* Fires which involve or are fuelled by liquids or liquefied solids; examples are petrol, oil, grease, paint and solvents.
- *Class C – flammable gases* Fires which involve or are fuelled by gas or gases; examples are propane, butane and methane.
- *Class D – combustible metals* Fires which involve metals in solid or powder form; examples are aluminium, magnesium and titanium.

  Note Many fires involve a combination of types of materials.

Examples of fire causation and sources

Electrical equipment

All sorts of electrical plant, equipment and tools can overheat and catch fire, short circuit (especially defective cables) or produce earth leakage fires. These potential fire problems can occur in motors, switch gear, transformers, lights, cabling, wiring, switches and fuse units. In fact almost every electrical component is a potential fire risk and electrical equipment is the main cause of fire both domestically and within industry. Proper design, installation, maintenance, inspection and testing – carried out by well-trained and qualified electrical staff – will reduce the risk from this source. Keeping electrical equipment clean and free from any build-up of dust, rubbish or waste is a vital preventive measure. This applies particularly to computer and electronic cabinets.

Hot work (burning, welding and soldering etc)

This is a common source of fire problems in many organisations. A stray spark or hot metal particle may start a fire. In particular off-cuts from burning operations can and do result in fires. Carelessly discarded hot welding rods and welding metals can be a source of ignition. Welding or burning of pipe lines, tanks, pressure vessels and drums, especially those which contain or have contained gas or flammable substances, can present dangerous explosion and fire risks unless adequate emptying, purging or spading operations have been carried out. Once again, identification of the hazards in each situation, followed by measures to minimise or eliminate the risk, is the answer if fires caused by hot work are to be avoided. Examples of such measures are:

- The use of effective hot work permits which allow for inspections before and after hot work tasks.
- Use of flash back arrestors on gas burning equipment.
- Ensuring equipment is in good condition, hoses are properly fastened and are not perished and gas bottles are secured.
- Limit hot work in production or warehouse areas to the absolute minimum (consider cold cutting).

Smoking

This is another common cause of fires.

All industrial concerns should have a clear smoking policy whether for health or fire reasons (or both). Fortunately a more health conscious workforce and health protection legislation is making smoking much less common and implementation and acceptance of No Smoking rules much easier. However, this can drive smokers into illicit smoking which is extremely hazardous. Smoking rules are a question of 'horses for courses'. Some industries because of their very nature – for example, oil and gas refineries, petrochemicals – have complete site bans, as do a number of industries associated with food production. In others, smoking areas may be the answer, provided the smoking areas created are themselves safe – non combustible, well ventilated and properly used. The question of smoking policies is dealt with at greater length in chapter 27.

Friction

This is a common cause of fire in some industries. The overheating of bearings and moving parts of plant and equipment can be a risk in practically any industry. Good-condition monitoring is the key to fire safety in these areas. Drive belts, though less common in industry today, can be the cause of friction, resulting in overheating and possible fire; poor adjustment may be the main problem.

Another main source of friction fires is conveyor belts. Millions of tonnes of materials are transported by conveyor. Fires travel at speed along conveyors and conveyor housings often because other services are located alongside the housing. When belts burn, they eventually part, falling back into transfer houses creating major fires. Stalled drums and rollers, and slipping belts are the main sources of friction. These can be cured by more regular cleaning and improved maintenance. The replacement of rubber belts by those of a non fire propogating specification would virtually eliminate serious conveyor fires.

Open flames

Open flames, as found in heaters, boilers (both domestic and industrial), furnaces, driers etc are still a basic cause of fire. The fuel is commonly gas or oil but may be wood, coal and in some specialised industrial equipment may be waste solvent or special gases or mixtures of gases. Fires are caused by poor housekeeping – combustibles too close to the heater, fires or explosions at the burner units, rupturing fuel lines, poor maintenance, clogged pilot lights, dirty burner units, dirty exhaust ducts and so on.

Everybody is familiar with the tar boiler and the fires that this piece of apparatus has caused. These are nearly all due to misuse – gas cylinders unsecured, often located too close to the boiler, damaged gas lines and quite often due to the fact that the tar is overheated and 'boils over'. Good standards of maintenance, common sense and 'risk awareness' are the solutions here.

Flammable liquids and gases

These are found in varying quantities in most industries and often in the home or garage.

- *Flammable liquids* The property of flammable liquids is that they change from liquid to gas at a given temperature much like water turns to steam in a kettle when heated. The more volatile the liquid the lower the temperature required to set this change in motion. The flammability of a liquid is judged by its 'flashpoint', the minimum temperature at which the liquid will produce enough gas to produce a flammable or explosive mixture in air. Thus, the lower the flashpoint of a liquid the more flammable it is. Those liquids with a flashpoint of between 32°C and 55°C are known as flammable, those with a flashpoint below 32°C are known as highly flammable and their use and storage are governed by the Highly Flammable Liquids and Gases Regulations.

  In some industrial uses flammable liquids are used under considerable pressures, which heats them, and when a fuel pipe is ruptured the liquid is forced out often at something approaching its automatic ignition temperature.

  It strikes a source of ignition and a serious fire results. Industrial hydraulic systems using mineral oil fluid are a major example. The key factors in fire safety with flammable and highly flammable liquids are:

  - always keep liquids in closed containers
  - keep them away from sources of heat and ignition
  - keep large stocks in separate flammable stores (for highly flammable liquids this is required by law)
  - avoid dispensing such liquids from drums into smaller containers, except in properly separated and equipped areas
  - make sure all containers of such materials are clearly marked (manufacturers are required to mark flammable and highly flammable materials)
  - make sure that these materials are not stored in glass bottles.

- *Flammable gases* Most people will be aware of propane, oxygen and acetylene cylinders, which require careful handling and storage (again required by law in the case of the highly flammable gases). However, perhaps rather less well known and recognised as a hazard are the billion or so aerosols produced in the UK every year. Halon gas was used as a propellant in these aerosols, but because of its effects on the ozone layer halons have been banned and most aerosol propellants are now highly flammable hydrocarbons – propane, butane etc. Many aerosols can become flame-throwers if used near a naked flame.

Hot surfaces

These heat sources will radiate or allow heat to be conducted to such a degree that when combustible materials or substances are placed on them or too close to them, ignition occurs followed by fire. In some industries hot surfaces will be around furnaces, heat exchanging equipment, flues and chimneys, canteen equipment (ovens, fryers and hot food units), soldering irons and around steam generation plant. Hot surfaces may be an inherent part of the production process.

Heaters

These are of four basic types; air heaters, dryers, stoves and hot water providers. Heating can be provided by electricity or fuelled by paraffin, oil, gas, wood, coal or coke etc. A heater can be of a fixed or portable design. The main cause of hazards with nearly all these types of heaters is either bringing combustible materials too close to the heat source or placing the heat source too close to materials. Other fire risks can occur because poorly maintained equipment causes leaking fuel, because of poor operation and control, because of poor ventilation and because the equipment is left unattended for long periods.

Spark generation

This may result from a variety of sources, including:

- Sparks can be generated from foreign metal objects falling into process machinery and causing sparks.
- At times spark generation can come from normal process operations eg the braking of moving machinery or reversing of moving machinery.
- Some tools and equipment give off sparks as a function of their operation, grinding tools being one of the best known examples. Metal cutting saws and concrete cutting discs are other examples.
- Electrical sparks can be generated by electrical moving power pick-up systems eg on overhead cranes and transfer cars by arcing.

In all these cases if the spark comes into contact with materials or flammable substances a fire can start some considerable time after the spark was generated. Should the spark fall into a cable tunnel or service trench where possibly there is an air flow, this will help the fire take hold, often hidden from view.

Though a rare source of ignition, sparks are produced by static electricity and this should always be considered. Conditions for static electrical spark generation are usually found in very dry and/or low humidity areas of plant and if flammable vapours or gases are present this could lead to fire or explosion.

Finally, sparks can be simply generated by the use of hand tools (eg hammer or chisel). So if the atmosphere could contain flammable vapours, or gases are present, non-spark generating tools and equipment must be considered.

Arson

Maliciously started fires are an increasing and worrying feature. The motive is usually vandalism or personal revenge. In many cases it can be the result of unlawful entry into work premises. Good security is one of the most effective prevention measures.

Sub-standard housekeeping

The term housekeeping is a well-used one throughout industry and commerce. Poor housekeeping is evidence of ineffective managerial control. Unfortunately many of the products of sub-standard housekeeping are combustible materials which can assist in the rapid spread of fire. Common examples are:

- leakages from containers, tanks and pipelines
- spillage of liquids and substances, especially oil and grease
- discarded rags, timber, packaging, waste paper
- uncontrolled tipping of waste and rubbish
- unrecovered equipment eg ladders, scaffolding boards and pallets
- discarded sealed containers
- blocked emergency exit doors and passageways
- incorrect storage and mixed storage of dangerous substances
- overfilled and mixed loads in skips.

The burning of rubbish in an uncontrolled manner or in unsuitable areas is a closely linked cause of fire.

Contractor operations

Though this topic does not strictly fall in the remit of causes and sources it would be remiss not to draw attention to the fact that contractor operations can result in major fire safety problems. Clearly many of the previous topics listed in this part of the chapter can and do apply to contractor operations of all types from large building and construction sites to the small specialist contractor brought in to carry out a repair. The fire risk usually, though not always, increases when contractors are on site. This is partly because contractors are not familiar with or may not understand in-house fire hazards or how they relate to the work being undertaken by the contractor. Contractors' employees may not have the same commitment to the prevention of fires as does the in-house workforce.

Time and cost factors placed on the contractor's task may lead to short-cutting of safe systems of work. Standards of fire safety may not have been specified to the contractor. Contractors may not have been instructed in fire-fighting requirements and emergency procedures linked to the task of work to be carried out. In many cases the amount of rubbish, combustible waste and flammable construction material can form a major fire hazard.

On the safety management of contractors more generally see chapter 15.

Other causes

There are other less familiar causes and sources of fire that need consideration. These include phosphoric materials and substances, oxygen enrichment and spontaneous combustion.

Spread considerations

Small fires, if uncontrolled in their early stage, become major fires or disasters. Once ignition has taken place the subsequent development (or fire spread), if not stopped in its early stages, can be extremely fast. In some cases the speed of fire spread can have fatal consequences. In the tragic Hillsborough football stadium disaster in 1985 the speed of fire spread was graphically displayed on our TV screens as news coverage illustrated how a small fire, unseen at first, developed into a disaster within minutes.

Fire spreads and develops (at times at an exponential rate) through heat transference. Heat transfer brings all combustible materials to a temperature where ignition can easily take place. Materials may reach temperatures where they will spontaneously burst into flames. The heat transference mechanism is facilitated by conduction of heat, flame radiation and heat convection. Another factor in flame and fire spread is burning vapour or gas arising from near-by combustibles, especially flammable liquids or gases. These quite often produce the burning and disintegration of articles such as plastic containers. Escaping flammable gases can add greatly to the speed of fire spread. Fires have been known to travel faster than a person can run, and where there are narrow vertical spaces, for example between high-piled stacks of materials forming a chimney, can produce temperatures of several hundred degrees at the top of these spaces within minutes of ignition.

The prevention of major fires and disasters must include the prevention of fire spread. In many workplaces this can be achieved at low cost, although sophisticated fire fighting and containment systems can and do require considerable expenditure. When set against possible losses, especially in high-risk operations, these costs can be seen as justified.

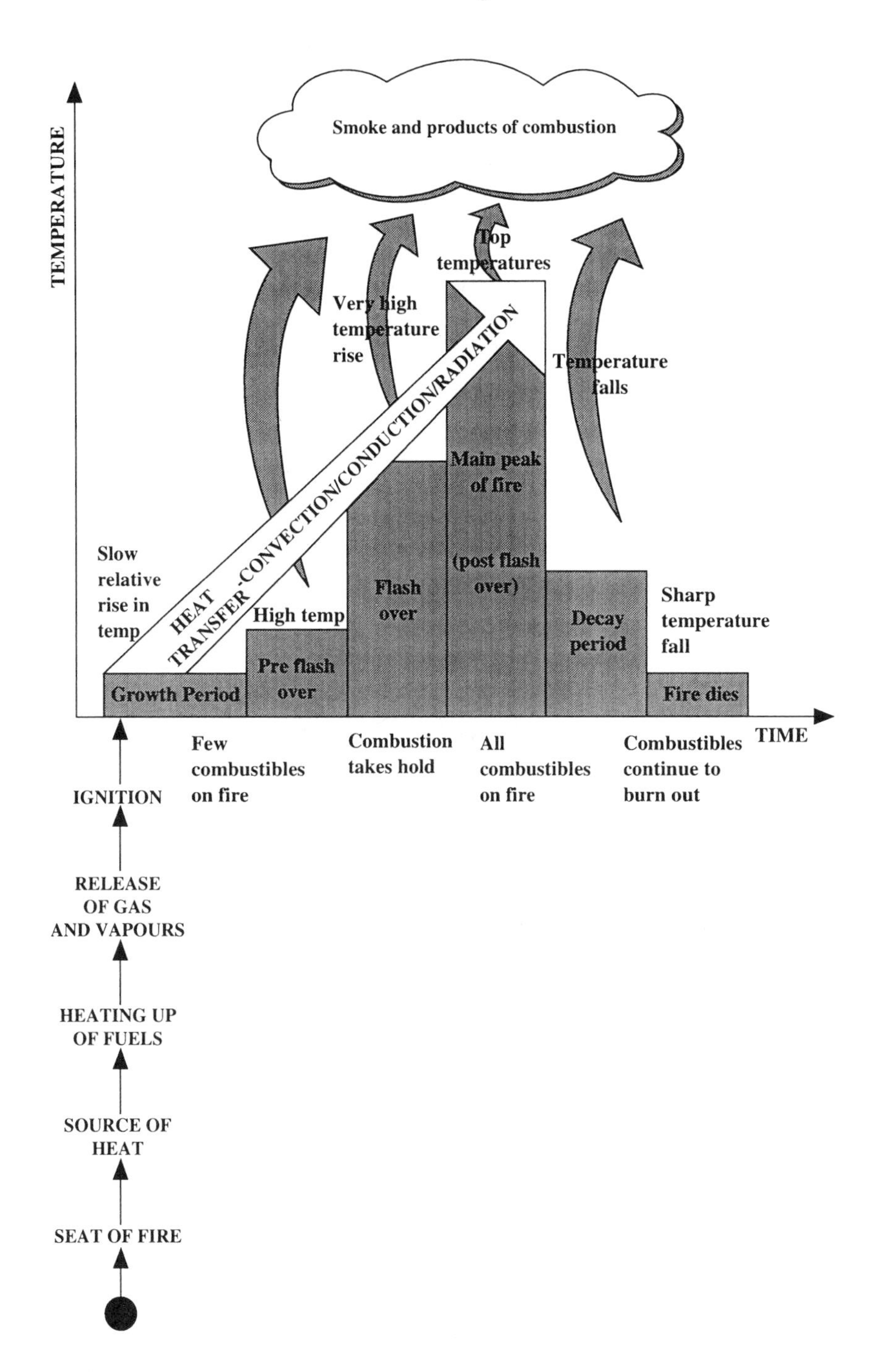

Figure 21.3 *Steps and stages in the development of a fire*

FIRE PRECAUTIONS: PREVENTION AND ACTIVE CONTROL

The steps and stages in the development of a fire are shown in figure 21.3. The measures that will effectively prevent fires or control the main effects are often referred to as fire precautions.

Whenever considering fire prevention and control, no matter how complex the situation, there are a few very basic principles which never change. For a fire to occur and develop there must be three elements present:

- heat (the ignition source)
- fuel
- oxygen (air)

These three elements make up what is known as the Fire Triangle.

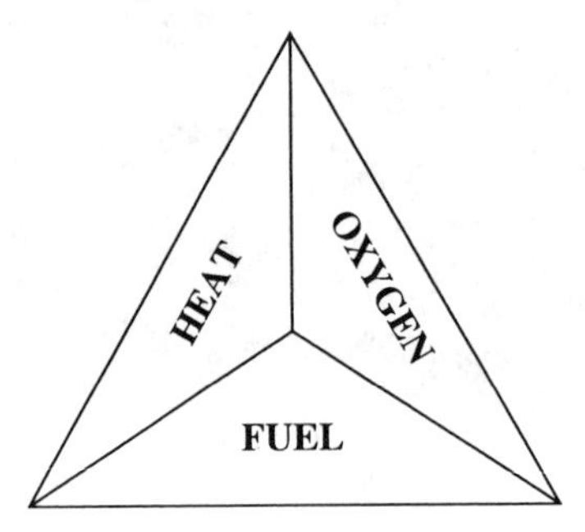

Figure 21.4 *Fire triangle*

Remove any one of these elements and fire cannot occur or develop. This is a very simple principle, but one that is extremely difficult to implement. Heat is often required or is generated in many production processes, fuel is needed and oxygen is present in the air.

Examples of removal of fuel

- improved housekeeping – less combustible rubbish
- changing combustible process fluids for safer ones
- use of non combustible building materials.

Examples of removal of heat (ignition source)

- control of hot work (burning welding etc)
- automatic cut-offs on process and domestic heating sources
- control of smoking.

These lists can be extensive and will be dealt with in more detail later.

Effective fire precautions should be aimed at major reduction or elimination of all types of fires and explosions. Fire precautions can be examined as follows:

- *prevention measures* – these are engineering and design features applied to stop the outbreak of fires and/or limit any consequential loss.
- *protection systems* – these cover a wide range of equipment, materials and devices which seek to minimise fire hazards by detecting, extinguishing and containing of fires (ie damage control)
- *preparation procedures* – these comprise proactive fire safety programmes together with pre-fire safety and emergency planning
- *managerial control policies* – these are programmes to ensure the integration of prevention, protection and preparation and to fulfil management's duties in respect of fire safety.

Key requirements are now presented under the above headings.

Fire prevention measures

Some measures can eliminate the possibility of fire completely, but the majority can only reduce the likelihood ie minimise the risk. Costs will usually be involved, and/or some effect on production systems and procedures, so decisions on implementation will involve weighing the likelihood of a fire and its consequences against those costs. Preventive measures can be considered under the following headings: electrical, mechanical, gas and flammable liquids.

Electrical measures

- Strict adherence to IEE and company installation standards and testing/ inspection specifications.
- Fitting of earth leakage circuit breakers will reduce the risk of fire in electrical equipment.
- Fitting of correctly rated fuses.
- Project design should ensure flammable liquids and gas-carrying pipelines do not run together or run alongside electrical cables, especially in cable tunnels or trenches.
- Project design should ensure that where cables may be subject to radiated heat or vulnerable to hot material spillage, cables should be re-routed, moved to a safer position, contained in protective conduit or placed in cable pipes.
- Consideration should be given in medium risk areas to the fire sealing of any cable laid through walls, roofs and floors etc and to dividing long cable runs by creating fire-check barriers in the form of one and a half metre wide bands of proprietory stopping material.
- Enhanced fire survival cables may need to be considered for certain applications.
- Open-topped cable trenches should be covered with steel or cast concrete

covers with closed dished hand or crane lifting points – or sand filled.

- Electrical equipment which can generate sparks (saws and grinders for example) should be adequately screened and shielded.
- Flameproof electrical equipment may be required for some applications.
- Electrical control rooms etc should be of wholly non-combustible construction.
- External transformers should be individually penned with power feed lines from the substation properly fire-sealed
- Internal transformers should not be oil filled, but should be solid state transformers or filled with fire-proof fluids.

Mechanical features

Examples include:

- Design project to dispense with cellars and basements for services whenever possible.
- Properly drainable drip trays or collection systems for oil equipment.
- Spark arrestors fitted to fork trucks' exhausts (or similar plant) in warehouse areas.
- Automatic fuel cut-offs on oil fired equipment.
- Use of non-combustible building materials including roof and wall linings.
- Adequate wall and bench protection for hot plates and cookers etc in mess-rooms
- Use of non combustible scaffold boards, use of non combustible pallets and containers etc.
- Combustion engines (especially generators) should not be located where any form of gas may be pulled into air intakes.

Gas safety

- Fit flame failure detection equipment to gas burning units.
- Ensure flash back arrestors and overpressure valves are fitted to all portable gas burning equipment eg burning guns, cutting and lancing torches etc.
- Consider the use of propane gas instead of acetylene.
- Provide ventilation equipment when solvents are being used
- Provide external gas bottle storage to comply with Highly Flammable Liquids and Gases Regulations.
- Check that major gas storage facilities are fitted with earthing and lightning conductor equipment.

Flammable liquids

- The major prevention measure in the case of flammable liquids is to consider whether use of the liquids is necessary. Could a non- flammable or less flammable product be used as a substitute eg water based solvents and high

water content hydraulic fluids?
- Can the flammable liquid be separated from sources of ignition or possible fire situations eg by locating it in another building?
- Provide safe spillage collection and provide non-drip/spill decanting and measuring equipment.
- Provide purpose designed safety containers and storage units, especially where transportation is required.
- Provide non-spark tools and equipment.
- Provide proper earth bonding for hazardous tanks, drums and vessels.

It can be seen that fire prevention is broadly about the design, purchase and use of fire-sensitive plant, tools, equipment and consumable materials. Management investment at this stage of fire safety is fundamental to an effective overall fire safety policy.

Fire protection systems

Fire protection systems can be divided into active and passive controls.

- *Active*
  - Detection and alarm systems
  - Fire extinguishers and appliances
  - Fixed fire fighting systems
- *Passive*
  - Structural fire systems
  - Means of escape
  - Emergency lighting.

Fire protection systems can be considered as the hardware of fire safety, the so-called first line of defence. However in dealing with means of escape and emergency lighting the objectives change from fire warning, fighting and containment to the protection of life. Reproduced in the Appendix to this chapter are some of the headings employed by the Loss Prevention Council. These provide an index of the sort of equipment and systems covered.

Detection and alarm systems

A fire detector is one (or more) sensors forming part of an automatic system which can give warning and indication of a fire condition. All detectors can be interfaced with other systems allowing automatic extinguishing discharge, plant shutdowns or door closing release mechanisms eg ventilation systems and air intake shutters. The basic types of detection units are:

- heat detectors
- smoke detectors and
- flame detectors.

Heat detectors can be divided into two main categories:

- Fixed temperature devices, which are designed to function in a way similar to that of a thermostat ie it responds to an increase in temperature.
- Rate of rise detectors which are designed to respond to a rapidly rising temperature or when a rising temperature has attained a pre-determined fixed temperature level.

Smoke detectors can be placed into three basic types:

- an ionisation smoke detector
- an optical smoke detector and
- an aspirating smoke detector.

Smoke detectors are usually much more sensitive and faster to respond than heat detectors, therefore they have to be carefully matched to their monitoring requirement, location and task. However smoke detectors are more likely than heat detectors to produce false indication alarms.

Flame detectors come in two basic types:

- ultra violet flame detectors; and
- infra red flame detectors.

It is important to note that warning and indication alarms are not normally activated directly by detectors but through electronic processing equipment. In essence these processors can sort out signals and indicate that the fire systems are:

- in normal condition
- in fault condition
- pre-warning of fire
- detecting fire.

Fire warning signals can also be raised from manual call points of the push glass system type. There are a number of key considerations that are important in relation to manual fire call points:

- the alarm point must require only one action to activate it
- it must not be reversible except by authorised persons
- alarm points must be located on exit routes
- there must be a sufficient number of alarm points to avoid excessive travel
- all alarm points must be effectively signed and located in well illuminated areas
- alarm point activation units should be linked to zoned fire warning signals which will then give a general warning around the area.

Alarm signals can be audible, visual or both. One important point about audible warning alarms is that they must be unique to the workplace and not be in any way similar to any other alarm devices. In addition it is vital to ensure audible warning alarms can be heard above normal workplace noise and be heard by everyone at all locations which could be at risk.

Two further points need consideration with detection and alarm systems – wiring and false alarms. Consideration should be given to the fault monitoring of electrical circuits ie open circuit, short circuit and power loss. Some critical circuits may need to be specially protected to prolong or protect them from fire damage. In some cases a radio-link may be the answer. The issue of false alarms needs to be addressed by any organisation with manual or automatic detection and alarm systems. Some 90 per cent of fire brigade calls are false alarms. Effective design and maintenance can help considerably in this area.

Fire extinguishers and appliances[5]

- *Portable fire fighting equipment* The types, quantities and locations of fire extinguishers required in the workplace will often be stipulated in the Fire Certificate or by the Fire Insurer. Smaller workplaces may need to make their own decisions as to what is required for their level of risk. Probably 75 to 85 per cent of industrial fires are brought under control by in-house fire fighting, mostly by use of portable fire extinguishers. Portable fire extinguishers usually contain one of the five extinguishing agents which are:
  - carbox dioxide in a black container (class B, C and electrical fires)
  - foam in a cream container (class B fires)
  - vapourising liquids (halons) in a green container (all classes of fire)
  - powder in a blue container (all classes of fire)
  - water in a red container (class A fires).

Carbon dioxide extinguishers work by displacing the oxygen supply (in the air) around the fire thus starving the fire. This type of extinguisher also places a thermal cooling on the flames further assisting extinguishing and reducing fire spread. The carbon dioxide in the extinguisher is in a liquid form under high pressure. These types of extinguishers are ideal for small fires of electrical apparatus and equipment as the gas leaves no damaging residues and ensures the extinguisher operator does not come into contact with any live electricity. Carbon dioxide will not cause any electrical malfunctions of electrical equipment. A further advantage of carbon dioxide is that it does not contribute to ozone layer depletion. However carbon dioxide extinguishers do have some disadvantages. If discharged in a confined space the gas and fumes are both toxic and an asphyxiant. During discharge the significant noise can unnerve the unwary operator and the discharge horn which becomes very cold during discharge can cause mild frostbite. Another disadvantage is that most standard carbon dioxide extinguishers are heavy by comparison with other types and this is not compensated for in the weight to fire extinguishing performance ratio.

In fire fighting operations the jet discharge from the horn should be directed straight at the centre of the fire or if the fire is in an enclosed unit, into the nearest opening.

Foam extinguishers fight a fire by forming a barrier between the fire and the surrounding air, smothering the fire by oxygen starvation. Foams can be used for class A and B fires. They are particularly suitable for class B fires (flammable liquids). However foams are not very effective if the flammable liquids are not retained or are a running liquid fire. They must not be used in fires involving live electrical equipment as some are likely conductors in certain exceptional conditions. The foam stored in these extinguishers can either be pressurised or powered by a gas cartridge.

When fighting flammable liquid fires in open containers, direct the discharge jet at an inside retaining wall and allow the foam to disperse and flow down onto the liquid fire surface. Where this is not possible, fan or sweep the discharge over the top of the liquid fire and allow the foam to cascade over the fire surface. Never aim foam discharge jets directly into the liquid fire. This will drive foam under the liquid fire surface and make it ineffective. It could also cause liquids and fire to be dispersed and splash over a larger area.

Halon extinguishers work via a complex chemical reaction with the fire/flame blocking out oxygen in the air around the fire. The storage of halons in extinguishers is as a liquid under pressure. On discharge it is first expelled as a liquid but rapidly evaporates into a vapour/gas. The main advantage of halon is that as it discharges in a liquid/vapour form its knock down speed on fires is very high. It can be used on both class A and B fires though is more applicable to the class B type. Further advantages are that it can also be employed on electrical fires and leaves no residues with any class of fire. Though toxicity of halon is classed as low, after any discharge into closed or confined spaces precautionary atmosphere checks should be undertaken before work areas are allowed to be re-occupied. Care must be taken not to use halon within confined space locations or in restricted or reduced air flows. Fire fighting with halons is best undertaken by fanning the discharge jet across the flames to allow a cascading effect to extinguish the fire. Halons are being phased out under the international Montreal Protocol Agreement due to the fact that halons deplete the ozone layer.

Powder extinguishers can be used on both class A and B fires dependent on the powder agent. Again the powder can be stored at pressure or discharged by a gas cartridge. The powder is likely to act on the fire in two ways. First, as a chemical inhibitor of oxygen and then by restricting the generation of fuel vapour. This combination of effects gives rise to a rapid knock down of the fire. In particular, powder is one of the few agents to have a positive controlling effect on running flammable liquid fires. While powder can be used with no electrical risk on live electrical equipment, it can result in extensive damage to electronic and electrical equipment of all types due to residues and discharged contents. The weight extinguishing ratio of powder is better than water or foam. Powder is best discharged over the fire in a cloud spray allowing contents to fall onto the fire. On running flammable liquid fires it can be used to slow and retard the spread of liquid by putting down a layer of powder in front of the spread.

Water extinguishers are still the most common type in use but are increasingly being replaced for general industrial applications by powder. These types of extinguishers are in the main applicable to class A fires. Water acts as a rapid cooling agent on the fuel of the fire and is normally very effective on class A fires. However it can be a hazard in class B fires and on electrical fires and should not be discharged onto either of these types. The water in the extinguisher is either stored at pressure for discharge or discharged by a pressurised gas cartridge. Water discharge jets should be directed to the base of the fire and fanned out across the centre of the flames. Hot spots within the fire should also be targeted. If the fire is spreading vertically, direct the water jet to the lowest point first and spray upwards.

The siting of portable fire extinguishers is very important. Extinguishers should be close to possible fire hazards. They should be highly visible and on general or emergency walkways. Selecting the correct size of extinguisher is an important consideration. How far will it need to be carried, what is the likely fire task and how many extinguishers need to be available? The type of extinguisher to be made available is a further consideration, for example halon for computer equipment, carbon dioxide for photocopy machines and powder for canteens etc.

When fighting fires with portable fire extinguishers some key points should be built into training. For example:

- only fight fire if you are trained to do so
- ensure before fighting a fire you are in a safe position – unrestricted emergency exit and an alternative emergency exit if there is a change of air or wind direction
- keep low to avoid smoke and heat and to aid vision
- always check that the fire brigade has been called
- ensure the fire is completely extinguished and not likely to re-ignite or smoulder.

Personal safety – never fight or continue to fight a fire if it is dangerous to do so eg:

- if the escape route could be cut off by fire or smoke
- if the fire continues to get out of control
- if there is a possibility of explosion due to gas bottles or other explosive substances.

Fire blankets – these are usually manufactured from a fibreglass type material and wall-mounted in a metal container. Fire blankets are mostly applicable in canteen operations, laboratories, small workshops and contractors' temporary accommodation.

- *Mobile fire fighting equipment* The need for mobile fire fighting equipment must be evaluated by management where it is considered that first response fire fighting equipment requires a large volume of extinguishing agent and/or high pressure delivery of agent. This type of fire fighting equipment is mainly in three forms:

- hose reels and water cannons
- trolley-mounted extinguishers
- fire engines or fire pumps.

Hose reels Though technically not truly mobile, hose reels do allow a limited mobile response to a fire in so much as the extinguishing agent discharge nozzle can be brought close to the location of a fire. Hose reels average some 30 metres in length but can be supplied in almost any length required. All hose reels are permanently connected to a water supply. To be effective in high buildings or plant, hose reels may need their own tanked water supply and pressurisation pump. Failure to maintain hose reel systems is a major disadvantage of this type of fire fighting equipment. Neglected systems fail at the very time they are needed most.

Water cannons Again, although not technically mobile, a water cannon may be an alternative to hosereels. A water cannon can deliver large volumes of water at high pressure throwing over large distances with variations to the configuration of jet/spray output. They are mobile in so much as (where required) design can allow a 360° operational turning circle and a 160° operating elevation arc. Because they are a powerful piece of equipment only trained and authorised staff should be allowed to use them. The disadvantages are they require much more water than hosereels and are more subject to obstruction.

Trolley-mounted fire extinguishers are another way of getting large amounts of extinguishing agent to a fire as a first response. They come in many shapes, sizes and configurations and are available for use with any of the five major extinguishing agents. In the main, these types of mobile extinguishers are mounted on two-wheel trollies which can normally be pulled by one person, though some are designed to be towed by a vehicle with a tow bar coupling. Again this type of mobile extinguisher is best used only by trained and authorised staff.

Fire tenders For a number of hazard and risk considerations an organisation may need to look at its own in-house fire tenders and fire fighting team as a first response to fire fighting. These machines can deliver large volumes of water or foam at high pressure. When connected to a water supply they can continue to deliver water or foam until (and even after) the fire brigade equipment comes onto site. A fire tender is a costly piece of fire fighting equipment requiring expensive maintenance and trained and fully equipped fire fighters. Fire tenders for industrial application can be simple trailer towed equipment, designed into four wheel drive vehicles or full-size fire brigade equipment.

Fixed fire extinguishing systems

Fixed fire extinguishing and control systems are possibly the best known fire equipment in commerce and industry after portable extinguishers. The term refers to a permanently installed protection system for the manual or automatic discharge of fire extinguishing or fire suppressing agents. They are again usually classified by the extinguishing agent; water, foam, gas or powder and in some cases further sub-divided by types of discharge.

- *Water systems* The most common of the fixed fire extinguishing/control systems are the ones using water. Systems using other extinguisher/suppressant agents tend to be for specialised applications.

Water systems are generally known as sprinkler or water spray systems. Without question, these are the most effective form of fire extinguisher/control measure available. Statistics over a period of many years show that where fires have occurred in sprinklered premises, they have operated and successfully extinguished the fire. Their effectiveness is illustrated by the fact that insurance companies, not normally recognised as philanthropists, are prepared to give premium discounts for sprinkler protection of between 50 to 70 per cent.

Sprinkler/water spray/installations generally have to be installed to a design specification known as the LPC/FOC 29th Edition Sprinkler Rules.[6] (LPC/FOC is the Loss Prevention Council/Fire Offices Committee). Previously this was solely an insurance body but now comprises officials from the Home Office, insurance industry, local authority and fire authorities. In some industries the installation of water sprinklers or water spray systems is a condition of obtaining insurance cover, in all other industries such protection means very much lower insurance premiums. Many local authorities insist on sprinkler protection in large office blocks, hotels, shopping malls etc.

Water sprinklers and water spray systems comprise a water supply, main sprinkler valve sets and a network of reducing diameter distribution pipework in which are numerous sprinkler or water spray heads. The water supplies may be town mains, private works mains (often incorporating a booster pump) or dedicated supplies in the form of tanks/reservoirs and pumps. These water supplies feed the distribution pipes which contain the sprinkler/spray heads via sprinkler valves which are non-return valves incorporating test facilities and automatic alarms. Depending upon the type of risk to be protected and the degree of exposure of the system to freezing, the water can either be held at the valves, usually by air pressure being introduced and maintained within the distribution pipework. This holds down a diaphragm within the sprinkler valves – release of the air pressure by a sprinkler head operating allows the diaphragm to lift and water to pass into the system.

In other systems the distribution pipework is full of water, held at bay by the sprinkler heads themselves which are sealed units. The sprinkler heads are heat activated and a range of operating temperatures of these heads is available depending upon the risk being protected. The 'general purpose' heads activate at 155°C. The design considerations of water supplies, pump sizes, pipe diameters, numbers and type of sprinkler heads will be dictated by the LPC/FOC

Sprinkler Rules and will depend upon the the type of risk to be protected. Effective density water discharge and droplet size have been calculated within these rules in respect of different types of risk. Approved equipment and approved installing contractors are also required by the rules. Disintegration of the quartzoid bulb or the metal strip by heat releases a seal and water (or air pressure followed by water depending upon the type of system) is released to form a spray pattern by striking the deflector plate on the head.

There are many variations of both wet and dry systems – individual sealed heads, a number of open sprays or projectors controlled by a sealed head, or total deluge of an area activated by detectors tripping the valve being the main ones. The type of discharge can vary from a sprinkler type of effect from sprinkler heads to high or medium velocity discharges often used in the control of fires involving flammable liquids. High velocity water will form an emulsion with liquids and prevents, for example, burning oil or spirit floating on the water, whilst medium velocity is much more of a control measure for cooling equipment and vessels around the seat of the fire.

The following are less common systems, used for specialised protection:

- *Foam fire fighting systems* These may be found in a variety of commercial and industrial forms. They include base and top basement injection into storage tanks (usually flammable liquids), top pouring, oscillating heads and conventional fixed spray or sprinkler modes. They can be employed both for internal and external application. The expansion properties of the foam are classed as low to high. Low expansion foam leaves a layer on the floor while high expansion foam is employed to fill the building or compartment in total if it is to do its job effectively. A range of medium expansion foams is available between these extremes.

- *Gaseous fire fighting systems* These mainly utilise one of two extinguishing discharge agents, carbon dioxide or Halon (mainly BTM 1301). The systems are usually activated by smoke detectors, occasionally by heat. They are used to protect sophisticated equipment – computers, electronic/electrical controls – and as a first knock-down protection for tanks and other vessels containing volatile liquids (often in these cases backed-up by sprinklers).

  In electrical areas or computer suites the activation is by tandem or double-knock smoke detection heads wired on separate circuits. One head will pick-up smoke at a very early stage and will sound an alarm, shut-off air conditioning fans, close dampers in ducts and so on. If the smoke increases in volume and two heads are activated this will trigger the gas release.

  CO_2 works on the exclusion of oxygen to extinguish the fire. Systems usually use about 30/35 per cent concentration. CO_2 at 30/35 per cent concentration will extinguish fires, but within a confined space can kill people, so foolproof lock-offs are required. Halon puts a chemical ‘spade’ in the combustion chain and requires only five – six per cent.

Halon is environmentally hostile and is gradually being replaced. FM200 is one of the better alternatives. Inergen is another, but this cannot be stored in liquefied form so banks of cylinders are required on installations of any size.

- *Powder fixed fire fighting systems* These are very rare in industry or commerce because of the cost of operation and maintenance. They do exist in a small number of places and are very specialist in design and operation. Hazardous liquids in road tanker filling bays is one application. Dry powder is mainly applied to local fire fighting systems but in certain hazardous situations powder flooding systems can be designed.

Other equipment

At this point it is appropriate to mention some items of equipment which are not strictly fire fighting systems themselves. However, they do contribute to the overall efficiency of fire fighting systems, and are in fact fixed pieces of equipment.

- *Fire hydrants* In large sites it may be necessary to have these inside the plant or works. In-works hydrants should whenever possible be on a separate ring main water supply and sited adjacent to roadways. Easy access in an emergency should be considered and clear high and low level marking of the hydrant site location is vital. They may be fed from works water reservoirs, tanks etc via pumps. Public street and industrial area roadways are served by a network of municipal water mains serving hydrant points.
- *Dry risers* – to assist in the speedy supply of water through high risk buildings or structures a second piece of fixed equipment known as a high rise dry main should be considered. A dry rise main comprises an inlet connection (clearly marked up) at floor level for fire brigade pumps to plug into or for a hose to be run and connected into the nearest hydrant supply. High rise fixed permanent pipe line systems run up through the building or structure and outlets (landing valves) are built into the pipe lines at each floor or level or at strategic locations. Fire fighting units would in the event of fire connect into the nearest outlet and once the main is charged from the ground level inlet, can quickly obtain a source of water for fire fighting without the need to run hoses from ground level to the fire location. For basement buildings, cellars, storage areas below ground level a similar system can be installed to carry a speedy supply of water to the fire areas. These systems are usually termed dry falling mains. In high fire risk buildings or plant both at high level and below ground these mains systems can be permanently charged with water. They are referred to as wet high rise or wet falling mains.
- *Foam inlets* These units are usually fitted to high fire risk plant or buildings especially where it could be hazardous for fire fighters to enter the area in a

fire condition. A foam inlet is simply a fixed pipe at ground level running through the wall of the structure to be covered and having a number of outlets into the risk area or areas. The inlet is fitted with a standard hose connection protected by a quick release cap. In the event of a fire within the structure fire fighters would connect to the inlet and pump the structure full or part full of fire fighting foam of the required type. This foam injection operation would smother the fire quickly with little damage to plant and equipment. This is a low-cost system ideal for oil-fired boiler rooms, pump rooms, hydraulic houses, cellars or basement areas and certain flammable storage units.

FIRE PRECAUTIONS: PASSIVE CONTROL

Buildings and structural fire systems

In the main, these fire safety systems comprise the following:

- fire doors and fire resistant glass
- fire barriers
- fire structural stability
- compartmental fire design
- fire separation management
- fire spread material inhibitors.

Structural fire systems should be designed to fulfil one or more of the following objectives:

- prevent collapse of a building or plant
- prevent flame/smoke spread over or along floors, walls or ceilings
- prevent fire/smoke spread within buildings or plants
- prevent fire spread beyond a burning building or plant.

Fire safety doors and glass

Fire safety doors are designed to hold and prevent the spread of fire or designed to prevent the spread of smoke. They can also be designed to undertake both roles. Average fire resistance of these types of door varies between 30 minutes and 4 hours. Treated or insulated timber doors form the usual material at the lower ranges while steel usually forms the material for the higher range prevention periods. Increasingly, modern timber fire doors are fitted with intumescent strips to either the fire door or frame. These strips expand at above 150°C and seal the gap between the door edge and door frame preventing fire spread. Where the prevention of smoke spread is required, smoke seals (similar to draft excluders) can be fitted even when intumescent strips are fitted. Considerable amounts of smoke can pass through a closed door before fire spread strips become fully operational. Some examples of the application of fire doors are as follows:

- doors to exit stairways, lobbies and main entrances
- doors to external fire exits and escape stairways
- doors in dead end corridors and passages
- crossing-corridor doors
- service shaft doors
- doors in fire separating barrier walls.

Where fire doors cannot be generally kept closed due to traffic or pedestrian flows they can be held open and fitted with a fire closing device. Closing devices are of two main types. One type has a fusible link (holding the door open) which melts at a pre-set temperature allowing the door to slide closed (on a sloping skid) pulled by the action of a falling weight. Another is an electro-magnetic unit holding the door open which is de-energised by a smoke detector and again pulled closed by a falling weight. A number of things can and do go wrong with self closing fire doors: careful evaluation needs to be undertaken before they are fitted and regular inspection carried out to ensure effective operation in the event of fire.

Normal types of glass do not give any resistance to fire as they shatter and fall out in the early stages when subjected to the heat of a fire. Where glass forms part of fire protection walls, ceilings or doors, fire resistant glass should be fitted. Wired glass is the most common low cost option. Recently unwired fire resistant glass in special frames has become available for this task though it is usually more expensive than wired glass. Specialist laminated and layered fire resistance glass is also available for high risk applications.

Fire barriers

Fire barriers can provide effective fire spread protection and minimise the time for re-start of production or operation. Fire spread in unseen concealed and confined wall, roof or floor cavity spaces and voids is a particularly hazardous situation in many undertakings. The slow build up of an undetected fire in these situations can spread considerable distances and be well established before it is seen or makes its presence known. Examples of the locations and areas of these types of fire are underground cable and service ducts, tunnels, conduits and pre-cast concrete or steel troughs. In addition air flow floor voids and underfloor ventilation trunking systems can be a major vehicle for fire spread. At or just above ground level other examples are open or closed cable racks, cable and pipeline ducts together with boxed-in cable runs, pipeline services and air shafts. Wall cavities, especially where electrical services can be found, present yet other unseen fire spread routes. At roof level in roof spaces and ceiling voids especially where there are heating or ventilation ducting, uncontrolled fire can be fast and dangerous.

All these fire problem situations can be controlled very effectively by installation and maintenance of fire barriers. They are of three main kinds:

- hard setting fire stopping cement
- solid fire retardant materials
- fire retardant packing material (temporary stopping).

Generally, fire barriers involve minimal cost.

Fire resistant stability

The structural fire resistant stability of buildings and plant in the event of fire is a specialist engineering subject. In outline it is the evaluation of the design and construction of buildings and plant. The object is that in the event of a fire a structure will maintain a stable condition for a reasonable period. It should not collapse nor should any load bearing element fail through initial exposure to heat or flame. This fire stability engineering includes the selection of fire-proof materials that will not distort or give way when exposed to high temperatures. It also includes engineering of key upright and cross structural members to support extra weight along escape routes in the event of fire, preventing total collapse. Each load-bearing element should be examined to determine its integrity when subject to fire and explosion. Load-bearing structure frames such as beams and columns, load-bearing walls, raised floors along with platform or gallery areas must all be examined for fire integrity. This aspect of fire safety design must above all ensure sufficient time for the safe escape of people and the provision of some protection to fire fighters.

Compartmentalisation

Fire compartmentalisation of building and plant is a fairly common industrial and commercial engineering feature. In essence it is simply the sub-division of buildings and plant by various civil engineering design methods to inhibit fire spread and to contain fire within a designated area. As the name implies, fire compartmentalisation is the enclosure of any space with fire resisting or retarding materials on all sides and where necessary at roof and floor level. Various fire resistance times may be applied depending on legal and commercial requirements. These are usually 30 minutes, 60 minutes, up to 4 hours and in special high risk situations longer.

Fire separation management

- *Separation of plant or units* Another design method which can be employed is the separation of high fire risk process departments or units from the main buildings or major areas of plant. Examples of these, both storage and operations are:
  - flammable liquid stores, paints, degreasers and cleaning substances
  - hazardous gas, propane, oxygen etc

- flammable raw materials storage, timber, rubber, plastics, oil etc
- flammable fuel storage, petrol, coke etc
- waste tip areas, rubbish, canteen waste etc
- finished goods store or warehouse products
- paint spraying units
- major adhesive operations
- industrial furnace or oven operations
- boiler rooms and pump rooms
- hydraulic pressurisation stations
- oil filled electrical transformers.

The prime purpose of the technique is to reduce the estimated maximum financial loss. This is the reduction of the maximum loss that is likely to occur in the event of a fire, assuming the failure of all other fire protection equipment and engineering that could be installed.

Increasingly, organisations are becoming more dependent on computer equipment. These facilities should be subject to a fire risk assessment in terms of both compartmentalisation and separation.

- *Separation walls* Where separation of high risk plant, equipment or units is not possible or not practicable fire separation walls can be employed to prevent fire and smoke spread. Walls which are common to two or more buildings or plant should be constructed or modified using fire resistant materials and constructed in such a way as to stop fire spread beyond the original unit. The construction must be of sufficient height to prevent fire spread to adjoining or adjacent walls or structures. In many cases the original walls of buildings and structures will (with little modification) be sufficient to be classified as fire separating walls.

Roofs which form an element of any fire separating walls must of course have fire separating design features to the same requirements as the walls to prevent the fire bypassing the fire spread wall and spreading through the roof structures to adjoining or adjacent units.

In circumstances of exceptional risk, walls can be engineered to act as blast walls in the event of explosion. The objective here would be to prevent or limit damage to adjoining premises.

Fire spread considerations

Whenever new building or plant or extensions are being planned or indeed when alterations or modifications are necessary, fire resistant/smoke retardant materials and furniture should always be considered. In particular wall, floor and ceiling or roof covering materials and textiles that can inhibit flame spread and retard fire should be examined. These materials must offer resistance to flame spread over their surfaces and retard heat and smoke. This fire protection feature is of most importance at and around the means of escape, doorways and stairwells for example. Performance of these materials is tested and classified into five classes. Class zero is a special application class, while class one is the best general class of materials.

Emergency preparation – means of escape

Means of escape comprise the protection of structures and the equipment whereby a safe route is created and is available for persons to travel from any point to a place of safety without the assistance of outside agencies. The golden rule of escape routes is that people should in all possible situations be able to turn their back on a fire and escape safely. However this may not always be practicable especially in premises with only one stairway or closed end corridors or in small single door rooms. Where two exits must be provided each should constitute a real alternative and not be so close to the other as to in practice provide only one real exit. In planning or designing means of escape three key points must be considered:

- overall travel distances
- the escape capabilities of exits and routes and
- people escape capacities.

Three elements of escape travel can be identified:

- travel within rooms or partitioned areas to an escape route ie protected corridor or passage
- horizontal travel along a stairway exit or final exit ie along corridors or passageways
- vertical travel down stairways inside or outside premises and then to a final exit.

Overall means of escape distances are made up of a combination of the above. The final escape exit must always lead to a place of safety. A final exit into a ground level enclosed area or a final exit onto an adjacent roof is not classified as a final escape exit to a place of safety.

The horizontal travel distances are distances covered to reach the nearest:

- final exit or
- door to a protected stairway or
- door to an external escape route.

As a rule of thumb, escape travel distances will be dictated by the type of premises, the occupational and operational purpose of the premises and the number of people at work or using the premises. Local fire brigade staff can advise on this subject. Regular checks need to be carried out to make sure that fire escape routes are never locked or obstructed, fire doors are not wedged open and self closing doors are free to close.

Emergency lighting

Emergency escape lighting is the provision of independent illumination for escape routes and escape route signs, together with the illumination of obstructions or hazards so as to ensure the safe evacuation of premises. Fires can and do lead to the failure of normal lighting systems which can make the use of escape routes difficult and at times dangerous. Provision of emergency escape lighting is necessary especially in windowless premises and those which are occupied during hours of darkness. Escape lighting must always be independent of normal electrical systems. Cables should be installed to fire resistant standards or be protected from fire. There are three main types of escape lighting systems:

- self contained lighting systems
- central battery lighting systems
- lighting systems using emergency generators.

Self contained lighting systems – these are the most commonly installed escape lighting systems. They are made up of a lamp, a battery, a mains supply charger and a changeover switching unit.

A central battery lighting system – this has a single location battery bank again charged up from the mains supply. The battery bank (in the event of fire) would supply power to the emergency lighting via a complete and independent secondary lighting circuit.

Lighting systems using emergency generators for escape lighting are usually not considered to be an adequate system by current standards because of the time it takes to run-up the generator units before they can provide adequate light.

Employees must be informed about escape routes, escape equipment and escape lighting so that they can become familiar with the procedures. Plastic 'how to escape in the event of fire' cards can be issued to employees and a short safety briefing given. Clear signs, reflective and where considered necessary, illuminated, must indicate the emergency escape direction, routes, fire exit doors and stairways.

FIRE SAFETY MANAGEMENT

Fire safety in any organisation will always come down to managerial control. Because many fire safety systems and procedures lie dormant and unused for months or even years at a time, their readiness can often become overlooked. The managerial role in ensuring the readiness of fire safety measures is all the more important as in many cases some of the hardware can be very costly to install.

Over the years many committees of inquiry into fire disasters have indicated that management deficiencies played very significant and in some cases a major role in the outcome of fires. An often quoted example is the King's Cross fire in 1987. The Fennell report on this disaster among many comments on managerial deficiencies stated "... London Transport at its highest level may not have given as high a priority to passenger safety in stations as it should have done."[7]

Management control of fire safety involves four main functions:

- planning
- documentation
- training
- inspection.

Planning

After conducting a detailed risk assessment, the managerial planning function should consist mainly of examining future requirements for fire safety equipment. Planning should evaluate replacement and servicing of current fire safety equipment. This will necessitate the provision of funding for any new or replacement hardware and equipment. Like all other managerial planning operations it could be subject to a rolling five year provision and finance plan together with engineering specifications and standards.

Documentation

Principal fire safety documentation comprises:

- a fire safety policy and
- an emergency procedure.

The fire safety policy document should set out how and by what means management intends to control fire safety within the organisation. After a senior management introduction and overview the policy amongst other things should detail:

- fire and explosive risks and risk areas
- line and functional responsibilities associated with fire safety
- details and locations of fixed fire fighting installations
- details of portable fire fighting equipment
- list of fire alarm call points
- location of emergency evacuation assembly points
- emergency evacuation roll-call procedure
- locations of hydrants, dry risers, foam inlets etc
- register of fire/smoke and escape doors and fire escapes
- list of fire barrier installations and cable fire stopping points

- inspection, monitoring and testing schedules
- maintenance standards.

The policy statement, along with relevant information, could be available to all members of the organisation. In large organisations it may be necessary to have a number of sub-documents covering fire safety. For example one on policy, one on standards, one as a register and perhaps one on inspection, testing and maintenance.

One document that all organisations need, however small, is an emergency procedure. This document should provide instructions on 'what to do in the event of an emergency'. It may be a wall or pocket card, but in larger organisations or those with hazardous operations a more comprehensive document is likely. Some of the components of an emergency procedure could be:

- examples of the types of emergencies covered by the document
- raising the alarm
- contacting emergency services
- emergency assembly locations
- emergency units and specialist organisational roles
- command and control duties
- communication methods and systems
- emergency control point location
- support equipment locations
- support services available.

An emergency procedure requires effective staff to make it work. They must be trained in all the skills required for fire safety and in particular in emergency procedures.

Training

Fire safety training has two constituents:

- information on fire safety and
- the practice of fire emergency procedures and the use of fire fighting equipment.

The frequency and content of fire safety training will depend on a number of factors. Some items of information should be presented to all employees while other, more comprehensive training will be necessary for those people with special fire safety roles or in high risk areas. People who are in buildings or plant with escape procedures will need to be briefed on the methods of escape, and this will need to be validated by practice and made the subject of records. Company employees who will be required to fight fires with local extinguishers, or take part in a fire fighting team, must be given the necessary skills training to ensure they can carry out their allocated fire fighting tasks.

Training might include the following information:

- fire safety legal framework
- action(s) in the event of fire
- means of raising alarms
- means of escape
- procedure for contacting the fire brigade
- action(s) on arrival of fire brigade and other emergency services
- location of portable fire appliances
- location and operation of fixed fire fighting systems
- components of fire
- fire safety and risk
- fire precautions
- fire systems, inspection, testing, monitoring and maintenance.

In addition, practical training is essential to ensure that when a fire occurs the relevant staff not only know what to do but can actually carry out the necessary procedures. The behaviour of company staff in fire emergency situations, especially escape behaviour, can in many cases be more important in the saving of lives than design of fire protection and fire fighting systems. The employees and managers of contractors should in certain situations be included in practical, fire prevention and procedure training, and where necessary be given information about fire safety topics relevant to their areas of work. Some of the general topics of practical fire training that need to be considered for contractors are:

- fire escape drills including alarm recognition
- handling and use of portable fire fighting equipment
- isolation and re-setting of automatic fixed fire fighting systems
- installing temporary and permanent fire barrier materials
- visual examination and understanding of fire safety signs
- emergency exercises including evacuations.

Inspection

A fire safety culture may be seen as a prerequisite of sound fire safety management. The concept of a safety culture is discussed in chapter 10. In the context of risks from fire, a safety culture involves:

- monitoring
- testing
- inspection and
- maintenance.

Monitoring

Monitoring is concerned, for example, with ensuring that there is no build up of flammable rubbish and wastes, especially in areas where ignition sources are present.

Other fire safety monitoring issues are:

- *Control of smoking*: compliance in controlled no smoking/no naked light areas should be monitored.
- *Burning, welding and hot work*: is it being controlled by implementing a hot work permit procedure?
- *Contractor operations* in hot work and high risk areas should be evaluated.
- *Company purchasing directives* should state that 'best standards' should be applied when buying or hiring heating and cooking appliances (especially portable appliances) and systems. This can be established through monitoring of engineering standards available to purchasing staff.
- *Monitoring of electrical wiring and installation of electrical equipment and appliances* against engineering fire safety standards and regulations will assist in preventing electrical fires.
- *Monitoring of process fire and explosion risks* The operation of boilers, central heating systems, furnaces, kilns, drying and heating units should be checked by instrumentation monitoring and by visual monitoring using trained staff.
- *Monitoring the storage, handling and use of flammable substances* Attention should be given to monitoring of, for example, gas and oxygen bottles and pipelines, paints, thinners, degreasing fluids, fuels, adhesives and cleaning fluids.
- *Security managers should review measures for the prevention of arson.* Security control systems and methods may need to be updated to combat this increasing problem.

Testing

Testing and training in testing is both a managerial duty and in certain cases a legal requirement. Equipment needing to be tested might include:

- fire alarms and evacuation alarms
- hydrants, foam inlets, dry risers, falling mains, hose reels and water cannons
- automatic closing fire doors
- emergency lighting and illuminated escape signs
- fire detection equipment
- sprinkler installations (sample heads) and fire gongs
- automatic start-up fire pumps
- fire fighters' emergency (electrical) switches
- automatic ventilation shut down
- lift emergency alarms.

Competent and trained staff should carry out the testing of fire equipment.

Inspection

Formal fire inspection schedules should be established to detect deficiencies.

There must be a system under which deficiencies are reported and rectification is given priority.

Inspections should cover the following, where applicable:

- fire safety signs, especially escape route signs, to ensure that they are in place and visible
- fire exits and fire escape stairways and routes, to make sure they are free from obstructions
- automatic fire systems to ensure that they are in auto-mode
- gaseous extinguishing bottles and water systems, to make sure they are charged (gauge readings)
- air pressure in spray systems (gauge readings) to ensure they are charged
- portable fire extinguishers, to check that they are charged, undamaged and are located at fire points in the required numbers
- fire stopping barriers to check they are sound and complete
- smoke and fire doors, to ensure they close properly
- heat shields and fire blankets, to check that they are in place
- access routes, to ensure they are clear for fire fighting teams and fire brigade appliances.

In addition, formal recorded housekeeping inspections should be carried out. Packing materials, timber pallets, spillage of flammable liquids, leaking gas lines, defective electrical cables, rubbish-strewn cable runs, overloaded skips and contractor cabins are some examples of housekeeping issues requiring regular inspection and control on a formal and recorded basis.

Maintenance

Competent and qualified engineering staff will be needed to undertake maintenance. A schedule of regular maintenance is important to ensure that work is carried out to a common, 'best practice' standard. Records of maintenance must match the schedule within a formal system. This should be under the control of a senior engineer who would need to be accountable for maintenance operations and routines, including rectification. The checking of maintenance records for in-house fire equipment would need to be part of this schedule. Importantly, maintenance schedules should list maintenance frequency and what maintenance activity is necessary on each occasion. Maintenance frequencies are usually weekly, monthly, quarterly, six monthly, annual and three to five yearly.

The keeping of a master register of fire equipment is important in the context of fire safety management and a useful training tool in any organisation. This document should list all fire safety equipment, facilities, major appliances and fixed systems etc. The location of these fire safety systems could be shown on detailed drawings and by site plans together with the basic outline of inspection, monitoring, testing and maintenance schedules.

APPENDIX: LOSS CONTROL INDEX

1. **Automatic sprinkler, water spray and deluge systems**
   Alarm valves
   Alarm and dry pipe valves
   Wet alarm valve stations
   Accelerators and exhausters for dry pipe valves
   Deluge valves and actuators
   Drop pipes (adjustable)
   Direct reading flow meters
   Multiples jet controls
   Pipe couplings and fittings
   Pre-action systems
   Control equipment
   Re-cycling systems
   Electrical alarm pressure switches
   Sprinkler heads
   Suction tanks for automatic pumps
   Vortex inhibitors
   Water flow alarm switches
   Water spray systems
   Water spray systems (medium and high velocity)
   Fire pumps

2. **Building products**
   Lining and composite wall and roof materials
   Firescreens and partitions
   Wall and floor penetration resistance seals
   Fire sealants and firebreak systems
   Lift landing doors
   Fire protection of structural steelwork
   Fire dampers

3. **Fire detection and alarm systems**
   Fire alarm equipment
   Fire alarm systems and components
   Automatic fire alarm systems
   Cables for fire alarm installations
   Central stations for fire alarm systems
   Signalling systems for central stations

4. **Firebreak doors and shutters**
   Installation requirements

Firebreak doors and shutters
Service and maintenance of doors

5. Portable fire extinguishers and fire hose reels
Portable fire extinguishers
Fire hose reels (water) for fixed installations

Notes

1. The Fire Certificates (Special Premises) Regulations, SI 1976/2003
2. The Building Regulations, 1991
3. On the Framework and Workplace Directives, see chapter 2. On the Management and Workplace Regulations see chapters 3 and 16.
4. The Fire Precautions (Places of Work) Regulations (as yet only in draft form)
5. The Fire Protection Association print an excellent pocket data sheet on this topic.
6. Loss Prevention Council/Fire Offices Committee Sprinkler Rules, 29th edition
7. Fennell Report on the King's Cross Fire.

22

Noise, Vibration and Radiation

NOISE

The importance of noise as an occupational health issue has achieved greater prominence in recent years. There have been two main reasons for this – the increasing numbers of successful claims against employers in respect of occupational hearing damage and the introduction of the Noise at Work Regulations in 1989. One of the landmark claims (*Berry v Stone Manganese Marine*) emphasised the importance of employees being made aware of the dangers of noise to their hearing and of employers both providing and enforcing the use of hearing protection.[1] These principles have since been incorporated into the Noise Regulations.

Hazards of noise

Noise in the workplace can cause a number of health and safety problems. Of these, occupational deafness is by far the most serious. Severe cases of occupational deafness are often accompanied by tinnitus (a continual ringing in the ears) although this also has other causes. Perforated eardrums are relatively rare (often caused by sudden very loud noises such as explosions or gunfire) and the eardrum usually heals up. Noise may also interfere with communication in the workplace and may be a cause of stress.

Hearing occurs through the action of sound waves on the eardrum. The eardrum activates the bones of the inner ear which exert pressure on the cochlea, a snail-shaped organ containing liquid. The motion of the liquid within the cochlea is detected by tiny hair cells which transmit the sounds to the brain.

It is these haircells which are permanently damaged by exposure to excessive noise. A useful analogy is to liken the hair cells to the grass on a lawn. The effect on the grass of the passage of a single person across the lawn will probably not be visible. However, several people walking across the lawn will bend back the grass to show a visible trail. although this may have disappeared by the following day. In contrast, if a lot of people walk across the grass on a daily basis the grass will become worn away, with their regular route indicated by a brown earth path.

Similarly the hair cells in the ear are capable of absorbing a certain amount of noise but higher levels produce noticeable effects – often described as a short term threshold shift. Longer exposure to high levels of noise will permanently damage the cells which, unlike grass, will never grow back.

Sound is made up of a mixture of frequencies and occupational deafness usually affects the frequencies at the higher end of the audible range before any others. The sufferer loses his ability to differentiate between various consonant sounds – he will often believe that the problem is others not speaking clearly rather than his own lack of hearing.

Some hearing loss occurs naturally through the ageing process and the hearing may be damaged in other ways eg through ear infections. However, skilled interpretation of audiometric checks (see pp248 and 285) can usually identify the differences between these and occupational causes, although leisure activities (eg shooting or exposure to loud music) may complicate the picture.

Sources of noise

Noise is generated by a vibrating source which may be either a vibrating surface or vibration in a fluid flow. There are many ways in which this vibration can be created – elimination or reduction of vibration is an important noise control technique (see pp525–7). Some of the more common sources of noise are:

- percussive impacts
  - — components within a machine
  - — machine tools upon workpieces
  - — materials/products in conveying equipment
- gas movement
  - — jet engines
  - — rotating equipment eg fans or cutter blades
  - — engine exhausts
  - — compressed air or steam
- human voices
- animal noises
- musical instruments.

Noise measurement

Units of measurement

The human ear can hear sounds at frequencies between 20 and 20,000 cycles per second or Hertz (Hz). It is most sensitive to those in the range 3,000 to 6,000 Hz, the frequencies of human speech.

Noise levels are measured in decibels (dB) but a correction is usually made to allow for the human ear's differing sensitivities to different frequencies. This is known as the 'A weighting' with noise levels corrected in this way shown as dB(A). The range of noises runs from zero (total silence) up to 140 dB(A) for very noisy situations.

The decibel scale is a logarithmic one so a 10 dB(A) rise eg from 80 dB(A) to 90 dB(A) in fact represents a ten-fold increase in noise. Consequently apparently small increases or decreases in the decibel figure results in significant changes in the noise level – a 3 db(A) increase approximates to a doubling of the noise.

Noises of a single frequency (pure tone noise) are relatively rare, usually produced by tuning forks or test instruments. Most noises are a mixture of a wide range of frequencies. Measuring instruments are capable of measuring the totality of noise they are exposed to, usually with the A weighting correction. More sophisticated instruments (see pp522 and 523) can be set to select noise from particular ranges of frequencies – known as octave band analysis.

Typical noise levels

Table 22.1 gives an indication of typical noise levels that can be found in given situations.

Table 22.1 *Typical noise levels*

dB(A)	
0	Threshold of hearing
10	Quiet whisper, leaf rustling
20	Very quiet room
30	Subdued speech
40	Quiet office
50	Normal conversation
60	Busy office
70	Loud radio
80	Busy traffic
90	Heavy vehicle nearby
100	Pighouse at feeding time
110	Woodworking shop
120	Chainsaws
130	Rivetting
140	Jet engine nearby

Daily personal noise exposure

Hearing is damaged by the total sound dose it receives. This depends not just on the noise level but also on the duration of exposure. The Noise at Work Regulations 1989 (see pp519–21) are based upon the concept of 'daily personal noise exposure' (usually represented as LEP,d) over an eight hour working day.

Two action levels are introduced based upon the daily personal noise exposure.

First action level – LEP,d of 85 dB(A)

Second action level – LEP,d of 90 dB(A)

A further action level the 'peak action level' is set at a peak sound pressure of 200 pascals but this is only of relevance for exposure to infrequent but loud impact or explosive noises, eg from guns, cartridge-operated tools or explosives.

Calculation of daily personal noise exposure is considered in more detail later but this is where appreciation of the implications of the logarithmic nature of the decibel scale becomes important. For example, doubling the sound level will halve the exposure time required to produce a particular daily personal noise exposure. The following exposures will all produce a daily personal noise exposure of 90 dB(A) – assuming the rest of the working day is relatively quiet.

dB(A)		
90	8 hours	
93	4 hours	
96	2 hours	
99	1 hour	
109	6 minutes)	a 10 dB(A) rise produces a
119	36 seconds)	ten-fold increase in noise

Instruments

There is a wide range of noise measuring instruments commercially available with differing facilities and levels of accuracy. Generally, equipment conforming to relevant British or IEC standards should be used. Instruments can be categorised into three main types:

- sound level meters
- integrating sound level meters
- personal dosemeters

Each of these types is described more fully on pp522 and 523.

Training

The Noise at Work Regulations require noise assessments to be made by a competent person. The topic of competence is discussed in some detail in chapter 7. As in other cases, the degree of competence necessary must match the demands of the situation. Some noise assessments may only require a limited degree of competence, built up through experience and private study whilst more complex noise exposures or use of more sophisticated equipment may necessitate attendance at formal training courses. See HSE *Noise Guide No. 6* for information about training.[2]

The Noise at Work Regulations 1989

The detailed requirements of the regulations and HSE guidance on them are contained in HSE *Noise Guides No. 1 and No. 2.*[3] Much other useful guidance has been published by the HSE, some of which is referred to elsewhere in this chapter.

Assessment

An assessment is required if employees are likely to be exposed to noise at or above the first action level (85 dB(A)) or the peak action level. A rough guide is that if people have to shout or have difficulty being understood by someone two metres away then an assessment is likely to be necessary.

Reduction of risk of hearing damage

Regulation 6 requires the employer to 'reduce the risk of damage to the hearing of his employees from exposure to noise to the lowest level reasonably practicable'. This duty applies even if an assessment is not required, although the risk of hearing damage below 85 dB(A) is small. At higher levels it may mean the employer has to take precautions over and above those required specifically by other regulations.

Reduction of noise exposure

Where exposure is likely to be at or above the second action level (90 dB(A)) or the peak action level the employer must reduce the exposure of employees, so far as is reasonably practicable, *other than by the provision of personal ear protectors*. Therefore at these higher levels the first duty should always be to look for 'reasonably practicable' means of reducing the noise rather than simply issuing hearing protection.

Ear protection

Where employees are exposed to noise at or above the first action level (85 dB(A)) they must be supplied with ear protectors by their employer if they request it. Once exposure reaches the second action level (90 dB(A)) or the peak action level then the emphasis changes. The employer must provide exposed employees with suitable personal ear protectors. (Types of protectors are reviewed in more detail on pp527–9).

Ear protection zones

Where zones reach the second or peak action levels they must be marked using standard signs (see below). Hearing protection is then compulsory for everyone who enters these zones, even though they may also be there for a short time. Since employers have a legal duty to enforce these standards they may be well advised to avoid designating compulsory hearing protection zones unnecessarily and use different signs (eg 'Hearing Protection Recommended') for zones which are noisy but below 90 dB(A).

EAR PROTECTION ZONE
EAR PROTECTORS MUST
BE WORN

Maintenance and use of equipment

Employers have a duty to ensure that equipment provided by them under the regulations (both ear protection and noise reduction equipment) is fully and properly used, "so far as is practicable". However, ear protection provided for those exposed to between 85 and 90 dB(A) is excluded from this obligation. Equipment must also be "maintained in an efficient state, in efficient working order and in good repair".[4]

Where the second action level is reached, employees must fully and properly use ear protectors and any other protective measures provided under the regulations. They must also report defects in equipment.[5]

Provision of information to employees

Employees exposed at or above the first or peak action levels must be provided with adequate information, instruction and training. This should include:

- the risks to hearing from noise
- their likely noise exposure
- how to obtain ear protectors
- how to use ear protectors and noise control equipment
- how to identify and report defects
- employees' own duties.

Methods used will depend upon the circumstances but could include:

- signs indicating noisy equipment and areas
- oral instructions and explanations
- leaflets and posters
- videos, films etc
- short training sessions or counselling.

Duties of manufacturers etc

Regulation 12 extends the existing duties of manufacturers and others under s6 of HSWA to require them to supply adequate information on the noise likely to be generated by articles they supply for use at work, where this is likely to reach the first or peak action level. The HSE's *Noise Guide No. 2* contains extensive guidance on this requirement, whilst *Noise Guide No. 7* describes procedures for noise testing machinery.[6]

Noise assessment

Its purpose

The purpose of the noise assessment is described in regulation 4(1). It can be simply represented as providing answers to the questions:

- Is there a noise problem?
- Who is exposed and to what extent?
- What needs to be done to comply with the regulations?
  - noise exposure reduction
  - providing ear protection
  - ear protection zones
  - informing employees.

There is a risk of organisations making mistakes similar to some made under the COSHH Regulations (see chapter 23) ie going too far into the technicalities and losing sight of the fundamental questions. Some employers have spent considerable amounts of time or money (in consultants' services) on producing vast quantities of noise data without ever correlating it and identifying what needs to be done. The assessment must be carried out by a competent person (see p518).

Information needed

In order to arrive at the likely daily personal noise exposure levels of employees, information must be obtained about the noise levels and duration of exposure in their workplace. This is not a simple process since in most workplaces noise levels and exposures vary. Several questions need to be asked:

- What are the noisy machines, processes, activities, situations?
- What are the noise levels associated with each of these?
- Do the levels vary?
- How long is noise emitted at each level?
- How long are employees exposed to each level?
- What is their position in relation to the noise source?
- Do groups of employees receive similar exposure?

These questions need not be answered in precise detail, but enough information must be gathered to identify whether the daily exposure level of employees has reached 85 dB(A), 90dB(A) or (more rarely) the peak action level.

Use of instruments

Each of the three main types of instrument can be useful in the assessment process.

- *Sound level meters* These give an instant read out of noise levels in dB(A). Where the noise level fluctuates a fairly accurate estimate can be made if the response time is set to 'slow'. They can be used to measure relatively uniform noise levels or to measure noise levels which might vary but where the time of exposure to each level is fairly well known. A calculation of the overall exposure levels of employees could then be made. HSE *Noise Guide No. 3* gives detailed guidance on such calculations.[7]
- *Integrating sound level meters* These meters can integrate the noise levels in a particular area over a whole day or a sample period, from which the LEP,d can be easily calculated. The more sophisticated instruments can print out a noise profile over the sample period, helping to identify the noisier activities or occurrences.

  Such instruments may be extremely helpful in identifying compulsory and recommended hearing protection zones, particularly if the workforce is fairly static. However, the information they provide is only of limited value in assessing the exposures of individual employees, especially those who move around a lot.
- *Personal dosemeters* These are designed to be worn by workers to determine their personal noise exposure over a sample period. They are particularly useful where the worker has to move between quiet and noisy areas in the course of his job. Microphones need to be placed in the wearer's hearing zone but at least 4cm from the head to avoid the reflection of sound from the body.

  Dosemeters provide the most accurate measurement of personal noise exposures – but only for that employee and that day or sample period. The difficulty in interpreting dosemeter results is in deciding how representative the measurement is and how far it can be extrapolated to provide assessments for other employees, other days or other ranges of duties. Detailed records of activities carried out and their duration are particularly important in this respect.

Both integrating sound level meters and personal dosemeters are prone to being sabotaged by exposure to abnormal noise and unexpected results from unattended instruments should be treated with caution.

Instruments are normally provided with a simple acoustic calibration device and checking should take place each time the meter is used. However, full laboratory calibration should also be carried out periodically. More detailed information on instruments is provided in the HSE's *Noise Guide No. 3*.[8]

Finalising the assessment

Once all the necessary information has been gathered and calculations made, the assessor should now be in a position to identify what needs to be done.

- *Noise exposure reduction* Where are the high levels of noise and what reasonably practicable means are available for their reduction, other than by using ear protection?
- *Ear protection* Which employees or groups of employees must ear protection be supplied to and where must its use be enforced? Who must ear protection be made available to on request (and how far does the employer want to go in encouraging its use)?
- *Ear protection zones* Where must compulsory ear protection zones be established? What signs are required and where should they be located? Who will enforce compliance within the zones? How will others who need to enter these areas obtain ear protection? Should the employer use signs to indicate areas where hearing protection is recommended but not compulsory?
- *Informing employees* Should individually noisy pieces of equipment be marked with signs or stickers? What general action needs to be taken to educate employees about the risks of noise and their exposure to it? Are there any specific noise-related training needs to address?

Assessment records

Regulation 5 requires an adequate record of the assessment to be kept until a new assessment is made. The assessment should record the workplaces, areas or jobs assessed, the date of assessment and the results. The record need not follow any prescribed form but might contain:

- noise exposure tables (identified by person, area, machine or activity)
- a plan showing noise exposures at various locations
- details of exposure times

R K JACKSON AND SON LIMITED
Northtown Works
NOISE ASSESSMENT REPORT

Dates of assessment: 7 and 8 September 1994 **Assessed by:** I Herd
Instrument(s) used: Smiths Sound Level Meter Model K42
Jones Personal Dosemeter Type C5

NOISE SURVEY RESULTS

Location	Noise Level dB(A) Average or Range	Typical Daily Exposure (hours)	Assessed Action Level			Comments
			0	1	2	
MACHINE SHOP						
Bandsaw	99	2			✔	Measurements showed no
Large grinder	94	1		✔		significant effects from
Power press	89	4		✔		machines on neighbouring
Milling machine	86	2	✔			operating or assembly positions.
Drilling machine	83	6	✔			There is little movement of
General background	75–80	8	✔			operators between machines.
PLASTIC SHOP						
Bench saw	99	2			✔	
Cross cut saw	97	2			✔	
Portable jig saw	101	1			✔	
Assembly bench						Dosemeter measured daily
– saw operating	92–95	3–5			✔	exposure at 90.7 dB(A)
– assembly only	70–80	8	✔			

Recommendations

1) NOISE EXPOSURE REDUCTION

Means of reducing noise emission should be investigated for
- the bandsaw in the Machine Shop
- the bench saw, cross cut saw and portable jig saw in the Plastic Shop

(Separate proposals will be prepared on this subject)

2) EAR PROTECTION
   - 2.1 Continue to provide ear muffs and ear plugs to all those who request them
   - 2.2 Ear muffs or ear plugs must be provided to
     - the operator of the bandsaw in the Machine Shop
     - all those working at any time in the Plastic Shop
   - 2.3 Use of ear protection must become compulsory for these activities

3) EAR PROTECTION ZONES
   - 3.1 Mark the bandsaw in the Machine Shop with a standard sign stating 'Ear protectors must be worn when operating this machine'
   - 3.2 Mark the entrance to the Plastic Shop with a standard sign stating 'Ear Protection Zone Ear protectors must be worn when any saw is operating'

4) INFORMING EMPLOYEES
   - 4.1 Mark the large grinder, power press and milling machine in the Machine Shop with signs stating 'Hearing protection recommended when operating this machine'
   - 4.2 Show all employees in the Machine Shop and Plastic Shop the video on noise (available from the Training Department)
   - 4.3 Brief the bandsaw operator, all Plastic Shop employees and their supervisors on the contents of this report, particularly stressing the new compulsory ear protection zone

Figure 22.1 *Noise assessment report*

- information on activities of peripatetic workers (including noise exposure levels and times of exposure)
- details of what action is required.

The example of a record form contained in HSE *Noise Guide No. 3* does not place sufficient emphasis on what needs to be done.[9] The completed sample record in figure 22.1 is more detailed in that respect, but it is intended as an illustrative extract rather than a complete report.

Assessment review

Like other types of assessment, noise assessments may need to be reviewed, particularly where changes have taken place. Circumstances justifying a review might include:

- installation or removal of equipment
- significant changes in workload or work pattern
- significant changes in machine speeds or materials used
- changes in building structure or machine layout
- significant alterations in employees' duties.

Periodic reviews should take place anyway to detect the cumulative effects of minor changes or gradual wear and tear. HSE guidance recommends two yearly checks for most kinds of machines. These may just involve spot checks rather than detailed re-assessments.

Noise reduction measures

There are various ways of reducing the exposure of employees to noise – through the specification of quiet machines or processes, by the reduction of noise generation or transmission and through the provision of ear protection. Regulation 7 requires that where the second or peak action level is reached then reduction of noise exposure should be by means other than ear protection. Specialist courses are available in noise control engineering and there are many publications on the subject, including some from the HSE.[10]

Specification

It may be possible to specify a less noisy type of process such as using welding instead of rivets or hydraulic piledriving instead of the use of impact methods, but in most cases specification will centre around the purchase of machinery. There may be a choice between products stated to be generating different levels of noise or there may be a need to specify performance standards. Here it is important to prepare the specification carefully, and in both instances the performance must actually be checked when the equipment is operating normally. One company is known to have specified noise standards for a large bandsaw being purpose built for them abroad. When the machine arrived it was found to meet the standard when it was running free, but not when it was cutting.

Regulation 12 requires that adequate information is supplied about likely noise generation – HSE Noise Guide No. 2 gives advice on this.[11] Procedures for noise testing machinery are covered in HSE Noise Guide No. 7.[12]

Reducing noise generation

Noise is generated by either a vibrating surface or vibration in a fluid, therefore if this vibration can be eliminated or reduced, then so will the noise. As well as considering the primary source of the vibration, account must be taken of other parts which may vibrate in sympathy (secondary vibration). Possible reduction measures include:

- *Vibrating surfaces*
  - — avoiding impacts or cushioning them (eg with plastic or nylon surfaces)
  - — replacing metal gears by nylon/polyurethane or by belts
  - — using isolating or anti-vibration mountings
  - — separating large vibrating surfaces from moving parts
  - — stiffening structural parts
  - — placing machines on vibration absorbing pads
  - — using damping materials on metal surfaces
  - — using mesh in place of sheet metal
  - — placing absorbent gaskets around doors and lids
  - — replacing rigid pipework with flexible materials
- *vibration in gases*
  - — choosing centrifugal rather than propeller fans
  - — using large diameter, low speed fans
  - — using large diameter, low pressure ductwork
  - — streamlining ductwork to avoid turbulence (this will also improve its efficiency)
  - — using effective silencers to reduce turbulence at exhausts
  - — using low-noise air nozzles, pneumatic ejectors etc at the minimum pressure necessary
- *good maintenance*
  - — replacing worn or badly fitting parts
  - — securing loose parts
  - — balancing rotating and other moving parts correctly
  - — providing good lubrication.

Attention should also be given to working practices by ensuring that noisy equipment is switched off when it is not in use.

Reducing noise transmission

Some of the measures already described will reduce the transmission of noise through equipment but the reduction of transmission through air is an important noise reduction technique.

- *Acoustic enclosures* Noisy equipment, both large and small, can be placed in acoustic enclosures. Important design considerations are:
  - — inner surfaces of the enclosure covered in sound-absorbent material
  - — openings in the enclosure minimised
  - — use of absorbent gaskets around doors, windows, service inlets etc
  - — avoiding the enclosure being in contact with vibrating parts.

  In some cases, particularly where the noise source is large, it may be preferable to provide an acoustic enclosure for the workers – either as a control booth or a noise refuge. The comfort of workers using such enclosures must be taken into account, especially in relation to ventilation, temperature and seating. Where the enclosure is used as a control booth the need for employees to go outside into the noisy environment should be minimised by situating all relevant controls within the booth and providing adequate visibility over the equipment being controlled.
- *Use of sound-absorbent materials* Noise transmission may also be reduced by the use of sound-absorbent materials on the walls and ceilings of large rooms to prevent noise reflection. These will be most effective if situated close to the noise source although there may also be considerable benefit from situating them close to operating positions. Use of portable absorbent screens may also be beneficial, particularly in protecting maintenance employees moving around in noisy areas.
- *Separation measures* Separation through distance can also be used to protect employees by situating them as far away from noise sources as possible or even in separate rooms. Noisy exhausts should always be directed away from workers and can be discharged further away by the use of flexible hose.

Ear protection

Choosing ear protection

There are several main types and many manufacturers of hearing protection equipment. In deciding which one to choose there are several factors to take into account but the most important of these is selecting a model which will actually be worn in practice. If the hearing protection is only going to be worn for 50% of the exposure period then the maximum reduction in total noise exposure that can possibly be achieved is 3 dB(A). Consequently comfort, convenience of use and compatibility with other PPE (eg helmets or eye protection) are extremely important criteria.

Levels of protection provided

Ear protectors do not reduce the noise equally across all the frequencies – generally they are more effective against the more important higher audible frequencies. The overall protection provided can be determined through a detailed calculation, using octave band analysis, of the noise exposure and attenuation data supplied by the manufacturer. This is described fully in HSE *Noise Guide No. 5*.[13]

Research work has shown that in many cases the actual attenuation achieved by ear protectors in practice can be considerably less than the figures specified by manufacturers. There may be many reasons for this – poor fitting, deterioration of equipment with age, unauthorised modifications or use with long hair, spectacles or jewellery. These should be reduced or overcome with good programmes of training, maintenance and replacement.[14]

Unfortunately no-one has devised a simple but accurate system of calculating attenuation figures. Most commercially available types will, if fitted reasonable well, provide attenuation of at least 15 dB(A) in most situations.

Cost factors

Another factor to take into account is the cost of the ear protection and here it is easy to be misled by the unit cost. Disposable plugs may appear a cheap option but they are likely to be used in large numbers compared to the reusable types of protection which can be expected to last much longer. However, whatever the costs, the vital factor is making sure the protection is worn – if it isn't then the money spent will have been wasted.

Types of hearing protection

There are three main types of hearing protection:

- *Ear muffs* (also called ear defenders) These are plastic cups which fit over the ears and are sealed against the head with cushions containing foam or a viscous liquid. The interiors of the cups are filled with soft plastic foam or some other sound-absorbent material. The cups are held in position by pressure bands, usually passing over the head. This makes their use with safety helmets difficult although the band can be positioned behind the neck or under the chin. However, some models are designed to be attached directly to the helmets and swing out of the way when not needed.

  Their efficiency is mainly dependent upon the tension of the headband and the condition and fitting of the seals. Headband tension may deteriorate with age or misuse. Use with spectacles, long hair or jewellery will diminish their effectiveness.
- *Ear plugs* Plugs fit directly into the ear canal. Some are intended to be reusable whilst others may be disposed of after a single use. They may be

attached to cords to prevent loss (especially important in food handling activities). Reusable plugs are usually made of plastic or rubber materials whilst disposable plugs consist of foam plastic or down material in a plastic membrane.

Use of plugs is ill-advised for anyone suffering from ear infections or irritations and some employees simply find the presence of anything within their ear unbearable. Hygiene is an important consideration as materials from the hands can easily be transmitted into the ear. Reusable plugs should be washed before they are used again.

Disposable plugs can generally be fitted with a minimum of training and will fit most sizes of ear. Most reusable plugs come in a variety of sizes. In some cases people need different sizes for each ear.

- *Semi-inserts* These consist of plastic or rubber caps which are held against the entrance to the ear canal by a headband. They combine some of the advantages and disadvantages of the other two types. It is particularly easy to slip them off and carry them round the neck on the headband. The headband may deteriorate with time and the caps need to be kept clean.

VIBRATION

Just as vibration can cause injury to the ears – through noise – so it can also damage other parts of the body. There are two main problems associated with vibration – hand-arm vibration syndrome (including vibration white finger) and whole body vibration. A proposed EC Directive on 'Physical Agents' will require action to be taken to deal with both types of problem.

Hand–Arm Vibration Syndrome (HAVS)

HAVS symptoms

The term 'Hand–Arm Vibration Syndrome' describes a group of symptoms and disorders in the fingers, hands and arms. These include effects on blood circulation, nerves, muscles and bones and are often very painful.

The most common condition in the group is Vibration White Finger (VWF) where the blood circulation in the fingers is affected, resulting in whitening of the fingers, usually accompanied by tingling and numbness. After attacks there is a painful throbbing return of blood flow and the colour of the fingers changes to red. Usually the effects start in the finger tips progressing gradually down the fingers. In severe cases the fingers may become blue-black and possibly even become gangrenous. Sensitivity in the fingertips may be lost, making delicate work difficult.

Many employees have brought successful damages claims against their employers for VWF[15] and it is a prescribed occupational disease under Social Security legislation as is another HAVS condition – carpal tunnel syndrome – which affects the wrist.

Sources of HAVS

HAVS usually arises from use of powered tools but may also be caused by holding a workpiece against a moving surface eg a grinding wheel. It is generally found that percussive action tools are worse than rotary tools. Use of chipping hammers, impact hammers, heavy duty hammer drills, portable grinders, chain saws etc are all potential causes of HAVS. The risk of conditions developing will be dependent upon factors such as

- duration of exposure
- frequency of rests
- energy level of the tool
- vibration frequency.

Possible precautions

There are a number of precautions available to reduce the risk of HAVS developing amongst employees. These include:

- modifying the process to eliminate or reduce the use of hand-held tools
- use lighter, higher speed tools to reduce low frequency vibration
- provide anti-vibration mountings
- use sleeves or grips of anti-vibration material eg sorbothane
- sharpen tools, dress grinding wheels to reduce the need for pressure
- maintain equipment in good condition
- use gloves to keep hands warm (however at low frequencies gloves may increase the transmission of vibration)
- reduce individual exposure by work sharing
- allow periodic work breaks
- advise exercise of fingers to improve blood flow
- carry out health surveillance to detect the early onset
- make employees aware of the risks and precautions.

Detailed guidance from the HSE on the subject is expected to be published shortly.

Whole Body Vibration (WBV)

WBV symptoms

This particularly affects vehicle drivers who can experience lower back pain or spinal damage. Other effects such as digestive disorders, loss of balance, loss of concentration and blurred vision have also been put forward. It is thought that the back and spinal effects are a combined effect of the increased wear and tear on the system induced by vibration and insufficient blood supply due to the vibration or prolonged spells in a sitting position.

Sources of WBV

The most severe intensities of vibration are usually found on wheeled, off-road vehicles during movement around sites. Dumpers, loaders, scrapers and site lift trucks have been measured as producing the highest vibration values in the driving position. Caterpillar tracked vehicles were generally lower. Whilst road tractor units and lorries were generally lower in vibration than off-road vehicles, the times spent driving such vehicles were usually much longer. Factors which must be taken into account in considering the extent of the problem are:

- the intensity of vibration in the driving position
- the duration of exposure
- seat design and availability of adjustment
- possible uncomfortable body position, due to the demands of the work
- the frequency of breaks or position changes.

Possible precautions

Three main principles of protection against vibration in vehicles were outlined in an informative review article in the *Health and Safety Practitioner.*[16] Some of this material may well form the basis for future HSE guidance. These principles were:

- reducing vibration at source, for example
  - Use of even surfaces (as far as reasonably practicable)
  - Speed related to surface conditions
- decreasing the transmission of vibration, for example
  - Seats mounted firmly on a rigid surface
  - Use of body or cab suspension devices
  - Appropriate choice of tyres
  - Provision of chassis suspension
  - Use of suspension cabs
  - Use of suspension seats – with adequate movement, adjustment and space
- improvement of the driver's position.

RADIATION

The intention here is to deal with all aspects of radiation, travelling through the electromagnetic spectrum and including particulate radiation. Some types of radiation will be included in the proposed EC Directive on Physical Agents.

Radio frequency and microwaves

Potential effects

The potential dangers from radiofrequency (RF) and microwave radiation are mainly those of deep burning to human tissues. Microwaves can produce the same deep heating effect in live tissue that they can produce in cooking. RF is used in a similar way to produce localised heating effects.

Possible sources

As well as microwave cookers these frequencies are used in drying and heating equipment and for some specialised welding operations. RF will also be present in various types of communication equipment.

Precautions

Microwave cookers are completely enclosed with interlocks ensuring the power is shut off if the door is opened. As far as possible other sources of RF or microwave radiation should be protected in the same way or at the very least provided with an adequate degree of screening. Regular checks should be made on the effectiveness of interlocks and the condition of door seals. It may be necessary to place signs on equipment if it is not immediately obvious as a source of radiation.

Infra-red radiation

Potential effects

Infra-red (IR) radiation can cause cataracts to the eyes and can redden or even burn the skin. IR is also a source of heat, therefore there is a danger of heat exhaustion arising due to long-term exposure.

Possible sources

Anything that glows is likely to be a source of IR, for example:

- furnaces or fires
- molten materials (eg metal or glass)
- burning or welding activities
- heat lamps
- some types of laser
- the sun.

Precautions

Eye protection with suitable filters is essential to protect the eyes from concentrated IR sources such as lasers or when carrying out burning or welding work. Protective clothing or, where possible, screens should be provided to protect the skin of employees who may be exposed to IR for extended periods. It may also be necessary to take steps to prevent heat stress, such as providing a cooling air-flow or ensuring the regular intake of fluids.

Visible radiation (including lasers)

Potential effects

Light in the visible frequency range can cause damage if it is present in a sufficiently intense form. The eyes are particularly likely to be affected but skin tissue may also be damaged. Indirect danger could also be created by employees being temporarily dazzled.

Possible sources

Any high intensity source of visible light could cause a problem. Lasers (which may also contain other radiation frequencies) are an obvious danger but so also are light beams, powerful light bulbs and the sun. In all cases danger may be due to direct or reflected radiation.

Precautions

Eye contact with any bright source of visible light should be avoided. Special precautions are likely to be necessary for the use of lasers (some of which may not be visible at all). These might include:

- effective screening to prevent direct viewing
- eliminating the possibility of reflection
- using interlocked enclosures
- controlling entry to areas by permit to work procedures
- providing warnings through standard signs.

The HSE have produced a Guidance Note PM19 *The Use of Lasers for Display Purposes*.[17]

Ultra violet radiation

Potential effects

Exposure of the eye to ultra violet (UV) radiation can produce conjunctivitis and a condition known as arc eye which feels as though sand or grit is present in the eye. UV may cause sunburn to the skin and is also a potential cause of skin cancer. Exposure to some chemicals (eg pitch) can predispose the skin to the effects of UV radiation. Ozone may also be produced by UV radiation, possibly in toxic concentrations.

Possible sources

There are many possible sources of UV to which people may be exposed at work:

- welding or other electrical arcs
- mercury vapour lamps
- tungsten halogen lamps
- crack detection equipment
- insect killers
- forgery detection devices
- some types of laser
- the sun
- sunbeds and sunlamps
- UV tanning and curing equipment.

Precautions

The normal use of some UV devices such as insect killers or forgery detection equipment usually involves low intensities or short duration of exposure. However, extended work or that involving higher levels of exposure may necessitate specific precautions. Enclosure or screening of the source provides the best protection but the use of PPE (clothing or eye protection with appropriate filters) or UV protection cream may be more practicable in some cases. For electric arc welding full face protection is necessary, involving a visor or hand-held screen with the correct filter, and precautions against passers-by being affected are often necessary.

UV devices should always be used with the correct type of lamp specified by the manufacturer and with the recommended filters always kept in place. HSE have produced guidance on use of such devices in tanning and in curing inks, varnishes and lacquers.[18]

Ionising radiation

Potential effects

There are a number of different types of ionising radiation – x-rays, gamma rays and particulate radiation (alpha and beta particles, protons and neutrons), each with their different penetrative power and differing effects on the body.

Effects can include nausea, vomiting and diarrhoea; burns, dermatitis and skin ulcerations; damage to cells and changes in the blood; cataracts; cancers of varying types and genetic effects.

Possible sources

Radiation can be produced by radioactive substances either present as a sealed source (where it is bonded within an inactive material or encapsulated within an inactive receptacle) or in its free form, often referred as an unsealed source. It may also be produced in radiation generators such as x-ray machines. Here the radiation can be turned on and off whereas radioactive substances provide a continuous source of radiation. Examples of where radiation may be found in the workplace include:

- Sealed sources:
  - radiography work
  - various gauges – thickness, level, density
  - static eliminators
  - smoke detectors
- Unsealed sources:
  - radioactive tracer work
  - luminous paints
  - radon gas (naturally occurring)
- Radiation generators:
  - x-ray machines
  - some laboratory analytical equipment
  - radar
  - some high voltage equipment

Precautions

The main piece of legislation affecting ionising radiation work is the Ionising Radiation Regulations 1985 which apply even to work with naturally occurring radon gas. The HSE have published several Approved Codes of Practice relating to the Regulations and many other sources of guidance are available.[19] Key features of the regulations include:

- advanced notification of certain types of radiation work
- restriction of exposure (as far as reasonably practicable) by
  - shielding, ventilation, containment
  - minimisation of contamination
  - safety features and warning devices
  - safe systems of work
  - adequate and suitable PPE
- controlled and supervised areas may need to be designated
- radiation protection advisers and qualified persons may need to be appointed
- local rules must be set down in writing
- radiation protection supervisors may need to be appointed
- employees must receive information, instruction and training

- radioactive substances must be controlled in respect of accounting for them and their storage, transport and moving
- hazards must be assessed and countermeasures introduced.

There are many other detailed requirements, particularly where controlled or supervised areas or classified persons have been designated.

Permission to store and dispose of radioactive substances is necessary from HM Inspectorate of Pollution (or their equivalents in Scotland and Wales).

Notes

1. *Berry v Stone Manganese Marine* (1971) 12 KIR 13
2. *Noise at Work – Noise assessment, information and control Noise Guides 3 to 8* – HSE 1990 (HS(G)56)
3. *Noise at Work – Noise Guide No 1: Legal duties of employers to prevent damage to hearing; Noise Guide No 2: Legal duties of designers, manufacturers, importers and suppliers to prevent damage to hearing. The Noise at Work Regulations 1989 – Guidance on Regulations* – HSE 1989 (formerly reference L3)
4. The Noise at Work Regulations, SI 1989/1790, regulation 10(1)(b)
5. *Ibid.*, regulation 10(2)
6. See notes 2 and 3, above
7. See note 2, above
8. See note 2, above
9. *Noise Guide No 3*, page 9 – see note 2, above
10. *Noise Guide No 4* – see note 2, above
    Noise from pneumatic systems – HSE 1985 (Guidance Note PM56)
11. See note 3, above
12. See note 2, above
13. See note 2, above
14. First-aid for hearing protection – *Health and Safety at Work* magazine, April 1994
15. See page 76: *Bourman v Harland and Wolff* – VWF is a foreseeable risk from 1 January 1978
16. Vibration in industrial vehicles and earth-moving machines – *Health and Safety Practitioner* magazine, May 1993
17. *Use of lasers for display purposes* – HSE 1980 (Guidance Note PM19)
18. *Commercial ultra violet tanning equipment* – HSE 1982 (Guidance Note GS18),
    Safety in the use of inks, varnishes and lacquers cured by ultraviolet light or electron beam techniques – HSE 1993
19. Some of the more important HSE publications are:
    Protection of persons against ionising radiations arising from any work activity: The Ionising Radiations Regulations 1985: Approved Code of Practice 1985 – HSE 1985 (COP16)

Exposure to radon: The Ionising Radiations Regulations 1985: Approved Code of Practice 1985 – HSE 1985 (COP23)
Dose limitation – restriction of exposure, Approved Code of Practice – HSE 1991 (L7)
Protection of outside workers against ionising radiations – HSE 1993 (L49)
A framework for the restriction of occupational exposure to ionising radiations – HSE 1992 (HS(G)91)

23

Hazardous Substances

HOW HAZARDOUS SUBSTANCES HARM THE BODY

The form of the substance

Substances, whether hazardous or not, can exist in several different forms and sometimes simultaneously – for example wherever liquids are present there will also be some vapour from the liquid. In considering the hazards associated with a substance all the forms in which it may be present must be taken into account.

Solid

Materials in their solid form rarely present problems, although a few may possibly be hazardous due to skin contact or ingestion.

Dust

Finely divided solid particles present rather more hazards to the body. The risk of inhalation must also be considered as well as the risks of skin contact or ingestion Particles between 0.2 and 5 microns (a micron is 10^{-6} metre) in size present the greatest dangers – particles larger than 5 microns are filtered out in the respiratory tract (as demonstrated by black spit) and those less than 0.2 microns do not settle out of the air in the respiratory system and are exhaled.

Particles in this 0.2 to 5 microns size range are invisible to the naked eye, so it is the dust that is normally invisible that causes the problems. However, visible dust will always be accompanied by smaller invisible particles and the proportion of invisible dust will increase with more processing, or more trampling on the floor.

Fumes

The term fume is used to describe solid particles which have condensed in the air from their gaseous state eg oxide particles which have formed from metals volatilised during welding. Fume particles are very small (usually less than a micron) and the main risk is of their inhalation.

Smoke

Smoke describes particles resulting from the incomplete combustion of materials and will often contain liquid droplets as well as dry particles. Again the main risk is from inhalation.

Liquid

Substances in their liquid form present risks of skin contact and also of ingestion. However, liquids will always be accompanied by a certain proportion of vapour and possibly also by mists or aerosols.

Vapour

Vapour is the gaseous form of a liquid, which presents risks through inhalation, although in certain cases contact with vapour may also affect the skin.

Mists and aerosols

These terms are usually used to describe the suspension of small droplets of liquid in the air. Mists can be generated in a variety of ways – chemical processes, use of cutting oils or spraying operations. Whilst the size of mist droplets is much greater than vapour molecules the risk is similar – mainly inhalation with a possible risk of skin contact.

Gas

Some substances exist completely in their gaseous form at normal temperatures. These present risks mainly through inhalation.

Entry routes

Hazardous substances can enter the body and cause harm by three different routes – inhalation, ingestion and the skin.

Inhalation

Inhalation of a hazardous substance may cause damage to the lungs or some other part of the respiratory system. However, inhalation is also a route by which the substance may enter the bloodstream and thus affect any part of the body's system or organs.

Ingestion

Although nobody would be expected to voluntarily eat or drink a hazardous substance, the possibility of inadvertent or accidental ingestion is always present. The consumption of food and drink in contaminated areas or with contaminated hands presents an obvious risk and smoking provides an excellent opportunity for contaminated fingers to come into contact with the lips.

The storage of hazardous liquids in unlabelled containers or, worse still, drinks containers is particularly dangerous. This has been the cause of many accidental poisonings through ingestion, some of which have been fatal.

Skin (and eyes)

Many substances present a direct threat to the skin and eyes through their corrosive or irritant effects. However a number of substances, particularly organic solvents, can enter the body by absorption through the skin. Once into the bloodstream or body tissue then either the solvent or materials dissolved in it can reach a vulnerable organ. Entry through cuts or cracks in the skin must also be considered.

The harm caused

The detailed review of the types of harm caused to the body by hazardous substances and the manner in which this harm is caused, is outside the scope of this book. However, some of the more common problems are outlined here. It is particularly important that those with management responsibilities recognise that a wheeze or a rash may be caused by substances used in the workplace rather than by some outside factor.

Respiratory problems

- *Pneumoconiosis* This is a general term used to describe irritation of the lung or impairment of the lung processes caused by the inhalation of dust. The nature of the symptoms, their extent and the possibility of recovery will vary from dust to dust. Some of the main types of pneumoconiosis are:
  - **Asbestosis** – see chapter 24
  - **Silicosis** – a form of pneumoconiosis due to exposure to free silica dust in a variety of industries – foundries, quarrying, potteries. As with asbestosis, silicosis results in shortness of breath through the effects of the silica particles on the lung tissue. Silicates (apart from asbestos and some talcs) do not cause the same problems as free silica.
  - **Byssinosis** – this is caused by exposure to cotton dust, sometimes leading to bronchitis and emphysema. It is not normally progressive once exposure ceases.
  - **Siderosis** – is caused by inhalation of iron oxide fume (eg in welding). It also is not considered to be progressive.

- *Respiratory irritation* Irritation of the nose, throat and lungs can be caused by a variety of substances. In some cases this will be due to the chemical effects of inhaling acid or alkali mists or gases such as chlorine or ammonia. Dusts may also cause irritation, particularly chromate dusts which may result in ulceration of the nasal passages. Inhalation of high concentrations of acid or alkaline gases (eg chlorine or ammonia) could produce very serious effects, including severe bronchial spasm resulting in death.

- *Asthma* Occupational asthma may be caused by inhalation of a variety of irritants or sensitising agents. What may cause an asthmatic condition for one worker may give no problems to another. Amongst the more likely causes are:
  - isocyanates
  - various hardening or curing agents
  - dyestuffs
  - platinum, nickel, chromium, or cobalt compounds
  - some hairdressing materials
  - wood dusts
  - grain dusts, flour, tea and coffee dusts
  - fish or crustacean products
  - mites and fungal spores
  - animal urine and droppings.

The symptoms of chest tightness and wheezing may occur at work but may also develop in the evening or at night, making detection difficult. Early removal of the employee from the suspected causative agent is generally the best approach. Those who have suffered from asthma previously (eg childhood hay fever) are more likely to be affected by occupational exposure to asthma causing substances. Smokers are also generally more susceptible.

- *Respiratory cancers* A variety of substances are known to cause cancer (carcinoma) within the respiratory system. Amongst these are:
  - pitch and tar fumes – carcinoma of the lung
  - wood dust (especially hard woods) – carcinoma of the nasal cavity or sinuses
  - nickel compounds – carcinoma at various sites
  - various organic solvents and other organic chemicals
  - asbestos (see Chapter 24)

- *Metal fume fever* This is a condition resulting usually from the inhalation of zinc or zinc oxide fumes. It is particularly likely to occur during the burning or welding of galvanised or zinc-primed metal. The symptoms are very similar to a bad attack of influenza but are usually over within 24 hours. Magnesium, copper and other metals may also produce similar effects.

Poisoning

There are many hazardous substances capable of poisoning many parts of the body in many different ways. It is only possible to list a selection of them here, separated into chemical types. The chronic and acute effects of the same substance may be different.

- *Acute* – resulting from short-term exposure to relatively large quantities of the toxic material, possibly resulting in death or serious injury.
- *Chronic* – resulting from smaller doses over a long period. The quantities exceed the body's ability to remove or detoxify the substances resulting in long term, often irreversible damage.

- *Metals and metal compounds*
  Lead – see Chapter 24
  Manganese – neurological effects, similar to Parkinson's Disease
  Mercury – tremors, lethargy, insomnia and even personality changes
  Beryllium – anaemia and other metabolic effects
  Cadmium – lung and kidney damage

- *Organic chemicals* Many organic chemicals, including most organic solvents, can produce some sort of detrimental effect within the human body. Typically organic solvents will attack the central nervous system giving a narcotic or anaesthetic effect. Solvent abuse is not just a modern phenonenon – there has been abuse and some tragic accidents involving industrial solvents over many years. Other effects vary from chemical to chemical with a wide range of organs and tissues prone to attack. The liver and kidneys together with the gastrointestinal system or even the blood itself are common targets. Some organic chemicals are also carcinogenic.

- *Inorganic chemicals* As with organic chemicals a wide variety of effects can be produced by inorganic chemicals. The poisonous effects of arsenic are well established with the nervous and digestive systems the targets in severe cases.

  Some chemicals exert an asphyxiant effect via the lungs, typically carbon monoxide which excludes oxygen from the haemoglobin in the bloodstream. Both hydrogen sulphide and hydrogen cyanide can also poison via the respiratory system to give a type of chemical asphyxiation – a characteristic in all three cases being the rapidity of the effect.

  Other inorganic gases such as carbon dioxide or nitrogen could also be the cause of asphyxiation, but generally through their exclusion of oxygen from the atmosphere rather than their toxic properties, although carbon dioxide is weakly narcotic at very high concentrations.

Skin effects

Work-related skin problems are much more common than is often realised.

Some skin conditions arise because of mechanical factors such as friction or physical factors such as heat or cold but many are due to contact with chemical substances.

- *Dermatitis – primary irritants* Many substances have a direct chemical effect on the skin, similar in some ways to their effects on the lungs. Strong acids and alkalis will cause direct damage to skin tissue, whilst exposure to dilute acids or alkalis or other substances such as organic solvents, soaps and detergents will remove oils and other natural constituents from the skin.

- *Dermatitis – sensitising agents* Other materials will produce an allergic reaction from the skin – something of a parallel to asthmatic effects within the lungs. The skin gradually becomes sensitised over a period of weeks or even months and eventually exposure even to a small quantity of the substance will produce a severe reaction.

  Amongst common sensitising agents are isocyanates, organic resins, coal tar products, chlorinated organic solvents and cutting oils, together with substances encountered in the food processing industry (although here mechanical, physical or biological factors must also be taken into account). As is sometimes the case with asthma, a substance which may cause a sensitised dermatitic condition in one employee will not necessarily produce any reaction in others.

- *Other skin conditions* Several substances are well established as potential causes of skin cancer. Amongst these are pitch, tar, soot, mineral oils and arsenic. The HSE leaflets depicting cancer of the scrotum, eyelid and hand as a result of contact with pitch, tar and mineral oils provide graphic displays of the effects[1]. Pitch also sensitises the skin to ultra violet radiation from sunlight producing a type of sunburn known as pitch burns.

Biological problems

Many occupational health problems are caused by contact with animals, birds or fish (including their carcases or products) or with micro-organisms from other sources. These include:

- *Anthrax* – this potentially fatal but now uncommon disease occurs through contact with infected cattle, although the long lived anthrax spores may be present in other sources.[2]

- *Brucellosis* – producing feverish symptoms from which recovery usually takes place within a couple of weeks, can be contracted from cattle, pigs or goats.

- *Allergic Alveolitis* – 'Farmer's Lung' from contact with mould or fungal spores present in grain etc.

- *Viral Hepatitis* – producing various symptoms including jaundice, usually from contact with blood or blood products.
- *Legionnaire's Disease* – a type of pneumonia caused by inhaling airborne droplets containing legionella bacteria. Possible sources include cooling towers, air conditioning units and hot or warm water systems.[3]
- *Leptospirosis* – 'Weil's Disease'. This is often associated with contact with rats' urine but may also be contracted from other small or large mammals.[4]

THE COSHH REGULATIONS

The Control of Substances Hazardous to Health Regulations 1994 (or COSHH Regulations as they are universally known) were first introduced in 1988 but have been amended several times, culminating in the 1994 version of the regulations. They are the main piece of legislation affecting the use of hazardous substances although separate regulations deal specifically with asbestos and lead (see chapter 24).

Substances hazardous to health

The full definition of 'substances hazardous to health' is contained in Regulation 2(1) and includes

- Substances specified as very toxic, toxic, harmful, corrosive or irritant in the approved list as dangerous for supply. (The regulations governing this list have since been replaced by the Chemicals (Hazard Information and Packaging for Supply) Regulations 1994 – or CHIP2 as they are known.)
- Substances for which a Maximum Exposure Limit (MEL) is specified in Schedule 1 to the Regulations or for which the HSC has approved an Occupational Exposure Standard (OES). (These terms are explained later in the chapter. Lists of MELs and OESs are published annually[5].)
- Biological agents
  (These are separately defined as "any micro-organism, cell culture, or human endoparasite, including any which have been genetically modified, which may cause any infection, allergy, toxicity or otherwise create a risk to human health".)
- Dust of any kind, when present at a substantial concentration in air.
- Any other substance creating comparable hazards.

COSHH applies to substances that have chronic or delayed effects such as carcinogens, mutagens and teratogens and given, the widely drawn nature of the definition above, it is prudent to treat the Regulations as applying whenever there are cases of doubt.

Duties under COSHH

Employers, including the self-employed, have detailed duties to their employees under the COSHH Regulations with the self-employed also having duties as if they were both an employer and employee. Regulation 3(1) extends these duties "so far as reasonably practicable" to "any other person, whether at work or not, who may be affected by the work carried on". This brings a variety of other people into consideration particularly in respect of the carrying out of assessments and the provision of adequate control measures. Those to consider are:

- contractors
- visitors
- joint occupants
- neighbours and passers-by
- members of the public
  - in buildings accessible to them
  - in public areas.

Precautions may need to be tighter in public buildings and spaces since the public cannot be expected to behave in as responsible a manner as employees, contractors or 'controlled' visitors. One prosecution under the COSHH Regulations resulted from an incident in which a child visiting a doctor's surgery left the control of his mother and drank a bottle of carbolic acid stored in an unlocked cupboard.

The COSHH regulations summarised

The main requirements of the regulations are summarised below, each being examined in greater detail later in the chapter.

(i) *Assessment* (Regulation 6)
A formal assessment of the health risks created by hazardous substances, based on the potential hazards and the degree of risk of exposure and the identification of necessary countermeasures.

(ii) *Prevention/control* (Regulation 7)
The prevention of exposure (by elimination or substitution) or the provision of adequate controls – through enclosure, extraction, ventilation, systems of work or personal protective equipment (PPE).

(iii) *Proper use of control measures* (Regulation 8)
Employees must make "full and proper use" of control measures and employers "take all reasonable steps" to ensure that they do.

(iv) *Maintenance, examination and test* (Regulation 9)
Control measures must be properly maintained and, in many cases, examined and tested at suitable intervals (in some cases the frequency of test and examination is specified).

(v) *Monitoring* (Regulation 10)/*Health surveillance* (Regulation 11)
It may be necessary in certain circumstances to monitor the degree of exposure to hazardous substances in the workplace and/or to carry out health surveillance of the employees exposed to those substances.

(vi) *Information, instruction and training* (Regulation 12)
Employees must be aware of the risks associated with the substances they are exposed to as well as the precautions that should be taken. Results of monitoring and health surveys must also be passed on to employees.

Some common COSHH mistakes

Experience of COSHH suggests that certain mistakes are commonly made by employers. These are dealt with next, prior to a detailed examination of what should be done to ensure compliance with the regulations.

The data sheet collection

On being asked for their COSHH assessments some employers produce a collection of manufacturers' or suppliers' data sheets, sometimes neatly bound and comprehensively indexed. Reviewing data sheets is an important early step in the assessment process but much more is required. It is essential to identify what the actual risks are in the workplace in order to identify the necessary control measures.

Spaceman syndrome

Because manufacturers and suppliers do not know all the ways in which their product might be used their data sheets often identify all sorts of precautions which *may* be appropriate dependent upon the circumstances. The slavish copying of these into *required* COSHH control measures can often result in PPE standards which are unnecessary, unrealistic and probably totally unenforceable.

Counting the trees

This is particularly common in large organisations where an undue emphasis is put upon assessing each and every hazardous substance present. In these situations assessment by process or activity (as described on p549) rather than by substance is far more appropriate – look at the type, density and shape of the forest rather than counting the trees.

Slaves to the form

There is no standard format for recording a COSHH assessment, despite what some travelling salesmen, or even safety practitioners, might suggest. Several different formats are illustrated in this chapter. Employers should pick the most suitable format for the circumstances. Regrettably overly-complicated standard forms are sometimes imposed within organisations.

Top secret file

The assessment is at the heart of the COSHH regulations. However, all the work involved in carrying it out will have been wasted if the results are not properly used. The necessary control measures must be introduced in the workplace and the workforce educated in the risks from substances they use and the precautions necessary. Far too many COSHH assessments stay hidden away on high shelves and in filing cabinets.

CARRYING OUT COSHH ASSESSMENTS

Who should make the assessment?

The answer to this question will vary depending particularly upon the scale of usage of hazardous substances. Many assessments will not need to be carried out by a 'chemical expert' or a health and safety specialist, although the assessor should always be conscious of the possible need for a second opinion or an expert view in some situations.

The assessor should have the following qualities:

- a basic understanding of chemicals and chemical principles
- some knowledge of the effects of chemicals on the body (as outlined earlier in this chapter)
- an awareness of the requirements of the regulations
- a knowledge of the work activities involving chemicals
- an enquiring mind.

If resources permit, assessment in pairs is recommended. Not only are two heads often better than one but an insider/outsider combination means that one person should know what's going on in a department or an area whilst the other might be more willing to ask questions and challenge why it should happen that way.

Finding out what substances are present

It is not just substances brought in from outside that need to be taken into account but all the substances which may be present in the workplace.

Raw materials and stores stock

Many substances can be identified from stocklists but reliance should not be placed on these alone. A tour of all the storage areas should be made, including maintenance cupboards where items may have been on the shelf for years but do not appear on the lists (or may have sneaked in via the back door).

Process materials

The products of a process, intermediates produced during it or the resultant by-products, waste or emissions may present greater problems than the raw materials. If the assessor's knowledge of the processes or activities is not good enough to identify all of these, advice will need to be sought from those with the relevant technical expertise or some documentary research undertaken.

Buildings and equipment

Hazardous substances may be present in the fabric of the buildings or equipment (eg painted or galvanised metal surfaces). They may also be due to some form of contamination (eg legionella bacteria or bird droppings).

What are the hazards from these substances?

There are various ways in which this information can be researched if necessary. Suppliers are legally required to provide information about their products and occasionally need to be reminded about this obligation or asked for further details. Sources of information are:

- manufacturers' or suppliers' data sheets
- information on container labels or on leaflets inside
- HSE publications including
  - — listings in Guidance Note EH40
  - — other Guidance Notes in the EH and MS series
  - — toxicity reviews
- reference books
- technical literature within the industry or activity
- occupational hygiene or occupational health specialists.

The extent to which the potential hazards need to be researched will depend upon what is already known about the substances and how much of it is present in the workplace. At the same time account should be taken of the control measures recommended in the literature, particularly HSE publications. However, as pointed out earlier, it is very much the assessor's role to decide what precautions are necessary in the specific circumstances of their own workplace.

Planning the assessment

Most of the preparatory work will have taken place in the office or in storage areas but the assessment proper must be conducted in the workplace. Larger workplaces will often need to be divided into manageable assessment units. For example a large garage could be split into the following units:

- cleaning and valeting
- body shop
- paint shop
- mechanical servicing
- breakdown and recovery work
- stores
- offices and showrooms.

The assessors should concentrate on the main activities and greatest risks within each unit. There may be a range of solvents and paints in use in a garage's paint shop but they are likely to be applied in a similar way using the same equipment. If the controls are effective for the most hazardous solvent and paint then, unless there are some different application methods, they are likely to be effective for everything else.

Similarly in an analytical laboratory there will be a wide range of chemicals available but many will be used only for a limited range of analytical procedures and some may not get used at all. The assessors should consider how each analytical procedure is performed and what precautions are necessary. Dealing with each chemical individually will be like 'counting the trees'.

In some workplaces occupational hygiene surveys may have already been carried out to determine the level of dust, fume, gas etc in the atmosphere. These should be reviewed as should other documents that could be of relevance, such as

- operating or task procedures
- training programmes
- health and safety rules
- personal protective equipment standards
- emergency procedures.

Assessing in the workplace

A proper assessment cannot be made without visiting the workplace, observing the working practices and conditions and talking to the people doing the work. It will not be necessary to see every activity or visit every location where the work is done, but the assessor will need to find out enough about what happens in the workplace to feel confident about making a judgement. The following checklists provide a guide to the approach to be taken.

Observe

The assessor should observe in the workplace:

- whether specified procedures are being followed
- what handling methods are adopted
- if the correct PPE is being used

- whether general ventilation is satisfactory
- what local exhaust ventilation is provided
- if the ventilation is operating correctly
- for visible dust and fumes in the atmosphere
- whether strong or unusual smells are present
- for spillages and leaks
- for accumulations of dust on surfaces and in corners
- arrangements for eating, drinking and smoking
- whether provision is made for storing PPE and clothing
- for unexpected substances present
- any unmarked containers.

Enquiries

Questions such as the following may be appropriate:

- Are conditions normally like this?
- What is it like when 'substance X' is being used?
- How much of 'substance Y' is used?
- How is it used?
- What precautions are taken with it?
- What are the respirators/gloves/suits like to wear?
- What happens if 'equipment Z' breaks down?
- What is done if there is a big spill?
- Where does the waste go?
- Where do the empty bags/drums go?
- How is the equipment/floor/ledges cleaned?
- What are the rules about eating/drinking/ smoking?
- What is the worst job that has to be done?
- Have employees ever felt unwell or had any health problems?
- How could things be improved?

Other considerations

The assessor should also consider what might foreseeably go wrong, for example:

- a fork lift holes a stored drum of chemical
- a lab technician drops a glass container
- a critical extraction system breaks down

Consideration should also be given to others who may be exposed to hazardous substances, for example:

- temporary workers
- contractors
- maintenance staff

- visiting administrative staff
- cleaners
- members of the public (especially children)

Identifying problems

It is possible that the assessor's observations and questions may confirm perfectly satisfactory working conditions but it is more likely that problems will be identified, for example:

- the written procedure is totally impracticable and is never used
- the largest units do not fit into the spray booth
- the mixer extraction has not worked for weeks
- some people get rashes from the respirators, so they don't wear them
- nobody realises the hazards from 'substance X'
- temporary workers are not properly trained
- nobody knows what to do if there is a leak
- employees sweep up the dust rather than walk to fetch the vacuum unit
- contaminated empty bags are put in the general rubbish skip
- nobody keeps children away during herbicide spraying
- contractors do as they please.

Forming conclusions

The assessor must now make conclusions as to whether the hazardous substances in use present a risk and, if so, whether existing control measures are adequate. Consideration must be given not just to the potential hazards of the substances but also to the quantity, frequency and manner of use. A sense of proportion needs to be maintained. Very often the scale of use of a substance is so small that a commonsense judgement can be made that it presents 'no significant risk'.

For example, correction fluid contains 1.1.1 trichloroethane, potentially a highly toxic solvent, but the quantities that most people use mean that it constitutes 'no significant risk'. (In an environment where solvent abusers may be present some simple security controls may be necessary.) However, use of larger quantities of the same solvent as a degreasing agent is likely to require an extensive range of controls.

A standard domestic-type bleach applicator (containing sodium hypochlorite) used by a cleaner in the toilets constitutes 'no significant risk', providing that the cleaner is aware of the risks of mixing bleach with acid-based cleaners to produce chlorine gas. However, if the public (especially children) have access to the premises the simple control measure of locking cleaning materials away should be taken.

Similarly the limited use of an aerosol containing a degreasing solvent or a lubricating agent by a maintenance fitter presents 'no significant risk' – even the below-average fitter can be expected not to discharge the aerosol up his own nose or into his own eyes. Here again, the use of the same or similar chemicals on a much larger scale will justify specifying quite detailed control measures.

Specifying control measures

Control measures are covered in much more detail later in the chapter but the choice is likely to be from the following 'menu'.

Elimination or substitution

Is it necessary to carry on using the substance? Can it be substituted by a less hazardous material?

Equipment-based measures

Design of the process and its equipment to minimise the risk created. The provision of local exhaust ventilation (LEV) or general ventilation.

Procedural measures

Specifying the working procedures to be adopted in using hazardous substances and, where necessary, for storage, cleaning and disposal of waste.

People-related measures

Identification and provision of the necessary personal protective equipment (PPE). Standards for eating, drinking and smoking. Arrangements for washing, changing, storing clothing. Ensuring that employees (and others) are provided with the necessary 'information, instruction, training'. Reduction of the numbers present in the area.

Recording the assessment

The COSHH ACOP states that there is no need to record assessments 'in the simplest and most obvious cases which can be easily repeated and explained at any time.'[6] In most cases it will be necessary to keep an assessment record of some type.

What to record

The sort of information that needs to be recorded is

- The workplace or operation to which the assessment refers.
- The activities or processes carried out.
- The substances present (cross-references to data sheet reference numbers may be made).

- The main risks present.
- Those who may be at risk.
- Other relevant information (cross-references may be made to existing procedures, previous occupational hygiene surveys, as appropriate).
- Significant observations.
- Other significant factors influencing the conclusions.
- Any further investigations needed eg dust or fume surveys.
- Control measures necessary
  — existing measures
  — recommended improvements or changes.

Format of the record

Recording all this information would require a fairly complex form and some organisations have produced forms capable of coping with this level of complexity. However, many situations are in practice much more simple and the assessment record can be kept in a correspondingly simple format. A COSHH assessment is not required to be in any standard format. The sample assessments illustrated in figure 23.1 to 23.4 all use different formats, each felt to be appropriate for the relevant situation.

Continuing the task

Review the recommendations

Many of the assessor's recommendations will need to be reviewed with other interested parties:

- there may be considerable cost implications
- major changes to work practices might be necessary
- significant extra training could be required.

Colleagues may have some good ideas about how the same objectives might be achieved through a simpler, cheaper or more effective route. Assessors must not be brow-beaten into changing valid conclusions because of their implications, but must also recognise that others can make useful contributions and be prepared to modify their assessments where appropriate.

Following things through

COSHH assessments must not become a 'top secret file'. They should be readily available to everyone involved. Moreover, the recommendations made as part of the assessment process need to be followed through. This may not be the assessor's role but management must appoint someone to check what has happened in practice. If major changes have taken place it may be necessary to carry out a fresh assessment, but in most cases annotating the assessment record with the action taken will suffice.

ASSESSMENT UNIT	PROCESS HALL MAINTENANCE
MAIN ACTIVITIES	Maintenance and repair work on fixed building installations, process vessels, overhead cranes and ventilation systems within the process hall. Most work takes place during process operations. A limited amount of bench work is carried out in an adjoining workshop. (Work on the building fabric is done by contractors.)
HAZARDOUS SUBSTANCES PRESENT	Data sheet numbers — Production substances: 13, 14, 16, 19, 20, 23, 26, 28 — Maintenance substances: 13, 46, 47, 48, 52, 53, 66, 77, 79, 80, 81, 83, 89
	Others — Graphlex (aerosol) Solvex (degreaser)
MAIN RISKS	Fume inhalation within the process hall, inside process vessels* and inside ventilation ducts* *Risk also of oxygen deficiency Skin contact with process solvents Control of process spills Welding and burning within the workshop Small-scale use of maintenance substances
THOSE AT RISK	Maintenance craftsmen, labourers and supervisors Temporary labourers and contractors (Production employees assessed in separate unit)
OTHER RELEVANT INFORMATION	Survey 13/94 showed all solvent vapours controlled within OES for production operatives Permit to work required for entry into process vessels or ventilation ducts (Air-fed RPE mandatory)
SIGNIFICANT OBSERVATIONS	Maintenance employees exposed to high vapour levels when changing agitator motors on operating vessels Some vessel lids left open by production Fume levels next to sump pumps in basement seem high Temporary labourer seen not wearing gloves Extraction on workshop welding booth good Some burning work being done on open bench
OTHER COMMENTS	All maintenance substances used in small quantities for very short duration. No significant risk from these.
EXISTING CONTROLS	LEV integrated into all process vessels Good general ventilation in process hall Permit to work system for vessel and ventilation system entry Good range of air-fed RPE available Solvent-resistant gloves provided for all employees Spill control kit immediately outside process hall Extracted booth in workshop for welding and burning
FURTHER INVESTIGATIONS	Carry out occupational hygiene surveys of vapour levels 1. During changing of agitator motors on process vessels 2. In sump pump area
RECOMMENDATIONS	1. Ensure regular examination and test of a) Process vessels LEV systems b) Workshop welding booth extraction 2. Ensure monthly examination and test of all air-fed RPE 3. Ensure production employees close all vessel lids 4. Improve induction for temporary labourers to take account of risks in this area, especially the need for gloves 5. Remind employees re use of extracted booth when welding and burning in the workshop

Signatures: R. Jones, A. Smith — Date: 10 October 1994

Next Assessment: February 1995 — Recommendation Follow up/~~Routine review~~

Figure 23.1 *Associated Products Ltd COSHH assessment*

SECTION *Car Valeting*	DATE *28 November 1994*	ASSESSOR *K Roberts*	SHEET *1 OF 2*

WORK ACTIVITIES

Valeting of new and second hand cars is carried out in one bay of the main garage. A separate enclosed area is used for steam cleaning. Ventilation in the area is good and there are no strong smells from the work. The three employees involved are all experienced in this work and there is no significant risk to other employees or visitors.

GENERAL RISKS AND CONTROL MEASURES

The small range of chemicals used are assessed in greater detail below. General risks are:

INGESTION All chemicals could be dangerous if ingested. All containers are correctly labelled (including hand sprays) and employees are aware of the importance of this.

SKIN AND EYES Most chemicals present some risk to the skin and eyes. Rubber gloves and goggles are provided and must be used for the chemicals indicated below. Employees are aware of the need to irrigate eyes or skin with water in case of contact. There is no history of skin problems.

INHALATION There is no significant inhalation risk, even of mist during steam cleaning.

CHEMICAL	METHOD OF USE	CONTROL REQUIRED
GLASS CLEANER Contains 25% isopropanol and ammonia	Applied from hand spray and buffed off with cloth. Slight smell. Only small quantities used.	None (no significant risk)
ALUMINIUM CLEANER Contains 20% Orthophosphoric Acid	Diluted with water (10:1) in hand spray containers. Sprayed on and then washed off with steam cleaner.	Gloves and goggles essential when handling concentrated solution. Do not mix with bleach.
CARPET CLEANER Slightly alkaline	Diluted with water (5:1) in upholstery cleaner tank and then further diluted (10:1) in machine during cleaning.	None (no significant risk)

Figure 23.2 *COSHH assessment ACME Garages Ltd*

COSHH ASSESSMENT

MAIN RISKS

1	Handling of various oils (engine, hydraulic, transmission, waste)
2	Handling of various greases
3	Oxy-acetylene burning and welding
4	Electric arc welding
5	Parts cleaning in proprietary unit (using paraffin based solvent)
6	Engine fumes
7	Use of various substances (listed on separate sheet) including chlorinated degreasing solvent and caustic materials. All used in small quantities for short durations.

CONTROLS

1 & 2	Rubber gloves provided. Worn for handling significant quantities of oil. (No history of skin problems amongst employees).
3 & 4	Limited burning and welding work, mainly of short duration. Large airy shop. Portable extraction unit available (but broken).
5	Unit includes brush applicator. Rubber gloves provided and worn.
6	Removed by natural ventilation. No evidence of fume build up when doors open. Door opened for a few minutes in winter if conditions necessitate.
7	Rubber gloves and goggles available when necessary.

RECOMMENDATIONS

1 & 2	Monitor employees to ensure gloves worn when appropriate
3 & 4	a) Repair portable extraction unit b) Examine and test extraction unit at least every 14 months (see reference material attached)
5	None
6	Monitor fume situation in winter months
7	a) Rubber gloves must be worn when handling substances indicated * b) Goggles must be worn when handling substances indicated°

Signed *B. Wright* *27 September 1994*

Figure 23.3 *Banger and Heap motor vehicle repairs*

HAGGLE & BILL, SOLICITORS
<u>COSHH ASSESSMENT</u>

HAZARDOUS SUBSTANCES

a) OFFICE MATERIALS
1. Type 433 cartridges for use with photocopier
2. VDU cleaning kits (containing isopropyl alcohol)
3. Correction fluid and thinners (containing 1,1,1,trichloroethane)
4. Spray adhesive (containing 1,1,1,trichloroethane)
5. Various solvent based marker pens

b) CLEANING MATERIALS
1. Bleach containing sodium hypochlorite
2. Metal polish
3. Furniture cleaner (containing 1,1,1,trichloroethane)
4. Acid descaling compound
5. Washing up liquid
6. Polish spray
7. Air freshener spray
8. Window spray
9. Scourer bleach

OVERALL ASSESSMENT
Subject to the comments below, the substances in use are of such a nature or used in such small quantities that they do not constitute any significant risk.

Cleaning materials are used in little more than household quantities and there are no special circumstances concerning the use of fairly standard office materials.

COMMENTS AND RECOMMENDATIONS
1. Wear disposable plastic gloves when clearing significant spillages of photocopier toner. (Relevant staff already aware.)
2. Avoid contact of bleach and acid descaler with skin and eyes. Rubber gloves must be worn when handling concentrated solutions or for lengthy work with dilute solutions.
3. In cases of skin or eye contact with the above, irrigate thoroughly with water.
4. Do not mix bleach with other cleaning materials (in contact with acids it will release chlorine gas).
5. Keep the cleaner's cupboard in the public waiting room locked except when access is essential.
6. Draw the cleaner's attention to points 2 to 5 above.

C. Sharp

23 October 1994;

Figure 23.4

Assessment review

Review of a COSHH assessment could become necessary because of changes in processes or activities or for other reasons, for example:

- introduction of new or modified equipment
- changes in substances used
- different methods of use (eg spraying replaces brush application)
- altered volumes of work
- new plant layouts (eg affecting ventilation or exposure of others)
- new health hazards become apparent
- the HSE introduce tighter MELs or OESs
- employees have reported work-related health problems
- health surveillance indicates that controls are not adequate
- new or improved control methods become available.

Changes can often creep up without anyone realising and periodic reviews of COSHH assessments should be carried out anyway. A review does not mean the whole assessment has to be repeated and revised. A reconsideration of the position in the light of the changes that have taken place, possibly together with an observation of process or activity, could conclude that the existing control measures are still adequate and that the assessment is still valid.

CONTROL MEASURES

The employer's prime duty under regulation 7 (see below) is to prevent the exposure of his employees to hazardous substances. If this is not reasonably practicable then the exposure must be 'adequately controlled'. Once a control measure is specified as necessary to control the risk, its use becomes compulsory and it must be 'maintained in an efficient state, in efficient working order and in good repair' (see regulations 8 and 9 below).

The maintenance requirements of the regulations and the COSHH ACOP are quite stringent and it is prudent for employers to make it clear in their assessments which measures are required to ensure adequate control as opposed to recommended as an optional additional precaution.

Duties under the regulations

Prevention or control of exposure (regulation 7)

As already stated, regulation 7 requires that exposure of employees to hazardous substances be prevented or, where this is not reasonably practicable, adequately controlled. So far as is reasonably practicable, the prevention or control of exposure must be by measures other than the provision of personal protective equipment (PPE). Thus there is a hierarchy of measures to be taken:

1. Prevention of exposure (eg elimination or substitution)
2. Control of exposure by means other than PPE
3. Control using PPE.

Paragraph 3 of the regulation specifies control measures which must be adopted in respect of carcinogens whilst paragraph 10 through Schedule 9 to the regulations introduces special provisions relating to biological agents. In both cases these are supported by Approved Codes of Practice.[6] Other parts of regulation 7 set standards for PPE and define what constitutes 'adequate control'. These aspects will be considered more fully later in the chapter.

Use of control measures etc (regulation 8)

Employers must 'take all reasonable steps' to ensure that control measures are properly used or applied. This means proactively policing what is happening in the workplace and taking appropriate remedial action where necessary. Employees have a duty to make full and proper use of control measures, PPE etc provided *and* to report defects forthwith to their employer. The employee is expected to:

- use the correct substances
- use the equipment provided
- follow the designated working procedures
- wear the specified PPE correctly
- make proper use of changing accommodation, PPE storage, washing, showering or bathing facilities
- adhere to rules on eating, drinking and smoking.

Maintenance, examination and test of control measures etc (regulation 9)

Employers have a general duty to ensure control measures are "maintained in an efficient state, in efficient working order and in good repair". Engineering controls must receive thorough examinations and tests "at suitable intervals". For LEV these are specified as at least every 14 months (more frequently for some processes – see Schedule 3 to the regulations).

Apart from disposable types, respiratory protective equipment (RPE) 'provided to meet the requirements of Regulation 7' (ie as a control measure) must also be examined and, where appropriate, tested at 'suitable intervals'. This might discourage employers from unnecessarily specifying respirators as control measures. If the employees are required to wear them then the employer must examine and possibly test them.

In addition to the specific requirements for LEV and RPE, the need for maintenance and/or examination of other control measures must not be overlooked, for example:

- examination of general ventilation fans
- inspection of vessels, pipelines etc
- checking the condition of process enclosures
- keeping other types of PPE (gloves, clothing, eye protection etc) in good condition
- cleaning and repairing warning signs.

Monitoring exposure at the workplace (regulation 10)

The exposure of employees to hazardous substances must be monitored where 'requisite' for ensuring maintenance of adequate control or otherwise protecting the health of employees. According to the COSHH ACOP these 'requisite' circumstances would be:

- when failure or deterioration of the control measures could result in a serious health effect
- when measurement is necessary to ensure that the MEL, OES or other working standard is not exceeded
- when necessary as an extra check on the effectiveness of control measures.

Schedule 4 to the regulations specifies processes where such monitoring *must* take place. Otherwise it is completely left to the employer's judgement, probably applied during the assessment process, as to whether monitoring is 'requisite'.

Such circumstances are likely to be infrequent although the need for a one-off occupational hygiene survey or a short programme of surveys to check on levels of risk or effectiveness of control measures will be much more common. Occupational hygiene surveys are referred to on pp561–3.

Health surveillance (regulation 11)

Regulation 11 will affect only a very small proportion of employers and their workforces. One of its purposes is to maintain the standards of health surveillance already applying to a number of processes (listed in Schedule 5 to the regulations). However, "where appropriate for the protection of the health of his employees" the employer must ensure they are under suitable health surveillance.

There are three main factors to be taken into account before deciding that health surveillance is 'appropriate'.

1. There must be an identifiable disease or adverse health effect associated with exposure to a hazardous substance.
2. Exposure must be such that there is a reasonable likelihood that it may occur.
3. There must be valid techniques for detecting the disease or effect. (Such techniques are described in chapter 13).

In most workplaces other control measures should ensure that health surveillance is not necessary. However, some surveillance may be appropriate in certain cases, for example to identify those individuals susceptible to a sensitising agent producing a dermatitic or asthmatic effect. Advice should be sought from an occupational health specialist before embarking on a programme of health surveillance. The Employment Medical Advisory Service (EMAS) is a good source of advice for those without in-house specialists. Several HSE publications also provide guidance.[7]

Adequate control

If a substance is one for which a Maximum Exposure Limited (MEL) has been specified in Schedule 1 to the regulations then its control (for inhalation risks) is adequate only if exposure is below the MEL and is reduced 'so far as reasonably practicable'. Where an Occupational Exposure Standard (OES) has been approved (by the HSC) then control is adequate if the OES is not exceeded or where it is exceeded the reasons are identified and rectified 'as soon as is reasonably practicable'.

To a certain extent a parallel can be drawn between MELs and OESs and the speed limit on the road. Just as the chances or consequences of an accident are only slightly greater at 31 mph than 29 mph, the chances of ill health at 101 ppm are only slightly more than at 99 ppm. However, the law says the limits should not be exceeded – the line has to be drawn somewhere.

MELs have been laid down for a relatively small number of substances. These are clearly the most dangerous, many being carcinogenic. Not only is the 'speed limit' enforced rigidly but levels must be further reduced 'so far as is reasonably practicable'. Most MELs are long-term exposure limits, averaged over an eight-hour reference period, but in a few cases short-term exposure limits have been set and these must not be exceeded over a fifteen-minute reference period.

An OES is the concentration of an air-borne substance at which, according to current knowledge, there is no evidence that it is likely to be injurious to employees inhaling those levels daily. Whilst a little more latitude is given to excursions above the OES, too much satisfaction should not be drawn from a single survey showing concentrations just within the standard. Normal day to day variations could easily take exposure above the standard and it would be sensible to have a comfortable margin within the OES. As with MELs both long-term and short-term OESs are listed. The HSE publish a guidance note annually listing MELs and OESs together with proposed changes and substances under review. The guidance note also contains much useful background information on how the figures are set.[5]

Occupational hygiene surveys

An occupational hygiene survey could be undertaken either to establish whether a hazardous substance is present in the atmosphere in sufficient quantities to constitute a risk, or to check that existing control measures are performing adequately. A survey may be carried out as a regular check or as a one-off exercise.

The occupational hygienist is a specialist in his own right and hygiene surveys can extend to other hazards besides hazardous substances (see chapter 13). There is much specialist literature on the planning and conduct of hygiene surveys including many HSE publications in their Methods for the Determination of Hazardous Substances (MDHS) and Guidance Notes – Environmental Hygiene (EH) series.[8] The intention here is to give an indication of the ways in which such surveys may be carried out.

Since the principal purpose is to determine how much of an airborne contaminant is being inhaled, logic would indicate that any sample should be taken from the breathing zone of the employee (or other person) – a so-called personal sample. Most samples are taken this way but sometimes this is impracticable – the sampling equipment may not be very portable or it may be incompatible with the employee's duties. In such cases it may be necessary to take a static sample, either by the source of the contaminant (a machine or activity) or from the general atmosphere of the workplace.

Generally the principle to be followed will be to sample a situation worse than the one the employee will actually experience – as he works near the contaminant source for only part of the time or is constantly entering and leaving the workplace. If the sample is below the relevant MEL or OES then the employee will actually be inhaling far less.

The range of survey instruments available is constantly expanding but broadly speaking they can be placed into the following categories.

Chemical indicator tubes

A measured volume of air is drawn through the tube by a manual or mechanical pump. The pollutant reacts with the chemicals in the tube to produce a colour change. The pollutant concentration is indicated by calibrated markings on the tube or through the intensity of the colour. Such tubes can be used for measuring gases, vapour or aerosols.

Direct reading instruments

Various electrically operated instruments are available for measuring gases, vapours and even dusts in the atmosphere. These will give a direct reading of the concentration and can also be used to set off alarms when specified concentrations are reached. Large plants may have fixed installations of such instruments but portable units are more common and are particularly useful for monitoring atmospheres in confined spaces.

Sampling pumps and filter heads

Small battery operated pumps which can be placed in employees' pockets or attached to their belts are used to draw polluted air through a filter head or some other absorbent medium situated in the breathing zone (usually attached to the lapel). The samples collected can then be weighed, chemically analysed or even counted under microscopes (in the case of asbestos fibres on filters) and related to the volume of sampled air (recorded by the pump) to produce a concentration of pollutant. These methods can be used for gases, vapours, mists or dusts, with different filtration or absorbent materials used in different situations.

Care must be taken in the interpretation of all hygiene survey results. The conditions sampled should be representative of those normally encountered and note must be taken of the accuracy of the survey method. Records may need to be retained for as long as forty years.[9]

Elimination or substitution

One way of avoiding problems caused by hazardous substances is to stop using them. Chemical cleaning methods could be replaced by mechanical methods, organic solvents could be replaced by less hazardous solvents or even water, but there are many factors to be taken into account before deciding that elimination or substitution is 'reasonably practicable'. However, all the relevant factors need to be taken into account when considering such a move.

Is the new product/method effective?

Does the replacement paint produce as good a surface finish? Will the alternative herbicide or insecticide kill off weeds or parasites as effectively?

What are the cost implications?

What are the unit costs of the new product and will more of it be used? Will it take longer to apply or to dry and will existing equipment need to be replaced or modified?

Could new risks be created?

Will there be a dust rather than a vapour problem? Could there be mechanical dangers involved or will access be more difficult? Are any extra manual handling problems introduced?

Nevertheless, there have been many successful examples of elimination or substitution:

- replacement of Halon fire extinguishers
- widespread use of water-based rather than solvent-based paints
- the virtually complete replacement of asbestos by a variety of other materials.

Equipment-based controls

Many control measures can either be incorporated into the equipment used for a particular activity or provided through the availability of other equipment.

Enclosure

The hazardous substance can be used within a totally or partially enclosed piece of equipment or its spread can be limited by the physical segregation of a process or activity – by permanent walls or temporary means (eg the use of plastic sheeting).

General ventilation

Dilution of an air-borne contaminant by fresh air is a perfectly legitimate means of controlling risks. This might be achieved simply by opening windows or doors or could involve the use of general ventilation fans mounted in the roof or walls. However, the wholesale replacement of workplace air can be very inefficient, particularly as the extracted air will probably have been heated and the fresh 'make-up' air is likely to be cold.[10]

Local exhaust ventilation (LEV)

LEV can be directly incorporated into process equipment, provided through extracted booths or available in the form of portable hoods or ducts. Thought must be given to the point of discharge of the LEV, otherwise a different group of people may be put at risk. Environmental considerations may require the extracted air to be filtered or cleaned by other means. The design of LEV systems is dealt with later.

Cleaning equipment

The brush and mop sometimes provide an excellent means of getting more of the hazardous substance into the working atmosphere, as does the use of an airline. Generally vacuum cleaning methods are to be preferred. These can be provided by permanent vacuum systems, mobile sweepers incorporating vacuum equipment or portable vacuum cleaners. Vacuum units must be fitted with suitable filters, otherwise the hazardous substance will be recirculated into the atmosphere.

Waste containers

The provision of suitable containers for waste materials is often neglected. This may be waste directly generated by the process, materials from cleaning operations or 'empty' bags or drums, often still containing significant quantities of the hazardous substance. Such containers will generally need to be covered and in some cases they may need to be sealed or to include LEV provision.

Emergency equipment

If there is a possibility of spillage occurring it may be necessary to keep equipment readily available for containing and then removing the spill. Major problems may be created if the substance enters water courses or drains. Absorbent granules may be all that is required but for larger spills the use of temporary barriers and even mobile pumps and pipes could be necessary. Such equipment should always be kept in positions unlikely to be affected by the spill.

Design and maintenance of LEV systems

There are many detailed publications available on the design of LEV systems including a well laid out booklet from the HSE.[11] Any system will have up to five components.

The inlet

This could be a booth or other enclosure (eg a fume cupboard) or a hood. Its design is critical to the effectiveness of the system. Several factors need to be taken into account. Dust particles are larger and harder to capture than gas or vapour as are pollutants which already have significant motion eg sprays or dusts from grinding. Employees need to be left with adequate room to work and they should not be positioned between the source of the pollutant and the LEV inlet. Account may also have to be taken of other air movement in the vicinity.

Air flow rates fall off quite dramatically as the distance from the LEV inlet increases – something often badly overlooked by the uninitiated in designing LEV systems. As a rough rule of thumb, for an open-ended duct the air flow rate just one diameter away will have dropped to 10% of that at the duct opening. Consequently the more the flow of air from unwanted directions can be prevented (eg by flanges, canopies, side shields) and the closer the inlet can be brought to the source, the more effective will be the control.

Ductwork

The air flow rate along the ductwork must be sufficient to stop particles within the air stream from settling out. The ductwork itself needs to be sufficiently strong and of the right type of material to withstand the abrasive and chemical attack it is likely to experience in use. Changes of direction should be gradual and kept to the minimum necessary, in order to avoid resistance to the air flow and abrasion at tight bends. Access panels within the ductwork system may be required for cleaning and inspection purposes and also possibly for testing flow rates.

Cleaning devices

It may be acceptable to discharge the contaminated air directly into the outside atmosphere but in these environmentally enlightened (and heavily regulated) times this is increasingly less likely. There is much technical literature available on the design of air cleaning devices, the choice being heavily dependent upon the pollutant and its physical and chemical properties.

The main types in common use are:

- fabric filters (often in the form of bags)
- cyclone separators
- electrostatic precipitators
- spray scrubbers (using water spray curtains).

With potentially flammable or explosive pollutants there may well need to be explosion relief or fire extinguishing systems incorporated into the cleaning devices and the ductwork.

Fans

As a general principle fans should be situated on the clean air side of the air cleaning device. Centrifugal fans are more commonly used in LEV systems since axial fans can overcome only limited resistance to flow. Axial fans are more common in direct extraction units mounted in walls or roofs. Air movers driven by compressed air may also be used in situations where flammable gases or vapours are present or it is difficult to obtain an electrical supply.

Discharge outlet

The cleaned air will need to be carried to a suitable point for discharge into the atmosphere. This will need to be chosen so as to not create any dangerous or unpleasant conditions outside and also to prevent the air re-entering buildings. Other matters to be taken into account are any possible damage to other structures from residual materials or residual heat and the structural stability of the discharge taking into account weather conditions, especially wind.

As noted earlier, Regulation 9 of the COSHH regulations requires LEV used as a control measure under the regulations to be maintained and to receive thorough examinations and tests at least every 14 months. An HSE booklet provides ample guidance on what maintenance might be appropriate and methods of examination and testing.[12] Common methods include:

- *Air sampling or monitoring* The LEV's effectiveness can be checked by monitoring pollutants in the workplace air.
- *Smoke generators or dust lamps* These can be used to monitor the direction and speed of air flow.
- *Static pressures* Static pressures can be measured at standard points in the LEV system. Changes from normal indicate potential problems within the system.
- *Air velocity measurement* Velocities at working positions or hood or booth

entrances can be compared with standard capture velocities.[11] Flow rates within ducts may also need to be checked.

The COSHH ACOP lists the information which must be recorded following LEV examinations and test. Adverse results need to be followed up and rectified.

Procedural controls

The importance of safe systems of work is stressed in chapter 12. Points to be particularly borne in mind in developing safe systems of work in the use of hazardous substances are detailed below.

The form of the substance

The form in which hazardous substances are used may be a big factor in ensuring that their risks are adequately controlled. Use of granules as opposed to powder, or wet methods rather than dry, will have a significant influence on the quantities inhaled.

Work methods

These may be informal or may be specified in written task procedures (see chapter 12). Important considerations here might include specifying methods of use (eg brushing rather than spraying), or detailing the sequence in which operations take place in order to minimise exposure. Sprayed or dipped materials may need to be left under LEV for a defined period of time to enable evaporation of solvents to take place.

Work location

This too may need to be specified in order to minimise exposure. It may be better to carry out some activities in the open air in order to provide dilution ventilation, whereas in other cases the influence of the wind may need to be avoided eg because it might whip up dust clouds.

Work may need to be carried out away from other people who may not possess the necessary knowledge of the risks or be equipped with the appropriate PPE. In some cases they may have to be kept out of operational areas by the erection of physical barriers or signs or need to be given verbal advice to keep clear. The exact position of the employee carrying out the task might have to be identified so that he avoids unnecessary exposure or gains maximum benefit from other control measures, such as LEV systems.

Housekeeping

The regular removal of accumulations of hazardous substance is an important control measure. Frequencies of housekeeping activities should be defined and there may also be a need to clearly specify the methods to be adopted.

Waste disposal

Procedures need to be established for the safe disposal of hazardous waste. Here environmental considerations, as well as the health and safety of workpeople, come into play. Some environmental aspects are reviewed in chapter 28. As well as taking account of waste products and by-products from the process, it will be necessary to consider contaminated packaging materials (sacks, bags, drums, barrels), unwanted raw materials and rejected product.

Emergencies

Procedures may also be required to take account of health, safety and environmental risks which may foreseeably arise. These may be relatively straightforward eg use of containment measures, the sealing off of affected areas, the use of emergency PPE, or the temporary closing down of a process. However, in other cases the degree of risk of serious and imminent injury may be such as to justify the establishment of the type of formal procedure envisaged by regulation 7 of the Management Regulations (see chapter 10).

People based controls

Regulation 7 of the COSHH regulations requires adequate control to be established, so far as is reasonably practicable, by means other than the provision of personal protective equipment (PPE). However there are many situations where the use of PPE or some other people based method of control (possibly in conjunction with other methods) is the only effective way of achieving the desired result. Several aspects of people based controls are reviewed here:

Types of PPE

Various types of PPE may be necessary as control measures depending upon the nature of the hazardous substance and the manner of exposure to it. Those most likely to be required are:

- respiratory protective equipment (see below)
- protective clothing
- protective footwear
- gloves
- eye protection (including face protection).

Factors to be taken into account when selecting PPE are:

- the protection it gives against the hazardous substance
- the working environment in which it is to be worn
- its suitability for the task being performed

- wearer comfort
- ease of maintenance.

PPE may need to comply with relevant British or European standards (see chapter 25).

All control measures must be 'maintained in an efficient state in efficient working order and in good repair'.[13] Consequently appropriate arrangements will need to be made for inspection and, where necessary, replacement of PPE together with its cleaning and any other maintenance. Employees also have a duty (under regulation 8(2)) to report defects in PPE whilst the employer (through regulation 8(1)) must "take all reasonable steps to ensure it is properly used ...".

Respiratory protective equipment (RPE)

RPE is a particularly important form of control measure against hazardous substances and it receives special attention in the COSHH Regulations. HSE have published detailed guidance on the subject in *Respiratory Protective Equipment – A practical guide for users*.[14]

RPE can be divided into two types – respirators, which filter the contaminated air before it is inhaled, and equipment providing uncontaminated air from a separate source. These can further be subdivided as follows:

- *Respirators*
  - simple filtered respirators
  - powered respirators
- *Air-fed RPE*
  - fresh air hose equipment
  - equipment fed from a compressed air system
  - self-contained breathing apparatus

Many further variations are possible upon these basic themes dependent upon the demands of the situation. Factors to be taken into account in selecting RPE include:

- *Hazard related*
  - availability of sufficient oxygen
  - form of contaminant (dust, gas, vapour)
  - nature of contaminant
  - concentration of contaminant
  - relevant occupation exposure standard
  - manner of exposure
  - other hazards present
- *Work related*
  - need for mobility, visibility, communication
  - compatibility with other PPE

— physical effort required
— medical fitness
— face size and shape
— presence of facial hair
— use of spectacles

It is important that the wearer's needs are not overlooked in the quest for high standards of protection. If RPE is not worn even for a short time then the level of protection it provides will be significantly reduced. RPE has to be approved by the HSE before it can be used.

When using respirators it is particularly important to ensure that the filter fitted is the correct one for the type of contaminant(s) the wearer will be exposed to. In the case of equipment fed from compressed-air systems, the quality of the air should be checked periodically against standards published in the HSE's RPE booklet.

All RPE provided as a control measure must be examined and tested on a monthly basis to comply with the terms of the COSHH ACOP. Disposable respirators are excepted from this requirement and in some cases the intervals may be extended as far as three months. Even where the use of RPE is only recommended as an additional precaution, some form of regular examination, maintenance or testing may still be appropriate.

The extent of examination and testing necessary will depend upon the type of RPE – self-contained breathing apparatus will require much more than simple respirators. The HSE's RPE booklet and the COSHH ACOP provide further information on this together with details of the records which should be kept.

Eating, drinking and smoking

It may be necessary to place restrictions on eating, drinking or smoking in the workplace, particularly where the hazardous substances in use present risks through ingestion. Employees may be advised of such restrictions as part of their training but in some cases it may be necessary to mark restricted areas with signs. Employers using these types of control measures should be aware of the requirement placed on them to ensure any restrictions are properly observed. There may also be a need to provide suitable locations where employees may eat, drink or smoke in an uncontaminated area, with washing facilities conveniently accessible.

Washing, changing and clothing storage

While there are general obligations on employers to provide adequate facilities for washing, changing and clothing storage (see chapter 16), there may be a particular need for them in relation to employees exposed to hazardous substances. Good standards of personal hygiene may be important to prevent ingestion, skin exposure or the spread of hazardous substances. Washing facilities, conveniently accessible from the workplace, should be provided in such cases. Measures will need to be taken to prevent the washing facilities themselves from becoming contaminated and to keep them clean on a regular basis.

Where employees have to change into or out of protective clothing then suitable facilities should be provided for them to do this in an uncontaminated area. Accommodation should be provided for protective clothing when not in use and also to keep employees' ordinary clothing free from contamination. Use of pegs or hangers in an uncontaminated area may be sufficient but in some situations the use of lockers may be necessary. Appropriate steps should be taken to keep changing and clothing storage accommodation clean.

Information, instruction and training

COSHH Regulation 12(1) states that:

> An employer who undertakes work which may expose any of his employees to substances hazardous to health shall provide that employee with such information, instruction and training as is suitable and sufficient for him to know:
>
> (a) he risks to health created by such exposure; and
> (b) the precautions which should be taken.

As in other cases, employees need to be aware of both the risks and the precautions they should take against them. What is 'suitable and sufficient' will depend on the circumstances but might include:

- the nature and degree of the risks
- factors that may increase the risk eg smoking
- relevant equipment-based controls and how to use them eg LEV
- relevant procedure-based controls eg work methods
- where relevant PPE must be worn
- where RPE is required and how to wear it correctly[14]
- any eating, drinking or smoking restrictions
- washing, changing and clothing storage arrangements
- how to obtain more information on substances
- what to do in cases of doubt
- emergency procedures.

It may be best to precede any specific training by giving employees a general appreciation of hazardous substances and how they harm the body (see pp 538–44) together with a broad understanding of how these risks can be controlled.

Paragraph (2) of Regulation 12 also requires employers to provide employees with information on:

- results of the monitoring of workplace exposure (forthwith if the MEL has been exceeded)
- collective results of health surveillance.

Notes

1. *Skin cancer by pitch and tar* – HSE (MSB 4)
   Skin cancer by oil – HSE (MSB 5)
2. *Anthrax: health hazards* – HSE 1979 (Guidance Note EH23)
3. *The control of legionellosis including legionnaires disease* – HSE 1993 (HS(G)70)
   The prevention or control of legionellosis (including legionnaires disease), Approved Code of Practice – HSE 1991 (L8)
4. *Leptospirosis* – HSE (IND(G)84)
5. *Occupational exposure limits 19* – HSE (published annually) – (Guidance Note EH40)
6. *General COSHH ACOP (Control of substances hazardous to health), Carcinogens ACOP (Control of carcinogenic substances)* and *Biological agents ACOP (Control of biological agents)* are published together in one booklet (reference L5) by HSE Books 1995. The booklet contains the COSHH Regulations and related Schedules in full. Paragraph 21 refers to the recording of assessments.
7. *Health surveillance of occupational skin diseases* – HSE 1991 (Guidance Note MS24)
   Surveillance of people exposed to health risks at work – HSE 1990 (HS(G)61)
   Health surveillance under COSHH: guidance for employers – HSE 1990
8. *Monitoring strategies for toxic substances* – HSE 1989 (Guidance Note EH42 (Rev.)) – provides general guidance on the subject of hygiene surveys.
9. Control of Substances Hazardous to Health Regulations 1994, regulation 10(3)
10. *Ventilation in the workplace* – HSE 1988 (Guidance Note EH22 (Rev.))
11. *Introduction to local exhaust ventilation* – HSE 1993 (HS(G)37)
12. *The maintenance, examination and testing of local exhaust ventilation* – HSE 1990 (HS(G)54)
13. Control of Substances Hazardous to Health Regulations 1994, regulation 9(1)
14. *Respiratory Protective Equipment – A practical guide for users* – HSE 1990 (HS(G)53)

24

Asbestos and Lead

INTRODUCTION

The main parts of the COSHH Regulations (Regulations 6 to 12) do not apply to work covered by:

- the Control of Lead at Work Regulations 1980
- the Control of Asbestos at Work Regulations 1987.

These regulations pre-date the COSHH Regulations and to a certain extent provided a model for them to follow. Asbestos presents particular dangers and is subject to other legislative controls.

ASBESTOS – THE DANGERS

The nature of asbestos

Asbestos is a term used to describe a group of fibrous crystalline silicates. These are mined from naturally-occuring deposits in places such as Canada, Russia and parts of Africa. The definition of asbestos in the Control of Asbestos at Work Regulations includes:

- Crocidolite (often called blue asbestos).
- Amosite (often called brown asbestos).
- Chrysotile (often called white asbestos).
- Fibrous actinolite.
- Fibrous anthophyllite.
- Fibrous tremolite.
- Any mixture containing any of these minerals.

Diseases caused by asbestos

There are three main diseases associated with asbestos, all of which may take many years to develop.

Asbestosis

The lung tissues are damaged by the formation of 'asbestos bodies' around inhaled fibres. The effects usually take many years to become apparent, symptoms including shortness of breath and a cough. The condition is untreatable and irreversible and its severity will generally relate to the level of exposure.

Lung cancer

Carcinoma of the lung is known to be caused by inhalation of asbestos fibres. There is considerable evidence that the risks are significantly higher for those who smoke. Sometimes lung cancer has occurred in people who have had very limited exposure to asbestos but the risks are greater for those with heavier exposure.

Mesothelioma

This is another type of cancer which occurs in the inner lining of the chest wall or the abdominal cavity. It is particularly associated with exposure to crocidolite or amosite.

Another condition, asbestos corns, may result from irritation caused by asbestos fibres entering the skin, particularly the hands. Such corns are not thought to lead to skin cancer.

Where asbestos might be found

Despite its well-recognised dangers, asbestos has many useful properties which led to its widespread use, particularly for fire and thermal insulation and as a binding agent within other materials. Many of these applications are now banned and new asbestos products must be clearly labelled. However, there is always the possibility of asbestos-containing materials being found in existing buildings and structures. The older they are, the greater the likelihood of asbestos being present. Asbestos products may be encountered in the following situations:

Asbestos lagging

Asbestos was once commonly used to lag pipes, ducts, boilers, process vessels etc.

Asbestos insulation board or blocks

- Fire resistant doors or partitions.
- Furnace lining materials.
- Electrical switchgear.

Asbestos cement products

- Corrugated or smooth roofing or wall sheets.
- Gutters and pipes.
- Decorative finishes.

Sprayed asbestos coatings

These may have been used for fire protection or to control noise or condensation. Metal girders, roof surfaces, wall surfaces, pipes, ducts or vessels may have been treated.

Asbestos tiles and felts

Asbestos has often been used as a binding agent in tiles used for ceilings, floors etc. Some of these are of the vinyl type which would not normally be associated with asbestos. Roofing felts may also contain asbestos.

Other asbestos-containing products

Asbestos may also be present in the following products:

- Brake lining and clutch facings.
- Gaskets and filters.
- Protective clothing.

THE CONTROL OF ASBESTOS AT WORK REGULATIONS 1987

Full details of the content of the regulations and their associated ACOP are contained in a HSE booklet.[1] A separate ACOP has also been published: *Work with asbestos insulation, asbestos coating and asbestos insulating board*.[2] What follows is of necessity a summary of the requirements of the regulations and those carrying out significant amounts of asbestos work are recommended to obtain copies of the relevant publications.

Duties (Regulation 3)

The regulations place duties on employers in respect of their own employees. Many requirements are also extended to protect others who may be affected by their asbestos work – contractors, visitors, occupiers, neighbours etc.

Action levels and control limits (regulation 2)

Some regulations come into operation only when the 'action level' is likely to be exceeded (over a 12 week reference period). The expected exposure time in hours is multiplied by the likely airborne fibre levels (in fibres per ml of air) and compared with the relevant action level. If only chrysotile asbestos is present the action level is set currently at 96 fibre-hours per ml whilst if any other types of asbestos are present the action level is reduced to 48 fibre-hours per ml. The likely airborne fibre levels may be known from previous similar work or they may be obtained from relevant publications such as the HSE Guidance Note *Probable asbestos dust concentration at construction processes.*[3] It may be necessary to aggregate different exposure periods at different exposure levels. The ACOP describes how this is done.[4]

Control measures must be sufficient to ensure that inhaled air is within specified control limits. These are set as follows:

- *For chrysotile asbestos alone:*
  — 0.5 fibres/ml averaged over any continuous period of 4 hours
  — 1.5 fibres/ml averaged over any continuous period of 10 minutes
- *If any other form of asbestos is present:*
  — 0.2 fibres/ml averaged over any continuous period of 4 hours
  — 0.5 fibres/ml averaged over any continuous period of 10 minutes.

Detailed guidance on measurement of airborne asbestos is contained in another HSE Guidance Note.[5]

Identification (regulation 4)

Employers must not carry out work which exposes or is liable to expose employees to asbestos unless either:

- they identify the type of asbestos before work starts, or
- they assume it is not chrysotile alone and treat it accordingly.

The identification may be carried out by:

- representative sampling and analysis
- reference to plans or records
- use of information from suppliers.

Compliance with these requirements of Regulation 4 will be assisted if the employer has already made a survey of his premises and identified those materials which are known to contain asbestos and those which are suspected to contain it. The latter can be subject to more detailed investigation if work on it becomes necessary or its condition begins to deteriorate.

Assessment and plans of work (regulations 5 and 5A)

Before carrying out work which exposes, or is liable to expose, employees to asbestos the employer must make an adequate assessment of that exposure including:

- identifying the type of asbestos
- determining the nature and degree of exposure
- setting out the steps to be taken to prevent or reduce the exposure to the lowest level reasonably practicable.

The assessment must be reviewed if there is reason to suspect it is no longer valid or there are significant changes in the work.

The ACOP states that assessments should be in writing except for straightforward and low level exposure situations such as:

- minor electrical or plumbing work
- handling products where the fibre is bonded
- small-scale work not involving cutting or breaking ceiling or floor tiles or asbestos cement
- fitting of gaskets.

Detailed guidance on the assessment process is available in the ACOP.[6] The approach is very similar to that for COSHH assessments.

For work involving removal of asbestos from any building, structure, plant or installation or from a ship (including its demolition) the employer must first prepare a suitable written plan of work which must be kept until the work to which the plan relates is completed. The plan of work must contain specified information including details of work methods and protection and decontamination equipment to be used.

Notification (regulation 6)

The enforcing authority (usually the HSE) must be notified in writing at least 28 days in advance (unless they agree a shorter time) of any work which is liable to exceed the action level. This does not apply to employers licensed under the Asbestos (Licensing) Regulations 1983 carrying out work in accordance with their licence (see p582). The ACOP states that only one notification is necessary under this regulation for all activities. Separate jobs need not be notified unless the nature of the work changes significantly from that originally notified. A sample notification form is included in the ACOP.[7]

Information, instruction and training (regulation 7)

Employers must ensure that adequate information, instruction and training is given to employees:

- exposed or liable to be exposed to asbestos

- carrying out work in connection with the employer's duties under the regulations.

The ACOP lists what might be appropriate for exposed employees, including:

- health hazards of asbestos
- increased risks to smokers
- how control measures can reduce hazards
- correct use and maintenance of control measures
- procedures for reporting defects
- work methods, hygiene procedures, waste disposal arrangements
- control limits and action limits
- exposure assessment and air monitoring
- medical surveillance.

An HSE leaflet *Asbestos and You*[8] may be helpful in providing employees with relevant information.

Employees and safety representatives must be allowed access on request to various specified records the employer is required to keep under the regulations. They must also be informed as quickly as possible if control limits are unexpectedly exceeded and given details of the reason and the action taken or proposed to rectify the situation.

Specialist training is likely to be necessary for those carrying out asbestos assessments, air monitoring work and the examination, testing or maintenance of plant or equipment required under the regulations.

Prevention or reduction of exposure (regulation 8)

Regulation 8, as amended in 1992, in effect requires employers to adopt a hierarchy of measures to protect their employees:

- prevention of exposure, if reasonably practicable
- substitution of asbestos by a substance of no risk or lesser risk, where this is practicable
- reduction of exposure to the lowest level reasonably practicable by measures other than respiratory protective equipment (RPE)
- provision of RPE of an approved type or to an approved standard, ie if prevention is not reasonably practicable, substitution must be attempted; if this is not practicable then exposure must be reduced etc.

Practical aspects of control measures are considered on pp583–5.

Use of control measures (regulation 9)

Employers must ensure, so far as is reasonably practicable, that any control measure, personal protective equipment (PPE) etc is properly used or applied.

Employees for their part must make full and proper use of them and report defects forthwith.

Maintenance of control measures (regulation 10)

Employers must ensure that any control measure, PPE etc is maintained in a clean and efficient state, in efficient working order and in good repair. Exhaust ventilation equipment must be regularly examined and tested at suitable intervals by a competent person. Records of maintenance and repair must be kept for at least five years.

The ACOP requires that exhaust ventilation equipment is inspected every seven days and thoroughly examined and tested by a competent person every six months. It goes into some detail about how such inspections or examinations and tests are to be carried out.[9] Much of this matches measures described earlier to comply with the COSHH Regulations. Formal maintenance arrangements are also likely to be necessary for PPE (especially RPE), cleaning equipment (particularly vacuum cleaners) and washing and changing facilities.

Provision and cleaning of protective clothing (regulation 11)

- *Provision of protective clothing* Employers must provide adequate and suitable protective clothing for employees exposed to asbestos unless no significant quantity of asbestos is likely to be deposited on their normal clothes. Clothing is not necessary where dust emissions are low and short-lived. Direct contact of clothing with asbestos materials must also be taken into account. The ACOP gives examples of situations where protective clothing is unlikely to be necessary, for example:
    - hand drilling of damp asbestos cement below head level
    - handling products where the fibre is bonded
    - minor electrical or plumbing work
    - work near low-dust processes
    - fitting of gaskets.
- *Disposal or cleaning* Protective clothing must be either disposed of as asbestos waste or adequately cleaned at suitable intervals. Adequate cleaning is likely to involve washing, although dry cleaning may also be necessary to remove other materials.
- *Cleaning arrangements* Cleaning must be carried out on the work premises or at a suitably equipped laundry. Clothing being removed from the premises (for cleaning, further use or disposal) must be packed in a suitable container which is appropriately labelled (see below).
- *Clothing contamination* If an employee's personal clothing becomes signif-

icantly contaminated by asbestos then it must be treated as if it were protective clothing (and cleaned or disposed of accordingly).

Prevention or reduction of the spread of asbestos (regulation 12)

Employers must prevent the spread of asbestos from any place where work with asbestos is carried out or, if this is not reasonably practicable, reduce it to the lowest level reasonably practicable. This regulation rather duplicates the requirements of Regulation 8. Measures such as extraction, enclosure, cleaning or decontamination may be necessary.

Cleanliness (regulation 13)

Employers have a general duty to ensure that premises or parts of premises where asbestos work is carried out, together with plant used in the work, are kept in a clean state. Thorough cleaning is particularly necessary on the completion of asbestos work. Manufacturing processes giving rise to asbestos dust must be in buildings designed and constructed to facilitate cleaning and be equipped with an adequate and suitable vacuum cleaning system – a fixed system if reasonably practicable.

The ACOP refers to floors, workbenches, external plant surfaces and washing and changing facilities being cleaned at least once a day with inside walls and ceilings cleaned at least once a year. Ledges etc should be cleaned regularly to prevent asbestos accumulation. Fixed vacuum systems are preferable for cleaning purposes. Both fixed and portable vacuum filters must have high efficiency filters suitable for asbestos. Dry brushing or sweeping of asbestos should not be allowed. The hosing or washing down of surfaces is acceptable providing the residues are properly disposed of. Wiping of surfaces with a well-damped cloth is also permitted, providing the cloth is then placed in a suitable labelled container.

Designated areas (regulation 14)

Employers must designate an area as:

- an asbestos area, where exposure in that area for the whole of an employee's working time would be liable to exceed the relevant action level
- a respirator zone, where the concentration of asbestos would be liable to exceed the relevant control limit

Such designated areas must be clearly demarcated and identified by suitable notices as an asbestos area, a respirator zone or both, as the case may be. Only employees required to be there for work reasons may enter or remain in these areas. Eating, drinking and smoking is not allowed in these designated areas. Arrangements must be made by the employer for employees to eat or drink elsewhere.

Air monitoring (regulation 15)

Adequate steps must be taken to monitor the exposure of employees to asbestos where this is appropriate for health protection. Suitable records must be kept for at least five years and, where health records are required under regulation 16, for at least 40 years.

Monitoring may be by personal sampling or static sampling and might be necessary to:

- check the effectiveness of control measures
- confirm the right level of PPE has been chosen
- check the cleanliness of areas when work has ceased.

The need for monitoring is a matter of judgement and depends to a certain extent what information is available from other sources. Both the ACOP and an HSE Guidance Note contain information on monitoring procedures and records.[10]

Health records and medical surveillance (regulation 16)

Where employees are exposed (or are liable to be exposed) to asbestos in excess of the action level the employer must:

- ensure a health record is kept for each employee
- ensure employees have a medical examination not more than two years before exposure begins and at intervals of not more than two years.

Health records must contain extensive details of the asbestos work performed as well as information on medical examinations. The ACOP sets out the requirements in full.[11] Records (or copies) must be kept for at least 40 years.

Washing and changing facilities (regulation 17)

Employers must provide employees exposed to asbestos with adequate and suitable facilities for:

- washing and changing
- storage of protective clothing required
- separate storage of personal clothing not worn during working hours
- separate storage of RPE required.

In the case of low and infrequent exposure, washing and changing facilities may be those used by other workers. Greater exposure may necessitate the provision of separate facilities or the use of specialist units with 'clean' and 'dirty' changing areas separated by showers. Provision of vacuum cleaners for cleaning PPE or containers for collection of used RPE and protective clothing may also be necessary.

Storage, distribution and labelling (regulations 18 and 19)

Raw asbestos or waste which contains asbestos must not be stored, received into or despatched from any place of work or distributed within a place of work (except in totally enclosed distribution systems) unless it is in a suitable and sealed container clearly marked to show that it contains asbestos. Persons supplying products for use at work which contain asbestos must ensure that they are appropriately labelled. The labelling requirements are defined in Schedule 2 of the regulations.[12]

OTHER ASBESTOS-RELATED LEGISLATION

The Asbestos (Licensing) Regulations 1983

These regulations require those undertaking 'work in which asbestos insulation or asbestos coating is removed, repaired or disturbed' to hold a licence issued under the regulations. The term asbestos insulation excludes asbestos cement, asbestos insulating board and bitumen, plastic, resin or rubber articles which contain asbestos.

The requirement for a licence does not apply where:

- The total work does not exceed two hours and no one person will spend more than an hour on it in any seven consecutive days.
- The employer (using his own employees) or a self-employed person is undertaking the work at his own premises and has given advance written notice (generally 28 days) to the enforcing authority (the HSE or local authority).

Licences under these regulations are issued by the HSE even if the local authority is normally the enforcing authority. The HSE may impose conditions on the licence and may subsequently vary or add to them. It may refuse to issue a licence and can also revoke existing licences, subject to certain conditions. All work, whether carried out under a licence or not, must still be in compliance with the Control of Asbestos at Work Regulations.

The Control of Asbestos in the Air Regulations 1990

These regulations control the discharge of asbestos dust into the atmosphere. Their main requirements are:

- the concentration of asbestos in air emitted through discharge outlets must not exceed 0.1mg per cubic metre of air
- these concentrations must be measured at least every six months

(These two requirements are the responsibility of the person having control of the premises.)

- significant environmental pollution must not be caused by the working of products containing asbestos or demolition work involving asbestos

(This requirement is the responsibility of the person undertaking the work.)

The Asbestos (Prohibitions) Regulations 1992

Under these regulations all forms of asbestos other than chrysotile are collectively described as 'amphibole asbestos'. 'Amphibole asbestos' is in effect 'banned':

- its importation in any form is prohibited
- its supply, or supply of products to which it has intentionally been added, is prohibited
- use of it, or products containing it, is prohibited (subject to very limited exceptions).

Spraying of any form of asbestos is also prohibited.

Slightly more leeway is allowed in the case of chrysotile asbestos, although the supply and use of specified products containing chrysotile is prohibited. These are listed in the Schedule to the Regulations.[13] The prohibitions do not apply to any product which was in use before 1 January 1993 (unless it was prohibited under the previous regulations) and to activities in connection with disposal of such products.

Control of Pollution (Special Waste) Regulations 1980

The disposal of asbestos is subject to these regulations which place obligations on the waste producer, the waste carrier and the waste disposer.

ASBESTOS CONTROL MEASURES

Much has been published on control measures in relation to asbestos. One HSE booklet in particular gives a comprehensive description of the precautions to be taken in work with asbestos insulation, asbestos coating and asbestos insulating board.[14] Regulation 8 of the Control of Asbestos at Work Regulations requires a hierarchical approach to be taken to the selection of control measures. In most practical situations more than one of the measures required by this and other regulations is likely to be necessary. All possible control measures need to be considered in making an assessment (as required by regulation 5) or in preparing a plan of work (as required by regulation 5A). Many of the control measures referred to here match those previously referred to under the COSHH regulations.

Prevention

Much money has been spent and much asbestos exposure has been unnecessarily caused by ill-considered removal of asbestos-containing materials from buildings etc. Even if these materials are beginning to deteriorate other alternatives to removal should be considered, such as the application of a suitable sealant or the enclosure of the asbestos material. Removal may ultimately still be necessary but this is far easier to arrange in a building undergoing major refurbishment or demolition than in occupied premises.

Where asbestos work is underway, it is often necessary to erect a temporary enclosure around the work to prevent the exposure of those not involved in the work. Such enclosures are often kept under negative pressure by exhaust ventilation (fitted with suitable filters). Openings into the enclosure need to be correctly designed to aid the effectiveness of the ventilation and prevent contamination. Regular checking of the integrity of the enclosure and the efficiency of the ventilation will be necessary (see regulations 10 and 15).

Such enclosures are likely to be 'designated areas' (see regulation 14) and other 'designated areas' may also need to be identified with the use of barriers and signs.

Substitution

Substitution of asbestos materials has been required by the Asbestos (Prohibitions) Regulations or encouraged by the expense involved in providing necessary control measures. Even where the use of chrysotile asbestos is still permitted, most manufacturers and users are actively seeking substitutes before further prohibitions are imposed.

Exposure reduction

These control measures are very similar to the equipment and procedural-based controls already referred to under the COSHH Regulations (see Chapter 23).

Equipment-based controls

- enclosure – partial or total
- general ventilation
- local exhaust ventilation
- cleaning equipment (see regulation 13)
- washing and changing facilities (see regulation 17)
- waste containers.

All of these must be maintained (see regulation 10) which may require inspection, examination and testing. Air monitoring to ensure their effectiveness may also be necessary (see regulation 15).

Procedure-based controls

- work methods eg: use of wet methods or use of hand tools rather than mechanical methods
- cleaning procedures (regulation 13)
- waste collection and disposal methods
- personal hygiene/decontamination procedures (see regulation 17)
- emergency procedures.

Personal protective equipment

Respiratory protective equipment (RPE) should always be the last line of defence, only to be relied upon if it is not reasonably practicable to provide other control measures which will bring the concentration of asbestos in air within the relevant control limit. All RPE must comply with EC or HSE standards.

Suitable protective clothing will also be required, apart from very light exposure situations (see regulation 11) and suitable footwear and head covering may also be necessary. Where head covering is used it should normally be attached to the main overall so that it fits closely to the head and neck under the straps of any RPE. Arrangements will need to be made for cleaning and maintenance of all types of PPE in use and decontamination facilities may also be required.

LEAD – THE DANGERS AND CONTROLS

Lead poisoning

Lead poisoning is a disease reportable under RIDDOR and is also 'prescribed' for social security purposes. Its early physical symptoms include headaches, tiredness, constipation and loss of weight. In more severe cases the central nervous system is more seriously affected with symptoms such as twitching, restlessness and talkativeness. For pregnant women there may also be effects on the development of their child.

The Control of Lead at Work Regulations require medical surveillance of all those employees whose exposure to lead is significant. This usually includes testing of blood lead levels, with high levels resulting in reference to a doctor or even suspension from lead work. The HSE have produced a leaflet *Lead and You*[15] which explains to employees the dangers of lead and procedures for their medical surveillance.

Sources of lead

Inhalation of lead fumes or dust is the main danger although ingestion is also possible where poor standards of personal hygiene exist. Although lead paint is little used now, its use was once common, making the burning of lead painted metal in the demolition and scrap industries a particularly risky activity. Old metal structures and ships, especially naval vessels, usually present significant problems.

Large scale smelting of lead and the manufacture of lead batteries are obvious potential problem areas but these activities are restricted to a few locations which are generally well-equipped. The small scale reclamation of lead which takes place in many small scrapyards (which may include the breaking up of old batteries) could present greater risks in practice if suitable control measures are not applied.

The pottery industry is now only allowed to use glazes which are very low in lead content (described as 'leadless' or 'low solubility' glazes) and the use of lead in vitreous enamelling is relatively uncommon. There will also be some exposure to lead from the use of lead solders and other work with lead piping in plumbing.

Lead alkyls are still added to petrol as an 'anti-knock agent' and the manufacture and addition of these substances presents risks as does the manufacture of other lead compounds. Exposure to lead from vehicle exhaust fumes on public roads is excluded from the application of the Control of Lead at Work Regulations but exposure in other places (eg garages) is not.

The Control of Lead at Work Regulations 1980

These regulations follow a similar pattern to the COSHH Regulations and therefore only their main features have been included here. Full details of both the regulations and their associated ACOP are contained in the HSE booklet *Control of Lead at Work.*[16] Lead is defined in regulation 2 as meaning lead, its alloys, compounds and substances or materials containing it, which is liable to be inhaled, ingested or otherwise absorbed. Under regulation 3 employers have duties to protect others at work on the premises in the same respect as their own employees.

Assessment and control measures

Where work may expose persons to lead, regulation 4 requires an assessment to be made to determine the nature and degree of exposure. The ACOP gives some assistance in this respect[17]. Under regulation 6, employers must, so far as is reasonably practicable, adequately control exposure by means other than the use of respiratory protective equipment (RPE) or protective clothing.

Where use of RPE or protective clothing is necessary then an assessment must be made to determine what is required and the equipment chosen must comply with relevant EC or HSE standards (regulations 7, 8, 8A and 8B).

Other precautions

Many of the other precautions required under these regulations match those required by the COSHH Regulations or the Control of Asbestos at Work Regulations. They include:

- Information, instruction and training (regulation 5).
- Washing and changing facilities (regulation 9).
- Eating, drinking and smoking (regulation 10).
- Cleaning (regulation 11).
- Duty to avoid spread of contamination (regulation 12).
- Use of control measures (regulation 13).
- Maintenance of control measures (regulation 14).
- Air monitoring (regulation 15).
- Medical surveillance and biological tests (regulation 16).
- Records (regulation 17).

Notes

1. *The Control of Asbestos at Work Regulations 1987 and Approved Code of Practice* – HSE 1993 (L27) (Henceforth referred to as the Control of Asbestos ACOP)
2. *Work with asbestos, insulation, asbestos coating and asbestos insulating board. Approved Code of Practice* – HSE 1993 (L28)
3. *Probable asbestos dust concentrations at construction processes* – HSE 1989 (Guidance Note EH35 (Rev))
4. *Control of Asbestos ACOP* – paragraphs 1 to 5
5. *Asbestos – exposure limits and measurement of airborne dust concentrations* – HSE 1990 (Guidance Note EH10(Rev))
6. *Control of Asbestos ACOP* – paragraphs 10 to 20
7. *Ibid.* – Appendix

8. *Asbestos and You* – HSE 1993 (IND(G)107)
9. *Control of Asbestos ACOP* – paragraphs 45 to 53
10. *Ibid.* – paragraphs 85 to 94; and see note 5, above
11. *Ibid.* – paragraphs 95 to 97
12. *Ibid.* – Schedule 2
13. The Asbestos (Prohibitions) Regulations, SI 1992/3067
14. See note 2, above
15. *Lead and You* – HSE 1986 (MSA1)
16. *Control of Lead at Work Approved Code of Practice* – HSE 1985 (COP 2)
17. *Ibid.* – paragraphs 13 and 14

25

Personal Protective Equipment

INTRODUCTION

Use and management of PPE: the four stages

The use and management of personal protective equipment (PPE) can be seen as involving four stages:

- assessment
- selection
- deployment
- control.

This chapter will examine these four stages. It will look at *why* PPE is required. This will include the important consideration of risk assessment. A PPE risk assessment is the examination of the need for PPE including the level of protection and the evaluation of equipment design. After a broad risk assessment, the next stage is the selection process. This is the *what* stage: it will examine what type of PPE is required. It must also evaluate the legal aspects of the PPE to be purchased. This stage must include consideration of the intended use of the PPE in the workplace, and the manufacturer's product standards. Careful consideration must be given to both of these aspects when selecting the equipment.

Ergonomics of design of the PPE in relationship to acceptability to the user is an important issue. The vexed question of the acceptability of PPE to its wearer and user, combined with compatability to the task, should be addressed at this stage. Cost effectiveness of the protective equipment to be purchased should also be considered here.

The deployment stage is about considering *who* should use the PPE that is to be supplied and who will enforce its use. Is there to be a clear senior managerial commitment to its proper use as specified by the organisation's policy on health and safety? In the same vein, consideration must be given to whether senior management will make proper provision for adequate supplies of PPE, including replacement supplies, and where applicable spares. The final stage in the PPE process is about *when*. When should certain PPE be used? Once PPE is in use how should management control it in terms of training, storage, accommodation, maintenance, replacement, monitoring and review?

What is personal protective equipment?

In the principal UK PPE regulations (the Personal Protective Equipment at Work Regulations 1992)[1] PPE can be examined under two basic headings:

1. *Protective work clothing which in the main is worn*, and for example includes full and part body protection, footwear, gloves, head protection and clothing for weather protection, along with clothing for high visibility protection.
2. *Protective equipment which in the main is used by a form of attachment to the individual* and includes safety harnesses, respiratory protection and breathing apparatus, lifejackets and eye protection.

Although the principal PPE regulations do not cover hearing protection this should be considered in an overview of PPE as it is detailed in the Noise at Work Regulations 1989.[2] To the requirements of the PPE and Noise Regulations must be added a number of other items of PPE. These are discussed later in this chapter. There are also a variety of miscellaneous items including:

- barrier creams
- dust suits
- knee pads
- lumbar supports
- personal danger boards
- thermal clothing (for artificial cold conditions)
- safety nets
- water operations protective clothing
- personal gas/oxygen monitoring alarms
- intrinsically safe torches.

On the border between PPE and work equipment are items such as hand held or helmet lamps, portable radios, life-buoys and life line equipment, along with safety tools such as slingers hooks and hand guarded chisels. However, for an organisation undertaking an in-depth examination of their current position on PPE a broader definition may be more appropriate than a legal one. It may therefore be sensible to consider these borderline cases as PPE. In broader terms an organisation undertaking a review could consider PPE as:

Safety clothing, safety devices, equipment or tools that are personal or provided as temporary to staff for the purpose of protecting against or reducing injury to or the ill-health of any individual when employed at work. This can also include protection of the health and safety of authorised visitors to the organisation who could be affected by the organisation's operations or conditions. It could also be protection that may be necessary as a public hygiene requirement.

RELEVANT LEGISLATION

The general legal framework

The legal framework encompassing PPE in terms of conditions of use and standards is now quite extensive and detailed. It is however, mainly covered by two sets of regulations which are:

- The Personal Protective Equipment at Work Regulations 1992 (henceforth the PPE Regulations).
- The Personal Protective Equipment (EC Directive) Regulations 1992 (henceforth the EU Standards Regulations).[3]

The PPE regulations were drawn up for the safe and effective application of PPE while the EU Standards Regulations are aimed at performance standards of PPE and the supply of PPE of all kinds. The combined PPE Regulations and Guidance should be referred to by those wishing to obtain more detail on the subject.

There are a number of other sets of UK regulations which are mainly outside of the scope of the PPE regulations but clearly have a bearing on the subject of PPE management. These are:

- Control of Substances Hazardous to Health 1994 (see chapter 23).
- Noise at Work 1989 (see chapter 22).
- Construction (Head Protection) 1989 (see chapter 26).
- Control of Asbestos at Work 1987 (see chapter 24).
- Ionising Radiations 1985 (see chapter 22).
- Control of Lead at Work 1980 (see chapter 24).

These regulations all contain a PPE requirement and should always be considered along with the main PPE Regulations. In particular the Construction (Head Protection) Regulations make it a legal requirement to wear PPE on construction sites and projects. All six sets of regulations have a direct effect on the use of the correct PPE in given conditions. In the main these sets of regulations follow the same pattern of application as those detailed in the PPE Regulations.

The Personal Protective Equipment at Work Regulations 1992

There are eight key regulations.

An assessment (regulation 6)

Employers are required to carry out an assessment in order to examine the risks to health and safety within the place of work which cannot be avoided by means other than PPE. There is a requirement placed on the employer to ensure that the correct PPE is provided for the risks encountered in the circumstances of use.

The provision of PPE (regulation 4)

This regulation simply states that employers have a legal duty to provide PPE to employees who may be exposed to a risk to their health and safety. The exemption is where such a risk has been adequately controlled. The guidance accompanying this regulation states that no charge can be made to employees for the provision of PPE. Interestingly, this regulation links the PPE at Work Regulations with the EU standards on PPE. It states that PPE shall not be considered suitable unless it complies with any relevant Community Directives which are applicable to an item of PPE.

Compatibility of PPE (regulation 5)

Compatibility is still a problem associated with certain types of PPE. The PPE Regulations require employers to ensure that each of a combination of items of PPE, when worn together, continue to give effective protection against the specific risks they are intended to reduce or remove. Some manufacturers' products make this difficult as they are not wholly compatible when used in conjunction with each other.

Use of PPE (regulation 10)

This places a duty on both the employer and employee. The employer must make sure that any PPE provided is properly used. Employees must use the PPE provided and ensure it is returned to its accommodation after use.

Maintenance and replacement PPE (regulation 7)

This formally sets out what responsible employers are already practising, ensuring that an employee's PPE is maintained in an efficient state, is in working order and of good repair. This includes being replaced and cleaned as and when appropriate.

Accommodation for PPE (regulation 8)

This regulation requires the employer to ensure that appropriate accommodation is provided for PPE when not in use. The guidance notes accompanying this regulation point out that the accommodation must protect the PPE. The design of the accommodation must prevent ready to wear PPE from being mixed with PPE awaiting repair, cleaning or maintenance.

Reporting of loss or defect (regulation 11)

This regulation places a second duty on the employee. It states that every employee who has been provided with PPE shall report to the employer forthwith any loss or obvious defect in that PPE.

Training, information and instruction (regulation 9)

This regulation places a clear requirement on the employer to ensure that employees supplied with PPE are provided with information, instruction and training that is both adequate and appropriate. It goes on to define the information, instruction and training that should be provided:

- Employees must be given knowledge of the risk or risks which the PPE is intended to counter.
- Employees are to know the manner in which the various types of PPE are to be used.
- Employees are to understand their duties to ensure their PPE is effective, in working order and of good repair.

Information, instruction and training shall not be considered adequate or appropriate unless it is comprehensible to the PPE user.

Some information, instruction and training will be simple while other types can be complex. Some employees will need theoretical training along with practical training. Fitting and wearing information could form a starting point for this practical training. The guidance note indicates that a higher level of training will need to be provided for managers, supervisors and the workers who undertake examination and testing etc of PPE or who train others in these tasks. An important aspect of training is the information and instruction linked to the protection limitations of each type of PPE.

A good working rule would be first to evaluate every employee (or general groups) in terms of the use of PPE and then consider their training needs. This type of approach will enable the organisation to determine the duration and content of training sessions and examine re-training frequencies as recommended by the guidance notes. As with all training systems the guidance suggests holding training records. Finally it calls for employers to ensure that employees who undergo PPE training understand what they are being taught. In practical terms this is validation of training which can be fulfilled by two simple and well tried methods. One is to get each employee to sign a brief record of training specifying a date and time along with a summary of the subject matter under main classifications, ie eye protection, safety hats etc. The other method is a simple multi choice questionnaire with no more than ten key questions on each aspect of the PPE. A training certificate could also be issued.

The PPE (EC Directive) Regulations 1992

The other element of the legal framework of PPE relates to product standards enshrined in UK regulations published by the Department of Trade and Industry. These are enforced through The Personal Protective Equipment (EC Directive) Regulations 1992. To fulfil the objective of free movement of goods throughout the European market the European Union has sought to harmonise technical standards. The European Union standards on PPE are but one example of this process. The regulations set out how manufacturers are to demonstrate that their PPE products meet various essential requirements. Products meeting these requirements will carry the CE (Communities European) mark, which means that they can be sold anywhere in the European Union. In brief, any PPE products sold in or imported for sale in UK must:

- satisfy safety requirements based on levels and classifications of protection
- satisfy examination and surveillance by an approved body or in some cases satisfy approved methods of manufacture
- carry CE marking and information
- be accompanied by product use instructions
- have an associated technical file available.

The basic health and safety requirement of these regulations is that the PPE must preserve the health and ensure the safety of users, and must not harm other people or property when properly maintained and used for its intended design purpose. The overall basic health and safety requirement is satisfied by an approval system referred to as 'attestation'. This is an approval method (outlined in figure 25.1) under three headings of types of approval:

1. simple design PPE
2. complex design PPE; and
3. PPE which is neither simple, nor complex.

Simple design PPE This includes PPE which relates to the following:

- superficial mechanical (body) effects as in gardening work
- weak cleaning materials with easily reversible results
- handling hot components not exceeding 50°C
- atmospheric agents of neither exceptional or extreme nature
- minor blows, impacts, vibrations etc whose effects cannot cause irreversible lesions
- sunlight.

Complex design PPE This is intended to protect against mortal danger and includes:

- filtering respiratory devices
- full atmosphere insulating respiratory devices
- limited chemical attack and ionising radiations protection
- emergency equipment for high temperatures (100°C or over air temperature) and low temperature (–50°C or less air temperature) protection
- PPE to protect against falls from heights;
- PPE to protect against dangerous voltage risks
- motorcycle helmets and visors.

Most of the industrial and commercial PPE used in the UK will fall mainly into the middle band, between simple and complex. The basic message is that PPE

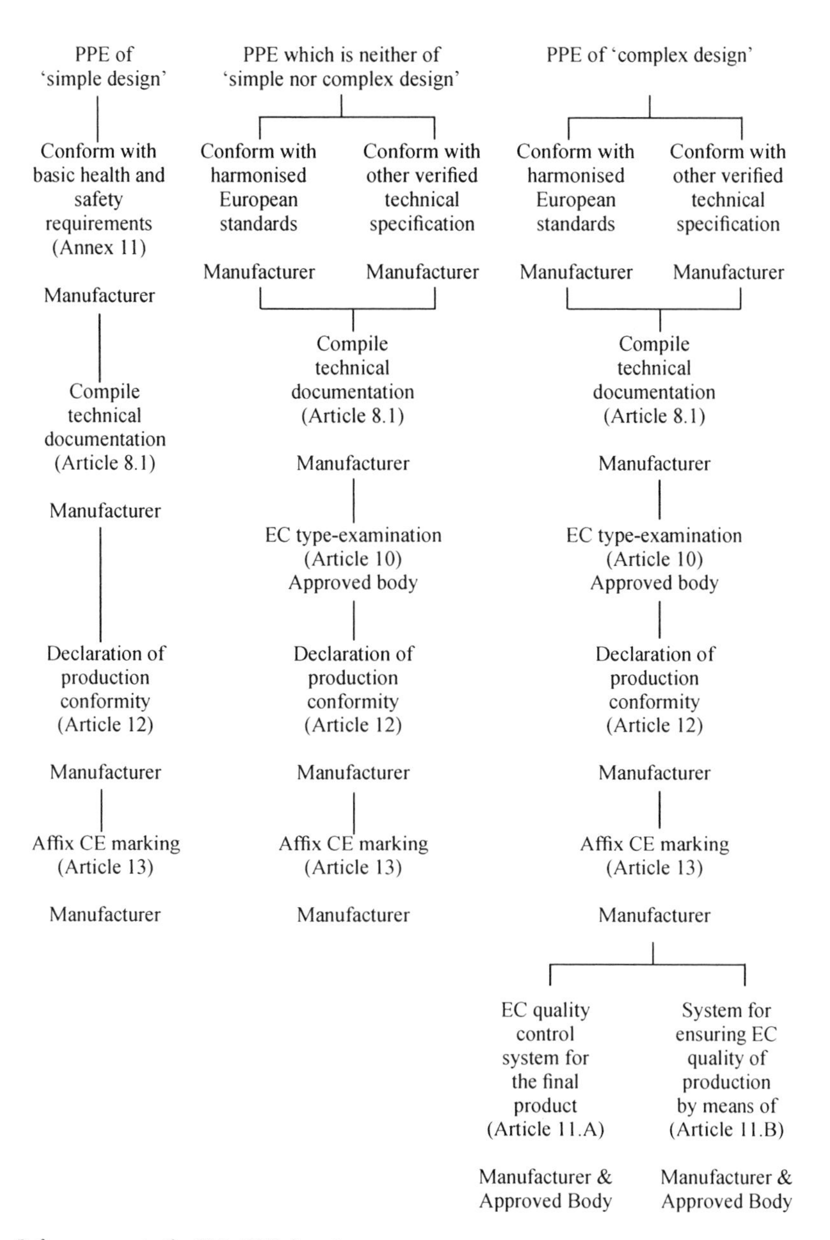

References are to the EU's PPE directive

Source: *PPE (EU) Regulations and Guidance*, © Crown Copyright reproduced with the permission of the Controller of HMSO.

Figure 25.1 *Attestation procedures– showing the responsibilities of manufacturers and Approved Bodies*

purchased by an organisation for its employees must comply with the CE standards. In all cases of approval, the manufacturer must affix the CE mark to both the PPE and its packaging. This must be followed by at least two figures: the year in which it was affixed, and where required, the approval body distinguishing number. Within the UK these PPE standards will be enforced by the Trading Standards Departments of local authorities.

Finally some of the main areas of PPE not covered by EU standards are:

- PPE for the armed services
- PPE for self-defence
- PPE designed for private use in adverse weather, damp, water or heat
- secondhand PPE.

RISK ASSESSMENT

Regulation 6 of the PPE at Work Regulations is closely related to the general requirement for a risk assessment contained in regulation 3 of the Management Regulations. PPE regulation 6 requires:

- an assessment of risk not avoided by other means
- definition of the characteristics required of PPE in order to be effective against that risk
- comparison of the characteristics of available PPE with the characteristics required of PPE.

While both regulations follow the same form, the associated guidance makes clear that the assessments are not a duplication. The guidance associated with regulation 3 states that PPE should be used as a last resort: PPE should be the final step in a hierarchy of safety control measures. The guidance note continues by stating that where the risks are sufficiently low, they can be considered to be adequately controlled. However, while it is clear what the HSE are stating, there could be some drawbacks.

EXAMPLE

Let us examine a task where, after a risk assessment has been undertaken, an organisation provides a lift trolley to move heavy blocks. The risk of the blocks slipping or falling when being manually handled has been eliminated. Therefore, no safety footwear is required or provided. But suppose the lift trolley is missing or becomes unserviceable, and employees need to keep the job going. They then revert to manual handling, and drop the heavy blocks onto their unprotected feet. The elimination of falling objects, or employees being struck by moving equipment, banging into fixtures, or even falling over causing head impact, is almost impossible. Therefore, while risk assessments are a legal requirement, experience sometimes dictates that (for example) footwear or safety helmets must be provided and worn even though the risk assessment finds the foreseeable risks are adequately controlled in normal circumstances.

Certain classes of PPE, such as safety footwear and safety helmets, should be regarded in most industries as standard issue. The legal requirement to wear safety helmets on building and construction sites is a prime example.

A risk assessment for PPE in any given task or work location should be seen as the 'why' element of PPE in the workplace. In essence, why should companies provide PPE? What PPE should be purchased and who should wear and use the PPE? The when element is when to use PPE both in principle and practice; it should be clearly underpinned by training, instruction and information for all employees who do or could use any form of PPE while at work.

A PPE checklist is set out in Appendix 1 to this chapter.

SELECTION OF PPE

Classification of PPE

Generally, PPE may be classified by body location. There are also combination types and a special category. The types of protection in the latter are not usually regarded as PPE.

General types by body location

- *Head protection* – eg safety helmets and hats.
- *Face protection* – eg shields, visors and screens.
- *Eye protection* – eg safety glasses and goggles.
- *Respiratory protection* – eg self contained and air line breathing apparatus, ori-nasal, positive pressure canisters and disposable respirators.
- *Hearing protection* – eg ear defenders and ear plugs.
- *Hand protection* – eg gloves and gauntlets, barrier creams.
- *Foot protection* – eg safety boots, shoes, socks and clogs.
- *Full body protection* – eg overalls, coveralls and work suits.
- *Part-body protection* – eg jackets, coats, trousers, aprons, shirts and waistcoats.
- *Arm protection* – eg oversleeves for lower or upper arms, full arm oversleeves, elbow pads.
- *Leg protection* – eg knee pads, spats, shin pads, over-leggings.
- *Back protection* – eg support belts and temperature control belts.

Combination types

In addition to the above, there are available a number of combinations of PPE. These combine different types of PPE functions. Many of the top manufacturers and suppliers of PPE make available double, triple and other combinations of PPE. These are designed to be compatible, each with the other, to facilitate the easy use of PPE by individuals who must wear a number of types of PPE simultaneously. Perhaps one of the best known is the forestry lumberjack combination. This is a safety helmet fitted with ear defenders and a steel mesh face shield visor with a weather resistant visor seal. The safety helmet is also fitted with a neck guard, sweatband and chin strap. Other examples are overalls in flame-resistant material with hoods and close fitting wrist and ankle adjustments. Another combination is a welder's eye and face shield: sometimes this is combined with a safety helmet.

Special types

There are a few types of PPE and some non-PPE items which can be listed under this heading. The list is not exhaustive but contains a number of the more common types found in general industry.

- safety harnesses and safety belts
- high visibility clothing
- personal danger boards (lock out tags)
- personal buoyancy equipment
- lifelines and lifebelts
- hand lamps
- tool belts
- barrier creams
- personal gas/oxygen monitors.

Standards

The standards applicable to PPE are detailed in Appendix 2 of this chapter. However, it should be noted that this is an ever-moving target as standards are constantly improving.

Types of PPE

Head protection

This can be in the form of a bump cap or much more commonly a safety helmet. Safety helmets come in many styles but all offer protection from impact and penetration from sharp objects. In addition, peaks on safety helmets can give limited protection if an individual falls or walks into obstructions, offering protection particularly to the nose and eyes. Safety helmets will of course also give all round protection to individuals when accidentally striking against fixed or moving objects. Some safety helmets will give limited protection against high or low temperatures and hot product splash. Wearing a safety helmet does not just protect against falling objects but can protect against a number of less likely hazards: the possibility of wearing one full-time should always be considered. One of the reasons why staff sometimes do not wear a safety helmet when it has been provided is that it does not fit securely. Consideration should always be given to purchasing safety helmets with a fitted chin-strap. Those wearing a safety helmet for the first time often complain about the weight of the helmet. This is usually a matter either of poor design or of a poorly adjusted head harness. If design and adjustment are correct, experience shows that people quickly get used to safety helmet weight. Purchasing of safety helmets with removable Terylene sweat bands is important for user acceptance.

Face protection

This is usually of three types: a clear face screen; a mesh shield; or a radiation shield. The face shields are manufactured in two sizes, one extending to below chin height while the other is at about chin height. The former can also be obtained with a chin-guard to prevent hazardous materials driving upwards to the face. Most face shields are bow-shaped and extend the face protection to the ears. Protection is offered against impact. Both shields can be chemical or molten-metal resistant. Face shields can be provided with tinted elements to protect against glare and reflective elements. They can also be provided to protect against heat. Obviously facial protection PPE will give protection to eyes and to some extent it is considered to be eye protection. However, face shields are only for face protection and should be employed in the main for that task unless a risk assessment demonstrates that they provide adequate eye protection.

Eye protection

This comes in three types: safety goggles, safety spectacles or glasses and eye shields. Safety goggles give the most complete all round protection as they fit closely to and on the face, forehead and nose, and are fitted with an adjustable headband allowing a close seal to be made around the eyes. Despite many developments over the last few years, wearer acceptability can still be a problem.

Impact protection and shielding are the most important safety issues to consider when selecting this type of PPE. The grade or class of impact protection should be known and evaluated. A number of types of safety goggles specify some protection against chemicals, dust, gas and molten metal. Other safety issues to consider are non-mist, no-fog, anti-mist, anti-scratch, anti-static and anti-UV requirements. Some safety goggles will allow the user to still wear prescription spectacles underneath. Other safety goggles offer 180° field of vision and some are of a lightweight design.

Finally, there are goggles designed specifically for welders. In addition to impact protection these provide anti-glare filter lenses. Welding lenses can be obtained with various grades of shading depending on what kind of welding is being carried out. These goggles are of two basic types. The first is a combined type which must be removed for normal vision. The second type is fitted with a flip-up front set of lenses for glare protection and a clear set of under fixed lenses for normal vision. This of course allows normal vision after or before welding without removing the safety goggles. In many cases this improves wearer acceptability. Lift up welding visors with eye protection are now commonly employed in welding operations. Developments in welding visor protection can provide eye and face protection integrated with a safety helmet; this is valuable in operations where both head and eye protection are necessary. Some welding visor protection can incorporate a high speed battery-powered automatic darking filter so that before an arc is struck the welder can clearly see the work area. This of course eliminates the requirement to nod-down or flip down (or up) the filter lens.

Safety spectacles are usually provided where the hazards found are of a lower level than those requiring safety goggles or face shields. Protection against general dust, especially wind-blown dusts is one of the main applications. There are many manufacturers of this type of eye protection and they come in all sorts of shapes, designs, colours and styles. Some are well-designed and shaped to reduce particle entry from cheek, forehead, and nose locations. Some are also fitted with adjustable arms (together with headbands) to obtain a close fit and reduce particle entry. Few safety spectacles these days are not fitted with side guards on the arms to prevent partial entry through the vulnerable area of the outer extremity of the eye.

Impact protection is an important consideration when purchasing safety spectacles as, in the main, the levels of protection are a grade or two below that of other types of eye protection. This is a vital factor in hazard identification and risk assessment. A simple problem found in a number of safety spectacles is the ergonomic design of the arm ear fitting. As a practical test anyone in an organisation who is responsible for ordering eye protection should be required, before purchasing, to wear the chosen safety spectacles for a few days. This test could be employed on a number of PPE items, would increase wearer acceptability and in the long-term would result in considerable savings in PPE expenditure. Some safety spectacles can be fitted with individual (medical) prescription lenses to protection grade standards. This has the advantage of increased wearer acceptability and offers protection to those requiring their ordinary spectacles during work.

Between safety goggles and safety spectacles come eye shields. They are not fully close-fitting like goggles but offer protection from particle entry which is better than that offered by a number of types of safety spectacles. The design is very similar to both spectacles and goggles but they are fitted with ear arms and can be used with a headband. On some types, both the angle and length of the arm are adjustable. These types are increasingly referred to as spoggles. They have the advantage of not misting and also of being easy to remove: this leads to them being more acceptable than goggles. Eye shields do not usually come up to the level of impact protection that the best safety goggles achieve. However, there are many applications, such as in workshops, where this type of PPE provides good eye protection. In some cases prescription spectacles can also be worn under eye shields.

Respiratory protection

The basic concept of respiratory protection equipment is detailed in both the COSHH and PPE Regulations. Respiratory protection must:

- be capable of adequately controlling the exposure and be suitable for the purpose that it is intended
- fully comply with the requirements imposed by the HSE in terms of protection standards.

Respiratory personal protection is issued mainly where risks of hazardous substances in the atmosphere are not fully controllable or will exceed the exposure limits laid down by the HSE. Only very rarely is respiratory protection used in a general manner. The prescribed tables of exposure are known as occupational exposure limits (OELs) (see chapter 23).

Respiratory protection is of four basic types:

- oxygen or air fed (known as breathing apparatus)
- power-assisted respirators (usually battery-powered)
- filter respirators
- disposable respirators.

The specific class of respirator to be employed will first depend upon the hazard classification. There are six principal types of hazard:

- dusts
- fumes
- gases
- mists
- vapours
- lack of oxygen to support life.

These hazards can affect the body in a number of ways, some of which can be fatal. (See chapter 13 for further information.)

With the exception of breathing apparatus all forms of respiratory protection require that sufficient oxygen exist at the workplace in order to support life (approximately 21% oxygen in the atmosphere).

Breathing apparatus comes in three main forms:

- self contained
- air line
- emergency escape sets.

Self contained breathing apparatus supplies pressurised purified air, usually to a full face mask with non-return exhaling valves to ensure a fully-sealed system. Where self contained back-carried cylinders are not required, or in difficult conditions, a purified compressed air feed can be connected to the breathing face mask. As noted, breathing apparatus is employed mainly when oxygen levels are low. It could however, be necessary in environments with high toxic hazards. Chemical resistance along with flame-retardant properties are important considerations.

Power-assisted respirators The second type of respirator (normally battery powered) is the power assisted type which has a fan (and motor) as an integral design feature within the filter. This type of respirator thus assists the user to breathe air in through the filter more easily. The battery is usually mounted on a belt. A heavy duty variation of this type of respirator carries a double filter and fan unit on a belt and feeds into the face mask via a flexible air tube.

Filter respirators Negative pressure (or non-powered) filter or cartridge respirators are essentially, as the name implies, systems whereby the air to be inhaled is first drawn through a filter. The type and level of protection required will depend on the type of filter to be employed on the equipment. Filters in single or twin units are usually fitted directly to the face mask. The face mask can be a full face mask or just an ori nasal mask (covering nose and mouth only). This is usually referred to as a half mask. When heavy duty filters are required they can be carried on a waist belt and a flexible air tube can be employed to connect the filter to the mask.

Disposable respirators The final and most commonly used personal respirator is the disposable respirator. This type of respiratory protection is a nose and mouth mask with one or two headbands. On some masks the headband is adjustable. A built in soft metal nose band allows the wearer to mould the respirator closely to fit the shape of their nose. Some are manufactured with an exhalation valve to reduce the build-up of heat and moisture inside the mask. This non-return valve also helps ease the breathing cycle. For especially hazardous applications some disposable respirators can be manufactured with flame retardant materials.

Finally, the design, selection and use of all forms of respiratory protection must be governed by a risk assessment carried out in line with regulation 6 of the COSHH Regulations 1994.

Hearing protection

Hearing protection comes in two main types, defenders or plugs. Defenders, referred to as ear muffs, are simply noise reduction cups fitting over the ears on a headband or mounted on fittings attached to a safety helmet.

Disposable ear plugs that can be directly inserted into the ear canal. They take the form of a pre-shaped plug, a mouldable plug or an expanding plug. Non-disposable ear plugs are solid reusable plugs. Some are linked together with a cord, others linked with a flexible chin band.

If noise levels are above 90db(A), hearing protection must be provided and worn which will reduce exposure levels to below 90db(A). If employees request hearing protection between 85db(A) and 90db(A) exposure levels, the employer must supply suitable hearing protection. (See chapter 22 on noise.)

Hand protection

Hand protection is an important area of PPE as most work tasks rely on the use of hands. Design and manufacturing standards are in three groups.

1. Hand protection of simple design which protects against minimal risk only.
2. Hand protection of intermediate design which protects against risks that are greater than minimal but less than mortal injury.
3. Hand protection of complex design which protects against irreversible or mortal injury.

There are gloves with or without cuffs and gloves with or without wrist grips. Gloves can also be non-slip and some kinds are disposable. Many kinds of gloves are liquid-proof or liquid-resistant. Other hand protection takes the form of gauntlets and mitts. For special applications, some gloves are manufactured with two fingers plus thumb/mitt cover. Under-gloves are also available for some applications.

When the type of hand protection has been selected it is vital to consider the 'risk-gaps'. There is a need to protect the gap between the glove and the sleeve of the jacket or coverall. The main risk is burns. Length and style of glove wrist protection is vital. The other risk is the possibility of hazardous substances getting down inside the glove. There are a number of liquid-resistant wrist or arm seals and quick-release fastenings which can overcome this problem. However if the communication gap between the purchasing department and the employees' supervisor is not bridged staff will continue to receive injuries from gloves which are poorly designed or unfit for their purpose. Other important points to be considered are size, fitting and material used. Ill-fitting gloves can cause wearer rejection. Most suppliers of quality gloves will have gloves available in small, medium and large sizes and may even provide special sizes when required. To ensure gloves are of the correct material for protection purposes a generic risk assessment on hand and lower arms will be necessary. For example, are there hazards present such as cutting edges, or sharp objects? Are hazardous substances present and are extremes of temperature a safety issue?

Acceptability of hand protection by wearers in the main centres on dexterity. If an employee finds it difficult to carry out a task because of wearing a glove there will be a temptation not to wear it. A fine and often difficult balance must be made between protection standards and user specification. However, with the increasing range of gloves available this is a reducing problem. Many manufacturers do and will design and manufacture to special user specifications.

The groups of hazards to be protected by hand protection and their design and manufacturing standards are listed as follows:

- *Chemical hazards*: protection against the permeation and penetration by chemicals to six levels of performance.
- *Cold hazards*: protection against three kinds of hazard: convective cold, contact cold and water.
- *Heat and fire hazards*: protection under four performance levels for flammability, contact heat, convective heat, radiant heat, small molten metal splashes and large molten metal splashes.
- *Impact cut hazards*: blade cut resistance to five performance levels.
- *Mechanical hazards*: abrasion, cut and puncture resistance to five performance levels.
- *Micro organisms*: resistance to penetration by micro organisms to six performance levels.
- *Static electricity hazards*: electrostatic discharge resistance to pass or fail standard only.

Foot protection

Foot protection takes three main forms. Safety shoes, safety boots or safety wellingtons with both men's and women's fittings and styles available. The main safety-design features and standards of safety footwear are as follows:

- toe impact steel protection to 100, 120, 160 or 200 joules (impact strength)
- heat resistant soles to 80°C, 120°C or 300°C
- acid and alkali resistant soles
- anti-static standards
- anti-slip extra grip
- steel pre-formed mid-soles
- oil resistant sole
- single, dual or triple density sole
- water repellent treatment
- shock absorbent heels and air cushion soles
- quick release fasteners
- additional upper foot steel protection.

In the UK over 30,000 accidents causing injury to feet are reported each year. These form 15% of reported industrial injuries. A foot injury usually means a person cannot get to work or carry out normal duties. Therefore investment in employees' safety footwear makes sound business sense.

As with hand protection, when the basic protection has been selected the 'risk-gaps' need examining before purchase is made. One gap is between the safety footwear and protective trousers. The ankle and possibly the lower leg may need protection eg against risk of sprain. The other gap is at the top or neck of the footwear which allows hazardous materials or substances to enter the shoe or boot.

Full and part body protection

Along with upper arm and leg protection, full and part body protection can be purchased in as many forms, colours, materials, styles and shapes as it is possible to imagine. In fact there will always be manufacturers who will make a number of types of PPE in these categories to most designs or specifications that the purchaser requires. Increasingly some of this PPE is being produced with corporate images, logos, badges and colours incorporated into the design. Gone or certainly going are the navy overalls. Basic considerations to be taken into account before purchase and use should be as follows:

- Is the PPE required full or part body protection?
- Does it need to be long lasting or disposable?
- Is there a requirement to protect against hot metal or does it need to be flame retardant?
- Will it be worn with any other types of PPE?
- Does it need to be chemical resistant or retardant?
- Does it need to be oil, grease or fat resistant or retardant?
- Is the PPE to be weatherproof?
- Is the PPE to be waterproof?
- Is it to protect against heat and high temperatures?
- Is it to protect against cold and low temperatures?
- Does it require a hood to provide neck, ears, head and limited face protection?
- What type of fastening is necessary, eg zips, buckles, buttons, velcro or quick release studs, and do these need give any flexibility?
- Does it require close fitting wrist bands, ankle bands, waist bands and neck bands?
- With which standards will the PPE need to comply?
- Is day time high visibility marking or material required?
- Is night time reflective markings or material required?
- Does the body protection need to be compatible with any other PPE to be used or worn?
- Does the body protection have flexibility of movement (climbing and bend-

ing etc) to prevent strain and sprain, especially back injury and torn muscles?
- Is the body protection manufactured from the best lightweight materials compatible with effective protection?

Back protection

This is a subject much to the fore with the arrival of the Manual Handling Operations Regulations 1992 (see chapter 19). Essentially, back protectors are very wide waist belts; some are air-inflated to give back support to employees in manual lifting operations. The belt moulds to the shape of the wearer's back and acts as a device for preventing muscular injury and lower back problems; in addition, it supports the abdomen. Some of these belts are designed to take hot or cold packs to adjust the wearer's body temperature when working in cold or hot conditions. This can reduce possible fatigue, helping the user to maintain correct lifting positions at work.

Protection against falling

Fall protection is of two basic types:

- harnesses
  - — full body
  - — chest
- belts
  - — restraining belts
  - — pole belts.

Many fatal accidents have occurred as the result of fall protection equipment not being provided or not being worn. Before deciding on the appropriate equipment to be employed, a risk assessment must be carried out. This will allow a safe work method to be established. Should it not be practicable to prevent the employee working within one metre of an edge from which they might fall, a full body safety harness with a lanyard should be used. However, the possibility of using a physical method to make the edge safe should be investigated first. When working on roofs, even from cat-ladders or similar, full safety harness should always be employed.

While a chest harness is more user acceptable than a full body harness, the full harness is safer on two counts. First, it spreads the fall impact between the upper and lower body contact points reducing friction and cutting injury, and results in less body/organ impact damage. Second under certain conditions with particular body shapes it is possible for a person to be torn out of a chest harness, especially if not correctly adjusted.

There is one special type of safety harness, a rescue harness. This is specifically designed to assist the rescue of personnel usually from underground areas or work at heights.

Most safety practitioners will not permit the use of safety **belts** with lanyards as more often than not their use is abused. This is because of the simple fact that many users do not understand that a safety belt is a restraining device only and a safety harness is a protecting device. Because of this problem it is always better to go for the higher standard and employ a full body safety harness.

Pole belts are wide support belts which can be clipped to anchor points allowing a person to stand on fixed steps or rungs with hands free to work. These are used, for example, on electric line poles, window ledges and man-hole access ladders.

The fitting of lanyards as anchor equipment is every bit as important as the safety harness or belt and is subject to a number of BS or EN standards: some are listed in Appendix 2. Lanyards with fittings must not exceed two metres in length. Some lanyards can be purchased with a built-in shock absorber: when activated the total lanyard length must not exceed 2.65 metres. Fall prevention equipment of all types should be the subject of a three-monthly inspection and examination, which must be recorded.

Personal danger boards

Alhough personal danger boards may not normally be considered PPE, they do protect against risks from moving machinery, electricity and pressurised systems. They are personal and they are a simple piece of equipment. Personal danger boards are a plastic board, 150mm x 75mm containing the owner's name and other identification. They can, where applicable, be equipped with a unique padlock which is chained to the board. This type of PPE would be employed by the individual to secure immobilisations and or isolations. The use of personal danger boards is dealt with in chapter 7.

MANAGEMENT CONTROL OF PPE

Management policy

Regulation 9 of the PPE Regulations makes clear the duty of the employer to the employee in terms of providing:

- information
- instruction
- training.

This regulation also makes clear that the information must be adequate and appropriate as well as comprehensible to the individual to whom it is provided. Training must include theoretical as well as practical elements.

What does the provision of suitable information, instruction and training involve? It should involve a simple, systematic and all-embracing approach to managerial control of PPE. For example, giving information about risks and the appropriate use of PPE is unlikely to be effective if employees are not instructed on the care, inspection and maintenance of PPE.

Before embarking on a programme of information, instruction and training, management policy should be drawn up and formally agreed with the organisation's employees. The policy must cover the following key points:

- why PPE needs to be used (a summary of risk assessments)
- when PPE is to be used (full time or when necessary?)
- who will need to use PPE (all employees, visitors, contractors?)
- what PPE will be used (legal requirements and CE standards).

The policy should specify what PPE has to be worn and the conditions under which it must be worn. It should not usually be left to individual employees to decide what PPE to use and when to use it. This issue can be made clear at PPE training sessions by reference to the legal requirement (regulation 10(2)) for employees to use and wear PPE supplied by an employer as a result of the employer's duty under the PPE Regulations.

To ensure that employees do not ignore their legal duty to use or wear PPE, a disciplinary code for the non-use of PPE could be drawn up by management and agreed by employees. The enforcement process will require monitoring by management. Other managerial control issues which should be detailed in the policy are:

- Who (individual or department) will be responsible for the purchasing of PPE?
- Who (individual or department) will be accountable for the issue of PPE?
- Who (individual or department) is responsible for supplying PPE training ?

Content of information, instruction and training

In terms of subject-matter, four types of PPE training could be undertaken. These are:

- existing general PPE
- special high risk PPE
- emergency PPE and
- new PPE.

The duration and the form that training should take will be dependent on the category of employee who is to attend and the type(s) of PPE to be covered. All types of PPE training are likely to cover the following:

- formal instruction as to who needs PPE and when PPE is to be used (ie company policy)
- information briefings (PPE 'tool-box talks')
- practical training (ie putting on PPE and wearing it)

- inspection and testing information (including demonstrations)
- maintenance instructions (cleaning, replacement and repair)
- legal duties of employees
- disciplinary code.

The specific subject content of PPE training will differ according to the category of employee to be trained: the principal categories are managers, supervisors and PPE users. Possible training content for each category is indicated below.

- *Management*
  - hazards and risks
  - legal duties
  - selection and purchasing
  - issue and spares system
  - training policy for PPE
  - control policy
  - review policy
  - disciplinary system
- *Supervision*
  - risk assessment
  - legal issues
  - type of PPE
  - issue of PPE
  - spares and replacement
  - training responsibilities
  - PPE checks (fault-finding)
  - the checking of working conditions
  - monitoring systems
  - records of issue
  - disciplinary system
- *PPE users*
  - risks
  - application of PPE
  - performance limits of PPE
  - legal responsibilities
  - fitting and compatibility of PPE
  - ergonomics of the PPE
  - care of PPE
  - accommodation and storage of PPE
  - inspection of PPE and observation of damage
  - maintenance, cleaning and repair
  - loss and damage-reporting system
  - spares and replacement system
  - system for recording the issue of PPE

— disciplinary system
— validation of training records.

A pocket card could be issued to users. Two types of cards are proposed but equally this sort of information could be included in safety manuals, unit trainer manuals or safe working procedures.

The first card proposed is a general statement of local standards and user/wearer requirements. This example is shown in figure 25.2. However each organisation will need to draw up and refine its own user/wearer standards list.

In this department the following PPE is applicable:

- Safety helmets are mandatory
- Hearing protection is mandatory in areas where signs indicate that it is necessary
- Safety footwear is mandatory in risk areas (again where signs indicate it is mandatory)
- Eye protection is a legal requirement in operations where signs so indicate
- Gloves are necessary on specified operations (see supervisor)
- Respiratory protection is essential on specified tasks
- Breathing apparatus is a legal requirement on specified jobs
- Safety harnesses are critical to some specified tasks
- Personal danger boards must be used to secure immobilisation and isolation
- Body protection must be considered in some specified areas

Figure 25.2 *PPE standing instruction*

The second card proposed is a PPE specification card which could be used to detail exactly what type of PPE is mandatory in a works, plant, department or for a particular task. In addition the card (on the reverse) could detail inspection and spares schedules together with replacement and storage instructions. An example of this card is shown in figure 25.3. The capital letters outline the pre-printed card data headings while the text suggests possible information to be included.

Other relevant documentation might be:

- a record, with dates and signatures, of the issue to individual employees of various PPE
- a validation record (containing dates, subject detail and signatures) of instruction, information and training.

Most manufacturers or suppliers of PPE will supply free, extensive and valuable literature on all aspects and on all types of PPE. Some can also supply useful training videos and may undertake the limited PPE training of company employees free of charge.

<table>
<tr><td colspan="3">PPE (Detail what PPE is covered by card eg safety helmet etc)</td></tr>
<tr><td>STANDARD
AND
TYPE</td><td>APPLICATION</td><td>WEARING AND FITTING</td></tr>
<tr><td>BS or BSEN standards and CE mark details. Manufacturer's name and type details. Include stores number</td><td>SUITABLE FOR
Details of hazards, risks
Specify when it must be worn, what location, jobs or tasks.
State mandatory, legal, essential, critical or necessary etc.
List signed areas.</td><td>Specify why PPE is required to be employed and how it is to be worn or fitted to meet accident or ill-health risks. Detail adjustment settings, size data and clearance settings.</td></tr>
<tr><td colspan="3">NOT BE USED FOR

</td></tr>
</table>

(Front of PPE card)

(Back of PPE card)

<table>
<tr><td colspan="4">INSPECTION AND TEST</td><td>DATE ISSUED ____________
ISSUED TO ____________</td></tr>
<tr><td>DATE</td><td>INITIALS</td><td>DATE</td><td>INITIALS</td><td>REPLACEMENT METHOD</td></tr>
<tr><td></td><td></td><td></td><td></td><td rowspan="7"></td></tr>
<tr><td></td><td></td><td></td><td></td></tr>
<tr><td></td><td></td><td></td><td></td></tr>
<tr><td></td><td></td><td></td><td></td></tr>
<tr><td></td><td></td><td></td><td></td></tr>
<tr><td></td><td></td><td></td><td></td></tr>
<tr><td></td><td></td><td></td><td></td></tr>
<tr><td></td><td></td><td></td><td></td><td>SPARES AVAILABLE</td></tr>
<tr><td></td><td></td><td></td><td></td><td rowspan="7"></td></tr>
<tr><td></td><td></td><td></td><td></td></tr>
<tr><td></td><td></td><td></td><td></td></tr>
<tr><td></td><td></td><td></td><td></td></tr>
<tr><td></td><td></td><td></td><td></td></tr>
<tr><td></td><td></td><td></td><td></td></tr>
<tr><td></td><td></td><td></td><td></td></tr>
</table>

Figure 25.3 *PPE pocket card*

APPENDIX 1: P.P.E. REVIEW CHECKLIST

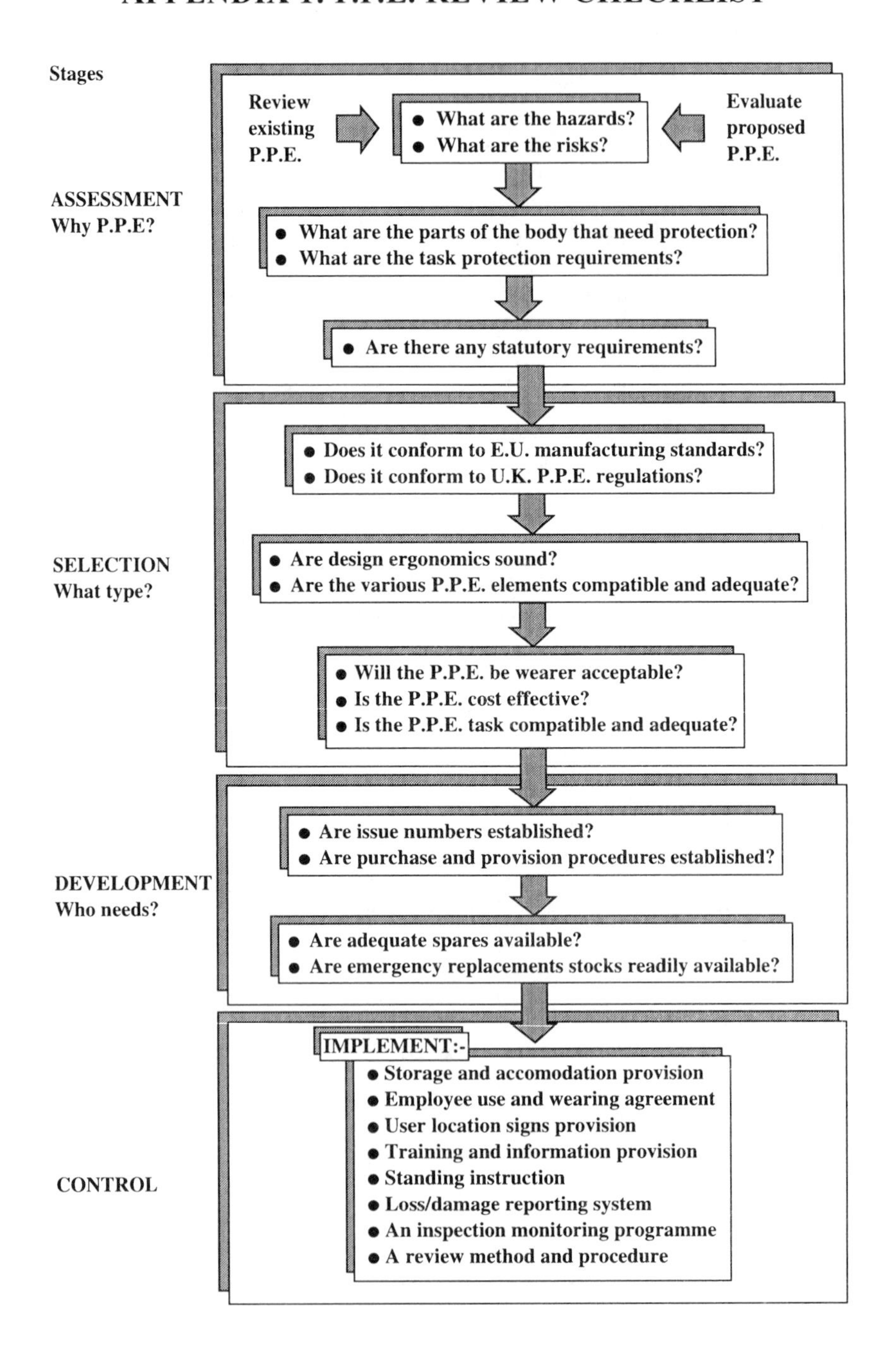

Stages
ASSESSMENT
Why P.P.E?
Review existing P.P.E.
• What are the hazards?
• What are the risks?
Evaluate proposed P.P.E.
• What are the parts of the body that need protection?
• What are the task protection requirements?
• Are there any statutory requirements?
SELECTION
What type?
• Does it conform to E.U. manufacturing standards?
• Does it conform to U.K. P.P.E. regulations?
• Are design ergonomics sound?
• Are the various P.P.E. elements compatible and adequate?
• Will the P.P.E. be wearer acceptable?
• Is the P.P.E. cost effective?
• Is the P.P.E. task compatible and adequate?
DEVELOPMENT
Who needs?
• Are issue numbers established?
• Are purchase and provision procedures established?
• Are adequate spares available?
• Are emergency replacements stocks readily available?
CONTROL
IMPLEMENT:-
• Storage and accomodation provision
• Employee use and wearing agreement
• User location signs provision
• Training and information provision
• Standing instruction
• Loss/damage reporting system
• An inspection monitoring programme
• A review method and procedure

APPENDIX 2 BRITISH AND HARMONISED EUROPEAN STANDARDS

Note

The reader of this list will understand that PPE standards identification is a constantly changing scene. A number of original British Standards have been replaced by harmonised European Standards. In the United Kingdom these standards are designated EN standards with a British Standard prefix ie BS EN. Standards prefixed with 'pr' are provisional standards at the time of going to print 1995.

Head protection

BS 3864:1989 *Specification for protective helmets for firefighters* (To be replaced by BS EN 443)
BS 4033:1966 *Specification for industrial scalp protectors (light duty)* (To be replaced by BS EN 812)
BS 5240 Part I:1987 *Industrial safety helmet – specification for construction and performance* (To be replaced by BS EN 397)

Eye protection

BS 1542:1982 *Specification for equipment for eye, face and neck protection against non-ionising radiation arising during welding and similar operations*
BS 2092:1987 *Specification for eye protection for industrial and non-industrial uses* (To be replaced by BS EN 166, 167 and 168)
BS 6967:1988 *Glossary of terms for personal eye protection* (To be replaced by BS EN 165)
BS 7028:1988 *Guide for selection, use and maintenance of eye-protection for industrial and other uses*
BS EN 169 *Personal eye protection: Filters for welding and related techniques: Transmittance requirements and recommended use*
BS EN 170 *Personal eye protection: Ultraviolet filters: Transmittance requirements and recommended use*
BS EN 171 *Personal eye protection: Infrared filters: Transmittance requirements and recommended use*
pr EN 165 *Personal eye protection: Vocabulary*
prEN 166 *Personal eye protection: Specifications*
prEN 167 *Personal eye protection: Optical test methods*
prEN 168 *Personal eye protection: Non-optical test methods*

Footwear

BS 953: *1979 Methods of test for safety and protective footwear*

BS 1870: Part 1:1988 *Specification for safety footwear other than all rubber and all plastic moulded compounds*
BS 1870: Part 2:1976 (1986) *Specification for lined rubber safety boots*
BS 1870: Part 3:1981 *Specification for polyvinyl chloride moulded safety footwear*
BS 2723:1956 (1988) *Specification for fireman's leather boots*
BS 4676:1983 *Specification for gaiters and footwear for protection against burns and impact risks in foundries*
BS 4972:1973 *Specification for women's protective footwear*
BS 5145:1989 *Specification for lined industrial vulcanised rubber boots*
BS 5462: *Footwear with midsole protection*
BS 5462: Part 1:1984 *Specification for lined vulcanised rubber footwear with penetration resistant midsoles*
BS 5462: Part 2:1984 *Specification for lined or unlined polyvinyl chloride (PVC) footwear with penetration resistant midsoles*
BS 6159: *Polyvinyl chloride boots*
BS 6159: Part 1:1987 *Specification for general and industrial lined or unlined boots*
BS 7193:1989 *Specification for lined lightweight rubber overshoes and over-boots*

The following will probably replace BS 1870 and BS 953:
prEN 344 *Requirements and test methods for safety protective and occupational footwear for professional use*
prEN 345 *Specification for safety footwear for professional use*
prEN 346 *Specification for protective footwear for professional use*
prEN 347 *Specification for occupational footwear for professional use*
prEN 381 *Protective clothing for users of hand held chain saws: part 3 Test method for boots*

Gloves

BS 697:1986 *Specification for rubber gloves for electrical purposes*
BS 1651:1986 *Specification for industrial gloves*
BS EN 421 *Protective gloves against ionising radiation to include irradiation and contamination*
prEN 374 (Parts 1 to 5) *Protective gloves against chemicals and micro-organisms*
prEN 381 (Parts 1 to 6) *Protective gloves for users of hand held chain saws*
prEN 388 *Protective gloves: Mechanical test methods and specifications*
prEN 407 *Protective gloves against thermal hazards*
prEN 420 *General requirements for gloves*
prEN 511 *Protective gloves against cold*
prEN 659 *Fire fighters' gloves: Protection against heat and flame*

Hearing protection

BS EN 352–1 *Hearing protectors – safety requirements and testing Part 1: ear muffs*
BS EN 352–2 *Hearing protectors – safety requirements and testing Part 1: ear-plugs*

Protective clothing

BS 1547:1959 *Specification for flameproof industrial clothing (materials and design)*
BS 1771: Part 1:1989 *Specification for fabrics of wool and wool blends*
BS 1771: Part 2:1984 *Specification for fabrics of cellulosic fibres, synthetic fibres and blends*
BS 2653:1955 *Specification for protective clothing for welders*
BS 3595:1981 *Specification for life-jackets*
BS 3791:1970 *Specification for clothing for protection against intense heat for short periods*
BS 5426:1987 *Specification for workwear and career wear*
BS 6249:1982 *Materials and material assemblies used in clothing for protection against heat and flame*
BS 6249: Part 1:1982 *Specification for flammability testing and performance*
BS 6629:1985 *Specification for optical performance of high visibility garments and accessories for use on the highway*
prEN 340 *General requirements for protective clothing*
prEN 342 *Protective clothing against cold weather*
prEN 343 *Protective clothing against foul weather*
prEN 366 *Protective clothing: protection against heat and fire: method of test: evaluation of materials and material assemblies when exposed to a source of radiant heat*
prEN 367 *Protective clothing: Method of determining heat transmission on exposure to flame*
BS EN 368 *Protective clothing – Protection against liquid chemicals – test method resistance of materials to penetration by liquids*
BS EN 369 *Protective clothing – Protection against liquid chemicals – test method resistance of materials to permeation by liquids*
prEN 373 *Protective clothing: Assessment of resistance of materials to molten metal splash*
prEN 381 (parts 1 to 6) *Protective clothing for users of hand held chainsaws*
prEN 393 *Life-jackets for personal buoyancy aids: buoyancy aids, 50 N*
prEN 394 *Life-jackets and personal buoyancy aids: additional items*
prEN 395 *Life-jackets and personal buoyancy aids: life-jackets, 100 N*
prEN 396 *Life-jackets and personal buoyancy aids: life-jackets, 150 N*
prEN 399 *Life-jackets and personal buoyancy aids, life-jackets, 275 N*
EN 412 *Protective aprons for use with hand-knives*

prEN 463 *Chemical protective clothing: protection against liquid chemicals: method of test: determination of resistance to penetration by liquids (Jet Test)*
prEN 464 *Chemical protective clothing: protection against gases and vapours: method of test: determination of leak tightness (internal pressure test)*
prEN 465 *Protective clothing: protection against liquid chemicals: performance requirements: type 4 equipment: protective suits with spray-tight connections between different parts of the protective suit*
prEN 466 *Chemical protection clothing: protection against liquid chemicals (including liquid aerosols): performance requirements: type 3 equipment: chemical protective clothing with liquid-tight connections between different parts of the clothing*
prEN 467 *Protective clothing: protection against liquid chemicals: performance requirements: type 5 equipment garments providing chemical protection to parts of the body*
prEN 468 *Chemical protective clothing: protection against liquid chemicals: method of test: determination of resistance to penetration by spray*
prEN 469 *Protective clothing for fire-fighters*
prEN 470 *Protective clothing for use in welding and similar activities*
prEN 471 *High visibility warning clothing*
prEN 510 *Protective clothing against risk of being caught up in moving parts*
prEN 531 *Protective clothing for industrial workers exposed to heat (excluding fire-fighters' and welders' clothing)*
prEN 532 *Clothing for protection against heat and flame: Method of test for limited flame spread*
prEN 533 *Clothing for protection against heat and flame: Performance specification for limited flame spread of materials*
prEN 702 *Protective clothing: Protection against heat and fire – Test method: Determination of the contact heat transmission through protective clothing or its materials*

Respiratory protection

EN 132 *Respiratory protective devices Definitions*
EN 133 *Respiratory protective devices Classification*
EN 134 *Respiratory protective devices Nomenclature of components*
EN 135 *Respiratory protective devices List of equivalent terms*
EN 136 *Respiratory protective devices Full face masks: requirements, testing, marking*
EN 140 *Respiratory protective devices half-mask and quarter-masks requirements, testing, marking*
EN 141 *Respiratory protective devices Gas filters and combined filters requirements, testing, marking*
EN 142 *Respiratory protective devices Mouthpiece assemblies: requirements, testing, marking*

EN 148–1 *Respiratory protective devices Threads for face-pieces: Standard thread connection*
EN 148–2 *Respiratory protective devices Thread for face-pieces: Centre thread connection*
EN 136-10 *Respiratory protective devices – full face masks for special use – requirements, testing, marking*
EN 137 *Respiratory protective devices – self-contained open-circuit compressed air breathing apparatus – requirements, testing, marking*
EN 145 2 *Respiratory protective devices – self-contained closed-circuit compressed oxygen breathing apparatus for special use – Part 2: requirements, testing, marking*
EN 148-3 *Respiratory protective devices – threads for face-piece Part 3: thread connection M 45 x3*
BS EN 371 *Respiratory protective devices – AX gas filters and combined filters against low-boiling point organic compounds – requirements, testing, marking*
BS EN 372 *Respiratory protective devices – SX gas filters and combined filters against organic compounds – requirements, testing, marking*

Emergency respiratory equipment

EN 400 *Respiratory protective devices for self-rescue – self-contained closed-circuit breathing apparatus – compressed oxygen escape apparatus – requirements, testing, marking*
EN 401 *Respiratory protective devices for self-rescue – self-contained closed-circuit breathing apparatus – compressed oxygen – escape apparatus – requirements, testing, marking*
EN 403 *Respiratory protective devices for self-rescue – filtering devices with hood for self-rescue from fire – requirements, testing, marking*
EN 405 *Respiratory protective devices for self-rescue – valved filtering half-masks to protect against gases or gases and particles – requirements, testing, marking*

Protection against falls from height

BS EN 341 *Personal protective equipment against falls from a height – descender devices*
BS EN 353-1 *Personal protective equipment against falls from a height – guided-type fall arresters Part 1: guided-type fall arresters on a rigid anchorage line*
BS EN 353-2 *Personal protective equipment against falls from a height – guided-type fall arresters Part 1: guided-type fall arresters on a flexible anchorage line*
BS EN 354 *Personal protective equipment against falls from a height – lanyards*

BS EN 355 *Personal protective equipment against falls from a height – energy absorbers*
BS EN 358 *Personal protective equipment against falls from a height – work positioning systems*
BS EN 360 *Personal protective equipment against falls from a height – retractable type fall arresters*
BS EN 361 *Personal protective equipment against falls from a height – full body harnesses*
BS EN 362 *Personal protective equipment against falls from a height – connectors*
BS EN 363 *Personal protective equipment against falls from a height – fall arrest systems*
BS EN 364 *Personal protective equipment against falls from a height – test methods*
BS EN 365 *Personal protective equipment against falls from a height – general requirements for instructions for use and for marking*

Source: *PPE at Work: Guidance on Regulations*, booklet L25, HMSO, 1992, pp36–9 © Crown Copyright reproduced with the permission of the Controller of HMSO.

Notes

1. Personal Protective Equipment at Work Regulations, SI 1992/2966
2. Noise at Work Regulations, SI 1989/1790 On noise more generally see chapter 22.
3. The Personal Protective Equipment (EC Directive) Regulations, SI 1992/3139

26

Construction and Demolition

Much of the legislation and good practice which applies generally to health and safety at work also applies to construction and demolition processes. However, the Workplace Regulations do not apply to construction sites (see chapter 16); moreover, construction work is covered by a separate body of law which specifically relates to health and safety in construction. The present chapter deals with this construction-specific legislation, with the more common hazards found in construction work and with the management of health and safety in this area. The specific legal framework is taken as the starting point.

CONSTRUCTION LEGISLATION

Construction (Design and Management) Regulations 1994

Introduction

The Construction (Design and Management) (CDM) Regulations require a systematic approach to be taken to the co-ordination and management of construction projects. This must start at the conception, design and planning stages, continue through work on site and extend into subsequent maintenance and repair. The regulations are based upon the interrelationship between several key roles (the same person or organisation may fulfil more than one role) and introduce some important new terms. These are:

- *Clients* These are the people for whom the project is carried out. They (or their formally appointed agents) must appoint competent people as planning supervisor, designer(s) and principal contractor and must ensure that sufficient resources are allocated to the project in order to comply with the law.
- *Planning supervisor* Those appointed to this role have overall responsibility for co-ordinating health and safety aspects in the design and planning phase. They must ensure that a health and safety plan is prepared, monitor health

and safety aspects of the design, advise the client on resource implications and prepare a health and safety file.

- *Designers* They must design so as to avoid, reduce or control risks as far as reasonably practicable. This duty includes future maintenance as well as the construction phase. They must identify any risks which still remain.
- *Principal contractor* In preparing and presenting tenders etc the principal contractor must take account of specific health and safety requirements. He must develop the health and safety plan (originally prepared by the planning supervisor) and co-ordinate the activities of all contractors and sub-contractors so that the plan and relevant health and safety legislation is complied with.
- *Health and safety plan* Such plans are expected to operate in two stages:
  - *Pre-construction* – bringing together relevant health and safety information from the client, designers, planning supervisor etc.
  - *Construction* – the practical steps which need to be taken to address these issues and also to deal with other risks which develop or are introduced as the project progresses.
- *Health and safety file* This is intended to be a maintenance manual for those who will inherit the project after completion. It should identify risks which may exist during future maintenance, repair, renovation or demolition and provide assistance in dealing with these risks.

The Regulations are accompanied by an ACOP which also details how various parts of the Management Regulations and the PUWE Regulations have particular relevance to construction projects.[1] Further guidance booklets have also been published by the HSE.[2]

Definitions

Several important terms are defined in regulation 2. These include:

- *Construction work* This is defined as "the carrying out of any building, civil engineering or engineering construction work". The definition goes on to give many specific examples of what this includes, for example:
  - construction, alteration, conversion, fitting out, commissioning, renovation, repair, upkeep, redecoration or other maintenance, de-commissioning, demolition or dismantling of a structure
  - the preparation for an intended structure
  - the assembly of pre-fabricated elements to form a structure (or its subsequent disassembly)
  - the removal of a structure or part of a structure and related product or waste
  - work relating to mechanical, electrical, gas, compressed air, hydraulic, tele-communications, computer or similar services.

This is not an exhaustive list – in cases of doubt reference should be made to the full definition.

- *Contractor* This means "any person who carries on a trade, business or other undertaking (whether for profit or not) in connection with which he
  (a) undertakes to or does carry out or manage construction work
  (b) arranges for any person at work under his control (including, where he is an employer, any employee of his) to carry out or manage construction work;"
- *Notifiable* "A project is notifiable if the construction phase
  (a) will be longer than 30 days; or
  (b) will involve more than 500 person days of construction work"

Other terms defined within the regulation are:

- Agent
- Cleaning work
- Client
- Construction phase
- Design
- Designer
- Developer
- Domestic client
- Health and safety file
- Health and safety plan
- Planning supervisor
- Principal contractor
- Project
- Structure
- Arrange
- Being reasonably satisfied.

Application (see figure 26.1)

The Regulations apply to construction work but with the following exceptions (regulation 3):

- where the local authority is the enforcing authority
- where the work is being carried out for a domestic client
- where the project is not notifiable *and* involves fewer than five workers.

Local authority exception

The HSE is the main enforcing authority for construction work. However, in premises normally inspected by the local authority (eg offices and shops) they also have jurisdiction over minor construction work. Such cases require the following criteria to be met:

- the work is carried out by people normally working on the premises; and
- the work is non-notifiable (see above and regulation 7)
- the work is entirely internal
- physical segregation or suspension of normal activities are not involved.

In these circumstances the work is completely removed from the scope of the CDM Regulations.[3]

Domestic client exception
This exception applies where the construction work is included in a project for a domestic client. "Domestic client means a client for whom a project is carried out not being a project in connection with the carrying on by the client of a trade, business or other undertaking (whether for profit or not)" (regulation 2(1)). However, the requirement to notify the HSE in writing (regulation 7) will apply if the "construction phase" of the project is longer than 30 days or involves more than 500 person days of construction work (regulation 2(4)); construction phase is defined in regulation 2(1)). Moreover, there is no exception from the requirements placed upon designers under regulation 13. Commercial developers who sell houses before the project is complete will remain clients to whom the Regulations apply (regulation 5).

Non-notifiable projects involving fewer than five workers
The Regulations will not apply where the client has reasonable grounds for believing that the project is not notifiable (using the 30 days/500 person days test) *and* the largest number of persons carrying out construction work at any one time is (or will be) fewer than five. However, the exception does not apply to the demolition or dismantling of a structure ie the Regulations do apply in such cases even if the work is non-notifiable and employs fewer than five workers. More generally, the exception does not apply to the duties placed upon designers under regulation 13.

Other questions of application
The Regulations apply to Great Britain but are extended beyond Great Britain by the same order that applies to HSWA, although not to the same extent.[4] The HSE is the enforcing authority for the Regulations (regulation 22). In general, the Regulations cannot be used as a basis for civil actions (ie breach of statutory duty – see chapter 4).[5] However, there are two exceptions:

- Regulation 10 – the duty of the client to ensure, so far as is reasonably practicable, that the construction phase does not start without preparation of a health and safety plan meeting the requirements of regulation 15.
- Regulation 16(1)(c) – the duty placed upon the principal contractor to "take reasonable steps to ensure that only authorised persons are allowed into any premises or part of premises where construction work is being carried out".

Duties of clients

"Client means any person for whom a project is carried out, whether it is carried out by another person or is carried out in-house" (regulation 2(1)). Every client has the following duties under the Regulations:

- to appoint a planning supervisor (PS) (regulation 6)
- to appoint a principal contractor (PC) (regulation 6)
- to satisfy himself that the PS, PC and designer are competent to perform their roles (regulation 8)

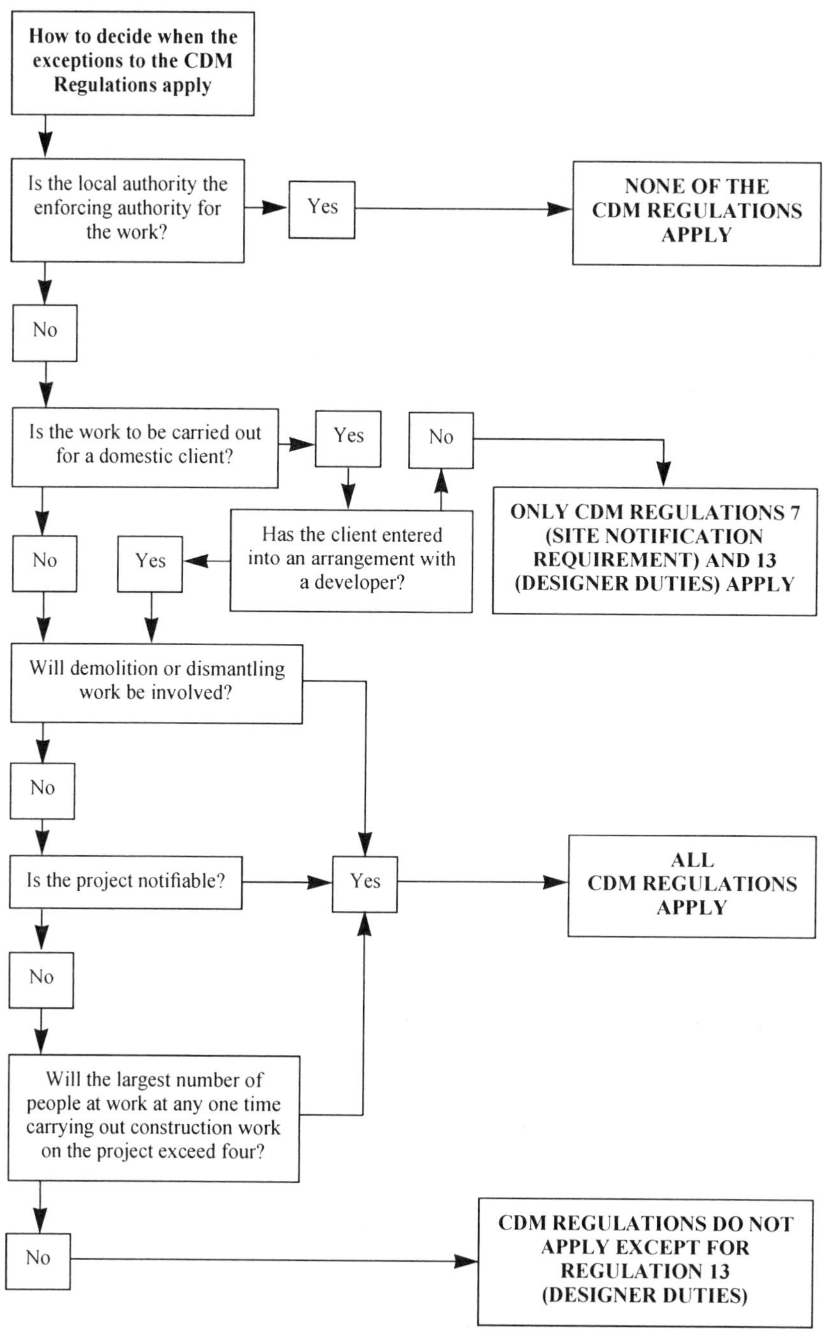

[Source: *HSC Guide to Managing Health and Safety in Construction*, HSE Books, 1995, p4
© Crown Copyright reproduced with the permission of the Controller of HMSO.]

Figure 26.1 *Application of the CDM Regulations*

- to satisfy himself that the PS, PC and designer will allocate adequate resources for health and safety (regulation 9)
- to ensure that the construction phase does not start without the preparation of a health and safety plan (HASP) (regulation 10)
- to ensure that certain information is available to the PS (regulation 11)
- to ensure that the health and safety (HAS) file is available.

Appointment of PS and PC

There must be a PS and a PC for each project. They should be appointed as soon as is reasonably practicable after the client has the information necessary to appoint them. There is no restriction on who may be appointed as a PS: it could be a company, partnership or individual. The PC must be a contractor (regulation 6(2)). The PS and PC could be the same person and could be the client. However, there is a requirement (in regulation 8) that the client shall not appoint a PS, PC or designer unless "reasonably satisfied" that they are competent to perform their statutory duties. The ACOP gives guidance on checking for competence and checking that the PS, PC and designer will allocate adequate resources for health and safety.[6]

Preparing a plan before construction

The client must ensure, so far as is reasonably practicable, that construction does not begin until a HASP complying with the Regulations has been prepared (regulation 10).

Information for the PS

The client is under a duty to provide to the PS the information necessary for the PS to perform his duties under the Regulations. The information should be provided as soon as is reasonably practicable and certainly before the work commences. The information relates to the state or condition of any premises at the site and should be that which the client can ascertain by making reasonable enquiries (regulation 11).

Availability of the HAS File

It is the client's duty to take reasonable steps to ensure that the information in the HAS File is kept available for inspection by any person who may need the information in order to comply with his statutory duties (regulation 12). At the end of the project, the PS delivers the HAS File to the client. The client's duty is to keep the information available for those involved in future work at the site.

Liability of clients

The client will be liable for breach of the duties mentioned above unless he or she appoints an agent or another client to act as the only client in relation to a particular project (regulation 4). The client must be reasonably satisfied that anyone he or she intends to appoint must be competent in terms of meeting his duties under the Regulations. Where the appointed person declares in writing that he will act as the client, the liability under the Regulations lies with him.

In the absence of such a declaration, the appointed person will be liable so far as within his authority while the client will also be liable.

Where a project is carried out for a domestic client and the client enters into an arrangement with a developer, the Regulations apply to the developer as if he were the client (regulation 5). This covers the situation where commercial developers sell domestic premises before the construction is complete but arrange for construction to be completed. In such a situation, the developer is to be treated as if he were the client.

Duties of designers[7]

The designer is under a duty to take reasonable steps to ensure that the client is aware of the duties placed upon clients under the CDM Regulations (regulation 13). "Design" includes drawings, design details, specifications and bills of quantities. "Designer" means "a person who carries on a trade, business or other undertaking" in connection with which he prepares a design or arranges for a person under his control (including an employee) to prepare a design (regulation 2(1)).

Designers must ensure that design considerations show adequate regard for the need:

- To avoid foreseeable risks to those carrying out construction or cleaning work and others who may be affected (this applies to the initial construction and to future construction and cleaning work).
- To combat such risks at source.
- To give priority to the protection of *all* persons who might be affected rather than merely the protection of each person, in order to secure the greatest benefit.

There is a duty to make available adequate information which might affect anyone carrying out construction or cleaning work or any person affected by such work.

The above matters are to be included in the design only if it is reasonable to expect it, and only to the extent that it is reasonably practicable. In addition, there is a duty to co-operate with the PS and any other designers to the extent necessary for each of them to fulfil their duties under the Regulations. Designers have responsibilities in relation to the use of an article *after* construction as a result of s6 of HSWA (see chapter 3).

Duties of the PS

As indicated, a project is notifiable to the HSE if it is longer than 30 days or involves over 500 person days. The responsibility for notification lies with the PS (regulation 7). The notice must be given in writing (unless the HSE approves otherwise) and contain the details set out below. HSE Form 10 (rev) can be used for this purpose. Where the project is carried out for a domestic client in the

- Date of forwarding
- Exact address of the construction site
- Name and address of the client or clients
- Type of project
- Name and address of the planning supervisor
- A declaration signed by or on behalf of the planning supervisor that he has been appointed as such
- Name and address of the principal contractor
- A declaration signed by or on behalf of the principal contractor that he has been appointed as such
- Date planned for start of the construction phase
- Planned duration of the construction phase
- Estimated maximum number of people at work on the construction site
- Planned number of contractors on the construction site
- Name and address of any contractor or contractors already chosen

[*Source:* Schedule 1 to the CDM Regulations 1994]

Figure 26.2 *Particulars to be notified to the HSE under regulation 7 of the CDM Regulations*

absence of any arrangement with a developer, the principal contractor has the duty of notifying the HSE.

Under regulation 14 the PS must ensure, as far as reasonably practicable, that designs are such that the designers have carried out their duties under regulation 13(2) and must take reasonable steps to ensure co-operation between designers.

The PS must be in a position to:

- Advise any client or contractor about the competence of designers and their ability to allocate adequate resources for health and safety.
- Advise any client about the competence of contractors and their ability to allocate adequate resources for health and safety, and to advise them as to whether a HASP complying with the Regulations has been prepared and construction may start.

The PS is responsible for preparing the HAS File. This should include the design information (under regulation 13) and any other information which it is reasonably foreseeable will be necessary to ensure health and safety at work. There is a duty to review the File, to ensure it is up to date and to deliver it to the client.

The central requirement of the Regulations, and the cornerstone of the PS's role, is the duty to ensure that a HASP has been prepared before construction commences (regulation 15). The HASP must contain:

- a general description of the work
- completion date (including completion dates for any intermediate stages)
- details of risks to the health and safety of persons carrying out construction work if known to the PS or reasonably foreseeable
- any other information known or obtainable by reasonable enquiry which a contractor would need to have to demonstrate his competence or his ability to make adequate provision for health and safety at work; or which a contractor would need to develop the HASP in accordance with his duties under regulation 15(4) and to meet statutory requirements as to welfare.

Guidance as to what should be contained in the HASP is given in figure 26.3.

1. **Nature of the project**
   - Name of client
   - Location
   - Nature of construction work to be carried out
   - Timescale for completion of the construction work

2. **The existing environment**
   - Surrounding land uses and related restrictions, eg premises (schools, shops or factories) adjacent to proposed construction site, planning restrictions which might affect health and safety.
   - Existing services, eg underground and overhead lines
   - Existing traffic systems and restrictions, eg access for fire appliances, times of delivery, ease of delivery and parking.
   - Existing structures, eg special health problems from materials in existing structures which are being demolished or refurbished, any fragile materials which require special safety precautions or instability problems.
   - Ground conditions, eg contamination, gross instability, possible subsidence, old mine workings or underground obstructions.

3. **Existing drawings**
   - Available drawings of structure(s) to be demolished or incorporated in the proposed structure(s) (this may include a health and safety file prepared for the structure(s) and held by the client).

4. **The design**
   - Significant hazards or work sequences identified by designers which cannot be avoided or designed out and, where appropriate, a broad indication of the precautions assumed for dealing with them.

(Continued over)

- The principles of the structural design and any precautions that might be needed or sequences of assembly that need to be following during construction.
- Detailed reference to specific problems where contractors will be required to explain their proposals for managing these problems.

5. Construction materials

- Health hazards arising from construction materials where particular precautions are required either because of their nature or the manner of their intended use. These will have been identified by designers as hazards which cannot be avoided or designed out. They should be specified as far as is necessary to ensure reliable performance by a competent contractor who may be assumed to know the precautionary information that suppliers are, by law, required to provide.

6. Site-wide elements

- Positioning of site access and egress points (eg for deliveries and emergencies)
- Location of temporary site accommodation
- Location of unloading, layout and storage areas
- Traffic/pedestrian routes

7. Overlap with client's undertaking

- Consideration of the health and safety issues which arise when the project is to be located in premises occupied or partially occupied by the client

8. Site rules

- Specific site rules which the client or the planning supervisor may wish to lay down as a result of points 2 to 7 above or for other reasons eg specific permit-to-work rules, emergency procedures

9. Continuing liaison

- Procedures for considering the health and safety implications of design elements of the principal contractor's and other contractors' packages
- Procedures for dealing with unforeseen eventualities during project execution resulting in substantial design change and which might affect resources

[Source: CDM Regulations 1994, ACOP, pp40–1 (Appendix 4) © Crown Copyright reproduced with the permission of the Controller of HMSO.]

Figure 26.3 *Contents of the health and safety plan*

Duties of contractors

As noted, the PC must develop the HASP. He must take such measures as are reasonable to ensure that the HASP contains the following features:

- arrangements for ensuring health and safety taking into account the risks involved in the construction work and any other work activity on site; and
- sufficient information for any contractor to understand how to comply with the statutory welfare provisions.

It is the PS's responsibility to ensure that the HASP is drawn up and available to prospective PCs at the tendering stage so that they are aware of the project's health and safety requirements (see ACOP, paras 77–8). The HASP then has to be developed by the PC: the plan is the foundation upon which the PC will base his health and safety management of the project. Among other things, it will include the PC's duties under regulations 16–18 (see below), the assessments prepared by contractors under the Management Regulations and the information about emergency procedures to be given to employees, again under the Management Regulations. Guidance is provided in the ACOP, paras 82–88.

Regulation 16 lays further duties on the PC:

- to take reasonable steps to secure co-operation between contractors
- to ensure, so far as is reasonably practicable, that contractors and employees comply with safety rules
- to take reasonable steps to ensure only authorised persons are allowed on site
- to ensure that the notice given to the HSE under regulation 7 is displayed
- to promptly provide the PS with information needed for the HAS File.

The PC is also given powers:

- to give reasonable directions to any contractor to allow the PC to meet his duties under the CDM Regulations
- to include safety rules in the HASP. However, these rules must be brought to the attention of those affected by them.

Under regulation 17 the PC must ensure, so far as is reasonably practicable, that every contractor is provided with comprehensible information about the risks to that contractor or anyone under his control. The PC must also ensure, again, so far as is reasonably practicable, that any contractor provides his employees with the information and training required by the Management Regulations (regulations 8 and 11 respectively).

There is a requirement in regulation 18 of the CDM Regulations for the PC to ensure that employees and self-employed persons are in a position to discuss health and safety matters and offer advice to him. There must also be arrangements made to coordinate the views of employees or of their representatives.

Finally, duties are placed upon all contractors to co-operate with the PC, provide the PC with necessary information, and comply with any safety rules or directions. Contractors must promptly provide the PC with information for the reporting of injuries, diseases and dangerous occurrences, as well as information that the PC should give to the PS for inclusion in the HAS File. No employer shall cause or permit any employee to work on construction work unless the employer has been provided with certain information, notably:

- the name of the PS
- the name of the PC
- the contents of the HASP (or any part(s) relevant to the work of the employee).

The same applies to self-employed persons. In both cases it will be a defence if the person can show he made all reasonable enquiries and believed he had been provided with the information or that regulation 19 did not apply because of one of the exceptions in regulation 3.

Other construction legislation

Section 127 of the Factories Act 1961 applied a number of provisions contained within the Act to 'building operations and works of engineering construction'.[8] Most of these provisions have since been repealed, superseded by more recent legislation applying to all work activities. However, five sets of regulations made under the Factories Act still govern operational requirements in respect of construction health and safety.[9] They are:

- Construction (General Provisions) Regulations 1961.
- Construction (Lifting Operations) Regulations 1961.
- Construction (Working Places) Regulations 1966.
- Construction (Health and Welfare) Regulations 1966.
- Construction (Head Protection) Regulations 1989.

There are also regulations governing safety standards for plant and equipment to ensure conformity with EU rules. Included here are regulations covering protection against noise, falling objects and roll-over.[10]

A central requirement of the General Provisions Regulations is that every contractor and every employer of workmen who normally employs more than 20 persons at any one time must appoint a safety supervisor (regulation 5). This person must be experienced and suitably qualified for the purpose. The appointment has to be made in writing. The safety supervisor's job is to advise the contractor or employer on observance of the legal requirements and generally to promote health and safety and monitor compliance with the law. The supervisor's name must be entered on the abstract of the Regulations which has to be displayed. The safety supervisor may be part-time, can be shared between contractors or employers and is allowed to have responsibility for more than one site.

The remainder of the legal requirements contained in the various sets of regulations are discussed below under subject headings:

- safety at heights
- excavations and confined spaces
- demolition
- tools and equipment
- site facilities.

Prior to this discussion, it is necessary to briefly describe the proposed Construction (Health, Safety and Welfare) Regulations.

Proposed Construction (Health, Safety and Welfare) Regulations

In May 1995 the HSC issued a consultative document on its proposals for a new set of regulations to consolidate the General Provisions, Working Places and Health and Welfare Regulations[11] and to give effect to the minimum safety and health requirements for construction sites as laid down in the EU's construction sites directive.[12] The intention is that health, safety and welfare on construction sites will be governed primarily by two complementary sets of regulations – the CDM Regulations and the proposed Health, Safety and Welfare Regulations – the former dealing with organisation and management, the latter with minimum standards. It is expected that the new regulations will be made in 1996.

The Construction (Lifting Operations) Regulations 1966 will remain. These are likely to be revoked in two to three years' time as the result of a separate consolidation of UK lifting legislation. In the meantime, parts of the Working Places Regulations relating to lifting operations[13] will be transferred to the Lifting Operations Regulations, which will be modified to apply to "construction work" rather than the old definitions (see next paragraph). Office accommodation and welfare facilities on construction sites will be brought within the scope of the Workplace Regulations 1992 (see Chapter 16).

The proposed regulations use the broad definition of construction work found in the CDM Regulations.[14] This combines the old concepts "building operations" and "works of engineering construction" and adds the installation or removal of services which are part of a structure and the installation or dismantling of fixed plant where there is a risk of falling two metres or more. The new regulations will apply to all construction work wherever it is carried out. However, some provisions, such as those relating to traffic routes, will apply only to construction sites (as distinct from construction work). A construction site is defined as a place where construction work is set apart, so excluding minor construction work in occupied premises.

The proposed regulations contain a number of new requirements.

Safe places of work

Every place of work shall, so far as is reasonably practicable, have working space that is adequate for the work being carried out. Suitable and sufficient steps shall be taken to ensure, so far as is reasonably practicable, that no person gains access to any unsafe place.

Prevention of falls

An intermediate guard rail or equivalent will be necessary to prevent falls from a height of two metres or more. Minimum standards for such rails etc will be laid down.

Falling objects

Where necessary, suitable and sufficient protection against falling objects will include provision of covered routes.

Traffic routes

Construction sites must be organised, so far as is reasonably practicable, to allow pedestrians and vehicles to move safely. Further provisions are set out to give effect to this requirement.

Doors and gates

Doors and gates need to be fitted with safety devices if necessary to meet the requirements laid down.

Excavating and materials-handling machinery

So far as is reasonably practicable, such machinery will need to be fitted with devices to protect the driver against crushing and falling objects.

Inspection of scaffolds

Scaffolding will have to be inspected prior to use and after exposure to certain conditions.

Fire precautions

New provisions are proposed in respect of fire-fighting devices, fire detectors and alarm systems, examination and testing, fire drills and signage. On fire safety generally, see chapter 21.

Emergency routes and exits

There will be a new requirement for emergency routes or exits to lead as directly as possible to a safe area. There will be new requirements to keep routes and exits free from obstruction and to provide emergency lighting. There are other detailed requirements.

Ventilation

There shall be a sufficient quantity of fresh or purified air. Provision is made for warning of failure of ventilation plant and for atmospheric testing.

Temperature

So far as is reasonably practicable, the temperature must be reasonable taking into account the purpose for which the place is used and any protective clothing provided.

Good order

Construction sites will need to be kept clean and tidy with the perimeter sign-posted.

Artificial lighting

The colour of artificial lighting must not adversely affect or change the perception of any health and safety sign or signal.

Office accommodation and welfare facilities

The Workplace Regulations[15] are to be applied to construction sites (see chapter 16).

SAFETY AT HEIGHTS

The Construction (Working Places) Regulations contain many detailed requirements relating to ladders, scaffolds and roofwork. Mobile elevating working platforms have become much more common since the passing of the regulations and they have also been included as a topic in this section as has the closely related subject of lifting operations.

Ladders

Regulations 31 and 32 of the Working Places Regulations govern the construction, maintenance and use of both ladders and folding step-ladders. All ladders must "be of good construction, of suitable and sound material and adequate strength". Many of the requirements of the regulations together with other points relating to the use of ladders are contained in the checklist in figure 26.4. Most of the checklist is also applicable to non-construction work.

- Ladders should only be used for access or for short-term light work
- Check ladders before use for defects eg twisted or cracked stiles, loose or missing rungs
- Don't use home-made ladders or ladders with makeshift repairs
- Don't paint ladders – the paint may hide defects
- Use ladders on a firm level surface
- Don't use loose packing material to gain extra height
- Rest the top of the ladder against a solid surface
- Use the ladder at a 75 degree angle (1 out for 4 up)
- Ladders must extend 1.05 metres above any landing place (unless other adequate hand-holds are available)
- Secure the ladder by its stile, preferably at the top
- If the ladder cannot be fixed a second person should foot it (this includes while it is being fixed)
- The base of the ladder may need protection against passing traffic
- Beware using ladders (especially aluminium ones) near electrical conductors
- Carry tools or equipment in shoulder bags or belt attachments – keep both hands free for climbing
- Beware of overreaching when working from ladders
- Only one person should be on a ladder at any one time
- Watch for wet, icy, muddy or greasy rungs

HSE Guidance Note GS31 contains much detailed advice on the SAFE USE OF LADDERS, STEP-LADDERS AND TRESTLES[16]

Figure 26.4 *Ladders checklist*

Scaffolds

The requirements of the Working Places Regulations were formulated with traditional scaffolding (often referred to as general access scaffolds) in mind. Recent years have seen the increasing use of tower scaffold systems to provide quick access both on construction sites and for general maintenance work. Such tower scaffolds are subject to the same legislative requirements as general access scaffolds. Many of these requirements together with other aspects of scaffolding safety are included in the checklists in figure 26.5.

Mobile elevating work platforms (MEWPs)

Such platforms (often referred to as cherrypickers) can be used to provide access in situations where use of ladders would be unsafe and scaffolding either uneconomical or impracticable. The platform itself must be provided with upper and lower guard rails and toeboards and the whole unit subjected to a

GENERAL ACCESS SCAFFOLDS

- Scaffolds must be competently designed in relation to anticipated loadings
- They must be erected, altered and dismantled only by competent workers
- Construct scaffolds on firm, level foundations
  – provide additional support as necessary
- Stabilise the structure with adequate braces and ties
- Fully board all platforms with properly supported scaffold boards
- Provide guard rails, mid rails and toeboards
- Provide safe access by properly secured ladders
- Avoid ground level tripping hazards and head level projections
- Carry out weekly inspections by a competent person
  (Form F91, the scaffold inspection register, lists items to check)
- Also inspect after alterations or bad weather eg strong winds

For further advice consult HSE Guidance Note GS 15
– GENERAL ACCESS SCAFFOLDS [17]

TOWER SCAFFOLDS

- Always follow the manufacturer's or supplier's instructions
- Erect the tower vertically on a firm, level surface
- Secure the tower to a rigid structure if it is likely to be subject to significant horizontal forces eg winds
- Lock the wheels and/or outriggers
- Provide safe access by properly secured ladders
  – normally inside the tower
- Board out the working levels
- Provide guard rails, mid rails and toeboards
- Inspect as for general access scaffolds – see above

For further advice consult HSE Guidance Note GS42
– TOWER SCAFFOLDS [18]

Figure 26.5 *Scaffolding checklists*

regular programme of inspection and maintenance to ensure that it remains in good condition.

- *Operators* Operators of MEWPs must be fully trained and familiar with the manufacturer's instructions for the type of platform in use. They should also know what to do if the platform fails in an elevated position.

- *Siting* MEWPs must be sited on firm level ground capable of supporting their load. Care should be taken that they are not placed on uncompacted backfill, hidden cellars, drainpipes etc. Their tyres should be fully inflated and outriggers or stabilisers used where necessary. Platforms should be positioned so that none of their parts protrude into traffic routes or should be protected from passing traffic. Access to unprotected electrical conductors or dangerous machinery should be prevented.

- *Methods of use* Platforms must always be used within their operating limits – weight loading, elevation and outreach. They are not intended for use in lifting or transporting goods other than for the work being done from the platform. They must not be used as a jack to support anything. Handling of sheet materials in windy conditions will be difficult and strong winds (over 30 mph) may affect the stability of the platform. Those using MEWPs could be made more secure by wearing harnesses attached to the platform.

Roofwork

The Construction (Working Places) Regulations contain detailed requirements affecting roofwork. Many fatalities each year involve roofs – not just in specialist roofwork but in other activities such as inspection, general maintenance work and cleaning. The main potential risks of roofwork are summarised below together with the precautions necessary to overcome them.

Falls through roofs

- *Risks* Many roofs are made of non-load-bearing materials such as asbestos cement or plastics. Other roofing materials may deteriorate over time eg rusted steel or rotted wood. Openings in roofs and skylights also present risks. Skylights are sometimes painted over or difficult to see in certain light conditions.
- *Precautions*

— Check the condition of the roof before venturing onto it.
— When in doubt always use crawling boards or roof ladders.
— Don't chance walking along lines of roof bolts or roof ridges.

Falling off

- *Risks* Falls may also occur from pitched roofs and flat roof edges or during gaining access to the roof from below. The risks of falling off are far greater during bad weather (eg wind, rain, ice or snow) or if the roof is covered with moss, lichen or other slippery material. Obstructions on flat roofs or in roof gulleys also present tripping hazards. Access to the roof may be more dangerous if the roof edge is curved or if the edge does not provide a stable surface against which to rest access ladders.
- *Precautions*

— Provide guard rails at the edges of pitched roofs and for work near the edge of the flat roofs.
— Use body harnesses if use of guard rails is not practicable.

— Plan safe systems of work for major roofwork projects.
— Use properly secured ladders for access (see earlier checklist).
— Use scaffolds or mobile platforms if ladders cannot provide safe access.

Falling objects

- *Risks* Slates, tiles or other roofing materials as well as dropped tools can present dangers to other workers and passers-by below.
- *Precautions*

— Provide toeboards or other edge protection (especially on pitched roofs).
— Do not throw or drop items from roofs.
— Provide enclosed rubbish chutes or lower materials in containers.
— Secure tools and materials as appropriate.

Other risks

- *Risks* Work on roofs can involve close approach to uninsulated electric cables, ducts emitting toxic or flammable gases or vapours or ducts or other items of plant hot enough to cause burns.
- *Precautions*

— Identify these types of risks before roof work starts.
— Establish appropriate safe systems of work.
— Control of access through permits to work may be necessary.

Detailed advice is provided in the HSE booklet *Safety in Roofwork*.[19]

Lifting operations

The Construction (Lifting Operations) Regulations contain many detailed requirements in respect of cranes, hoists and other lifting appliances together with chains, ropes and other lifting gear. These cover the construction, standards of installation, testing, inspection, maintenance and use of such items. Some of the more important requirements are summarised below together with other key aspects relating to lifting operations. One vital ingredient is training – whether employees are driving cranes, operating hoists, acting as signallers or banksmen or slinging loads it is essential that they have received an appropriate degree of training and are competent to carry out the task.

Cranes

Cranes must be positioned on firm level ground capable of supporting the weights involved (see the earlier section concerning the siting of MEWPs). All cranes must be thoroughly examined at least every 14 months and both tested and thoroughly examined at least every four years with the appropriate certificates retained. They must also be inspected weekly by the driver or other competent person with the results being recorded in Form F91. Cranes with a safe working load of more than one tonne must be fitted with an automatic safe load indicator which must also be inspected weekly. Crane drivers should have a clear view of lifting operations or, if this is not possible, be provided with signallers or banksmen as appropriate. A standard system of signalling exists and should be used in such cases.

Hoists

The term hoist refers to any lifting machine with a carriage, platform or cage whose movement is restricted by a guide or guides. The lifting appliance itself is subject to weekly inspection by a competent person (recorded in Form F91) and to requirements for testing and thorough examination. The latter requirements differ between mechanically and manually powered appliances.

Precautions are also necessary to prevent people being struck by the hoist or falling down the hoistway. These include:

- enclosing the hoistway where people may be struck or fall down eg working platforms or window openings
- fitting secure gates at ground level and on all landings
- opening gates to load or unload only when the hoist is at rest
- not allowing significant gaps between the hoist platform and landing platforms.

Devices to support the hoist platform in the event of a failure of the hoist rope or other hoisting gear should be provided as should automatic devices to prevent over-run beyond the hoist's highest intended point of travel. Operation of the hoist should only be possible from one position from which the operator should be able to see all the landing levels. More detailed requirements apply where hoists (or other lifting appliances) are used to carry people. Several HSE publications provide further advice on hoists.[20]

Small lifting equipment

Care is also important in the use of manually powered lifting appliances (such as pulley blocks and gin wheels) and chains, ropes and lifting gear (chain slings, rope slings, hooks, shackles, eye bolts etc). All such equipment is subject to requirements for testing and thorough examination by a competent person. For chains, ropes and lifting gear thorough examinations must be six-monthly.

Pulley blocks and gin wheels must be properly secured to poles or beams of adequate strength to support the block or wheel and its load. Hooks should be designed to prevent displacement of the load or be fitted with safety catches. The lifting equipment must also be effectively secured to the load and the load lifted in such a way as to prevent it swinging. Here the role of the person slinging the load is vital.

SAFETY BELOW GROUND

Excavations

The Construction (General Provisions) Regulations contain several requirements relating to excavations, shafts, earthworks and tunnels. Although most of the comments in this section relate to excavations, many of them can and must be applied to other below ground situations. The main risks in respect of excavations are of the sides of the excavation collapsing or of people, materials, vehicles or nearby structures falling into the excavation.

Excavation collapses

Collapses of trenches and other excavations continue to be the cause of many fatal accidents. Just a cubic metre of soil weighs over a tonne. Its weight may cause instant death or result in suffocation before the weight can be removed. Excavations of less than 1.20 metres depth may not need to be supported and the same may be the case for deeper excavations if their sides have been firmly battered back to a safe angle. In considering whether support of excavations will be necessary account must be taken of the possible effects of the weather (particularly rain and snow) and of passing vehicles and other activities in the vicinity of the excavation. In cases of doubt the excavation should always be supported.

Support may be provided by timbers, sheets, props, piles or proprietary support systems. The installation, alteration or dismantling of support for excavations must be carried out by competent and experienced workers under the direction of competent supervision. Support equipment must be "of good construction, sound material, free from patent defects and of adequate strength". It must also be inspected by a competent person on each occasion before being taken into use.

All excavations must be inspected by a competent person at least once on each day during which people work in them, with trenches deeper than two metres being inspected before the start of each shift. A thorough examination of the excavation must be carried out by a competent person at least every seven days and after events which may have affected the support (eg material falls or blasting work). Records of such examinations must be kept in Form F91.

Falls into excavations

- *Falls of people* Excavations more than two metres deep must be protected with suitable barriers or securely covered. This is particularly important in respect of excavations in public places. Barrier tape is not acceptable for this purpose. Safe access by ladders or other means must be provided for persons entering or leaving the excavations. Suitable lighting may also be necessary in the area of the excavations.

- *Falls of materials* Loose materials should not be stored close to excavations and it should be remembered that any storage in the vicinity may affect the stability of the excavation. It may be necessary to provide protection at the edge of the excavation to prevent materials falling in and use of head protection by those within the trench may also be required.

- *Falls of vehicles* Barriers or baulks must be provided to keep vehicles out of areas where there are excavations. These will need to be much more substantial than those provided to safeguard people and they will generally need to be marked to ensure they are visible to drivers. Lighting may again be necessary. Where vehicles must approach the edge of excavations (eg to tip materials) suitable stops must be provided to prevent them going over the edge.
- *Collapses of structures* The excavation may undermine the foundations of nearby structures increasing the risk of their collapse into the excavation or in other directions. Extra support may be required to prevent this happening. Specialist advice may have to be sought in cases of doubt about the quality of the foundations or the extent of the support necessary.

Buried services

There are a variety of underground services which must be considered before carrying out excavation work. These include gas, electricity, water, drainage, fuel, telephone and cable TV. Damage to these services involves the risk of injury to those excavating and disruption or even danger to users of the services or passers-by. A safe system of work should be established which in some situations may involve use of a permit to work (see chapter 12).

Research and locate

Checks should be made through plans, drawings and contact with the owners of services to establish whether services are buried on site and if so, to find their location and depth. Public utilities generally offer a free telephone enquiry service. Consideration should be given to whether the service needs to be isolated for work to continue safely. It must be recognised that plans are not always accurate (in respect of location or depth) and use of a cable or pipe locating device may be necessary. Such detectors should be used only by trained operatives and they cannot locate plastic or other non-metallic pipes. The positions of services may be indicated by marker posts or by marker tape underground, approximately 300mm above the service. Tiles or slabs are also sometimes used to protect services underground.

Digging methods

Trial holes should be dug by hand to establish the position of the service more accurately. Spades and shovels should be used as picks or forks may penetrate the service. Digging should be careful with a watch kept for marker tape, tiles or slabs as well as the service itself. Further use of detector devices may be necessary. If unidentified services are found then work should stop until they can be identified. It may be dangerous to make assumptions – a rusty conduit may contain a live cable and services will not necessarily run in straight lines. Mechanical excavators or power tools should not be used within 0.5 metres of gas pipes or electric cables. It may be necessary to provide temporary support for service pipes exposed during excavations.

Emergencies

Damage to buried services must be reported immediately to the service owner. In the cases of electricity, gas or fuel lines people should be kept clear of the area and in the latter two cases smoking and other sources of ignition should be excluded from the area. Evacuation of surrounding areas may need to be considered. Where contact is made with live electrical cables no attempt should be made to remove equipment from the area. Drivers of excavators touching live cables should jump clear of their vehicles to avoid creating a live to earth connection.

Backfilling

Light materials should generally be used to backfill around services. Tipping of hardcore could cause damage. Care should be taken during compacting to avoid damage and also to ensure that backfill under pipes is properly compacted to prevent later settlement causing damage. Warning tape or tiles should be placed 300mm above the service and plans updated where appropriate.

HSE booklet HS(G)47 *Avoiding Danger from Underground Services* contains more detailed advice on the subject.[21]

Confined spaces

Although there are many confined spaces involved in work underground, similar conditions may also exist above ground, eg in work inside tanks, roof voids or even small rooms. Toxic gases, vapours or fumes may be present or there may be insufficient oxygen. Some gases or vapours may also create flammable or explosive atmospheres. Access into and out of confined spaces may be difficult creating particular problems in cases of emergency. Cramped working conditions may also increase the risk of contact with machinery, electrical cables or service pipes.

High risk situations

All confined spaces represent potential sources of danger but situations requiring particular care include:

- where work involves paint spraying or large scale use of adhesives
- burning or welding work in confined spaces
- use of petrol or diesel engines in or near the confined space
- presence of chemicals, slurries or similar hazardous substances
- work in sewer systems
- work in excavations in rubbish tips or other contaminated land
- rusting or chemical processes depleting oxygen concentrations
- natural processes in soils or rocks producing methane or carbon dioxide, or depleting oxygen.

Safe systems of work

Safe systems of work should always be established before activities in confined spaces commence. Application of permit to work procedures (see chapter 12) may be particularly helpful in ensuring that risks are properly identified and appropriate precautions taken. Use of well trained employees is important so that they are capable both of recognising dangers for themselves and of implementing the necessary precautions. It may be possible to avoid the need for entry to be made into the confined space but if the work cannot be done in any other way then precautions may need to include:

- removing residual substances before work starts
- testing the atmosphere before entry (for toxic or flammable gases or vapours, or lack of oxygen)
- monitoring the atmosphere as work continues
- purging the atmosphere or improving the ventilation (*never* use oxygen to 'sweeten' the air)
- improving access (both for entry and exit, and also within)
- providing appropriate respiratory protective equipment (RPE) (a separate air supply is necessary if there is oxygen deficiency)
- providing other necessary PPE
- training employees in the use of the RPE and PPE
- using flameproof or similarly protected lighting
- using hand or air-powered tools
- prohibiting smoking
- requiring those in the space to wear rescue harnesses with lifelines attached
- providing people outside to carry out a rescue and/or raise the alarm (they too must be fully trained and equipped with necessary RPE and PPE).

DEMOLITION

Both the CDM Regulations and the other older Construction Regulations apply to demolition work. Part X of the General Provisions specifically covers demolition with regulation 39 requiring the appointment of a competent supervisor, experienced in demolition work. As in other fields of work the health and safety of employees and of other persons must be taken into account but the risks to neighbours, passers-by and trespassers are probably greater in demolition work than in construction activities. The principal risks of demolition include:

- falling objects
- unintended collapse of structures
- falls
- fire and explosion
- presence of live services (gas, electricity, water etc)
- contaminated materials (from asbestos, chemicals, radioactive substances, biological waste etc).

The risks of specific demolition operations should be identified before work starts and the necessary precautions planned and then implemented. Preparation of a method statement is a good means of ensuring that this is done systematically although a permit to work procedure could also achieve this, particularly for smaller projects. Many of the precautions detailed in other parts of this chapter and elsewhere in this book are equally applicable to demolition work but some specific areas where attention may be necessary are detailed below.

Site security

Only those directly involved in the demolition work should be allowed onto the site. If space allows, a buffer zone should be created around the perimeter. The site should be clearly marked using warning signs and preferably fenced, although the use of marker tape may be acceptable in some circumstances. The use of hoardings around the site as well as providing increased security can also protect passers-by from falling materials. The erection of protective fans around structures may also be necessary to provide protection against falling objects. At critical stages of demolition work the use of sentries or patrols could be required.

Removal of materials

Consideration must be given at an early stage to how materials are to be removed from the site. Identification and removal of contaminating materials (eg asbestos) should normally take place before the main demolition starts, otherwise a large volume of material could become contaminated. Thought should be given to whether lower levels of the structure are capable of supporting the weight of materials coming down on to them from work above. The use of chutes to convey debris safely to ground level is particularly recommended.

Service isolation

Services present on site should be identified before work starts and, where necessary, arrangements made for their isolation. Purging of gas or other pipelines may also be required before work may safely start.

Structure support

Either the structure being demolished or neighbouring buildings or structures may need to be supported to prevent their premature or unintended collapse. Specialist advice could well be necessary on this aspect of the work. The sequencing of demolition activities may be important in preventing collapse.

Demolition methods

Methods of demolition which keep workpeople remote from the position of demolition are generally preferable. Use of a crane and swinging ball or a long reach machine achieve this objective although the vehicle cabs should always be sufficiently protected against falling materials. Explosive methods might also be used but this is certainly an area where specialist advice is essential, particularly in relation to the direction and distance of fall of debris. If hand methods of demolition are to be used then the provision of safe access for those involved is critical as is their protection from falling materials. Safety helmets and other types of PPE will certainly be necessary.

TOOLS AND EQUIPMENT

Legal requirements

The Provision and Use of Work Equipment Regulations 1992 (PUWER)[22] apply to virtually all tools and equipment used in construction. The regulations came into operation on 1 January 1993, although in the case of work equipment first provided for use prior to that date some regulations will not apply until 1 January 1997. In such cases the older legislation replaced by PUWER will still apply. The guarding requirements contained in the Factories Act, Offices, Shops and Railway Premises Act, Construction (General Provisions) Regulations, Woodworking Machines Regulations, Abrasive Wheels Regulations and many other pieces of legislation will all eventually be totally replaced by PUWER.

The definition of work equipment includes 'any machinery, appliance, apparatus or tool and any assembly of components ...'. In the construction context this includes hand tools, powered machinery, cartridges and cartridge tools, scaffolding, site vehicles and many other pieces of equipment. Those regulations which came into immediate effect contained requirements of relevance to all types of work equipment which are summarised below.

Work equipment must be:

- suitable for its purpose
- selected with regard to working conditions and health and safety risks
- used only for suitable operations and in suitable conditions
- conform with relevant EC directives.

Account must also be taken of those using the equipment:

- adequate health and safety information provided
- adequate training provided
- use of some types of equipment restricted.

These requirements mirror much of what was previously contained in other specific legislation or widely accepted as good practice.

Hand tools

Some of the general principles of PUWER are particularly relevant to the use of hand tools ie:

- the correct tool must be chosen for the job
- tools must be maintained in good condition
- employees must be trained in their use.

Examples of the practical application of those principles are given in the checklist below:

- hammer heads sound and properly secured onto shafts
- shafts free from splinters and cracking
- chisels, drifts, punches, wedges etc with sound cutting or driving edges or tips
- heads free from mushrooming or splintering
- files with correctly fitted handles
- scissors or shears used in preference to knives
- knives preferably with partly guarded or retractable blades
- correctly fitting spanners with square jaws
- screwdrivers with correctly sized and shaped tips
- blades and drill bits kept sharp
- cutting actions away from the body.

Power tools

The general topic of guarding of machinery is dealt with in greater detail elsewhere in this book but the requirements of regulation 11 of PUWER are of particular relevance in considering power tools in use on construction sites. The regulation requires employers to take effective measures to prevent access to any dangerous part of machinery (or a rotating stock bar) or to stop its movement before any part of a person enters a danger zone. A hierarchy of measures is specified for achieving this ie the employer may only use the second measure if it is not practicable to achieve safety by the first method, and so on down the list. The measures specified are:

1. fixed enclosing guards
2. other guards or protection devices
3. jigs, holders, push-sticks or similar protection appliances
4. information, instruction, training and supervision.

In practice this matches the approach taken in the Woodworking Machines Regulations, most of which will be replaced by PUWER. Taking a circular saw as an example, the portion of the saw blade below the bench is enclosed by a fixed guard. However, enclosure of the blade above the bench would not be practicable and it is normally protected by an adjustable top guard and riving knife. Nevertheless some of the blade must still be exposed, necessitating the use of a push-stick and the operation of the saw by employees who have received sufficient training. Compliance with the standards set out in the HSE booklet giving guidance on the Woodworking Machines Regulations[23] is likely to achieve compliance with PUWER.

The HSE have also published advisory material on other types of power tools which present significant risks in the construction industry such as chain saws and abrasive wheels.[24] Both of these types of tools provide good examples of where the training of operators is essential. The majority of the Abrasive Wheels Regulations will largely be superseded by PUWER but the requirement under regulation 9, that abrasive wheels may only be mounted by fully trained and competent persons who have been appointed in writing by their employer, still remains.

Many tools in use on construction sites are powered by electricity. Electrical safety is dealt with in some detail in chapter 20. Risks on site can be reduced by the use of cordless tools or those operating from a 110V supply. Such systems are centre-tapped to earth so that the maximum shock potential is 55V. If mains voltage does have to be used then suitably rated residual current devices (RCDs) or similar protection should be used.

Electrical equipment should be selected according to the environment it is expected to operate in, which in the case of a construction site may involve rough use and exposure to wet and other adverse conditions. Equipment should also be subject to the same types of user checks, formal visual inspections and combined inspections and tests as equipment elsewhere (see chapter 20). The table opposite sets out the frequencies recommended by the HSE for such activities.[25]

	User check	Formal visual inspection	Combined inspection and test
110V portable hand tools, extension leads, site lighting, movable wiring systems and associated switchgear	Weekly	Monthly	Before first use and then 3-monthly
240V portable and hand held tools, floodlighting and extension leads	Daily or every shift	Weekly	Before first use and then monthly
Fixed (non-movable) 240V equipment	Weekly	Weekly	Before first use and then monthly
Residual current devices (RCDs)	Daily or every shift	Weekly	

Cartridge operated tools

There is considerable potential for injury in the use of such tools. Fixings may penetrate through surfaces because too strong a cartridge has been selected or the structure contains voids or low density materials. Ricochets may also occur due to working on hard materials or to obstructions within the material. It is essential that operators are trained in the use of the equipment and also in the care and control of cartridges. Colour blind operators may not be able to distinguish between the colour coded identifications on cartridges. The checklists below contain some of the key safety aspects relating to the storage of cartridges and the use of cartridge tools.

Cartridge storage

- Store in a secure, labelled, waterproof container.
- Keep different strength cartridges in separate compartments.
- Keep cartridges for immediate use in a carrying case.
- Return unused cartridges and cartridge tools to secure storage.

Use of cartridge tools

- Check the material of the structure being worked on (also check behind the structure).
- Select the correct cartridge (carry out a trial fixing if necessary).
- Wear eye protection, hearing protection and a safety helmet.
- Carry tools barrel down.
- Do not load until immediately before use.
- Do not use force when loading.
- Keep others clear of the work area (use signs or marker tape where appropriate).
- Always follow the manufacturer's instructions, particularly in the event of a misfire.
- Beware of the recoil (never work on an unsecured ladder).

Gas burning and welding equipment

The health hazards associated with burning and welding are dealt with elsewhere in this book but the risks of fire and explosion from such work must also be taken into account. The checklists below summarise some of the key safety features in the storage of gas cylinders and the use of burning and welding equipment.

Cylinder storage

- Store cylinders in a secure and well ventilated compound.
- Keep flammable gases separate from oxygen cylinders.
- Preferably store empty cylinders separate from full ones.
- Secure cylinders in the vertical position.

Use of equipment

- Use suitable flashback arresters and pressure regulators.
- Ensure satisfactory joints are made at all connections.
- Secure gas cylinders in the vertical position.
- Check the equipment for damage and leaks before use.
- Use appropriate PPE (eye protection, gloves and possibly body protection).
- Keep combustible materials clear of the work area.
- Check drums, tanks, pipelines etc for flammable materials before working on them (purging or cleaning may be necessary).
- Keep oil and grease away from fittings (especially on oxygen equipment).
- Do not leave equipment unattended when lit.
- Remove equipment from confined spaces during breaks and at the end of each work period.

Personal protective equipment (PPE)

The Personal Protective Equipment at Work Regulations apply to construction activities. They require an assessment to be made to establish what PPE is required to protect against risks which cannot be avoided by other means. This subject is covered in greater detail in Chapter 25 but some of the types of PPE which may be required to protect against risks present in construction are detailed in the table below.

PPE	Types of Risk
Head protection	Falling objects Impact with fixed objects
Eye protection	Impact eg work with power tools and some hand tools Dust, chemical splashes, sprays, molten metal splashes Burning or welding
Foot protection	Falling objects Sharp objects piercing Chemicals
Hand and arm protection	Abrasions from manual handling Hot or cold materials or conditions Chemicals Vibration
Protective clothing	Cold, heat, bad weather Traffic (high visibility clothing) Chemicals Use of chain saws Drowning (life jackets etc)

Hearing protection or respiratory protection may also be necessary to comply with the Noise at Work or COSHH Regulations (see chapters 22 and 23).

The Construction (Head Protection) Regulations 1989 require the wearing of suitable head protection unless there is no risk of head injury other than by a person falling. The employer's duty is to provide suitable head protection and to make sure it is worn. Employees must wear the protection and report any damage or loss. There is an exemption for turban-wearing Sikhs.[26]

SITE FACILITIES

Transport, access and egress

The General Provisions Regulations lay down various requirements about workplace transport whether on water, rail or road.

The Working Places Regulations require, so far as is reasonably practicable, that there shall be safe means of access and egress. Every place where a person works must be kept safe.

Welfare facilities

First-aid

Construction is subject to the Health and Safety (First-Aid) Regulations 1981 (see chapter 16).

Shelter, drying facilities etc

The Construction (Health and Welfare) Regulations require the provision of adequate and suitable accommodation for taking shelter in bad weather, and for depositing non-work clothing (regulation 11). Dry facilities for taking meals and boiling water must be provided. There are other detailed requirements.

Washing facilities and sanitary conveniences

If an employee works on a site for more than four consecutive hours there must be adequate and suitable washing facilities (regulation 12). There are various other detailed requirements (in the Health and Welfare Regulations) dependent upon the numbers employed, including sanitary conveniences if more than 25 people are employed on a site (regulations 13 and 14).

Notes

1. HSC, *Managing Construction for Health and Safety*, Construction (Design and Management) Regulations 1994, ACOP, HSE Books, L54, 1995
2. See: HSC, *Guide to Managing Health and Safety in Construction*, HSE Books 1995; and HSE, *Health and Safety for Small Construction Sites*, HS(G)130, HSE Books, 1995
3. For further details see the ACOP accompanying the CDM Regulations, pp38-9 (Appendix 3) see note 1
4. Regulation 20, The Health and Safety at Work etc Act 1974 (Application Outside Great Britain) order SI 1995/263
5. Regulation 21
6. ACOP, paras 32–46

7. See: HSC, *Designing for Health and Safety in Construction*, HSE Books, 1995
8. The definitions of "building operations and works of engineering construction" are contained in s176(1) of the Factories Act 1961
9. Construction (General Provisions) Regulations, SI 1961/1580; Constructing (Lifting Operations) Regulations, SI 1961/1581; Construction (Working Places) Regulations, SI 1966/94; Construction (Health and Welfare) Regulations, SI 1966/95; Construction (Head Protection) Regulations, SI 1989/2209
10. Construction Plant and Equipment (Harmonisation of Noise Emission Standards) Regulations 1988; Falling Object Protective Structure for Construction Plant (EC Requirements) Regulations 1988; Roll-over Protective Structures for Construction Plant (EC Requirements) Regulations 1988
11. It is proposed to revoke the Working Places and Health and Welfare Regulations in their entirety and to revoke most of the General Provisions Regulations. The Construction (Health and Welfare) (Amendment) Regulations, SI 1974/209 will be revoked in their entirety
12. EU Temporary or Mobile Construction Sites Directive 1992, OJ L:245/13. The minimum requirements are set out in Appendix IV to the Directive
13. Working Places Regulations, regulations 19 and 20
14. SI 1994/3140, regulation 2(1). The Factories Act 1961, s176(1) will be amended so that building operations and work of engineering construction will mean construction work as defined in the CDM Regulations
15. SI 1992/3004
16. *Safe Use of Ladders, Step-ladders and Trestles*, HSE, Guidance Note GS31, HSE Books, 1984
17. *General Access Scaffolds*, HSE, Guidance Note GS15, HSE Books, 1982
18. *Tower Scaffolds*, HSE, Guidance Note GS42, HSE Books, 1987
19. *Safety in Roofwork*, HSE, HS(G)33, HSE Books, 1987
20. *Construction Goods Hoists*, HSE, Construction Information Sheet No. 13, HSE Books, 1988
    Safety of Rack and Pinion Hoists, HSE, Guidance Note PM24, HSE Books 1981
    Inclined Hoists Used in Building and Construction Work HSE, Guidance Note PM63, HSE Books 1987
21. *Avoiding Danger from Underground Services*, HSE, HS(G)47, HSE Books, 1989
22. Provision and Use of Work Equipment Regulations, SI 1992/2932
23. *Woodworking Machines Regulations 1974, Guidance on Regulations*, HSE Booklet L4, HSE Books, 1991
24. *Training and Standards of Competence for Users of Chain Saws in Agriculture, Arboriculture and Forestry*, HSE, Guidance Note GS48, HSE Books, 1990; *Safety in the Use of Abrasive Wheels*, HSE, HS(G)17, HSE

Books 1992; *Training Advice on the Mounting of Abrasive Wheels*, HSE Guidance Note PM22, HSE Books, 1983

25. See HSE, *Health and Safety for Small Construction Sites*, HS(G)130, HSE Books, 1995, pp42-44 dealing with electrical equipment and other electrical risks
26. Employment Act 1989, s11

27

Social Factors and Health and Safety at Work

The purpose of this chapter is to examine the health and safety issues which arise out of the interaction between work activities and certain social factors. These factors are:

- the scheduling of work (eg long hours and shiftwork)
- smoking, drinking and drugs
- pregnancy and childbirth.

A brief note on AIDS is included. More general issues of health and hygiene, such as what employers can do in the form of health education, promotion and treatment, are addressed in chapter 13. The present chapter commences with a brief analysis of the problem of stress.

STRESS

Stress can be seen as a response made by people to the demands placed upon them. A certain amount of stress is beneficial in that it helps motivation: the problem occurs when the demand is too great or prolonged, or is insufficient.

The existence of stress as a workplace health and safety issue is increasingly being recognised. Stress is specifically referred to as a factor to be taken into account in complying with the 1992 Regulations on Display Screen Equipment and the HSE published a major report in 1993 on *Stress Research and Stress Management.*[1] Stress-related illness may be evidence of a breach of the general provisions in s2 of HSWA. Moreover, in *Walker v Northumberland County Council* it was found that the second nervous breakdown suffered by a social services manager was a result of overwork caused by the employer's inadequate resourcing of his area of work.[2] This was held to be a breach of the employer's duty of care in that there was a failure to provide a safe system of work. It seems quite likely that there will be increasing use of the law in relation to stress-related illness, and employers would be prudent if they gave the issue more consideration. In particular, employers need to recognise that they:

- owe a duty of care to their employees not to cause them psychiatric damage through the volume or character of the work they are required to do
- must provide a safe system of work
- must take reasonable steps to protect the employee from reasonably foreseeable risks
- are just as liable for mental injury as they are for physical injury.

The sources of stress are numerous and may lie outside or within the workplace, but wherever the source, the effects are likely to be apparent in work performance. There may well be a link with other health-related problems such as alcohol or drug-abuse or smoking.

Some of the principal work-related sources of stress are:

- impossible deadlines
- overwork or underwork
- change
- promotion or lack of it
- racial or sexual harassment
- bullying or other victimisation
- fear of redundancy
- long or unsocial hours
- lack of control over the work
- role ambiguity or conflict
- poor relationships at work including management-worker conflicts.

The highest-rated work factor causing stress is job loss.[3]

Measures to alleviate stress within the workforce are likely to need to involve both the occupational health service and the senior management of the organisation. Such measures might include:

- controlling the causes of stress (some of which may require considerable organisational change)
- training employees to cope more effectively with stress
- counselling activities (providing relevant assistance where appropriate).

WORK SCHEDULES

The effects of adverse work schedules and shiftwork are increasingly coming into focus as occupational health issues.

Potentially adverse work schedules

There are a number of types of work schedules which may produce adverse effects in those called upon to work them, but not necessarily in all cases. One person's stress is another's challenge, a schedule criticised by one employee as inflexible may be welcomed by another as predictable.

Excessively long hours of work

Excessively long periods at work, whether in single spells or with only short recovery periods, are certainly likely to produce adverse effects in most people. Long work periods in the medical profession and the security industry have attracted attention in recent years.

Although hours of work in some sectors, particularly transport, have been subject to increasing legislation the longstanding controls on the hours of work for women and the under 18s were removed by the Sex Discrimination Act 1986 and the Employment Act 1989. However, the general requirements of HSWA mean that the employer cannot totally ignore hours of work as a health and safety issue. Moreover, an employer requiring an employee to work excessive hours may be in breach of his duty of care (see chapter 4). In *Johnstone v Bloomsbury Health Authority* the High Court refused to accept the employer's application that the case of a junior hospital doctor – relying on the above argument – should be struck out.

Excessively demanding work

The Manual Handling Operations Regulations 1992 (see chapter 19) mean that in the vast majority of cases excessive physical demands of work schedules must be taken into account. The most obvious effect of excessive mental demands at work is stress (see above).

Shiftwork

Many workers have to come to terms with shiftwork either temporarily or permanently. A short spell working shifts can be a good education for a manager in appreciating the problems that his employees face – for them it becomes a lifestyle rather than just a work schedule.

Possible adverse effects

Shiftwork affects the natural rhythms of the body, controlling body temperature and other metabolic activities such as the digestive system. These, together with the many external factors (family, friends, neighbours etc) can affect the shift worker's ability to sleep, especially at times other than during the night.

The build-up of fatigue together with the unsocial hours being worked begin to have an impact upon the social life of the shift worker – he becomes less willing or less able to enjoy social activities with his family or friends. These effects and the physical effects (fatigue, digestive problems etc) often build up gradually and insidiously.

Coping with shifts

Many factors govern the ability of employees to cope with shifts. Some of these are associated with the job itself – jobs that are well-rewarded, interesting and meaningful are much more likely to appeal. Physical work is generally easier to cope with than mental work. If the employee is employed in a monitoring or reactive capacity, there may be a risk of falling asleep.

Research has shown that body rhythms adapt better to non-rotating shifts (eg permanent night shift) than rotating ones. In the case of rotating shifts, the lower the frequency of rotation then the greater the body's chances of adapting.[4]

Younger people usually cope better with shifts than older ones and men better than women, who may be affected by their reproductive cycles. People differ in their physical make-up – some need less sleep; some are 'night-people' as opposed to 'morning-people'.

Individuals' own social circumstances also influence their ability to cope with shifts. For example, patterns that are compatible with those of partners or the demands of children will be more acceptable.

Frequent rotations of shift or sudden changes of shift schedules should be avoided, particularly where these involve night work. Apart from the disruptions to the body rhythms that these bring, a lack of predictability or reliability is more likely to dislocate the employee's social life. Finally, employees may need to create for themselves a home environment more conducive to daytime sleeping.

Effects and costs

The direct and indirect effects of adverse work schedules and shiftwork are generally well recognised, but the magnitude of the costs associated with them is probably not so well appreciated.

Accidents

Physical and mental fatigue are obvious causes of accidents, the costs of which were reviewed in chapter 6. A study of near-miss train accidents in Japan showed that 82% occurred between midnight and 0800. Major disasters such as Bhopal, Chernobyl and Three Mile Island occurred during the same period.

Poor products

Tired workers are much more likely to produce sub-standard and defective products with all the associated costs. Even if employers have these cost data available, they are unlikely to have details of the time of day when poor products were produced or of the time the employee producing them had spent at work.

Poor decisions

The potential for poor decisions due to fatigue among the medical profession is quite awesome and the public are never likely to be aware of the true extent of the problem. Similarly poor executive decisions made under conditions of personal fatigue are never likely to be apparent to a wider audience.

Ill-health

Many people become ill through the physical and mental effects of fatigue although the medical certificates do not always say so. Gastro-intestinal problems, headaches and other disorders may be listed as the cause of absence. Inter-related problems such as abuse of alcohol or drugs may complicate the picture.

Social

Continuous fatigue or the disruptive effects of shiftwork may well affect the ability to enjoy life and relate well to family and friends. Whilst the initial impact is upon the employee, there may eventually be a significant cost to the employer through indirect effects eg a lengthy absence after a marital breakdown or the sudden departure of a valued employee for pastures new.

SMOKING, DRINKING AND DRUGS

Smoking

Tobacco smoke contains various substances which can cause cancer or other health problems.[5] It may cause more distress among those with respiratory disorders, such as asthma, than among most non-smokers. There is strong evidence that women who smoke during pregnancy place their babies at risk. The HSE recommends that pregnant women should also avoid passive smoking.[6] Passive smoking is defined as the inhalation of environmental tobacco smoke by non-smokers.

The main issues appear to be:

- Are employers under a legal duty to restrict or ban smoking because of the risks to the health, safety and welfare of their employees and/or others?
- What should be in a smoking policy, how should it be introduced and can the introduction of a smoking policy be unlawful?
- Can dismissal for non-compliance with a smoking policy be unlawful?
- Are there any legal dangers in discriminating against smokers eg by restricting recruitment to non-smokers?

Although there are no definitive rulings from the courts of the UK, the evidence in relation to the first of these issues is accumulating. Medical research has now established a foreseeable risk. Therefore, it appears that employers have a duty to consider the risk to their employees which arises from smoking, and to control that risk. This is both a duty at common law – the duty to take reasonable care (see chapter 4) – and a general statutory duty under HSWA, s2, especially s2(2)(e) (healthy working environment). There are also specific statutory duties. The Workplace Regulations require "a sufficient quantity of fresh or purified air" (regulation 6).[7] They also require "suitable and sufficient rest facilities", including rooms and areas in which non-smokers are protected from the discomfort of tobacco smoke (regulation 25). Moreover, there must be "suitable facilities" for pregnant women and nursing mothers: clearly these would need to be smoke-free (again, regulation 25).

In the Australian case of *Scholem v New South Wales Health Department*, the plaintiff claimed that 13 years of exposure to tobacco smoke at a Sydney Health Centre had made her asthma worse. She was awarded damages on the ground of her employer's negligence in not banning smoking in the workplace. There was also a criminal offence because of breach of the Shops and Factories Act[8] In the UK, in an action for damages, a non-smoker accepted £15,000 from her employer as an out-of-court settlement.[9]

There are a number of good reasons for introducing a smoking policy in addition to the legal duty to do so. Action on Smoking and Health (ASH) cites the following:

- improvement of staff morale eg because of better working conditions
- reduced absenteeism
- reduced cleaning and redecoration costs
- reduced ventilation and air conditioning costs
- improved productivity and efficiency
- possible reductions in fire and other insurance premiums (as early as 1992, at least one company, Legal & General, offered employers who ban smoking, a discount on the cost of insuring their workforce against death and disability)
- improved corporate image.

(See: *How to Achieve a Smoking Policy at Work*, London: ASH, 1988).

Care needs to be taken when introducing a smoking policy and ASH suggests a 5-step approach:

1. *Set up a working party* to review current practice and to determine the objectives
2. *Raise the issue among the workforce*: that is, the issue of where people smoke; circulate information about the health hazards of smoking and encourage feedback.
3. *Consult the workforce* This might involve the use of surveys, ballots, meetings and discussions with trades unions and/or staff associations.

4. *Draft the policy* The principle underlying it should be that non-smokers have the right to breathe air that is free from tobacco smoke. The options can be set out. Clear rules need to be laid down about where smoking is permitted (if at all).
5. *Implement the policy* A lead-in time of 12 weeks is recommended by ASH. Some employers have opted for a phasing-in of the policy eg an evolutionary period before the implementation of a complete ban. Other employers feel the total ban is too draconian, and ban smoking in common areas but allow it in private offices and/or designated smoking areas. Many policies provide assistance, such as literature, counselling or group sessions for smokers in order to help them give up the habit.

The HSE recommends a number of practical steps to protect non-smokers.

- improving ventilation so that smoke is more effectively removed from the working environment
- segregating smokers, and non-smokers in separate rooms where possible
- letting employees in each working area decide whether smoking should be allowed there
- discouraging or banning smoking in common areas such as reception, corridors, lifts, open plan work areas, conference rooms, all or parts of canteens, etc
- banning smoking in all parts of the premises except in areas designated as smoking areas
- restricting smoking (either throughout the workplace or in common areas) to certain times of day
- agreeing a total ban on indoor smoking
- encouragement and help for smokers who wish to give up smoking, including advice on counselling.

(Source: HSE, *Passive Smoking at Work*, 1988)

In general, the law provides support for employers in this area. Not only can an employer introducing a smoking policy be seen as fulfilling his general duty under s2 of HSWA, but the introduction of such a policy is likely to be within an employer's discretion. This was held to be so in *Dryden v Greater Glasgow Health Board*, where a lifetime smoker resigned in response to the introduction of a smoking policy. Moreover, it was held that the employee had no implied contractual right to smoke. However, this does not mean that the introduction of a smoking policy will always be lawful.

A policy introduced in such a way as to breach the relationship of mutual trust and confidence between employer and employee would allow an employee to resign and claim compensation for an unfair constructive dismissal. Undue haste, and lack of consultation and support might be constituents here.[10]

Whether an employer can lawfully dismiss an employee who is in breach of smoking policy will depend on the facts of the case. If handled reasonably, the employer is likely to be able to defend against an action for unfair dismissal just as on any other disciplinary grounds. Refusal to employ smokers does not appear to contravene any UK laws, although conceivably it could be indirect discrimination if the proportion of smokers in any particular racial group or in either of the sexes was very high. Moreover, while it may be an employer's duty to ensure that the air in the workplace is free from tobacco smoke, it is not necessary to employ only non-smokers in order to fulfil that duty. Possibly such a recruitment practice might be tested against the European Convention on Human Rights.[11] Apparently, in 1991 over 6,000 US employers were refusing to employ smokers.[12]

The use of materials such as those available from the Health Education Authority should be combined with personal counselling of employees to encourage them to give up smoking. This is especially important for those exhibiting adverse physical symptoms which may have been identified through lung function testing and may be causing actual absence. In some workplaces employees are assisted to give up smoking by the provision of nicotine patches or chewing gum. The occupational health service may also have a significant role to play in the development of a smoking policy within the workplace.

Drinking and drug abuse

Here too, much can be done to publicise the dangers of alcohol and drugs and to counsel employees who have problems. However there are added difficulties due to the possibility of unacceptable or even dangerous behaviour of those at work whilst affected by drink and drugs. A balance has to be struck between helping employees with problems and ensuring adequate standards of behaviour and performance and this is best done within the framework of an alcohol and drugs policy. The HSE booklet *Drug Abuse at Work – a Guide for Employers* provides much useful guidance on formulating and implementing such a policy.[13]

The term drug abuse refers to the use of illegal drugs and to the misuse, whether deliberate or unintentional, of prescribed drugs and substances such as solvents. Drug abuse may affect work performance even if the abuse takes place outside the workplace. The HSE guide gives examples of signs of *possible* drug abuse (eg an increase in short-term sickness absence) and reminds employers that drug abuse is expensive for the abuser, so may result in theft or other forms of dishonesty. Various savings can flow from having a successful policy on drug abuse, including reduced absenteeism, increased productivity and the reduced risk of accidents.

The HSE recommends that the employer's policy on drug abuse at work should be part of the overall health and safety policy and that generally it should be in writing.

Under the policy employers should:

- clearly define drug abuse;
- explain that drug abuse leads to health problems, including addiction, so that it is necessary to identify abusers quickly and provide the help they need;
- encourage employees with drug problems to seek help voluntarily;
- provide or signpost the necessary help and professional advice;
- stress the confidentiality of any discussions and subsequent treatment;
- make clear that absence for treatment and rehabilitation will be regarded as normal sickness;
- recognise that relapses may occur;
- enable an employee to return to the same job after treatment or, where this is not feasible or advisable, try to provide suitable alternative employment;
- explain that if help is refused and/or impaired performance continues disciplinary action is likely;
- explain that dismissal action may be taken in cases of gross misconduct;
- state that trafficking will be reported immediately to the police. There is no alternative to this procedure;
- provide for the policy to be monitored and reviewed regularly in consultation with workplace representatives.

In developing the policy an employer will need to:

- take account of local community services which can help with advice, particularly for smaller companies. The Health and Safety Executive's Employment Medical Advisory Service may be able to advise on what is available;
- seek the commitment of line managers and employees to the policy by fully consulting workplace representatives, including any safety representatives appointed under the Safety Representatives and Safety Committees Regulations 1977, at all stages of developing the policy;
- agree with workforce representatives what action to take where an employee:
  - admits to needing help
  - is found abusing drugs
  - has a relapse after receiving treatment.

(Source: *Drug Abuse At Work: A Guide for Employers*, HSE, IND(G)91L (revised) 1992 © Crown Copyright reproduced with the permission of the Controller of HMSO.)

Figure 27.1 *Employers' policies on drug abuse at work*

The HSE guidance goes on to deal with implementation of the policy, including the question of screening, and the legal aspects of drug abuse are also covered. The guide contains a useful description of commonly-abused drugs, along with information about how they are taken and an indication of their harmful effects.

As with smoking and drugs, the act of drinking or being drunk at work may affect not only the individual concerned but also others. There may be health and safety implications regardless of whether the drinking reflects alcoholism or simply the consumption of an excessive amount of alcohol. The difference will probably lie in the way the problem is dealt with: alcoholism tends to be treated as a question of illness while drunkenness is likely to be considered as misconduct. Much of what the HSE recommends in relation to a drug abuse policy is relevant to the question of alcoholism.

PREGNANCY AND CHILDBIRTH

Legal framework

The EU Pregnant Workers' Directive has resulted in new provisions for the health and safety of pregnant workers, those who have recently given birth and those who are breastfeeding.[14] Some of these provisions have been brought into effect through amendments to the Management Regulations and to the EP(C)A. In addition, the Maternity (Compulsory Leave) Regulations are relevant here, and the Sex Discrimination Act. Finally, the general provisions of HSWA will apply, and the Workplace Regulations require the provision of rest facilities.

Risk assessment and avoidance

The Management of Health and Safety at Work (Amendment) Regulations 1994 insert new requirements into the original 1992 regulations (see chapter 3).[15] The 1994 provisions apply to any new or expectant mother, defined as an employee who is pregnant; who has given birth or miscarried in the previous six months; or who is breastfeeding. The first requirement is for the employer to carry out an assessment of the risks to such workers from exposure to certain agents, processes or working conditions. These are set out under the following headings but the list is not exhaustive:

- physical agents
- biological agents
- chemical agents
- certain other agents
- certain industrial processes
- underground mining work.

The HSE booklet *New and Expectant Mothers at Work*[16] provides guidance on the major areas of risk including:

- lifting and handling tasks
- noise
- shocks, vibration and movement
- harmful substances
- radiations
- temperatures
- chemical and biological agents
- working conditions (eg VDUs).

A risk assessment must be carried out if the employees include women of child-bearing age and the work could involve risk, by reason of the condition of a new or expectant mother, to her health and safety or that of her baby, as the result of exposure to substances, processes or working conditions.

The second requirement is that where the results of the assessment reveal a risk, the employer must take measures to ensure that the worker's exposure to the risk is avoided. If this is not possible, the employer must alter the working conditions and/or working hours of the particular employee, providing that that employee has notified the employer of her condition, and done so in writing.

Where risks cannot be avoided

If it is not feasible or reasonable to adjust the employee's working conditions or hours, the employer will be required to suspend the employee with pay. The first option, however, will be to offer suitable alternative work if there is any. There is no requirement to offer alternative work if it does not exist, but where it does exist a failure to offer it and subsequent dismissal will amount to automatically unfair dismissal. In the absence of suitable alternative work, a failure to suspend with pay will have the same effect.[17]

Suitable alternative work means work that is suitable for the employee and appropriate for her in the circumstances. The terms and conditions must not be "substantially less favourable" than her normal terms and conditions. If the employee unreasonably refuses suitable alternative work, she will lose the right to be paid during her maternity suspension. Maternity suspension is suspension in consequence of risks arising out of a maternity risk assessment or nightwork.

Where a suitable alternative job is not offered to an employee, or an employee is not paid during suspension, they may complain to an industrial tribunal.

Nightwork

Where a new or expectant mother performs nightwork and a medical certificate shows that it is necessary for her health and safety that she should not do so, the amended Management Regulations state that the employer must suspend her from work. Again, where there is suitable alternative work, this should be offered instead of the worker being suspended.

Compulsory maternity leave

Under the Maternity (Compulsory Leave) Regulations 1994, an employee entitled to maternity leave is not permitted to work during the two weeks beginning with childbirth.[18] An employer in breach of this requirement will commit a criminal offence, whereas a breach of the maternity leave provisions of the EP(C)A will be a civil matter. Where the employee works in a factory, the Factories Act 1961 prohibits return within four weeks.

Sex discrimination

Special treatment afforded to women in connection with pregnancy or childbirth is not a breach of the Sex Discrimination Act 1975.[19] Moreover, acts done for the purposes of protecting women from risks specifically affecting women, or protecting them as regards pregnancy or maternity, are not unlawful where they are done to comply with existing statutory provisions. Nor are they unlawful if done to comply with relevant statutory provisions as defined in HSWA.[20] Finally, the Employment Act 1989 lists a number of specific provisions relating to the protection of women at work: compliance with any of these is not unlawful sex discrimination.[21]

Rest facilities

The Workplace Regulations require that suitable facilities for rest shall be provided for any person at work who is a pregnant woman or nursing mother.[22]

AIDS

The body's normal defences against illness may break down following infection by a virus known as HIV. If a particular form of cancer or serious infection develops as a result, the individual is said to have Acquired Immune Deficiency Syndrome (AIDS). Not all individuals who become infected with HIV will necessarily develop AIDS.

The virus is passed on by sexual intercourse with an infected person or by taking infected blood into one's own bloodstream. Transmission through breast milk is also a possibility. Infection is not spread through the air or by touch. Nor is there any danger from handling objects which have been used by an infected person.

There is no risk where there is no direct contact with the blood, semen or other body fluids of infected individuals. Thus, for most employees there will be no risk. However, there are certain jobs, particularly in the health services, which may involve some risk because of contact with these things. In such circumstances, HIV infection should form part of an employer's risk assessment. Guidance is available from the HSC and HSE.[23]

Notes

1. Health and Safety (Display Screen Equipment) Regulations, SI 1992/2792. See Annex B, para 5 (p42) of the combined regulations and guidance booklet, L26 *Stress Research and Stress Management*, HSE, 1993. See also HSE free leaflets: *Mental Health at Work* (IND(G)59) and *Mental Distress at Work* (IND(G)129). Priced publications by the HSE in this area include: *Managing Occupational Stress: A Guide for Managers and Teachers in the Schools Sector*; and *Workplace Health and Safety in Europe*. Another useful publication is: *Survey of Stress at Work*, MIND, 1991.
2. *Walker v Northumberland County Council*, [1995] IRLR 35 High Court
3. *The Safety and Health Practitioner*, February 1993, p20
4. On the literature generally see: *Shiftwork, Health and Safety: An Overview of the Scientific Literature 1978–90*, HSE, 1992
5. *Fourth Report*, Independent Scientific Committee on Smoking and Health, London: HMSO, 1988. According to the Royal College of Physicians, 350 million working days are lost each year in the UK as a result of sickness, and of these, some 50 million are attributable to cigarette smoking. (*Personnel Management*, Factsheet 27, March 1990.)
6. *Passive Smoking at Work*, HSE, IND(G)63L, 1988. See also: *Smoking Policies at Work*, London: Health Education Authority
7. The Workplace Regulations are dealt with in Chapter 16
8. This case was widely reported in the UK national press: see, for example, *The Guardian*, 28 May 1992, p9
9. *Bland v Stockport Metropolitan Council* See: *The Independent*, 28 January 1993
10. For guidance on the law of unfair dismissal generally see Lewis, P, *Practical Employment Law*, Oxford: Blackwell, 1992
11. Council of Europe, 1950
12. *Financial Times*, 19 August 1991
13. *Drug Abuse at Work – A Guide for Employers*, London: HSE, 1992. Publications by Alcohol Concern may also be useful.
14. Pregnant Workers' Directive, 92/85/EC
15. Management of Health and Safety at Work Regulations SI 1992/2051, amended by SI 1994/2865
16. *New and Expectant Mothers at Work*, HSE, 1995
17. EP(C)A, s60

18. SI 1994/2479, regulation 2
19. By virtue of s2(2)
20. SDA s51
21. Employment Act 1989, s4 and schedule 1
22. Regulation 25(4)
23. See: Department of Employment/HSE: *AIDS and Employment*; and HSC, *AIDS: The Prevention of Infection in the Health Services*

28

Environmental Aspects

The focus of this chapter is not upon environmental matters in general: rather, the aim is to consider those environmental issues which arise out of, or are otherwise associated with, health and safety at work. After an introduction which identifies the environmental pressures facing business, the chapter concerns itself with air and water pollution and its control, waste on land, hazardous substances and statutory nuisance (including noise). A final section examines management's environmental strategy requirements.

INTRODUCTION

The environmental pressures on business are unlikely to recede. Regulation by law (requiring investment in prevention and clean-up); shifts in the burden of taxation towards pollution and resource-use; and increasingly environmentally sensitive consumers are all likely to make it worthwhile for a company to have an environmental policy.

'Green consumerism' has helped to put environmental auditing on the corporate agenda and has encouraged eco-labelling and life-cycle assessment – an examination of the product's environmental impact from the beginning to the end of its life (or to resurrection if recycled). Life-cycle assessment can lead to life-cycle design: designing a product to minimise its impact throughout its life. Environmental auditing is used not only to determine how a company stands in relation to compliance with the law, but also how it stands in relation to competitive and consumer environmental pressures. Flowing from increasing analysis of businesses' environmental performance is a growth in corporate environmental reporting.

There are other factors at work which mean that environmental issues now raise serious financial and policy questions for business. For example, some firms are making environmental demands upon their suppliers. There is also the question of insurance – perhaps a fifth of Lloyds' recent losses are environmentally-related, for example, oil spills. Furthermore, contaminated sites raise large questions of liability. Altogether, these and other factors suggest that environmental failure will have serious financial implications for businesses and their insurers.

Other developments of relevance here include the growth of eco-labelling schemes at national and supranational, notably European Union, level. In the UK, there is the environmental standard BS 7750. Germany's 'take-back' laws – forcing companies to recycle used packaging and old computers – are an example of the direction in which the law might travel. There may also be links internationally between competition law and environmental controls in order to stop businesses where environmental regimes are lax from benefitting through reduced costs and increased competitiveness.

The objects of national environmental policy were set out in a White Paper in 1990.[1] These include:

- improvements in the quality of air and water
- the setting of acceptable standards (such as maximum emission levels)
- the provision of incentives to industry
- the use of control mechanisms to enforce policy.

POLLUTION

Water

Regulation of the abstraction of water is to be found in the Water Resources Act 1991. However, this has little relevance to health and safety at work, and it is the pollution of water that is now considered. Public Health Acts[2] control the discharge of effluent and pollution into the network of sewers. River pollution is subject to the Water Resources Act 1991 (henceforth, the 1991 Act) and control by the National Rivers Authority (NRA). The legislation sets out a number of offences which can be avoided only by obtaining a consent from the NRA.[3] Consents may have conditions attached to them. The offences include allowing any poisonous, noxious or polluting matter or any solid waste to enter any controlled waters; and more generally discharging any trade effluent or sewage effluent in contravention of a relevant prohibition (the latter relates to NRA prohibition notices).

Air

The Clean Air Acts 1956 and 1968 make an occupier liable for the emission of 'dark smoke'.[4] (See Figure 28.1) The local authority can designate smoke control areas. Since smoke can be treated as a public nuisance, abatement and prohibition orders are possible under the Public Health Act 1936.[5]

FIRM FINED OVER FIRE

XYZ Ltd were fined £50 after admitting breaking clean air laws. Magistrates at Anywhere heard how council environmental health officers saw clouds of black smoke billowing from one of the company's sites.

An enquiry found that a worker had spilled diesel oil and then set it alight in a bid to get rid of it. The company had no written policy for dealing with spillages of this kind. The foreman was temporarily absent and the worker had taken it on himself to get rid of the oil.

The company was ordered to pay costs of £40.

This case occurred prior to the increase in the standard scale of fines in 1992.

Figure 28.1 *Clean air legislation in operation*

The HSWA imposes general duties on persons in control of prescribed premises to prevent emission into the atmosphere of noxious or offensive substances.[6] The premises prescribed and the substances defined as noxious or offensive are listed in regulations under HSWA.[7] The duty extends to rendering harmless and inoffensive any substances which are emitted. The standard required is "best practicable means", which includes the way the plant is used, the supervision of any emissions and the use of any preventive devices. Best practicable means is a high standard, exceeded only by an absolute requirement. Enforcement of this provision is the responsibility of the Secretary of State for the Environment and not the HSC.[8]

The Environmental Protection Act 1990 (the 1990 Act) introduced a two-tier system of control. Processes covered by the Act are laid down in regulations.[9] Processes listed in Part A are subject to integrated pollution control (IPC – see below) by HM Inspectorate of Pollution (HMIP): Part B processes remain the responsibility of the local authorities, whose remit does not extend beyond air pollution. This system will run alongside the HSWA provisions until all the prescribed processes have been incorporated into the 1990 Act system.

Integrated pollution control

Integrated pollution control (IPC) is a scheme of authorisation, control and enforcement of processes capable of polluting the environment. A process falls within this scheme if it releases to air, water or land any substance capable of causing harm to living organisms (including humans) which are supported by the environment.[10] Processes become subject to these controls only when they have been prescribed by the Secretary of State for the Environment.[11] Once a process is prescribed, persons are prohibited from carrying on that process unless authorised by HMIP or the local authority and in accordance with the conditions of the authorisation.[12] The procedure for gaining authorisation is set out in regulations.[13] (See Figure 28.2) A condition which has to be met if there is to be authorisation is that a person must use the "best available techniques not entailing excessive cost" (BATNEEC) in order to prevent environmental damage as a result of using the process.[14] The Department of the Environment's guidance note helps to define the components of this:

- **best** means the best technology available anywhere in the world
- **available** includes patented processes if they are available under licence
- **techniques** include operational matters (such as numbers and qualifications of staff, training and supervision) as well as the design, construction, layout and maintenance of buildings
- **not entailing excessive cost** involves balancing the additional costs against the resulting reduction in emissions.

STATUTORY NOTICES

Environmental Protection Act 1990
Part I
Application for Authorisation

ABC Ltd, of High Street, Anytown, has applied for an authorisation from A Metropolitan Borough Council to continue to operate a coating and printing process AND a rubber compound production process.

A copy of the application is available for public inspection free of charge at:

A Metropolitan Borough Council Office

Written representations about the application may be sent to A Metropolitan Borough Council at the above address within 28 days of this notice.

Figure 28.2 *Notice of application for authorisation in respect of prescribed processes*

Breach of a condition of an authorisation is a criminal offence and in criminal proceedings the operator has the responsibility to demonstrate that the BATNEEC were used. Where there is a release of substances into more than one environmental medium, the standard is the "best possible environmental option", that is, one resulting in the minimum overall environmental pollution.

WASTE ON LAND

Public Health and Control of Pollution Acts

Disposal of waste may constitute a breach of public health legislation. The local authority has powers to abate nuisances and/or to remove noxious matter. The Public Health Act 1936 includes as a statutory nuisance water (such as a stagnant pond) which is either a nuisance or prejudicial to health.[15] The enforcement procedures are now contained in the EPA 1990. The Public Health Act 1961 gives local authorities the power to remove rubbish from land if it is seriously detrimental to the amenities of the neighbourhood, but only after notifying the owner and the occupier of the land of their intention.

The Control of Pollution Act 1974 provided a more general scheme for dealing with waste, the provisions now being contained, with additions, in the 1990 Act, and in particular in the Controlled Waste Regulations 1992.[16] The 1974 Act defined waste very broadly and in relation to the person disposing of the material. "Controlled waste", which is subject to regulation, is defined as household, industrial and commercial waste, or any such waste. These terms are themselves defined. The general scheme is that the unlicensed disposal of controlled waste is an offence. The offences extend to depositing controlled waste on land or causing or knowingly permitting such disposal, and the use of any plant or equipment (or causing or knowingly permitting its use) in the disposal of controlled waste or in dealing with controlled waste.

The Environmental Protection Act 1990

The definitions of waste and controlled waste are similar to those in the 1974 Act, but with more detail.[17] The 1990 Act makes it an offence to deposit, treat or dispose of waste without authority and in addition, a new duty of care is introduced.[18] This applies to anyone who produces, imports, stores, treats, recycles or disposes of controlled waste. A person who has waste has a duty to stop it escaping. If waste is handed on to someone else, it must be secured. It should be transferred only to a person legally authorised to remove it. A written description of the waste must be handed over, and a transfer note completed and signed. Anyone receiving waste will need to be legally authorised to do so. The recipient should receive a written description and complete and sign a transfer note. A code of practice provides guidance on what is required.[19]

Persons authorised to collect and dispose of waste are council waste collectors, registered waste carriers, exempt waste carriers (such as charities and voluntary organisations), holders of waste disposal or management licences and those exempt from the need to hold such a licence. When waste changes hands, the paperwork has to be done by both parties: both need to complete and sign the transfer note. The written description and the transfer note can be one document: a model form has been published with the code of practice, but other forms can be used providing they meet the legal requirements. The written description must contain as much information as is necessary for someone else to handle the waste safely. The transfer note, to be completed and signed by both parties, must include:

- What the waste is and how much there is.
- What sort of containers it is in.

- The time and date the waste was transferred.
- Where the transfer took place.
- The names and addresses of both parties.
- Details of the category of authorised person of each party.
- If either or both of the parties, as a waste carrier, has a registration certificate, the certificate number and the name of the council that issued it.
- If either or both of the parties has a waste licence, the licence number and the name of the council has issued it.
- Reasons for any exemption from the requirement to register or have a licence.

Both parties must keep copies of the transfer note and description for two years. They may have to prove in court where the waste came from and what they did with it.

The Control of Pollution (Amendment) Act 1989 aims to make it more difficult for the carriage and disposal of waste to be carried out illegally. It is illegal for a person to transport controlled waste to or from anywhere in Britain in the course of business unless they are a registered carrier. The Controlled Waste (Registration of Carriers and Seizure of Vehicles) Regulations 1991[20] require registration with the waste regulation authority as a carrier of controlled waste. The authorities are:

- In England
  - Metropolitan areas –
    The London Waste Regulation Authority
    The Greater Manchester Waste Disposal Authority
    The Merseyside WDA
    or District Council in other metropolitan areas
  - Non-metropolitan areas -
    The County Council
- In Wales
    The District Council
- In Scotland
    The District or Islands Council

Registration must be sought if waste is to be carried by road, sea or inland waterway. It can be refused if the application is not made in conformity with the rules, or if the applicant or some other person connected with the business has been convicted of certain offences and the waste regulation authority considers that registration would be undesirable. There is a right to appeal against such a refusal.

The problem of packaging waste

The EU has produced a draft directive on packaging with the following aims:

- to reduce the quantity of packaging and packaging waste;
- to reduce its toxicity
- to promote recycling.

The ultimate aim is to reduce the amount of packaging waste disposed of in landfill tip sites. In future, these are likely to be available only as a last resort.

Within five years of its introduction, the directive will ban the production of any packaging which cannot be recycled. Other objectives are: within ten years, to recover 90 per cent by weight of packaging waste; to recycle 60 per cent of the amount recovered; and to limit to 10 per cent the amount of waste permanently disposed of by landfill or incineration.

As far as the management process is concerned, this means not only that there will have to be systems for recovering waste but also that packaging will have to be produced to specific standards (eg dealing with biodegradability and the presence of hazardous substances). For the consumer's benefit, a system of marking the packaging to indicate what it is made of will be needed.

Apart from cost, a number of other problems are likely to be generated by these proposals. One is the fact that recycling degrades the material. Another is what happens to recyled material when it becomes unusable. A further issue is whether there is a market for recycled material.

From a management view, it is perhaps desirable to distinguish between what is not possible and what is not liked, and use this as a starting point for a comprehensive programme of waste management. Germany has already laid down standards of waste management which are quite close to those of the EU, and these apply not only to German businesses. While this can be said to distort trade within the EU, it seems doubtful whether the EU will disapprove of something so close to its own environmental policy.

HAZARDOUS SUBSTANCES

This section does not deal with the control regime in respect of the escape of hazardous substances leading to serious emergency nor with 'mainstream' health and safety at work issues. These are dealt with in the chapters on emergencies (chapter 10) and hazardous substances (chapter 23). Rather, the aim is to deal with the day-to-day environmental control of the production, transportation and use of hazardous substances.

Food safety

The Food and Environmental Protection Act 1985 deals with the contamination of food. It allows the appropriate Minister to make emergency orders where there has been or may have been an escape of substances. Such orders may be made where the escape is likely to create a human health hazard through food consumption, or where the substances are or may in future be derived from anything unsuitable for human consumption. It is an offence to contravene an emergency prohibition, although certain actions may be permitted by the Minister. Appropriate Ministers also have powers to direct that food shall not be consumed if it is considered unsuitable.

The Act provides the Minister with other powers, namely:

- to make regulations to protect the health of humans, creatures and plants
- to safeguard the environment
- to secure safe, efficient and humane methods of pest control[21]
- to make information available in relation to these matters.

The Minister has powers to prohibit, approve, consent and generally review in these areas.

The Act provides for codes of practice on the advice of the Advisory Committee on Pesticides.[22] Inspectors under the Act are given wide powers.[23] As in HSWA, directors and other officers of the body corporate can be personally liable for offences committed.[24] The defence of due diligence may be argued in respect of alleged offences under the Act.[25]

The Food Safety Act 1990 generally provides for local authorities to be 'food authorities' for the purposes of the Act. It is an offence to render food so as to be injurious to health or to sell food not complying with food safety requirements. "Authorised officers" of the food authorities may serve improvement and prohibition notices and may seize food suspected of being in breach of the Act. Much of the detail is in regulations under the Act.[26]

The Act defines food in a way that is wider than conventional usage: it includes anything used as a food ingredient, animals eaten live (eg oysters), drink, slimming aids, dietary supplements and water used in food. The Act covers a broad range of commercial activities relating to food itself, as well as crops and live animals – defined as "food sources" – and the derivation of food from them. It also covers articles that food comes into contact with, such as wrapping materials and manufacturing plant (eg a mixing vat). The activities relating to food itself comprise all the operations involved in selling, possessing for sale, delivering, preparing, labelling, storing, importing delivering, preparing, labelling, storing, importing and exporting food.

The main offences under the Act are:

- Selling or possessing for sale, food which does not comply with food safety requirements (see below).
- Rendering food injurious to health eg by intentionally adding something to it or extracting something from it.
- Selling, to the purchaser's prejudice, food which is not of the nature, substance or quality demanded eg selling cod as haddock.
- Falsely or misleadingly describing or presenting food ie false statements or pictures.[27]

The food safety requirements are that food must not have been rendered injurious to health, must not be unfit and must not be so contaminated that it would be unreasonable to expect it to be eaten. Injurious to health means that it would harm a substantial part of the population. Food would be unfit if, for example, it was toxic. It would be unreasonable to expect food to be eaten if, for example, it contained a rusty nail.

Day-to-day enforcement is carried out by local authorities. Trading standards officers deal with the labelling and composition of food, environmental health officers with hygiene and food which is unfit for human consumption. The system of improvement and prohibition notices mentioned earlier is used in relation to food premises which are insanitary. (Food premises should be registered.) An emergency prohibition notice can be used where there is an imminent risk to health. In parallel, where the problem is more widespread, central government can make emergency control orders to require various steps to be taken. These powers are in addition to those contained in the 1985 Act (see above).

Due diligence is the general defence under the Act. Where it is held that an offence has been committed, the fine in the Magistrates' courts can be as high as £20,000 eg where there has been a breach of the food safety requirements. Fines are unlimited in the Crown Court and prison sentences are a possibility both there and in the Magistrates' courts. Additionally, injured consumers may sue at common law, proceed under the Consumer Protection Act 1987 or be awarded compensation by a criminal court where they have suffered as the result of a criminal act.

Other aspects

The planning regime

The planning regime is to be found in the Planning (Hazardous Substances) Act 1990. The keeping of any hazardous substance on, over or under land (with the exception of small quantities) will require the consent of the hazardous substances authority (the local authority). Application procedures are similar to those in respect of planning permission.

Genetically modified organisms

The Environmental Protection Act 1990 imposes a system of control to prevent or minimise any damage to the environment which may arise from the escape of a genetically modified organism (GMO).[28] The Act defines "organism" and "genetically modified". Damage to the environment is specified as being the presence in the environment of GMOs with a capability for harming living organisms supported by the environment.[29]

The Secretary of State has various powers which can be exercised by regulations. There is a prohibition against importing or acquiring, releasing or marketing any GMOs without a risk assessment and notice to the Secretary of State.[30] Failure to comply with these procedures leads to an offence: in certain cases the onus is on the accused particularly to show that appropriate techniques were used to comply with the requirements laid down in consents and for monitoring.[31] The Secretary of State is required to keep a register of notices, consents etc.

Miscellaneous controls

The 1990 Act also provides the Secretary of State with powers to restrict the importation, use, supply or storage of injurious substances or articles.[32] It allows the Secretary of State to obtain information about potentially hazardous substances and provides for a register of contaminated land.[33] New substances have to be notified under the Notification of New Substances Regulations 1993.[34]

Supply, labelling, packaging and transport of dangerous substances

The Chemical (Hazard Information and Packaging for Supply) (the CHIP 2) Regulations 1994 govern the supply, classification, labelling and packaging of dangerous substances and preparations.[35] Among other things the regulations require suppliers to classify chemicals according to their principal hazards and provide specific information about such hazards and any relevant precautions on labels and in safety data sheets.

The road carriage provisions formerly in the original CHIP Regulations 1993 are now to be found in the Carriage of Dangerous Goods by Road and Rail (Classification, Packaging and Labelling) Regulations 1994.[36] The carriage of dangerous goods by rail is covered by the Carriage of Dangerous Goods by Rail Regulations 1994. Together with other railway legislation they are designed to ensure that existing health and safety standards are maintained, and where necessary, improved following privatisation.[37]

NUISANCE

Statutory nuisance

The law relating to statutory nuisance is now contained in the EPA 1990. The local public health authority is able to deal with nuisance if they are satisfied that a statutory nuisance exists or is likely to occur or recur. An abatement notice is served, against which there is a right to appeal within 21 days.[38] Failure to comply with an abatement notice can lead to a fine of up to £20,000 in the magistrates' courts.

Statutory nuisance is defined as:

- premises which are themselves prejudicial to health

- smoke, fumes, gas or noise emitted from premises so as to be prejudicial to health or a nuisance
- dust, steam, smell or other effluvia on business premises where these things are prejudicial or a nuisance
- accumulation or deposits prejudical or a nuisance
- any animal kept in such a place or manner as to be prejudicial or a nuisance
- any other matter declared by legislation to be a statutory nuisance eg a stagnant pond under the Public Health Act 1936 – see above, pp670–1.

Smoke, fumes, gas, dust, steam, smell or other effluvia emitted from premises subject to the Alkali Etc Works Regulations Act 1906 – being replaced by EPA Part I – are the responsibility of HMIP and the HSC. The defence of "best practicable means" is available in respect of premises, smoke, dust, steam, smell or other effluvia, accumulations or deposits, animals and noise, except where the nuisance is caused by smoke from a chimney.

Best practicable means is interpreted as follows:[39]

- Practicable means reasonably practicable taking into account local circumstances, the current state of technical knowledge and the financial implications.
- The means to be employed include design, installation, maintenance and manner and periods of operation of plant and machinery and design etc of buildings and structures.
- The test is to apply only as far as it is compatible with any duty imposed by law, with safety and safe working conditions and with the exigencies of any emergency.

Noise

The principal legislation governing 'public' noise is the Control of Pollution Act 1974, although the enforcement procedures are contained in the EPA, 1990. There is also specific legislation eg the Civil Aviation Act 1982 deals with aircraft noise.

Noise from construction sites can be controlled by the local authority by serving a notice specifying how the work should be done (including the plant and machinery), hours of work and level of noise.[40] Relevant codes of practice must be taken into account, as must the need to secure best practicable means to minimise noise levels.[41] Appeals lie to the magistrates within 21 days.[42]

An area can be designated as a noise abatement zone. Noise levels have to be recorded in a register.[43] The local authority can then issue a noise reduction notice. Rights of appeal lie to the magistrates within three months.

Civil liability

In addition to the statutory provisions, there is the possibility of individuals pursuing civil actions against those creating a nuisance. Similarly where negligence is involved. In certain cases there may be possible an action for breach of statutory duty.

The rule in *Rylands v Fletcher* is of importance in this area. This holds that there is strict liability ie liability without the need for fault, where there is an escape from land which causes damage (here, the escape of water to a mine on adjoining land).

As far as negligence is concerned the general principle is that of the duty of care as set out in *Donoghue v Stevenson* (see chapter 4). However, the need for the risk to be foreseeable has been emphasised recently in a water pollution case. Seepage of a toxic substance from tanks at industrial premises travelled through the ground and entered a watercourse some considerable distance away. It was held that this risk was not reasonably foreseeable, so no liability arose (*Cambridge Water*).

MANAGEMENT'S ENVIRONMENTAL STRATEGY

This section focuses upon the strategy needed by management in order to deal with environmental matters and in particular those matters which are subject to regulation by law.

New businesses or projects

If the business is a new one or is involved in a major development, then depending upon the nature of the business or development an environmental impact assessment (EIA) may be desirable or necessary. EIA is the process of evaluating the environmental impacts likely to arise from a project. Many countries list projects which require an EIA by law or which require an EIA if an environmental authority demands one. An environmental authority will review the EIA documentation prior to making a decision to approve or reject the project on environmental grounds. The approval may be unconditional or be subject to conditions such as compulsory self-monitoring or the use of specific technologies.

BS 7750

Essentially, this addresses the need for companies to have an environmental policy. There are a number of stages an employer should follow to qualify for BS 7750:

1. Conduct a preparatory review of existing management practices, taking into account relevant legislation.
2. Establish an environmental policy – this is a *public* statement of commitment to environmental improvement and should cover matters such as pollution and waste control.

3. Derive environmental objectives and targets.
4. Designate a responsible person to ensure that an environmental management programme is set up, and that the policy and objectives are being met.
5. Draw up an environmental management manual.
6. Identify training needs.
7. Conduct an environmental audit.
8. Carry out an environmental review to check that the policy is up-to-date.

This is a specification for environmental management systems. It lays a foundation for effective environmental management but does not guarantee it.

Effective environmental management

Three principal stages can be seen as comprising effective management in this area:

- enhancing environmental awareness within the organisation
- undertaking environmental audits
- developing an environmental improvement strategy.

The first of these requires key employees to be involved in identifying the issues to be confronted. Employees more generally should be made aware of the organisation's current environmental performance. A plan should be drawn up setting out the organisation's environmental aims, objectives and targets, and this information should be communicated to employees. Environmental audits should monitor plans and performance against various targets in order to reveal areas of risk and weakness. Measures for improvement need to be drawn up and priorities established in order to ensure practical and continuing implementation of the policy.

The Institute of Management has set out environmental guidelines for managers and provides an example of an organisational environmental policy (see Figure 28.3).

Environmental awareness training is part of an effective environmental policy. Managers and other employees should be trained, depending upon their position, in:

- environmental issues, such as liability for waste
- environment legislation
- the company's environmental organisation, such as – in a large firm – the functions of an environmental services department
- key areas of environmental policy, including waste management and the control of effluent, noise and air pollution.

Ultimately such a policy will need translating into detailed, practical steps which can be taken by employees. These steps are often quite simple, for example, putting a net over a skip when it is travelling or locking tip and waste areas when not in use.

- Recognise its obligations to its owners, employees, suppliers, customers, users, society and the environment
- Allocate to a director or senior manager responsibilities that include environmental issues.
- Educate and train employees in environmental matters, as appropriate.
- Make the most effective use of all natural resources for the benefit of the organisation and the overall public interest.
- Progressively reduce the amount of waste, avoid harmful pollution and make serious efforts to find ways of reprocessing or converting waste materials into useful products.
- Reduce environmentally related risks.
- Market products, services and processes which create the minimum environmental damage and can be used safely.
- Be willing to exercise influence and skill for the benefit of the community within which the organisation operates.
- Ensure that all public communications are true and unambiguous, with full disclosure of environmental, health and safety issues.
- Respect the interests of neighbours and the world community.

Reproduced with permission from the Institute of Management.

Figure 28.3 *Institute of Management example of an organisational environmental policy*

The DTI role

The Department of Trade and Industry recognises that business must respond to environmental pressures and in May 1989 launched its Environmental Programme. The DTI's Environmental Unit brings together all of the Department's widespread interests in environmental issues. Details of what the Environment Unit does, and of practical ways in which it can help business by offering advice, encouraging best practice and supporting research and development, can be found in its publication *The Environment: A Challenge for Business*.[44]

Notes

1. *This Common Inheritance: Britain's Environmental Strategy*, HMSO, Cm 1200, 1990. See also: *Report of the Royal Commission on Environmental Pollution*, HMSO, 1994
2. Public Health Acts 1936, 1937 and 1961
3. The offences are set out in s85 of the 1991 Act: consents are provided for in s88
4. This term is defined, technically, in s34 of the 1956 Act
5. 1956 Act, s16 provides for smoke to be classified as a public nuisance
6. HSWA, s5
7. The Health and Safety (Emissions into the Atmosphere) Regulations 1983/943 as amended by 1989/319
8. Control of Industrial Air Pollution (Transfer of Powers of Enforcement) Regulations 1987
9. Environmental Protection (Prescribed Processes and Substances) Regulations 1991/472, amended by 1992/614
10. 1990 Act, s1
11. 1990 Act, s2
12. 1990 Act, s6
13. The Environmental Protection (Applications, Appeals and Registers) Regulations 1991/507
14. 1990 Act, s7
15. Public Health Act 1936, s259
16. The Controlled Waste Regulations SI 1992/588, issued under the EPA 1990 and the Control of Pollution (Amendment) Act 1989
17. 1990 Act, s75
18. 1990 Act, ss33 and 34
19. *Waste Management – The Duty of Care, A Code of Practice*, HMSO, 1991
20. SI 1991/1624
21. See: Control of Pesticides Regulations, SI 1986/1510
22. See: Control of Pesticides (Advisory Committee on Pesticides) (Terms of Office) Regulations, SI 1985/517; and 1985 Act, ss 16-17
23. Food and Environment Protection Act 1985, s19
24. 1985 Act, s21
25. 1985 Act, s22
26. For example: Food Labelling Regulations 1984 as amended; Food Hygiene (General) Regulations 1970 as amended: Imported Food Regulations 1984; and various regulations on milk and dairies, food composition, the use of food additives and packaging materials, and registration of premises. A full list of regulations issued under the 1990 Act is available from the Ministry of Agriculture, Fisheries and Food.
27. Note also the detailed regulations on labelling: The Food Labelling Regulations 1984

28. EPA 1990, s106
29. 1990 Act, s107
30. 1990 Act, s108 ff
31. 1990 Act, ss 118-19
32. s140
33. ss142 and 143
34. SI 1993/3050
35. SI 1994/3247. See *The Complete Idiot's Guide to CHIP 2*, IND(G)181(L), HSE 1995.
36. SI 1994/669; CHIP Regulations, SI 1993/1746
37. The Carriage of Dangerous Goods By Rail Regulations, SI 1994/670. See also: Railways (Safety Case) Regulations and Railways (Safety Critical Work) Regulations 1994
38. Statutory Nuisance (Appeals) Regulations SI 1990/2276
39. 1990 Act, s79(9). Guidance on the application of the definition to noise can be found in codes of practice issued under the Control of Pollution Act 1974, (CPA) s71.
40. CPA, ss60-1
41. These are codes issued under CPA s71: see The Control of Noise (Code of Practice for Construction and Open Sites) Order SI 1984/1992 and Control of Noise (Code of Practice for Construction and Open Sites) Order, SI 1987/1730
42. Control of Noise (Appeals) Regulations, SI 1975/2116
43. See: Control of Noise (Measurement and Register) Regulations, SI 1976/37
44. *The Environment: A Challenge for Business*, DTI, 1991

Index

abatement notice 676
Abrasive Wheels Regulations 644, 646
access control 341
Access to Health Records Act 1990 86
Access to Medical Reports Act 1988 86, 90
accident investigation
contractors 343–4
employee involvement 303
procedures 185–9
techniques 189–91
accidents
works transport related 390–1
workshops and garages 402–3
accidents, causes of
environment 117
equipment 116
immediate and basic 112–17
loss causation model 112, 113
materials 116–17
people 116
transport 416–19
Accident Prevention Advisory Unit (APAU) 118
acoustic enclosures 527
Action on Smoking (ASH) 658
Advisory Committee on Pesticides 674
Advisory Committee on Safety, Hygiene and Health Protection at Work 39
agency 28
agriculture 352
AIDS 91, 664–5
air-borne substances 292, 293, 561, 564
air monitoring 580 *see also* air-borne substances
air pollution 668–9
alarm systems, fire 493–5
alcohol *see* drinking
Alkali Etc Works Regulations Act 1906 677
amosite (brown asbestos) 574
'Anthropometric Consideration' 378
appeal system (legal matters) 30
arc-eye 449, 533
arson 487
articulated tractor unit 406
asbestos 293, 562, 573–85
Asbestos (Licensing) Regulations 1983 577, 582
Asbestos (Prohibitions) Regulations 1992 583, 584
Asbestos and You, HSE leaflet 578
asbestosis 540, 574
asthma 541
audiometric testing 284–5, 516
audits, health and safety 126, 219–29, 235–48, 312–13 *see also* inspections
Avoiding Danger from Underground Services, HSE booklet 641
'A weighting' 517

Blackspot Construction, HSE Report 218–19
blocking plates 384
blood tests 285
body protection 605
bomb threats 201, 205, 206–8 *see also* suspect packages
breathing apparatus 602 *see also* respiratory protective equipment
bridges 399, 403
British Approvals Service for Electrical Equipment in Flammable Atmospheres (BASEEFA) 456
British Examining and Registration Board in Occupational Hygiene (BERBOH) 290

Index

British Occupational Hygiene Society 290
British Standard publications 454
British Standard EN60529 455
British Standard 4547 481–4
British Standard 5304 368, 374
British Standard 5345 456
British Standard 6651 455
British Standard 7750 678
buffer zones 386
Building Regulations 1991 479
Building (Scotland) Act 1959 34
bulldozer 408
burden of proof 32, 86, 87
buried services 640
byssinosis 540

Careers Services 280
carpal tunnel syndrome 423, 530
Carriage of Dangerous Goods by Road and Rail (Classification, Packaging and Labelling) Regulations 1994 676
cartridge operated tools 647–8
'Centre Tapped Earth' transfer 458, 646
Chemical (Hazard Information and Packaging) Regulations 1993 (CHIP) 102, 544
Chemical (Hazard Information and Packaging) Regulations 1994 (CHIP 2) 544, 676
chemical indicator tubes 562
childbirth *see* pregnancy
Children and Young Persons Acts 1933–69 34
chiropodist 288
chrysotile (white asbestos) 573, 576
circuit conductor 450, 458–9
Civil Aviation Act 1982 677
civil law 24–6
  contract 27, 28
  tort 24–5, 26–7, 52
  trust 24
Civil Liability (Contribution) Act 1978 86
Clean Air Act 668
client (Construction (Design and Management) Regulations) 619, 620, 622, 623, 624
Coal Mines Act 1842 50
code of practice, definition 54–5
combustibles 483, 487
commendation 126
common law 21–2, 52
communication on health and safety matters 144–52
Community Charter of Fundamental Social Rights *see* Social Charter
compactors 408
Companies Act 1985 56
competent person 54, 60, 137, 273, 518, 630, 631, 638, 639, 644
  electrical safety 473–4
  requirement for competence 137–8
competent staff *see* competent person
computerised safety communication 309
conductors, electrical 456–9
confined spaces 641–2
  entry into 261
connectors, electrical 459
Construction (Design and Management) Regulations 1994 325, 349, 619–30
  client 619, 620, 622, 623, 624
  designer 619, 620, 622, 625
  notifiable 621, 622, 623, 625–6
Construction (General Provisions) Regulations 1961 33, 131, 353, 407
Construction (Head Protection) 1989 591, 630, 649
Construction (Health, Safety and Welfare) Regulations, proposed 631–3
Construction (Health and Welfare) Regulations 1966 353, 630, 650
Construction (Lifting Operations) Regulations 1961 630, 631, 637
Construction (Working Places) Regulations 1966 630, 631, 633
construction industry 352–3
Consumer Protection Act 1987 58, 84, 675
Continuing Professional Development (CPD) 142
contract of employment 21
contract order 335
contracts 27, 29

contractors 146, 160, 238, 267–8, 320–46, 403
  approved status 334
  communications with 334–40
  duties 624–30
  fire safety 487–8, 510
  handbooks 335–6
  induction 308, 339–40
  permits to work 342–3
  pre-work meetings 336–9
  relevant legislation 320–5
  review meetings 344
  safety monitoring 313
  selection of 326–34
  services 273–4
  statutory duties 57–8, 62
  vicarious liability 83
  work completion reports 344–6
Control of Asbestos at Work Regulations 1987 573, 575–82, 587, 591
Control of Asbestos in the Air Regulations 583
Control of Industrial Major Accident Hazards (CIMAH) Regulations 1984 200, 215
  requirements outlined 203–4
Control of Lead at Work, HSE booklet 586
Control of Lead at Work Regulations 1980 573, 586, 591
Control of Pollution Act 1974 671, 677
Control of Pollution (Amendment) Act 672
Control of Pollution (Special Waste) Regulations 583
Control of Substances Hazardous to Health (COSHH) 58, 71, 77, 154, 221, 271–2, 294, 352, 404, 424, 544–61, 591, 602, 649
  carrying out assessments 547–58
Controlled Waste (Registration of Carriers and Seizure of Vehicles) Regulations 672
Controlled Waste Regulations 671
corporate manslaughter 52, 71
cost considerations 314
co-tenants 160 *see also* shared premises
County courts 29, 30, 31
Court of Appeal 30, 31
courts, system of 29
Courts and Legal Services Act 1990 35
cranes 412–13, 637–8
Criminal Justice Act 1991 34
criminal law 23, 25–7
crocidolite (blue asbestos) 573, 574
cross audits 247
Crown Court 29, 30, 31, 70
cumulative trauma disorder 423
cyclists 399

danger, definition 368
danger zone (machinery), definition 368, 383, 384
Deacon, Andrew 118
deafness, occupational 515
decibels 517
demolition 642–3
Department of Trade and Industry (DTI) 96
Deregulation and Contracting Out Act 1994 96
dermatitis 282, 286, 543
design and manufacture, safety of moving machinery 368–9
designated areas 580, 584
designer (Construction (Design and Management) Regulations) 619, 620, 622, 625
Diploma in Safety and Health 142 *see also* Diploma in Safety Management, NEBOSH Diploma
Diploma in Safety Management 142 *see also* NEBOSH Diploma, Diploma in Safety and Health
direct reading instruments 562
Disablement Advisory Service 280
discipline, safety 92–3
discussion workgroups, safety 311
Display Screen Equipment (DSE) Regulations 1992 434–45, 653
Diving Operations at Work Regulations 1974 358
dosemeter, personal 522–3
drinking, alcohol 660–2
drivers
  code of discipline 397

general safety policy 403–5
locomotive 413–14
monitoring of standards 396–7
selection and training 395–6, 415
drug abuse 660–2
Drug Abuse at Work, HSE booklet 660
due diligence 71, 452, 454
dump trucks 401
dumpers 406, 531
duty of care, employers' 27–8, 51, 103, 653–4
see also employers
action for breach of 74–5
delegating liability 87–8
general duties to employees 56–7
negligence 80–1 *see also* negligence
owed to non-employees 57–8
reasonable care 75–6
safe fellow workers 79–80
safe plant, equipment and tools 79
safe premises 77
safe system of work 77–8
travelling 81
types of action 74–5
vicarious liability 81–3

ear protection 519, 523, 527–9, 603
ear protection zone 519–20, 523
earth-moving machines 408
eating 570
electricity
burn 449, 457
conductors 450, 457–9
dangers of 447–9
earthing 457–8
electric shock 447–9, 457, 472
explosion 449
fire 449
flammable or explosive equipment 456
isolation 460–1
nature of 446–7
protective equipment 453–4
Electricity at Work Regulations 1989 352, 447
an outline 450–74
emergency 568, 632
controller 218–19
lighting 506–7
means of escape from fire 505–6
off-site 204
on-site 204
plans 304
procedures 204–15
emergency services 208
employees 54
competence requirements 138–40
duties under HSWA 451
protection in health and safety cases 89–91
refusal to work on grounds of lack of safety 92–3
safety attitudes 297–8
safety behaviour 298–9
safety culture 299
safety involvement 306–14
statutory duties 59, 63
employers
audit and inspection legal duties 221
duties under COSHH 545
duties under HSWA 451–4
RIDDOR requirements 194–8
Employers' Liability (Compulsory Insurance) Act 1969 63, 81, 85
Employers' Liability (Defective Equipment) Act 1969 33, 76, 79
Employment Appeal Tribunal (EAT) 91
Employment Medical Advisory Service (EMAS) 66, 280
Employment Protection (Consolidation) Act 1978 89–91
employment protection in health and safety cases 89–91
employment status 28
enforcement liaison officer 67
enforcing authorities 194, 621, 623
Environmental Hygiene (EH) 562
environmental impact assessment (EIA) 678
environmental policy 678–9
Environmental Protection Act 1990 58, 669, 671–2, 675, 676
environmental strategy 678–80
equipment isolation 257–60
ergonomics
approaches 421–2

meaning of 420
principal sources of risk 422
problems 420–1
Ergonomics at Work, HSE leaflet 420
Essentials of Health and Safety at Work, HSE book 159
European Communities Act 1972 40–1
European Court of Justice 36, 38
European Health and Safety Agency 48, 101–2
European Union 23, 36–49, 349
directive on Manual Handling of Loads 424
fourth health and safety action programme 45, 48, 101
Framework and Workplace directives 479–80
legal system 36–42
social policy 39, 102–3
evacuation 209–11
excavations 639–40
excavators 408, 640
explosives 479
eye protection 308, 436–7, 533, 599–601, 613, 649
eyesight test for VDU users 434, 436–7
see also vision testing

Factories Act 1844 50
Factories Act 1961 51, 63, 64, 75, 103, 138, 217, 350–1, 353, 393, 424, 630, 644
Factory Act 1833 50
Factory Act 1937 51
Factory and Workshop Act 1901 51
Faculty of Occupational Medicine 279
falling, protection against 606–7, 617–18, 632
Fatal Accidents Act 1976 86
fatalities, reporting of 196
five-step approach 107, 158, 165, 166
Five Steps to Risk Assessment 158
fire 201, 205, 477
barriers 503
causation 483–8
classification 481–3
compartmentalisation 504
electrical prevention measures 491–2
flame detectors 493, 494
means of escape 505–6
prevention and active control 490–502
protection systems 493–502
safety doors and glass 502–3
safety knowledge 480–1
safety management 507–12
spread considerations 488, 505
fire fighting equipment
blankets 497
dry risers 501
fixed systems 498–501
foam inlets 501–2
hose reels 498
hydrants 501
tenders, in-house 498
water cannons 498
fire fighting extinguishers
carbon dioxide 495
foam 495–6
halon 496
powder 496
trolley mounted 498
water 497
Fire Officers Committee (FOC) 499
Fire Precautions Act 1971 68
fire protection 599
fire triangle 477, 490
first-aid 357–60
flammable
gases 483, 485, 492
liquids 483, 485, 492–3
substances 483, 486
foot protection 604–5, 613–14, 649
'Four Cs' 122–3, 128
fumes 539

general register 351
generic assessment *see* risk assessment
genetically modified organism (GMO) 275
goggles, safety *see* eye protection
grandparents, safety 309
group management safety surgeries 307
guard 368
adjustable 377
automatic 377–8

distance 377
enclosed 376–7
fixed 374
interlocking 377
perimeter/area 374–6
guarding characteristics 370
guarding devices 378–83
guidance notes 54–5

handbooks 147–8 *see also* manuals
contractors' 335–6
hazard
definition of 155
reporting books 301–2
tags 311, 319
telephones 302
Hazard and Operability Studies (HAZOP) 274
head protection 598–9, 613, 649
Health and Safety at Work etc Act 1974 22, 26, 27, 28, 32, 34, 52, 54, 56–61, 68, 69, 74, 96, 119, 268–9, 272, 283, 322–3, 372–3, 390, 622
Health and Safety at Work etc Act 1974 (Application Outside Great Britain) Order 1995 34
Health and Safety at Work (Northern Ireland) Order 1978 34
Health and Safety Commission 21, 23, 66
constitution, duties and powers 64–5
review of legislation 95–101, 103–4
Health and Safety (Display Screen Equipment) Regulations 1992 154
Health and Safety (Emission into the Atmosphere) Regulations 1983 158
Health and Safety (Enforcing Authority) Regulations 1989 67
Health and Safety Executive 21, 66
constitution, functions, structure and working of 65–6
notices and prosecutions 67–9
powers of inspectors 69
risk assessment guidance 159
health and safety file 624, 626–7
Health and Safety (First-Aid) Regulations 1981 357–60
Health and Safety in Motor Vehicles Repair, HSE booklet 171
health and safety plan 624, 626–30
health and safety policy 119–21, 135, 139, 147
Health and Safety Policy Statements (Exception) Regulations 1975 34
health and safety specialists 129–32, 144, 157, 188, 266, 278–9
needs of 141–2
health education 289–90
health questionnaire 285–6
health surveillance 560–1
Herald of Free Enterprise 52, 71, 103
High Court 21, 29, 30, 31
hoists 638
hot work 483–4
How to Achieve a Smoking Policy at Work 658

immobilisation, principles of 383–8
importers 58
improvement notice 67, 111
Incident Report of the Month competition 194
Index of Protection (IP) 455
Industrial Robot Safety, HSE booklet 374
Industrial Training Act 1964 91
Industrial Tribunals 91
infra-red radiation (IR) 532–3
ingestion, of hazardous substances 540
inhalation, of hazardous substances 539, 541
injuries
electrical 448
ergonomic 422–3
meaning of major 195
moving machinery 388
not reasonably foreseeable 87
not sustained in the course of employment 88
reporting of 196
treating services 288
inoculations 288–9
inspections 125, 312, 465
aims 222–3
classification of 223–6
electrical 479–80
fire safety 510–12

in practice 235–42
key points 247–8
scheduled 227–30
Inspectorate of Pollution, HM 669
Institute of Management 679
Institute of Occupational Hygienists 290
Institution of Electrical Engineers' Regulations for Electrical Installations (IEE Wiring Regulations) 454–5, 456, 458, 459, 460, 461, 464
Institution of Environmental Health Officers (IEHO) 143
Institution of Occupational Safety and Health (IOSH) 141–2, 143
insulation, double 458
integrated pollution control (IPC) 669, 670
International Loss Control Institute (ILCI) 118
International Safety Rating System (ISRS) 118–19
interpretation, statutory 32
Interpretation Act 1978 32
ionising radiation 534–6
Ionising Radiation Regulations 1985 535, 591
iron oxide fume 540
isolation 384, 460–1, 470
equipment 257–60

ladders 633–4
lanyard 606–7
lasers *see* visible radiation
law *see* torts, contracts
civil 24–5
correct application of 28–9
criminal 23
European Union 23
extent 26
relation between criminal and civil 25–6
types 23
Law Reform (Contributory Negligence) Act 1945 87
Law Reform (Personal Injuries) Act 1948 84
lead
poisoning 585–6
sources of 586
Lead and You, HSE leaflet 586
legal framework
of health and safety at work 52–3
tomorrow's 103–4
legal personality 27–8
lifting operations 637–8
lift trucks 408–9
lighting 461
emergency 506–7
litigation 84–8
loaded goods vehicle 395, 410
loading shovels 409
local authority
health and safety enforcement 67, 351
notification of emergency 208
local exhaust ventilation (LEV) 559, 564, 567, 571
design and maintenance of systems 565–7
Locomotive and Wagons (Used on Lines and Sidings) Regulations 1906 413
locomotive design 414–15
Loss Prevention Council 493, 499
LPC/FOC 29th Edition Sprinkler Rules 499
lung cancer 574
lung function testing 284
Lyme Bay 52, 71, 103

Maastricht Treaty 36–7, 39–40
magistrates' courts 21, 29, 69
management
commitment and interest 133
communication 144–52
responsibility for health and safety 128–9
risk assessment 157
Management of Health and Safety at Work (Amendment) Regulations 1994 662
Management of Health and Safety at Work Regulations 1992 46, 59, 61–3, 138, 140, 153, 154, 164–5, 170, 200–1, 202, 221, 295, 323–5, 392, 629
Management Regulations *see*

Management of Health and Safety at Work Regulations 1992
manslaughter 71, 103 *see also* corporate manslaughter
manual handling
avoidance of 425
carrying out risk assessment 426–9
key action points 429
nature and scale of problem 423
reducing risks 429
Manual Handling Operations Regulations 1992 123, 154, 424–33, 606, 655
manuals
driver training 396
health and safety procedures 148
risk assessment 160, 166
safety 307
manufacturers 159–60
statutory duties 58, 272, 521
master register of fire equipment 514
Maternity (Compulsory Leave) Regulations 662, 664
maximum exposure limit (MEL) 292, 544, 558, 560, 561, 571
measuring health and safety performance 124–5
mechanical anchors 384
medical practitioners 279
Memorandum of Guidance on the Electricity at Work Regulations 1989 454
mentor 309
mesothelioma 574
metal fume fever 541
Methods for the Determination of Hazardous Substances (MDHS), HSE booklet 562
mines and quarries 351–3
Mines and Quarries Act 1954 71, 351–2
Mines and Quarries (Tips) Act 1969 352
Mines and Quarries (Tips) Regulations 1971 352
mobile crane *see* cranes
mobile elevating work platforms (MEWPs) 634–6
mobile plant 373 *see also* vehicles
mobile work platforms, power operated 411–12
Montreal Protocol Agreement 496
moving machinery
definition of 367
hazards of 373–4
types of safeguards for 374–83
unguarded 372
multiple penalties, example of 70
musculoskeletal disorders 288, 423

National Examination Board in Occupational Safety and Health (NEBOSH) 141–2, 278
National Rivers Authority (NRA) 668
national vocational qualification (NVQ) 269–70
near-miss reporting 302–3
example 316
negligence
contributory 87
duty of care 80–1, 364
New and Expectant Mothers at Work, HSE booklet 662
newsletters 149
noise 284–5, 294
assessment 521–5
hazards of 515–16
measurement 516–18
reduction measures 525–7
sources of 516
Noise Guides, HSE booklets 518, 521, 522, 523, 525, 526, 528
Noise at Work Regulations 154, 163, 352, 515, 517, 518, 519–21, 649
non-employees, statutory duties 57
notice boards 148–9
notices 63
notifiable (Construction (Design and Management) Regulations) 621, 622, 623, 625–6
Notification of New Substances Regulations 1993 676
nuisance, statutory 676

occupational exposure standard (OES) 293, 544, 558, 560, 561, 562
occupational health
advice 289

outlined 276–8
provision of service 278–80
screening 281–6
Occupational Health Nursing Certificate 279
occupational hygiene surveys 560, 561–3
occupational hygienist 132, 290–1, 562
occupiers, duties of 83–4
Occupiers' Liability Acts 1957 and 1984 81, 84
octave band analysis 528
Offices, Shops and Railway Premises Act 1963 (OSRPA) 51, 64, 138, 351, 644
offshore oil and gas safety 66
Offshore Safety Act 1992 69
Ohm's law 447–8
one-to-one interviews 308
operation of moving machinery, safe 369–73

pallets 404, 409
parking areas 398, 405
pass-card systems 342
pedestrian traffic 397
permit to work 259, 471, 640
common mistakes and problems 266–7
contractors 342–3
designated situations 261–2
design of form 263–4
inadequate cancellation 267–8
issue and cancellation sequence 262–3
issuer 264–5
transferring 267
where permits may be required 261
why use 260–1
personal danger board 384, 389, 591, 598, 607
personal injury, definition of 54
personal protective equipment (PPE) 161, 272–3, 534, 535, 558–9, 566, 571, 578–9
British and Harmonised European Standards for 613–18
for construction work 649
management control of 607–11
provision and cleaning of 579–80
types of 568–9
Personal Protective Equipment at Work Regulations 1992 154, 591–3, 649
Personal Protective Equipment (EC Directive) Regulations 1992 593–6
physiotherapy 288
pin trucks *see* lift trucks
Piper Alpha oil platform disaster 69, 103
Planning (Hazardous Substances) Act 1990 675
planning supervisor, construction 624, 625–7
pneumoconiosis 540
poisoning 542
pollution 668–70
portable appliance tester (PAT) 464
portable equipment, maintenance of 461–6
portable machinery 373
portable tools 373
posters, health and safety 149
posture 422
power failure 202
practicable, meaning of 53, 57 *see also* reasonably practicable
pregnancy 662–4
Pregnant Workers' Directive 662
pressure systems and transportable gas containers 363–4
Pressure Systems and Transportable Gas Containers Regulations 1989 137, 221, 363
principal contractor 622, 629–30
prohibition notice 67, 111, 675
Provision and Use of Work Equipment Regulations 1992 (PUWE) 53, 369–72, 392–3, 424, 620, 644–6
Public Health Act 1936 670–1, 676
purchasing controls
hazardous substances 271–2
personal protective equipment (PPE) 272–3
contractors' services 273–4

quarries *see* mines and quarries

residual current devices (RCDs) 646
respiratory
  problems 540–1
  protection 601–2
respiratory protective equipment (RPE) 569–70, 571, 578 *see also* breathing apparatus
risk assessment 61
  acting on 167–71
  carrying out 157–64
  display screen equipment 435
  fire safety 481
  formats for records 165–7
  hazardous substances 547–58
  manual handling operations 424–33
  manufacturers 369
  model assessments 163–4
  moving machinery 373–4
  personal protective equipment 596–7
  practical examples 171
  purpose and meaning of 153–4
  railway operations 418
  recording 164–7
  tankers 410
  visual display units 435
risk, definition of 155, 368
road/rail crossings 418, 532
Road Transport and Traffic Acts 393
Road Transport in Factories and Similar Places, HSE guidance note 393
roadwork repairs 400
Robens Committee of Inquiry 48, 51, 217
roofwork 636–7
room dimensions 356

safeguards 368, 374–82
safe systems of work 250–1, 640, 642
safe working practices 77, 310, 474
safe working procedures 310
safety attitude 233, 297, 307–8
safety briefings 308
safety committees 59, 151, 296, 301 *see also* safety meetings
safety counsellor 301–2, 302–3
safety culture 122, 133, 299, 311–12, 314
safety device, definition 368
safety infrastructure 295, 299–304
safety inspections 124, 125, 227–30 *see also* risk assessments, safety sampling, safety tours
safety inspectors, voluntary 313
Safety of Machinery BS 5304 368, 374
safety manuals 307
safety meetings 300–1 *see also* safety committees
safety observations 232–5
Safety of Moving Machinery British Standard 5304 368, 374
safety plan, annual 303–4
safety policy statement 119–21, 133
safety representatives 59–61, 157, 295–6
Safety Representatives and Safety Committee Regulations 1977 59, 295, 301
safety responsibility, areas of 312
Safety in Roofwork, HSE booklet 637
safety sampling 230–1 *see also* risk assessments, safety inspections, safety tours
safety signs 360–1
Safety Signs Regulations 1980 360
safety spectacles *see* eye protection
safety supervisor 630–1
safety surgeries 307
safety survey 226–7, 305–6
safety tours 232 *see also* risk assessments, safety inspections, safety sampling
sampling pumps and filter heads 562
sanitary conveniences 356
scaffolds 634
scissor lift platforms *see* mobile work platforms
scrapers 408, 531
Selecting a Health and Safety Consultancy, HSE booklet 129
self-employed 54, 61, 154, 162, 195, 322, 323, 324–5, 353, 451, 545
serious and imminent danger, meaning of 201–3
setting health and safety standards 123–5
Sex Discrimination Act 655, 662, 664
shared premises 62, 204 *see also*

co-tenants
shiftwork 655–7
siderosis 540
signing-in systems 341
silicosis 540
Single European Act of 1986 36, 42
Single European Market 46
'six-pack' 41, 96
smoke detectors 494
smoking policy 484, 570, 657–60
Social Action Programme 44–5
Social Charter 39–40, 43–4, 45
Social Policy Protocol 39
sound-absorbent materials 527
sound level meters 522
sound level meters, integrating 522
spark generation 486–7
speed limits 399
speedometer 404
sprinkler systems 499–500
standards of care 53
statute law 22–3
statutory duty, breach of 365
stress 653–4
Stress Research and Stress Management, HSE report 653
substances hazardous to health 544 *see also* Control of Substances Hazardous to Health
suggestion schemes 150
suite danger board 389
suppliers 83–4, 159–60, 272
Supply of Machinery (Safety) Regulations 1992 369, 374
suspect packages 201, 208 *see also* bomb threats
suspension from work
  on maternity grounds 90
  on medical grounds 90–1
  right to receive pay 91
Suzy Lamplugh Trust 362

tankers 409–10
task observation 310
task procedures, written 251–7
telescopic boom *see* mobile work platforms
temperature 355
temporary employees 324, 321, 325
tendinitis 423
tennis elbow 423
termination after medical screening 286
tinnitus 515
tipper lorries 411
tools 644–8
torts 26
tractors and trailers 410
trading standards officers 67, 596
traffic routes 397–9, 632
trailers (flat backs) 411
training 132–3, 140, 161, 162, 167, 251, 257, 261, 300, 313
  accident investigation 303
  audits and inspections 244
  drivers 396, 404
  emergency procedures 215–16, 304
  fire safety 509–10
  first-aid 360
  hazardous substances 571
  induction 139, 140
  locomotive drivers 415
  managers and supervisors 143–4
  manual handling assessors 125
  personal protective equipment 593
  safe systems of work 268–70
  safety representatives 296
  VDU users 437
transport maintenance standards 400–3
transport training programme 396
transportable gas containers *see* pressure systems
Treaty of Rome 1957 36–7
triangle of fire *see* fire triangle

ultra violet radiation 533–4
Unfair Contract Terms Act 1977 84
urine tests 285
user checks 463, 465

vapour 539
vehicle
  design 393–4
  load considerations 405–6
  loading and unloading 417–18
  passenger carrying 409
vibration 529–31

in gases 526
vibrating surfaces 526
vibration white finger 76, 422, 529
vicarious liability 27, 81–3, 84–5, 88
violence to staff 202, 361–3
visible radiation 533
vision testing 284 *see also* eyesight tests
visitors 146, 160, 210, 239, 322–3, 341
visual display unit (VDU) 434–45
visual indication signals 388–9

wagons, railway 415
waste disposal 568, 670–1
waste packaging 672–3
water pollution 668
Water Resources Act 1991 668
welding equipment 648
whole body vibration (WBV) 530–1
Woodworking Machines Regulations 644
work completion reports, contractors' 344–6
workgroups, safety discussion 311–12
workplace, definition of 349
Workplace (Health, Safety and Welfare) Regulations 1992 (the Workplace Regulations) 55, 349, 353–7, 392, 650
work-related upper limb disorder (WRULD) 422–3
work schedules 654–7
workshops and garages 402–3
Workstation Minimum Requirements, HSE Guidance 435

zone classification system 456